Radiology of Parasitic Diseases

Hongjun Li
Editor

Radiology of Parasitic Diseases

A Practical Approach

Springer

Editor
Hongjun Li
Capital Medical University
Beijing You An Hospital
Beijing
China

ISBN 978-94-024-0909-3 ISBN 978-94-024-0911-6 (eBook)
DOI 10.1007/978-94-024-0911-6

Library of Congress Control Number: 2016959213

Printed on acid-free and chlorine-free paper

This Springer imprint is published by Springer Nature
The registered company is Springer Science+Business Media B.V. Dordrecht
The registered company address is Van Godewijckstraat 30, 3311 GX Dordrecht, The Netherlands

Foreword One

Jianping Dai
Vice President of Chinese Medical Association
On July 15, 2015

The expansive territory and complex climate and geological environment have bred numerous types of parasitosis in China, which is one of the countries subject to severe epidemics of parasitosis. Besides, parasitosis has remained as a major long-term public health concern in China. Along with the progress of globalization and economic integration and rocketing development of science and technology, the ongoing profound changes of human living environment and behaviors exert tremendous impacts on the occurrence and prevalence of infectious diseases, which is characterized by continual emergence of new infectious diseases and recurrence of traditional infectious diseases, posing threats to human beings. As the morbidity or mortality of patients with parasitosis is primarily associated with complications, early diagnosis and differential diagnosis of complications is key to the survival quality and term of patients with parasitosis. Meanwhile, an important link of complication diagnosis and differential diagnosis is imaging examinations.

At present, Prof. Li has led to publish the Practical Radiology of Infectious Diseases (English and Chinese version). Under the requirements of readers and clinical demands, Prof. Li pooled together nationwide experts and clinical resources to compile the Radiology of Parasitosis, which presents a more detailed introduction of the fundamental radiological theories and typical cases of various parasitotic diseases.

From conception to completion of the manuscript, members of editorial board have been trained for three times, and a special team was convened and dedicated to compilation and composition. Lasting over 2 years, this book encompasses two parts, with ten chapters, over 300,000 characters and 500 figures. The concise, well-structured, and systematic knowledge is reader friendly, with convenience for searching and reading. It well demonstrates the primary characteristics of radiology of parasitosis, providing valuable guidance for diagnosis of parasitosis-associated diseases and efficacy evaluation. All the firsthand data in the book lay a solid foundation for further research in radiology of parasitosis. This treatise opens up a new area, serves as a complement for medical radiology of China, and also offers an essential reference for prevention, medical treatment, and researches in this field.

I feel much honored to compose this foreword since this book incorporates great endeavors and wisdom of the author, with comprehensive and systematic contents as well as a highly readable style. I believe this book will improve the public cognition to parasitosis, facilitate academic communication and exchanges, as well as press ahead the cause of prevention and treatment of parasitosis.

Foreword Two

Ke Xu
Director, Medical Imaging Research Institute
of China Medical University
Director membership, Society of Radiology
of Chinese Medical Association
On July 16, 2015

Around the world, parasitic diseases not only undermine the health and life quality of patients but also incur tremendous losses to social economic development. However, despite the current rapid development of radiological techniques and continuous advent of new methods and theories, there is no treatise on comprehensive and systematic illustration of imaging changes of parasitotic diseases, which does not reflect the significance of imaging to diagnosis of parasitosis, and application of latest and most advanced radiological techniques to diagnosis of parasitosis. On such accounts, Prof. Li, together with professors from departments of medical radiology of dozens of hospitals in China, integrated multicenter resources and systematically summarized the spectrum of parasitic diseases to jointly compile this book, Radiology for Parasitosis, which will fill the blank of radiology for parasitosis at domestic or abroad. Proceeding from disease cases, this book describes the etiology, pathology, clinical manifestations, imaging presentations, and differential diagnosis of each disease and makes commentaries based on the experience of the author. This monography aims to comprehensively and systematically elaborate the imaging changes of common parasitotic diseases and relevant complications so as to provide guidance and reference for department of radiology and clinical physicians.

This book comprises two parts with ten chapters, about 300,000 characters and over 500 images in an informative and reliable style. I wish and believe that the publication of this book will actively promote the treatment and prevention of parasitotic diseases and the cause of radiology.

Foreword Three

Xiaoyuan Feng
Chief Editor Chinese Journal of Radiology
Former direct member, Society of Radiology
of Chinese Medical Association
On July 16, 2015

The expansive territory and complex climate and geological environment have bred numerous types of parasitosis in China, which is one of the countries subject to severe epidemics of parasitosis. Besides, parasitosis has remained as a major long-term public health concern in China. Along with the progress of globalization and economic integration and rocketing development of science and technology, the ongoing profound changes of human living environment and behaviors exert tremendous impacts on the occurrence and prevalence of infectious diseases, which is characterized by continual emergence of new infectious diseases and recurrence of traditional infectious diseases, posing threats to human beings. As the morbidity or mortality of patients with parasitosis is primarily associated with complications, early diagnosis and differential diagnosis of complications is key to the survival quality and term of patients with parasitosis. Meanwhile, an important link of complication diagnosis and differential diagnosis is imaging examinations.

From conception to completion of the manuscript, lasting over 2 years, a special team was convened and dedicated to compilation of relevant domestic and foreign resources and composition. This book encompasses two parts, with ten chapters, over 300,000 characters and 500 figures. The contents cover fundamental theories of parasitosis and radiology for special diseases, which is typical of case-based introduction and discussion, with convenience for searching and reading. This book displays the primary radiological characteristics of parasitosis, which is highly valuable for diagnosis of parasitotic diseases and efficacy evaluation.

Radiology for Parasitosis presents a new vision for domestic and foreign clinical and imaging counterparts and expands a new area for medical radiology, which complements and improves current medical radiology of China, and is a critical reference book for prevention, treatment, and researches of parasitosis. I am very gratified about the efforts and achievement of Chinese scholars in radiology of parasitosis and honored to write this foreword.

Foreword Four

Zhengyu Jin
Director membership, Society
of Radiology of Chinese Medical Association
On July 16, 2015

Parasitic diseases are infectious diseases caused by parasites latent in human and animal bodies and are widely distributed, diverse, and greatly detrimental. Parasitic diseases have remained as a common public health concern around the world.

In light of the concepts of evidence-based medicine, imaging examination is an important method for diagnosis of parasitotic diseases and efficacy evaluation and is an essential link to enhance the outcomes of prevention and treatment of infectious diseases.

Currently, despite the current rapid development of radiological techniques and continuous advent of new methods and theories, there is no treatise on comprehensive and systematic illustration of imaging changes of parasitotic diseases, which does not reflect the significance of imaging to diagnosis of parasitosis, and application of latest and most advanced radiological techniques to diagnosis of parasitosis. On such accounts, Prof. Li, together with professors from departments of medical radiology of dozens of hospitals in China, integrated multicenter resources and systematically summarized the spectrum of parasitic diseases to jointly compile this book, *Radiology for Parasitosis*. Proceeding from disease cases, this book describes the etiology, pathology, clinical manifestations, imaging presentations, and differential diagnosis of each disease and makes commentaries based on the experience of the author. This monography aims to comprehensively and systematically elaborate the imaging changes of common parasitotic diseases and relevant complications so as to provide guidance and reference for department of radiology and clinical physicians.

This book encompasses two parts, with ten chapters, over 300,000 characters and 500 figures. The well-structured contents are user friendly, which is convenient for reading and searching. This monography exploits a new area for Chinese medical radiology and is a complement and improvement of current medical radiology, which is highly significant. It is also an important reference book for prevention, treatment, and research of parasitic diseases.

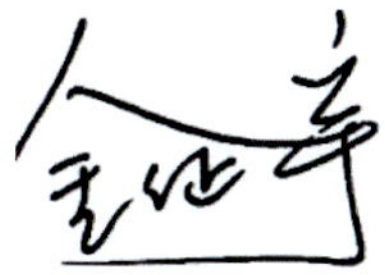

Preface

Radiology for parasitosis is a discipline that studies the imaging characteristics and evolution patterns of infectious diseases induced by protozona, helminth, arthropods and annelids, and mollusk lurking in human bodies.

The expansive territory and complex climate and geological environment have bred spawned numerous types of parasitosis in China, which is one of the countries subject to severe epidemics of parasitosis. Besides, parasitosis has remained as a major long-term public health concern in China. Along with the progress of globalization and economic integration and rocketing development of science and technology, the ongoing profound changes of human living environment and behaviors exert tremendous impacts on the occurrence and prevalence of infectious diseases, which is characterized by continual emergence of new infectious diseases and recurrence of traditional infectious diseases, posing threats to human beings. Prevention and treatment of infectious diseases once ignored are now coming to limelight. Since the lethal causes of patients with infectious disease are closely associated with occurrences of complications, early diagnosis and differential diagnosis of complications are critical to the life quality and survival of patients. Meanwhile, as primary methods for diagnosis and differentiation of complications, imaging examinations such as CT, X-ray, and MR play vital roles in prevention and treatment of parasitic diseases. Lack of systematic theories, technical standards, and diagnostic guides of radiology for parasitosis as well as the urgent clinical demands underpin the researches and compilation of this book.

Moreover, popularization of knowledge about radiology for parasitosis has not already only confined to special hospitals of infectious diseases, but the departments of radiology in comprehensive hospitals also confront the need of popularization of the knowledge of a new discipline—radiological diagnosis and differential diagnosis of infectious diseases.

However, systematic studies about clinical radiology of parasitosis are still absent around the world, and no systematic theoretical standards guide clinical practices. The previous sporadic valuable data, whether lost or scattered, need to be collected, summarized, analyzed, and compiled to fill the blankness of radiology for parasitosis in systematic theories, which will promote standardized diagnosis of parasitosis-associated complications, enhance knowledge of diagnosis and differentiation of parasitosis complications, and boost the effects of prevention and treatment of complications related to infectious diseases. It is of great significance for elevating the life quality and survival of patients with infectious diseases. Based on a multicenter study integrating nationwide experts and clinical resources, the author compiles this landmark monograph about clinical imaging of parasitosis, which fills the blank of the systematic blank of imaging about parasitosis and enriches medical imaging.

Moreover, this monograph incorporates the format styles of both Chinese and Western books, which includes large quantities of classical cases and highlights key points with arguments and text along with abundant pictures.

Major features: This book is composed of two parts, including ten chapters with over 500 images. It is close to clinics and of high practical and reference value, which is suitable to healthcare professionals at all levels. 95 % imaging materials are firstly displayed and boast independent and complete intellectual property rights.

The publication of this book consolidates the foundation for clinical imaging of parasitosis and exploits a new area.

The academician Liu Yuqing, a pioneer in radiology in China, is very gratified about the publications of the serial monographs about infectious disease imaging by Prof. Li Hongjun and indicated that such academic achievements represent great advances of infectious disease imaging in China, usher in a new area, and enrich medical radiology.

There might exist some errors in this book and your kindly comments are highly appreciated for the improvement of this book.

The compilation of this book is sponsored by clinical medicine development special fund of Beijing Municipal Administration of Hospitals (project No.: ZYLX201511).

Brief Introduction

Parasitic diseases are caused by parasites that parasitize in humans and animals with worldwide distribution. The species of parasites are diverse and the parasitic diseases seriously threaten the health of humans and animals, which remain to be a common public health concern across the world. The research team led by Prof. Li Hongjun, the editor of this book, firstly and systematically summarized the parasitic diseases spectrum from the perspective of radiology. By screening over 100 classic cases with clinical and radiological diagnosis, contrast studies were performed by comparing their radiological and pathological data to unveil the pathological evolution and the nature of the diseases based on the radiological findings. The studies were intended to provide valuable diligent scientific reference for the diagnosis, treatment, control, and scientific research of parasitic diseases and their complications.

Instead of simply reviewing and introducing, we integrated reviews and case studies in this book. The readers will firstly read the basic theories with following illustrations by clinical classic cases. In such a way of reading, we expect a more favorable reading effect than atlas, reviews, and headline guides. The whole book consists of two parts with ten chapters. The first part is a general introduction, which elucidates such fundamental theories as the etiology, epidemiology, and imaging examination techniques of parasitic diseases. The second part focuses on specific diseases and falls into four individual chapters covering protozoal diseases, helminthiasis, nematodosis, and trematodiasis. Each chapter introduces relevant diseases which are expounded respectively by several cases. Each case comprises complete history of disease, laboratory tests results, pathological figures, and imaging data (ultrasound, X-ray, CT, and MRI). In addition, each case is characterized by comprehensive clinical data, classic radiological images, detailed differential diagnosis, and discussion, which unravels diagnosis of similar imaging signs.

This book is targeted at physicians at the department of radiology and clinical departments as well as medical students.

Contents

Contributors

Honorary Editors: Jianpign Dai, Ning Li

Chief Editor: Hongjun Li

Associate Editors: Wenxiao Jia, Jingliang Cheng, Wenya Liu, Yi Xiao, Li Li, Quansheng Gao

Members of Editorial Committee

Yanhua Yu, Jingxu Ma, Yunling Wang, Hong Wang, Jian Wang

Jian Wang, Lu Wang, Jiamin Deng, Lili Kong, Lina Gu, Fangfang Dai

Dongwu Bao, Dongying Bao, Wei Xing, Shengxiu Lv, Meiji Ren

Bailu Liu, Hanqiu Liu, Xin Liu, Guangbin Jiang, Jun Xu, Haiqing Sun

Fuhua Yan, Qun Lao, Xuhua Yang, Xiumeii Yang, Yuxin Yang, Baiyan Li

Hongchi Li, Hongchen Li, Hongyan Li, Ming Li, Ping Li, Xueqin Li, Hongjun Li, Wenxiao Jia, Jingliang Cheng, Wenya Liu, Yi Xiao, Li Li, Xiaobiao Zhang, Lili Jiang, Zhiyang Jia

Ting Li Hua He, Jifeng Zhang, Na Zhang, Xiaochun Zhang, Lihong Zhang Puxuan Lu, Tingting Chen, Li Gou, Wenbin Luo, Qingxia Zhao, Yuxin Shi, Chunhui Jiang, Jinli Lou, Qi He, Xingxing Yuan, Wanjun Xia

Yan Dang, Meiling Xu, Quansheng Gao, Yulin Guo, Yinglin Guo, Hui Guo, Yonghua Tang Kui Huang, Yi Jiang, Song Shu, Zhiyan Lu, Yu Guan

Tingting Fan, Jiangfeng Pan, Jin Xu, Jing Ning, Wenyan Gao

Academic Consultants

Chunshui Yu, Zhenchang Wang, Peijun Wang, Guangming Lu, Baozhong Shen

Xiaoyuan Feng, Shiyuan Liu, Kuncheng Li, Hui Zhang, Min Chen, Zhengyu Jin, Chunwu Zhou, Weijian Jiang, Huishu Yuan, Ke Xu, Youmin Guo, Changhong Liang, Ping Han, Bin Song, Mengsu Zeng, Gaojun Teng

Contributing Institutions

City Public Health and Clinical Center, Affiliated to Fudan University, Shanghai, China

Shanghai Changzheng Hospital, Shanghai, China

Ruijin Hospital, affiliated to School of Medicine, Shanghai JiaoTong University, Shanghai, China

Tianyang People's Hospital of Guangxi

Longtan Hospital, Guangxi Zhuang Autonomous Region

General Hospital, Ningxia Medical University

Chengdu Municipal Public Health and Clinical Medical Center

Academy of Military Medical Sciences

The Third Affiliated Hospital, Suzhou University, Suzhou, Jiangsu, China

Zhongnan Hospital of Wuhan University, Wuhan, Hubei

Hangzhou Children's Hospital

The First Affiliated Hospital, Zhengzhou University, Zhengzhou, Henna, China

Municipal Sixth People's Hospital, Zhengzhou, Henan, China
Dengzhou Municipal People's Hospital
Family Planning Guiding Center, Wolong District, Nanyang, Henan, China
Baihe Town Hospital, Nanyang, Henan, China
City TCM Hospital, Nanyang, Henan, China
Nanyang Medical College
The First Affiliated Hospital of Nanyang Medical College
The Second Affiliated Hospital, Harbin Medical University, Harbin, Heilongjiang, China
City Public Health Medical Rescuing Center, Chongqing, China
Affiliated Huashan Hospital, Fudan University
Beijing You'an Hospital, Affiliated to Capital Medical University, Beijing, China
Jinhua Municipal Central Hospital, Jinhua, Zhejiang
The Third People's Hospital of Shenzhen
City Kangan Hospital (former City Hospital for Infectious Diseases), Mudanjiang, Heilongjiang, China
Taiping People's Hospital, Daowai District, Harbin, Heilongjiang, China
Taihe Hospital of Shiyan city
Tianmen First People's Hospital
The First Affiliated Hospital (Xinan Hospital), the Third Military Medical University, Chongqing, China
The First Affiliated Hospital, Xinjiang Medical University, Urumqi, Xinjiang Uygur Autonomous Region, China
The Second Affiliated Hospital, Xinjiang Medical University, Urumqi, Xinjiang Uygur Autonomous Region, China
The People's Hospital, Xinjiang Uygur Autonomous Region, China
Municipal Sixth People's Hospital, Xinjiang Uygur Autonomous Region, China
School of Foreign Languages, Southern Medical University, Guangzhou, China
Linyi People's Hospital, Linyi, Shangdong, China
Yantai Yeda hospital, Yantai, Shangdong, China

Introduction of Chief Editor

Hongjun Li
Director member, infection radiology special committee,
society of radiology, Chinese Medical Association
Director member, infection radiology society,
Chinese Association of STD/AIDS Prevention and Control
Team leader of Infectious Disease Imaging Management,
Chairman of Radiology Committee on Infectious and Inflammatory Diseases,
Society of Infectious Disease Hospital Management,
Chinese Hospital Management Association
Standing committee membership,
Beijing Medical Radiology Society,
Chinese Medical Association
July 8, 2015

Hongjun Li (M.D., Prof.) is a 49-year-old radiologist with an educational background in the UK. Currently, he is also a supervisor for the master's degree program in radiology. Professor Li is now offered the special government allowance from the State Council in China in recognition of his outstanding contributions to the field of medicine. He is also recognized membership in the Ten-Hundred-Thousand talent program in China at the "hundredth" level in medicine. Meanwhile, he has achieved membership as one of the 215 high-level academic leaders in Beijing.

Research direction: Radiology of Infections and Infectious Diseases

Current Positions:

Director, Department of Radiology, Beijing You'an Hospital, Capital Medical University, Beijing, China

Deputy Dean, Department of Medical Imaging and Nuclear Medicine, Capital Medical University, Beijing, China

Chief Editor, *Radiology of Infectious Diseases* (Elsevier platform)

Social Affiliations

Chinese Medical Science and Technology Award expert review committee member

National Study Abroad Foundation-funded project expert review committee member

Beijing Natural Science Foundation project expert review committee member

Director member, infection radiology special committee, society of radiology, Chinese Medical Association

Team leader of Infectious Disease Imaging Management, Society of Infectious Disease Hospital Management, Chinese Hospital Management Association

Committee membership, Abdominal Specialty Group, society of radiology, Chinese Medical Association

Committee membership, Beijing Medical Radiology Society, Chinese Medical Association

Committee member, Radiological Society of South Korea

Director, Imaging Quality Control Center of Fengtai district, Beijing

Expert membership, experts' pool for Differential Diagnosis of Occupational Diseases in Beijing

Editorial board membership for 14 journals such as *Chinese Medical Journal (CMJ), Journal of Clinic Infections & Therapy, Chinese Journal of Psychiatry, Journal of Clinical Radiology, Chinese Journal of Medical Imaging Technology, Journal of Magnetic Resonance Imaging, Radiologic Practice, Chinese Journal of AIDS&STD, Journal of Clinical Hepatobiliary Diseases*

Clinical interests: imaging diagnosis and differential diagnosis of hepatic diseases, contagious diseases and infectious diseases (such as AIDS-associated diseases); realizing clinical staging of HBV-associated hepatocellular carcinoma based on noninvasive multimodal imaging biomarker and localization, quantitative and qualitative diagnosis of imaging classification. Main directions: (1) multimodal imaging classification diagnosis of hepatocellular carcinoma associated with viral hepatitis; (2) multimodal molecular imaging of hepatic fibrosis and liver cancer associated with viral hepatitis; (3) evaluation of early hepatocellular carcinoma biological behaviors based on optic/PET/CT/MR; (4) functional and molecular imaging biomarkers of cerebral dementia associated with HIV-1; (5) biomolecular imaging studies of early HAND based on multimodal magnetic resonance; and (6) biomolecular imaging studies of early cerebral injuries in SIV-1-infected Rhesus monkey based on multimodal MR/PET.

Academic accomplishments: In recent 5 years, Prof. Li has directed or participated in eight national, provincial scientific research projects as well as three international collaborative projects, with the accumulative scientific research grants up to RMB 10 million yuan (current research funds RMB 8 million). He has contributed to 110 articles, including 33 articles published at Science-Citation-Indexed (SCI) journals and 110 articles in core journals. Prof. Li has obtained 27 sponsors, respectively, from National Natural Science Foundation, Beijing Natural Science Foundation, Yangfan Project and International Publication Foundation, National Natural Science and Technology Publication Foundation, and scientific research funds of Ministry of Health as well as special publication funds. Besides, he has edited and completed 19 treatises, including six in English, which are published and promoted around the world and exert strong academic influences at domestic and abroad. He was conferred seven provincial awards such as National Medical Science and Technology Award (first person) and applied 16 national patents and intellectual property rights (first person and owner).

Outstanding contributions: Prof. Li pioneers to establish the first systematic spectrum of legitimated infectious diseases from the perspective of medical radiology. His contributions shed light on and improve the fundamental theories about the radiology of infectious diseases and clinical applications. Additionally, his academic achievements further enrich the theories of medical radiology and comb and constitute the technological standards and diagnostic guide for radiology of infectious diseases. His findings exploit a new area for medical radiology and command the leading academic level around the world, laying a solid foundation for development of radiology of infectious diseases. Moreover, Prof. Li created *Radiology of Infectious*

Diseases (Elsevier), the first international journal about radiology for infectious diseases, and serves as the chief editor. Besides, his series of treatises about radiology of infectious diseases are the first books in such field published by Springer (6 treatises), of which *Radiology of HIV/AIDS* registers a download of over 20,000 chapters since publication. Furthermore, Springer PG commended that *Radiology of HIV/AIDS* had recorded the highest downloads within a short term in recent years, and the renowned professor of University of Washington (one of the experts formulating infectious disease diagnostic standards of US NIH) spoke highly of this monograph and published a book review on the internationally influential journal of *Clinic Infectious Diseases* (IF: 9.146), which improves the academic position of Chinese radiological diagnostics. He completed the global first unique specimen bank and data bank of HIV donor corpses in terms of tomography and radiological and pathological comparison. In addition, he built the first preliminary resource bank for radiological big data tissue specimen studies about the 39 Chinese legal infectious diseases, consolidating the basis for international cooperation and establishment of radiological diagnostics standards.

Part I

Basic Theories on Parasitic Disease

1 Introduction to Parasitic Disease

Yunling Wang

Parasitic disease is a type of infectious disease caused by parasites in human body such as protozoa, worms, arthropod, annelida and mollusc. Parasitic disease prevails in warm and humid areas of tropical, subtropical and temperate regions. Particularly in developing countries, the prevalence of parasitic disease constitutes a threat to human life and health. Among the 10 globally important tropical diseases in the Special Program for Tropical Diseases jointly enacted by the United Nations Children's Fund (UNICEF), the, United Nations Development Program (UNDP), the World Bank and the World Health Organization (WHO), seven are parasitic diseases, including malaria, schistosomiasis, lymphatic filariasis, onchocerciasis, leishmaniasis, African trypanosomiasis and chagas disease.

Parasitic disease is a public health issue across the world. In some developed countries, some parasitic diseases also prevail. In China, the large area is geographically across the frigid, temperate and tropical zones, with greatly diverse natural conditions. In the National Guideline for Agricultural Development (1956–1967), five major parasitic diseases were proposed to be eliminated, including schistosomiasis, malaria, filariasis, black fever and ancylostomiasis, with death in thousands of people. After the foundation of People's Republic of China in 1949, the government of China has achieved great success in controlling the prevalence of parasitic diseases. Currently, parasitic disease is still prevalent and threats to human health. With the social development, the environment and life style have been changing accordingly. In addition to improved traffic conditions and increased personnel mobility, the prevalence of parasitic disease shows according changes. Furthermore, due to increased organ transplantation, increased occurrence of cancer and AIDS as well as increased use of immunosuppressor, the population with immunodeficiency or compromised immunity is increasingly large. Therefore, opportunistic parasites are increasingly important pathogens. The asymptomatic infections caused by such opportunistic parasites as toxoplasma gondii, cryptozoite, isopsoriasis, cyclospora and strongyloides stercoralis see increasing occurrence.

The changes of natural conditions also contribute to the distribution and prevalence of some parasites. Due to the greenhouse effect and the subsequently increased global temperature, the areas with anopheles and other vector insects are enlarging. It has been speculated that the prevalence of schistosomiasis, trypanosomosis, dengue and yellow fever would exacerbate. The more patients with parasitic disease present challenges to the clinicians. During clinical management of parasitic disease, accurate diagnosis is of great importance and the first step to relieve the sufferings. The compiling of this book is intended to facilitate the accurate diagnosis of parasitic disease and thus to provide basis for treating the diseases.

Y. Wang
The Second Affiliated Hospital, Xinjiang Medical University,
Urumqi, Xinjiang Uygur Autonomous Region, China
e-mail: 1079806994@qq.com

H. Li (ed.), *Radiology of Parasitic Diseases*, DOI 10.1007/978-94-024-0911-6_1

2 Etiology of Parasitic Diseases

Yunling Wang

2.1 Definition of Parasite and Host

The pathogen of parasitic disease is a type of organisms that live by parasitizing to its host. Parasitism refers to the condition that two different natural organisms live together to form certain interrelationship. One is parasite and the other is host. The parasite, an organism that temporarily or permanently lives in or on the surface of the host. Via parasitizing, the parasite is supplied by the host and impairs the host. Therefore, the parasite usually benefits from its interrelationship with the host. However, the host does not benefit from the interrelationship but only provides nutrients and a living space for the parasite. Parasite is a pathogenic organism threatening the health of both human and animals, including virus, rickettsia, bacteria, fungi, protozoa, worm, and arthropod. Parasite refers to the animal parasites including single-cell protozoa and multicellular invertebrates that lead a parasitic life.

2.2 Classification of Parasite and Host

2.2.1 Parasite

According to duration and adaptability of the parasite-host relationship as well as specific ecological environment, parasite shows great diversity and can be categorized into the following groups.

2.2.1.1 Obligatory Parasite

The entire life circle behaves as parasite such as trichina and tapeworm. The eggs or larva of some worms may live independently, but during their infective stage, they live within the host to guarantee the development from larva into adult, such as roundworm and hookworm.

Y. Wang
The Second Affiliated Hospital, Xinjiang Medical University, Urumqi, Xinjiang Uygur Autonomous Region, China
e-mail: 1079806994@qq.com

2.2.1.2 Facultative Parasite

During larva and adult stages, a parasite can not only live its life circle independently, but also can live in the host, such as strongyloides stercoralis.

2.2.1.3 Accidental Parasite

A parasite, during its infective stage of life circle, accidentally gains its access into an organism other than its common host. For instance, fly maggot may gain its access into human organ, cavity or lumen.

2.2.1.4 Opportunistic Parasite

Some parasites show no obvious pathogenicity and their host are commonly asymptomatic. However, when the immunity of the host is compromised, such as patients with AIDS and patients receiving long-term treatment by hormone or anti-neoplastic drugs, the parasite may multiply abnormally with increased pathogenicity, causing obvious clinical symptoms and signs. In severe cases, it may lead to death. Such parasites include toxoplasma and cryptosporidium.

2.2.1.5 Ectoparasite

A parasite, such as louse, that lives permanently on the surface of its host is known as ectoparasite. Such parasite also includes mosquito, bedbug and ticks that only temporarily suck blood on the surface of their host and then leave when they are full. Ectoparasite is also referred to as temporary parasite.

2.2.1.6 Endoparasite

A parasite, such as trichomonas vaginalis, schistosome and malaria parasite, that lives in the intestinal tract, organs, tissues, blood and cells of the host is known as endoparasite.

2.2.2 Host

2.2.2.1 Definitive Host

The host that is parasitized by a parasite at its adult stage or sexual reproduction stage is known as definitive host. For

H. Li (ed.), *Radiology of Parasitic Diseases*, DOI 10.1007/978-94-024-0911-6_2

instances, human is the definitive host of paragonimus westermani.

2.2.2.2 Intermediate Host

The host that is parasitized by a parasite during its larval stage or asexual reproduction stage is known as intermediate host. In the cases of at least two intermediate hosts, they can be classified as the first or second intermediate host according to the sequence of being parasitized. For instance, paragonimiasis is the first intermediate host of Paragonimus westermani, while stone crab is its second intermediate host.

2.2.2.3 Reservoir Host

Some animals can act as host for parasites that finally live within human body during its parasitic stage. For instance, dog and cat can be host for adult paragonimus westermani that may finally parasitize human. According to epidemiological study of parasitic disease, these animals are known as reservoir host of paragonimus westermani that parasitize humans.

2.2.2.4 Paratenic Host/Transport Host

When larva of a parasite gains its access into the body of its unusual host, it fails to develop into adult parasite but keeps its larva state for a long period of time. However, when the larva gains another opportunity into its usual definitive host, it can further develop into adult. During such a process, the unusual host is known as paratenic host/transport host. Wild boar, chicken and duck may act as paratenic host/transport host of paragonimus westermani.

2.2.3 Life Cycle of Parasite

The life cycle of parasite refers to the entire course of growth, development and reproduction of one generation. It includes access route of parasite into the host during its infective stage, migration of the parasite to the usual parasitic site in the host, way of leaving the host as well as host species and environmental conditions necessary for its development.

2.2.3.1 Route of Parasitic Infection

Parasite must develop into its infective stage before its invasion into human or other vertebrates as its host. Only in such a way, the parasite can continue its survival, development and reproduction in its host. Different parasites have varied infective stage and their routes of access into host are also diverse. The common routes of access include the following:

Infection via Mouth

Parasite in its infective stage can gain its access into human body via mouth along with food, water, contaminated fingers or other utensils, which is the most common route of infection. For instances, by intake of infective eggs, intake of undercooked freshwater fish with metacercaria of clonorchis sinensis and intake of pork with cysticercus can cause parasitism by roundworm, enterobius vermicularis, liver fluke, taenia solium and toxoplasma.

Infection via Skin

Contacts to soil contaminated by filariform larva of hookworm or water contaminated by schistosoma japonicum cercariae may cause their direct and active access to human via skin.

Infection via Skin by Vector Insect

Parasite develops into its infective stage in vector insects. When the insect bites, stings or sucks blood on the skin, the parasite gains its access into the body of host via skin. For instance, sporozoites of malaria parasite and filariform larva can gain their access into human body via skin when mosquitoes carrying such parasites bite.

Contact Infection

In its infective stage, parasite in oral cavity, vagina or body surface of human body can infect other person via direct or indirect contact, such as entamoeba gingivalis, trichomonas vaginalis and sarcoptes mites.

Inhalational Infection

The infective eggs of enterobius vermicularis in dust can be inhaled and then swallowed to cause parasitism.

Infection via Blood Transfusion

Human malaria parasite in its intraerythrocytic asexual form can cause parasitism via blood transfusion.

Retrograde Infection

Enterobius vermicularis lay eggs around the anus of the host where larva may develop within several hours. Subsequently, larva may gain its access into the intestinal tract and further develop into adult.

2.2.3.2 Parasitic Site of Parasite

After parasite gains its access into its host, in some cases, it directly arrives to its parasitic site. For instances, the infective eggs of enterobius vermicularis with larva and cysticercus of taenia solium may directly develop into adult in the intestinal tract of the host and then parasitize the host there after being swallowed. However, in some other cases, the parasites may follow certain procedures to migrate in the host before reaching its usual parasitic site. For instance, larvae of roundworm and hookworm migrate from one organ to another in the host before reaching their usual parasitic site. The parasitic sites in human body can be roughly categorized into intravascular, intralymphatic, within blood or body fluid, intracellular, within tissues, respiratory tract, gastrointestinal tract and within other organs. Some parasites are rather exclusive in terms of parasitic site in the host, while others can live in multiple organs and tissues. Certain parasitic site that different parasites need in the host is the result of long-term inter-adaptation between parasite and its host after an evolving process.

2.2.3.3 Reproductive Pattern of Parasite

When parasite reaches its parasitic site and continues its development into its mature stage, it begins to reproduce its offsprings. Different species of parasites show varied reproductive patterns.

Asexual Reproduction

Asexual reproduction is a common reproductive pattern of parasite protozoa, including binary division such as amoeba trophozoite and leishmanial; multiple fission such as schizogony of plasmodium; gemmation such as endodygony by trophozoites of toxoplasma gondii. Polyembryony often occurs in some worms in their larva stage. That is to say, asexual reproduction, also known as larva reproduction, is a reproductive pattern of parasite in its larva stage.

Sexual Reproduction

Sexual reproduction is a common reproductive pattern of worm, with a female adult worm and a male adult worm to mate and the female adult worm to lay eggs or deliver larva. Most of the eggs or larvae develop into their mature stage in an external environment such as soil or intermediate host before they gain access into the host to develop into adult.

Alternation Generation

Some parasites have to experience both asexual and sexual reproductive pattern to complete a full life circle of one generation. That is to say, an asexually reproduced generation alternates with a sexually reproduced generation, which is known as alternation generation, such as plasmodium, toxoplasma gondii and fluke.

2.2.4 Nutrition and Metabolism of Parasite

Parasites live most/all of their life circle in the physiochemical environment of their host. Therefore, nutrition and metabolism of parasites depend on the host, and interactions of different degrees occur between parasite and its host.

2.2.4.1 Nutrition

The nutrients that parasite need are from substances that have been digested or semi-digested in intestines of the host. The intake of nutrients is, on one hand, via temporarily or permanently formed specific morphological structure such as pseudopod of protozoon, cytostome, cytopharynx, microcytostome or other organellae. On the other hand, the intake of nutrients can be directly via cytomembrane such as protozoon. Tapeworm, without alimentary tract, has to take in nutrients via cortex of its body wall. Most of the flukes have incomplete alimentary tract and take in the needed nutrients via integument of their body wall.

2.2.4.2 Metabolism

Energy Metabolism

The energy of parasite is mainly supplied via aerobic metabolism and anaerobic metabolism. Aerobic metabolism can completely oxidize metabolites and produce more energy. However, anaerobic metabolism, also known as glucose metabolism, produces lactic acid and pyruvic acid after anaerobic glycolysis of glucose, with less energy obtained. Lactic acid commonly can be utilized by the host to synthesize glycogen, which is subsequently oxidized into carbon dioxide and water.

Glucose Metabolism

Glucose metabolism can be achieved via three approaches: anaerobic glycolysis, aerobic metabolism of tricarboxylic acid cycle and shunt of phosphopentose metabolism. A great number of parasites, especially those parasitizing in the alimentary tract, live in an environment with no or low level of oxygen and their energy is mainly supplied via glycolysis. For instances, amylase in entamoeba histolytica can resolve starch that parasites have stored into glucose, and then produce energy via glycolysis. Some tapeworms in their larva stage can store large glycogen and obtain energy via glycolysis and tricarboxylic acid cycle. All nematoids can transform glucose or glycogen into pyruvic acid via glycolysis, which is then reduced into lactic acid or produces acetyl Co-A, followed by the metabolic process of tricarboxylic acid cycle.

Metabolism of Protein and Amino Acid

When glucose is unavailable, parasite can obtain energy from protein metabolism. Experiments have shown that most parasites produce more ammonia when glucose is insufficient, indicating active catabolism of protein. Amino acid that parasite obtains from the host enters into metabolic pool for synthesizing different tissues or participating in metabolism. Amino acids in metabolic pool commonly participate in the routine metabolic pathways of protein whose products include ammonia, urea and uric acid. Most of these products are toxic, playing significant role in the pathogenesis of parasitic disease.

Lipid Metabolism

Lipid in parasite includes glyceride, free fatty acid and phospholipid. Nonsaponifiable substance, such as sterol and related substances, can also be included. Lipid metabolism sometimes concurs with glucose metabolism, and sometimes with protein metabolism.

2.2.5 Classification of Parasite

The number of existent animal species in the world ranges from 1.5 million to 4.5 million, and more than 1 million have been described and categorized. The number of parasitic protozoa, worms and arthropods is also considerable. According to animal classification system, parasite predominantly distributes in Platyhelminthes, Nemathelminthes, Acanthocephala and Arthropoda in kingdom of animalia, and Phylum sarcomastigophora, Phylum apicomplexan and Phylum ciliata in subkingdom of gnotozoa.

Animal classification has seven levels, including Kingdom, Phylum, Class, Order, Family, Genus and Species. In addition, the animal classification system also includes some intermediate levels such as Subphylum, Subclass, Suborder, Superclass and Superorder.

Kingdom	Animalia
Subkingdom	Protozoa
Phylum	Sarcomastigophora
Class	Rhizopodea
Order	Ameoebina
Family	Entamoebidae
Genus	Entamoeba
Species	Entamoeba histolytica

Binomial system is applied for nomination of parasite. That is to say, a technical name of parasite include two parts, with the first part identifying the genus and the second part identifying its species. In some cases, the second part is followed by its subspecies. The name of species or subspecies is prior to the first name of the identifier and the year of its identification (the publication year of its identification report). And the technical name of parasite should be italicized in any printed document.

3 Pathophysiology of Parasitic Disease

Yunling Wang

Parasitic infection refers to the asymptomatic condition of the host after infected by parasite, while parasitic disease refers to the symptomatic condition after infected by parasite. Parasitic disease shows diverse clinical manifestations due to the degree of inter-adaptation between the host and the parasite. Factors affecting the clinical manifestations include the virulence of different parasite species, the quantity of parasites, the ability of parasites in escaping from defense responses of the host, as well as the nutritional status and immunity of the host. Host responses to invasion of parasite can be systemic or disseminated, otherwise local responses of the organ or tissue affected by parasites. Generally speaking, systemic responses caused by invasion of protozoon are more common than those caused by worms. However, local responses are usually caused by necrosis and inflammation of the infected organ or tissue, with formation of abscess or granuloma.

3.1 The Most Common Symptoms and Signs of Parasitic Disease

3.1.1 Fever

Fever is the most common symptom of many parasitic diseases. The body temperature and duration of fever are usually related to species of parasite, parasites load and the immunity of host. For instance, the onset of malaria is induced when schizonts mature and rupture to release schizozoites of plasmodium and their metabolites in erythrocytes, along with fragments of erythrocytes, into the bloodstream. Specific protein response is, therefore, triggered at the same time as schizonts of plasmodium at the erythrocytic stage increase.

Y. Wang
The Second Affiliated Hospital, Xinjiang Medical University, Urumqi, Xinjiang Uygur Autonomous Region, China
e-mail: 1079806994@qq.com

3.1.2 Anemia

Many species of parasites can cause anemia of the host, whose pathomechanisms differ. For instance, when adult hookworms parasitize in the small intestine and suck blood, the host suffers from chronic blood loss. Anemia is, therefore, caused due to deficiency of iron and protein, which is commonly diagnosed as hypochromic microcytic anemia. In another case, plasmodia parasitize and destroy erythrocytes to cause hypersplenism, suppressed hematopoiesis, and immune induced erythrocytic hemolysis, which further leads to anemia. Leishmania donovani parasitize in splenic macrophages to cause hyperplasia of mononuclear phagocytic system, leading to hypersplenism. And erythrocytic hemolysis induced by immune mechanism can also cause anemia.

3.1.3 Diarrhea

Diarrhea is one of the common symptoms of parasitic disease. The underlying mechanisms include mechanical damages to the parasitic site, toxic effect of metabolites and secretions of parasites, and allergic reactions induced by the parasites. Parasitic diarrhea can be attributed to a series of mechanisms, mostly caused by local intestinal lesions and accompanying intestinal dysfunction.

3.1.4 Anaphylaxis

Anaphylaxis is a common allergic response after infection of parasites. The allergen from parasites stimulates the host to produce specific IgE antibody, which has affinity to target cells and adheres to the surface of mast cells and basophils. When the same allergen re-enter the host, it bind to the IgE antinbody to degranulate the mast cells and basophils. The histamine, heparin, proteolytic enzyme, chemokine and others factors are released from the granules to cause anaphylactic shock.

H. Li (ed.), *Radiology of Parasitic Diseases*, DOI 10.1007/978-94-024-0911-6_3

3.1.5 Malnutrition and Impaired Development

Parasites live in or on the host and obtain nutrients directly or indirectly from the food, metabolites or tissues of the host for their growth, development and reproduction. Nutrients of the host are therefore greatly consumed. For those individuals with good nutritional status before parasitic infection, the adverse effects are not obvious. However, for those individuals with relatively malnourished status, malnutrition, severe malnutrition and even hypoproteinemia may occur. Some parasitic infections can even cause deficiency of specific nutrient. For instance, parasitic infection by diphyllobothrium latum can cause deficiency of vitamin B12, which adversely affects hematopoiesis of the host to lead to severe anemia.

3.1.6 Hepatosplenomegaly and Other Liver Damages

Hepatomegaly is a common sign of parasitic liver damage. For instance, when acute egg granulomas of schistosoma japonicum are formed, the liver commonly shows acute inflammatory swelling, with significant tenderness. Clonorchis sinensis that parasitizes in the hepatobiliary duct can cause hepatomegaly, predominantly the left liver lobe, as well as cholestasis and other symptoms and signs. Hepatic amebiasis induced by entamoeba histolytica is a common but severe complication of intestinal amebiasis. It's characterized by abscess formation, with chronic onset, hepatomegaly and accompanying hepatalgia, long-term irregular fever, anemia, and malnutrition. Echinococcosis granulosa and alveolar echinococcosis are generally referred to as hepatic echinococcosis, which is a parasitic liver disease with space occupying lesion. The parasites damage the spleen via directly parasitizing in the spleen, destroying normal splenic tissues to cause diseases such as splenic echinococcosis and splenic microfilarial granuloma. Parasitic infection can also lead to portal hypertension, which blocks splenic venous reflux to cause congestion, tissue hyperplasia and fibrosis.

3.1.7 Impaired Central Nervous System (CNS)

During certain stages of their life cycle, some protozoa and worms can invade into the brain and spinal cord of the host to impair the CNS. Parasitic CNS damages can be categorized into three types: space-occupying lesions such as amoebic brain abscess, cysticercosis and cerebral schistosomiasis, encephalitis and cerebral meningitis. Migration of some worm larvae into the brain may also cause inflammation that is characterized by infiltration of acidophilic granulocytes, which is known as acidophilic meningitis or meningeal encephalitis.

3.2 Opportunistic Parasitic Infection

When the immunity of host is compromised or deficient, such as patients with AIDS, long-term users of hormones or anti-neoplastic drugs, the parasites show enhanced productivity and pathogenecity to cause obvious symptoms or signs. The disease commonly exacerbates and even progresses to death. Such parasites are generally known as opportunistic parasites, commonly including pneumocystis jirovecii, cryptosporidium, toxoplasma gondii, isospora and strongyloides stercoralis.

3.2.1 Pneumocystosis

Human infection of pneumocystis jirovecii is found across the world, and most patients are asymptomatic. Since the identification of AIDS, the cases of pneumocystosis have been increasing, with consistent geographical distribution with AIDS, which is the most common opportunistic infection in patients with AIDS. Patients with AIDS complicated by pneumocystosis commonly experience general malaise, dry cough, body weight loss, progressive dyspnea, and fever. The characteristic sign is cotton like spots on the retina and the condition may progress into weak ventilation and respiratory distress syndrome, with symptoms and signs of interstitial lung disease (ILD). The patients may experience decreased lung capacity, decreased diffusion of carbon monoxide (CO) and increased air flow rate. The terminal state may be pulmonary heart disease, and the patients may die from respiratory failure.

3.2.2 Toxoplasmosis

Toxoplasmosis is a zoonotic parasitic disease with worldwide occurrence. Its pathogen is toxoplasma gondii and the disease is one of the most common opportunistic diseases in patients with AIDS whose infection rate is around 40 %. Clinically, the patients experience encephalitis and meningeal encephalitis as well as seizures, hemiplegia and mental disorders. Most patients die within months. Other parasitic site include eyes and lungs. In terms of eyes, the infection is mainly bilateral retinochoroiditis. In addition, patients may also sustain systemic responses and multiple organs injuries.

3.2.3 Isoporiasis

The common pathogen of human isoporiasis is isospora belli and it parasitizes in the cytoplasm of the intestinal epithelial cells. Via intake of mature oocysts, the individuals are infected. The disease is relatively endemic, which prevails mainly in Africa. After infection of isospora, the patients may experience severe diarrhea, enterogastritis, nausea, anorexia or even diffuse abdominal pain. The patients may also sustain long-term chronic diarrhea and weight loss, which are rarely accompanied by coccidiosis pulmonary infection.

3.3 Parasitic Infection and Neoplasms

In 1898, it was first reported a case of schistosomiasis complicated by rectum cancer. After that, the relationship between parasitic disease and neoplasms have attracted much scholarly attention. According to literature reports, schistosoma japonicum can be complicated by colorectal cancer, liver cancer, gastric cancer and breast cancer, while bilharziasis can be complicated by bladder cancer and cervical cancer. Moreover, complications of schistosomiasis mansoni include follicular lymphoma of the spleen, liver cancer, prostate cancer, and cervical cancer. And cholangio cancer is a complication of clonorchis sinensis infection. Opisthorchiasis can be complicated by cholangio cancer. Lymphoma is a complication of filariasis. Colorectal cancer is a complication of amebiasis. Nasopharyngeal cancer is a complication of malaria in males, and cervical cancer is a complication of trichomoniasis. Studies have shown that parasitic diseases are pathologically related to occurrence of neoplasms because they can induce inflammatory responses and cell proliferation, induce chromosome abnormalities, impair DNA methylation and weaken its self repair, induce genic mutation and immunosuppression. However, considering the large number of parasite species and their widespread geographical distribution, however, the relationship between parasitic diseases and occurrence of neoplasms remains to be further studied.

4 Pathogenesis of Parasitic Diseases

Yunling Wang

The pathogenic mechanism of parasitic infection varies according to species and quantity of parasites as well as parasite-hcost adaptation and host responses. The damage caused by parasites can be either confined within the parasitic site or extend into other parts in host. The damages are commonly caused by mechanical pressure, capturing nutrition, toxins, responses and anaphylaxis of host to the stimulation from parasites as well as passage for other pathogens invading the host.

4.1 Mechanical Damage

Mechanical damage refers to the damages or lesions to cells, tissues or organs of the host that are caused by mechanical pressure from parasites during the infection process. For instance, local skin lesions can be caused by invasion of mites or fly larvae into the skin of host. And dermatitis may be caused by penetration of cercariae of schistosoma japonicum into the skin, with local spots of papule and pruritus, which is an immediate and delayed allergic response. Hookworm dermatitis is a common hookworm diseases and the patients experience sensations of needling, burning and itching within dozens of minutes after invasion of hookworm larvae into the skin. Congestive spots and papule may develop thereafter, which may further progress into local redness, swelling and blisters. When such larvae as roundworm and hookworm move and pass through the organs (especially lungs), they may penetrate the pulmonary capillaries to cause vasculitis as well as embolism and rupture of pulmonary capillaries. Therefore, local cell infiltration and spots of hemorrhage may be produced. In some serious cases, pulmonary hemorrhage and edema or even secondary bacterial infection may occur. Especially when the parasites are large in size and quantity, the effects may be extremely dangerous. For instances, in the cases of larvae of echinococcus granulosus parasitizing in human liver, the patients are asymptomatic at first. However, along with their growing bigger in size, they compress liver tissues and other abdominal organs to show symptoms due to obvious compression and stimulation. The disease caused by parasitizing of Echinococcus in human skeleton is known as osteohydatidosis, whose occurrence is commonly found in the pelvis, central vertebra, and metaphysis of long bones. The patients are vulnerable to fracture or osteomiosis due to its destruction to bones.

Large quantity of parasites in the intestinal canal may cause intestinal obstruction. For instance, adult roundworms parasitizing in the intestinal lumen are likely to move into various organs with their opening in the intestinal wall, such as bile duct, pancreatic duct, and appendix, to cause biliary ascariasis, ascaris pancreatitis, and appendicitis.

4.2 Capturing Nutrition and Developmental Disorders

All the growth, development and reproduction of parasites are supplied by their hosts. And larger quantity of parasites need more nutritional substances from their hosts. For instance, roundworms, one of large-sized parasites, take in semi-digested food in the intestinal lumen of their hosts, whose metabolites are toxic irritant to the hosts. It captures nutritional substances and damages intestinal mucosa, leading to dysfunctional digestion and absorption, which further cause malnutrition of the hosts. Hookworms impair functions of human body by sucking blood from their human hosts. The patients sustain long-term chronic blood loss as well as losses of protein and iron that further causes anemia. When hookworms suck human blood, they bite the intestinal mucosa and the blood exudates. Meanwhile, hookworms often bite different sites at the intestinal mucosa and the previous bite wound may continue its bleeding before coagulation, which aggravates anemia. Tiny protozoon, giardia,

Y. Wang
The Second Affiliated Hospital, Xinjiang Medical University, Urumqi, Xinjiang Uygur Autonomous Region, China
e-mail: 1079806994@qq.com

H. Li (ed.), *Radiology of Parasitic Diseases*, DOI 10.1007/978-94-024-0911-6_4

parasitizes in the human upper digestive tract that affect its host via different approaches. Its anterior half part of ventral trophozoite is depressed to form a sucking disc like lacuna, which facilitates its adherence on the surface of intestinal epithelial cells. In the cases with large quantity of giardiae, they cover extensive area that primarily acts to absorb nutrients and, therefore, interfere the nutritional absorption by the host intestine. The loss of nutrients in a large quantity causes malnutrition.

4.3 Toxic Effects

Many species of parasites can produce toxic substances such as secretion, excrement, and decomposition of dead bodies, all of which have toxic effects on hosts. Entamoeba histolytica Schaudinn is one of commonly found protozoa parasitizing in human intestinal ileocecum. When it invades into human intestinal mucosa and liver, it can secrete histolytica enzyme to dissolve tissues and cells. In addition to capturing nutrients from human body, it invades tissues to cause necrosis, which further develop into ulceration of host intestinal wall and liver abscess. Trypanosoma cruzi is a blood flagellate that parasitizes in human smooth muscle cells and myocardial cells in a large quantity. In addition to releasing neurotoxic substances, it can adhere to the autonomic ganglion cells to impair nerves controlling muscular motor and cardiac contraction. Eventually myocarditis as well as esophageal and colonic hypertrophy and dilation develop. In China, filariasis caused by bancroftosis and malayi once prevailed, with their adult worms parasitizing in human lymphatic system to produce many microfilariae. The microfilariae move along with peripheral blood flow and their adult worms split to produce large quantities of toxic substances after death. These toxic substances can stimulate host and cause lymphangitis and subsequent hyperplasia and obstruction. The patients with bancroftosis may experience chyluria due to of urinary and abdominal lymphatic obstruction. However, the patients with malayi may experience local tissue swelling and thickening due to superficial lymphatic obstruction, which eventually develop into elephantiasis of local skin and tissues.

4.4 Inflammatory and Allergic Responses

The excrements, secretions and sheddings from parasites show antigenicity, which can stimulate the organism to produce specific antibodies and sensitized lymphocytes. Therefore, the organism is immunized. When the organism is stimulated for the second time by the same antigen, an abnormal or pathological immune response is produced, which is known as hypersensitivity reaction or allergy. Allergy has the same mechanism with immune response, both of which are actually specific immune response to certain antigen. However, allergy manifests as tissue damage and physiological disorders while immune response acts as a physiological defense exerting beneficial effects on organisms. The underlying mechanism of allergy involves substance stimulation and reactivity of organism. Substance stimulation is a prerequisite condition for inducing antibody to produce allergy, which rarely occurs in those with allergic constitution.. Based on the mechanism of allergy and its clinical manifestations, allergy can be classified into type I, type II, type III and type IV, which, in turn, is known as immediate type, cytotoxic type, immune complex type and delayed type.

Further Reading

Bowels J, Blair D, McManus DP. A molecular phylogeny of the human schistosomes. Mol Phylogenet Evol. 1995;4(2):103–9.

Chen J, Liu Y, Dai G. Enzymes of Entamoeba histolytica and their detoxification effects. Chinese J Parasitic Dis Control. 1996;9(1):40.

Jelinek T, Schulte C, Behrens R, et al. Imported Falciparum Malaria in Europe: sentinel surveillance data from the European network on surveillance of imported infectious diseases. Clin Infect Dis. 2002;34(5):572–6.

Jia S. Potential impact of global warming on spreading of malaria. Chinese J Parasitic Dis Control. 2004;17(1):63–4.

Office of National Survey on the Important Parasitic Diseases. National survey report on the important parasitic diseases in China. Chinese J Parasitol Parasitic Dis. 2005;233(5):332–40.

Rajan TV, Porte P, Yates JA, et al. Role of nitric oxide in host defense against an extracellular, metazoan parasite, Brugia malayi. Infect Immun. 2009;64(8):3351–3.

Shanshan Z. Hypersensitivity. In: Zhenzhou long. Medical immunology. Beijing: Peoples' Health Publishing House; 1998. p. 152.

Vadas MA, David JR, Butterworth AE, et al. Functional studies on purified eosinophils and neutrophils from patients with Schistosoma mansoni infection. Clin Exp Immunol. 1980;39(3):683–94.

WHO. Schistosomiasis and soil. transmitted helminth infections. preliminary estimates of the number of children treated with albendazole or mebendazole. Wkly Epidemiol Rec. 2010;81(16):145–64.

Wu W, Sun D. Progress in globally dliminating lymphatic filariasis. Chinese J Parasitic Dis Control. 2005;18(1):64–6.

Xu L, Yu S, Shuhui X. Distribution and damage of human parasites in China. Beijing: Peoples' Health Publishing House; 2000.

Zhou X, Wang T, Wang L. Epidemiological analysis of schistosomiasis in China. Chinese J Epidemiol. 2004;25(7):555–8.

Zhou S, Yuan W, Yuan F, et al. Malaria in China, 2007. Chinese J Parasitol Parasitic Dis. 2008;26(6):401–3.

5 Radiological Modalities for Diagnosis

Ping Li, Jifeng Zhang, Shanshan Yu, Yinglin Guo, and Xuhua Yang

5.1 X-ray Examination

5.1.1 X-ray Examination

X-ray examination is the most traditional, most widely used and most important modality in diagnostic imaging. Since the discovery of X-ray by a German physicist Wilhelm Röntgen in 1895, along with deeper understandings about properties of X-ray and the gradual improvement of X-ray apparatus, X-ray examination is playing an increasingly important role in clinical diagnosis of various diseases.

X-ray examination can be categorized into routine examination, special examination and contrast examination.

5.1.1.1 Routine X-ray Examination

Routine X-ray examination includes fluoroscopy and radiography. In a fluoroscopic examination, an image enhancement TV system is used with obviously enhanced contrast of the image and favorable imaging effect. During fluoroscopy, the patients can change their position via rotation so as to change the perspective for dynamic changes of the examined organ. So far, radiography is still the most commonly used modality in diagnostic imaging, with more favorable spatial resolution and density resolution than fluoroscopy. It can develop the tissues with large variances in density and thickness, and can also develop the lesions with minute variances in density and thickness. However, it is incapable of demonstrating dynamic changes of the examined tissues, organs and lesions.

5.1.1.2 Special X-ray Examination

Special X-ray examination includes tomography, soft X-ray radiography and high voltage radiograph, all of which are rarely used in clinical practice. To examine a special organ such as mammary gland, soft X-ray radiography is applicable to conduct mammography.

5.1.1.3 Contrast X-ray Examination

The contrast is to increase the density variance between different tissues as well as between normal and pathological tissues. The contrast X-ray examination is mainly applied for more favorable demonstrating different tissue structures or pathological changes with poor natural contrast. For contrast X-ray examination, a substance with higher or lower density than the target tissue is introduced into the tissue or its surrounding space to produce density difference for imaging and the following observation.

5.1.2 Selection of X-ray Examinations

Based on the full understandings about indications and contraindications as well as advantages and disadvantages of each X-ray examination, an appropriate X-ray examination should be selected according to the clinical primary diagnosis and clinical diagnosis. Generally speaking, safety, accuracy, conveniency, and economical cost should be taken into account.

5.2 Computed Tomography

Computed tomography (CT) is a diagnostic technique that integrates computer and X-ray.

P. Li (✉) • J. Zhang • S. Yu
The Second Affiliated Hospital, Harbin Medical University, Harbin, Heilongjiang, China
e-mail: 523371675@qq.com

Y. Guo
Taiping People's Hospital, Daowai District, Harbin, Heilongjiang, China

X. Yang
City Kangan Hospital (Former City Hospital for Infectious Diseases), Mudanjiang, Heilongjiang, China

H. Li (ed.), *Radiology of Parasitic Diseases*, DOI 10.1007/978-94-024-0911-6_5

5.2.1 Characteristics of CT Images

A CT image is composed of a certain quantity of pixels which are of various shades from black to white and are matrix arranged. These pixels show the X-ray absorbance index of the corresponding voxel. Images produced by different CT devices are of different pixel size and quantity. Smaller and more pixels constitute more exquisite image that indicates higher spatial resolution.

CT image is shown by different grey levels, which reflect the X-ray absorbance index of organs and tissues. Black shadow indicates low absorbance area, namely the low-density area, such as lungs. White shadow indicates high absorbance index, namely the high-density area, such as skeleton. Compared with X-ray images, CT images show high density resolution, which is the prominent advantage of CT.

5.2.2 CT Examination

5.2.2.1 Plain CT Scan

Plain CT scan refers to common CT scan with no use of contrast enhancement or radiography. In clinical practice, plain CT scan is firstly performed for patient when CT is necessary for the diagnosis.

5.2.2.2 Contrast Enhancement Scan

After the intravenous infusion of water-soluble iodinated contrast material, the blood iodine concentration increases, with following iodine concentration variance in organs and lesions. The density variance renders more favorable development of the lesions on the image.

5.2.2.3 Image Post-processing Technique

Spiral CT scan requires short scanning time and imaging time, but with large scanned area and thin layer thickness for scanning, to harvest consecutive sections images. After post-processed on a computer, 2-dimensional or 3-dimensional images and angiographic images can be reconstructed on any orientation. The commonly used techniques include:

Image Reproduction

The image reproduction technology can be categorized into three types, surface reproduction, maximum intensity projection and volume rendering. By using image reproduction technique, 3-dimensional CT images can be obtained, and thus observation from any perspective via rotation is possible.

CT Virtual Endoscopy

Simulation technology is a computer based technology, which has been integrated into CT or MRI to be developed into CT virtual endoscopy. The integration of volume data and computer based simulation, such as navigation and roaming in lumen, is capable of simulating the process of endoscopy, gradually demonstrating the lumen from one end to the other. The demonstration of lumen looks real with pseudo-color processing. Such procedures include virtual intravascular endoscopy, virtual bronchoscopy, virtual laryngoscopy, virtual cholangioscopy, and virtual colonoscopy, all of which show favorable demonstrating effect.

CT Perfusion Scan

After intravenous infusion of mass water-soluble iodinated contrast agent, consecutive scanning of interested organs such as brain, liver, kidney and heart is conducted along a certain direction to obtain multi-frame images. According to the density changes of the images at different time points, a time-density curve of each pixel can be drawn to calculate the parameters such as peak time, average transit time, regional blood volume and regional blood flow volume of the contrast agent when it reaches the lesion. After color coded processing, 4 parameter diagrams can be obtained. By analyzing these parameters and parameter diagrams, we can understand the hemodynamics within capillaries of the interested organ, namely blood flow perfusion. Therefore, CT perfusion scan is a type of functional imaging.

5.2.3 Application of CT in Clinical Diagnosis

5.2.3.1 Diagnosis of Diseases in Central Nervous System

CT examination has high diagnostic value and is widely used in the diagnosis of central nervous system diseases, such as intracranial neoplasms, abscess and granuloma, parasitic diseases, traumatic hematomas and cerebral injuries, cerebral infarction and hemorrhage, intraspinal neoplasms as well as intervertebral disc herniation.

5.2.3.2 Diagnosis of Head and Neck Diseases

CT is also of great value in the diagnosis of head and neck diseases, such as intra-orbital space-occupying lesions, paranasal sinus cancer in its early stage, small cholesteatoma of middle ear, destruction and dislocation of auditory ossicle, slight destruction of inner ear bony labyrinth, congenital dysplasia of ear, and nasopharynx cancer in its early stage. However, obvious lesions can be defined by plain X-ray, and CT examination is, therefore, unnecessary.

5.2.3.3 Diagnosis of Thoracic Diseases

As high-resolution CT has been applied into clinical practice, CT shows its superiority in the diagnosis of thoracic diseases. Contrast enhanced scan is commonly selected to define the presence of lump or enlarged lymph nodes at the mediastinum and hilum as well as bronchial stenosis or obstruction. It also facilitates the diagnosis of primary and metastatic mediastinal neoplasms, lymphonode tuberculosis, and lung carcinoma of the central type. The low dosage radiation scanning is even applicable in the screening of lung cancer. In addition, interstitial and parenchymal lesions within the lungs can also be favorably demonstrated. CT scan can favorably reveal the lesions that are undetectable by plain X-ray, such as the lesions overlapping with heart or major vessels as well as pleural lesions, diaphragmatic lesions, and lesions at the thoracic wall.

5.2.3.4 Diagnosis of Cardiac and Major Vascular Diseases

CT examination is of great importance for heart and major blood vessels, especially the major blood vessels. For the heart, CT is mainly applied to diagnose pericardial diseases and to show the cardiac chambers and walls. Because the scanning time is usually longer than a cardiac cycle, the legibility of images as well as their diagnostic value are limited. However, CT examination can favorably reveal calcification of coronary artery and cardiac valves, calcification at the major vascular wall and the changes of arterial aneurysm.

5.2.3.5 Diagnosis of Abdominal and Pelvic Diseases

CT is mainly applied in the diagnosis of diseases occurring in liver, gallbladder, pancreas, spleen, peritoneal cavity, retroperitoneal space, urinary and reproductive system, especially the space occupying lesions, inflammatory lesions and traumatic lesions. CT also has great value in examining the gastro-intestinal lesions that invade outwards as well as their adjacent and distant metastasis. However, it can not replace barium radiography, endoscopy and pathological biopsy in the diagnosis of gastro-intestinal lesions.

5.3 Magnetic Resonance Imaging

Magnetic resonance imaging (MRI) is a medical imaging technique using signals produced by resonance of nucleus in magnetic fields to reconstruct images of human body. In recent years, MRI has been developing rapidly and improving greatly, with capabilities of examining all body systems and worldwide application.

5.3.1 MRI Technology

MRI is different from CT scan because it requires transverse, sagittal and/or coronal imaging, as well as T_1WI and T_2WI. Therefore, appropriate impulse sequence and scanning parameters should be selected. The spin echo (SE) technique with multi-layer and multi-echo is commonly applied. Time parameters for scanning include echo time (TE) and repetition time (TR). T_1WI can be performed by using short TR and short TE, while T_2WI by using long TR and long TE, with the time recorded in milliseconds. According to the length of TE, T_2WI can be categorized into intense, moderate, and slight. The intensity of signal by T_2WI facilitates qualitative diagnosis of the lesions.

The SE impulse sequence commonly selected during MRI requires relatively longer periods of time in scanning and imaging. In order to overcome these disadvantages, imaging techniques such as gradient echo sequence and fast SE sequence have been developed with significant achievement in recent years, and have been widely applied in clinical practice. In addition, techniques such as fat suppression, water suppression and magnetic resonance angiography have also been developed to obtain more MRI data for analysis and diagnosis.

5.3.1.1 Fat Suppression in MRI

Adipose tissue has high proton density with short T_1 time but comparatively long T_2 time, which, therefore, is demonstrated as extremely high T_1WI signal and relatively high T_2WI signal. On the commonly adopted SE T_2WI images, the signal intensity of adipose tissue is further increased. Such properties of adiopose tissue provide favorable natural contrast for detection of lesions, such as lumps growing in subcutaneous tissue or bone marrow cavity. The bone marrow tissue or subcutaneous tissue is demonstrated with extremely high T_1WI signal, while the lump has an obviously longer T_1 value than adiopose tissue and thus is demonstrated with relatively lower signal. Therefore, the obvious contrast greatly facilities detecting the lesion.

From another perspective, these properties of adipose tissue may also decrease the quality of MR images, and thus affect the detection of lesions. Therefore, the significance of fat suppression technique in MRI are as the following: (1) to decrease motion artifact, chemical shift artifact and other related artifacts; (2) to suppress the signal from adipose tissue to increase the tissue contrast on images; (3) to optimize the effect of contrast enhanced imaging; and (4) to identify the presence of adipose tissue within lesion, because the T_1WI signals of adipose tissue, protein containing liquid and hemorrhage are high and thus fat suppression technique helps in differential diagnosis. For instances, neoplasms containing mature adipose tissue in kidneys are usually angioleiomyolipoma, and lesions with fatty degeneration in liver is usually well differentiated hepatocellular carcinoma or hepatocellular adenoma.

Fat suppression techniques in MRI are varied, which, but in general, are mainly based on the following two mechanisms: (1) chemical shift of fat and water; (2) longitudinal relaxation difference between fat and other tissues.

Based on the above-mentioned properties of adipose tissue, many techniques can be applied in MRI for fat suppression. MRI apparatus with different field intensities should be equipped with different fat suppression techniques, and MRI apparatus with the same field intensity can also be equipped with different fat suppression techniques according to the different body parts for examination, different purposes of examination and different imaging sequences adopted. The commonly used fat suppression techniques include: (1) spectral fat saturation; (2) STIR (Short TI Inversion Recovery); (3) SPIR (Spectral Presaturation with Inversion Recovery); (4) Dixon technique; and (5) presaturated zone technique.

5.3.1.2 Magnetic Resonance Chemical Shift Imaging (MRSI)

Chemical shift imaging is also known as in-phase/out-of-phase MR imaging. It is based on the chemical shift effect of the protons in fat and water molecules, and integrates nuclear magnetic resonance and chemical shift to analyze specific nucleus and its compounds for non-invasive studies of biochemical metabolism of living organism. MRSI combines the advantages of MRI and spectrum to show the changes of biochemical metabolism of the interested substance on MRI with spectral curves. Therefore, a comprehensive diagnosis combining anatomic morphology with biochemical change can be made.

Compared with common T_1WI (by in-phase MR imaging), out-of-phase MR imaging is characterized by: (1) obviously attenuated signal of tissue with mixed water and fat with the attenuation degree exceeding spectral fat saturation; (2) no obviously attenuated signal of pure adipose tissue; (3) crispening effect.

Currently, chemical shift MR imaging is more commonly applied to examine organs in abdominal cavity in clinical practice. Its application is commonly intended to:

1. differentially diagnose lesions at adrenal gland. Due to the common presence of lipid within adrenal adenoma, its signal intensity by out-of-phase imaging obviously decreases. However, chemical shift MR imaging helps in diagnosis of adrenal adenoma, with a sensitivity of 70–80 % and a specificity of 90–95 %.
2. diagnose and differentially diagnose fatty liver. Chemical shift MR imaging shows a higher diagnostic sensitivity than routine MRI and CT for the diagnosis of fatty liver.
3. detect the presence of fatty degeneration in the focal hepatic lesions. The patients with fatty degeneration in the focal hepatic lesions are more commonly diagnosed with hepatocellular adenoma or well differentiated hepatocellular carcinoma.
4. others. Chemical shift MR imaging is also helpful in the diagnosis and differential diagnosis of renal or hepatic angioleiomyolipoma.

5.3.1.3 MR Water Imaging

As MRI technology advances, MR Water Imaging has also gained wide application during recent years, which provides extremely valuable data for the diagnosis of diseases occurring in organs containing liquid.

The principle of MR Water Imaging is comparatively simple, mainly based on the properties of water with long T_2. In all the tissues and organs in human body, the T_2 value of aqueous components, such as cerebrospinal fluid, lymph, bile, gastrointestinal fluid and urine, is far higher than that of other organs. If T_2WI sequence with very high T_2 weight is selected, namely a very long TE (above 500 ms), the horizontal magnetization vectors of other organs almost completely detenuate to show extremely low or no signal. However, due to the long T_2 value of the aqueous components, a fairly large horizontal magnetization vector can be maintained. Therefore, the collected signals are virtually all from the aqueous components, and the technique is thus known as MR Water Imaging.

MR Water Imaging has gained comparatively wide application in recent years and now the commonly used techniques in clinical practice include:

MR Cholangiopancreatography (MRCP)

MRCP is the most commonly used MR Water Imaging technique in clinical practice. Its indications include cholelithes, choleangial neoplasms, choleangitis, pancreatic neoplasms, chronic pancreatitis, variation and malformation of cholepancreatic ducts.

MRCP can be performed using GRE sequence or FSE sequences. Currently, the new-type MRI apparatus is commonly equipped with single-shot FSE (SS-FSE) T_2WI sequence or half-fourier acquisition single-shot turbo-SE (HASTE) T_2WI sequence. MRCP is commonly performed by 3D or 2D consecutive thin-layer scanning and 2D thick-layer projecting scanning.

MR Urography (MRU)

MRU is also one of the commonly used MR Water Imagings in clinical practice. And its main indications include lithangiuria, kidney pelvic and/or calyceal neoplasm, ureteral neoplasm, bladder neoplasm, urinary tract obstruction of other reasons as well as variation and malformation of urinary system.

MRU adopts the similar sequence and scanning technique with MRCP, and the key points for image analysis also resembles to those of MRCP.

MR Myelography (MRM)

MRM has been increasingly applied in clinical practice in the past few years, with its imaging effect being close to myeloid iodinated contrast imaging. By integration of MR imaging to myelography, it has basically replaced myeloid iodinated contrast imaging. Its main indications include intraspinal neoplasm, spinal canal malformation, spinal nerve sheath lesion, spinal degenerative diseases, and spinal trauma.

MR Sialography

MR sialography is often used to detect lesions in the parotid duct. It is usually performed by using high-resolution 3D True FISP sequence or 3D FSE sequence.

MR Inner Ear Water Imaging

With the lymph in cochlea or semicircular canal as the natural contrast agent, MR inner ear water imaging produces images to examine lesions in membranous labyrinth. It is often performed by using high-resolution 3D True FISP sequence or 3D FSE sequence.

5.3.1.4 MR Angiography (MRA)

MRA has been one of the routine MRI techniques for radiological diagnosis of diseases, with advantages of noninvasiveness, conveniency, low cost, and generally no use of any contrast agent compared with DSA. Currently, MRA are performed by using time of fly (TOF), phase contrast (PC) or contrast enhancement MRA (CE-MRA).

MRA is clinically applied to examine the following blood vessels.

Cerebral or Cervical Blood Vessels

MRA is applied to detect lesions of the cerebral or cervical blood vessels, such as cerebral/cervical arteriostenosis or occlusion, arterial aneurysm, and vascular malformation.

Pulmonary Artery

MRA can also be applied to detect pulmonary artery embolism and pulmonary arterio-venous fistula. For pulmonary artery embolism, CE-MRA can more favorably reveal vascular embolism of subsegmental and above level arteries. For arterio-venous fistula, CE-MRA can reveal the feeding artery and draining vein.

Aorta

MRA is also applied to detect aortic aneurysm, aortic dissection, and aortic malformations.

Renal Artery

It is also used to detect renal arteriostenosis.

Mesenteric Vascular Vessel and Portal Vein

It is mainly used to examine mesenteric vascular stenosis or thrombus, portal hypertension and its collateral circulation.

Limbs Blood Vessels

It is mainly used to detect limbs vascular stenosis, arterial aneurysm, thromboangitis and vascular malformation.

5.3.1.5 MR Diffusion-Weighted Imaging (DWI)

DWI is currently the only noninvasive examination to detect the water diffusion in living organisms in vivo.

DWI is clinically applied in the diagnosis and differential diagnosis of hyperacute cerebral infarction. On DWI, the brain tissues in the cases of hyperacute and acute cerebral infarction are demonstrated with high signal. Compared with routine T_1WI and T_2WI, DWI can reveal the abnormal signal of infracted brain tissue earlier.

It should be noted that some other cerebral lesions may also show high signal on DWI, such as active lesion of multiple sclerosis, some neoplasms, hematoma and abscess. Therefore, differential diagnosis is important. In addition to brain lesions, DWI can also be applied to detect lesions in such organs as liver, kidney, mammary gland, spinal cord, and bone marrow. DWI might provide information for the diagnosis and differential diagnosis of lesions in these organs. However, relevant clinical knowledge and experience are still insufficient, and more clinical studies are needed.

5.3.1.6 Perfusion-Weighted Imaging (PWI)

PWI is one of the brain functional MR imagings, providing micro hemodynamic information within tissues. The mechanism and technique of PWI are complex, and it remains to be improved in clinical application. Herein, we briefly introduce this imaging technique. MR PWI can be performed by using first-pass contrast-enhanced MRI or arterial spin labeling.

The followings are often studied in clinical practice:

Brain PWI

Brain PWI is performed by the most commonly using single-shot GRE-EPI T_2WI sequence to explore the blood supply to cerebral ischemic lesions and brain neoplasms.

Myocardial Perfusion

Myocardial perfusion is performed by the most commonly using super fast spoiled GRE T_1WI sequence or multiple shots IR-EPI T_1WI sequence. It is mainly adopted to explore myocardial ischemia. PWIs conducted in the quiescent condition and loaded condition, respectively, can detect the reservoir of myocardial perfusion, which is helpful for the early detection of myocardial ischemia.

Kidney Blood Perfusion

MR renal perfusion imaging is a noninvasive examination to semi-quantitatively measure the hemodynamics of physiological and pathological kidney conditions with no renal toxicity. Combined with conventional MRI imaging sequences may more accurately reflect the morphology and function of the kidney.

Liver Blood Perfusion

Liver perfusion is a noninvasive, fast and safety method. It can provide both morphological information and hemodynamic characteristics of liver, which is helpful for accurate diagnosis. So, it plays an important role in clinical practice.

5.3.1.7 MR Spectroscopy (MRS)

MRS is currently the only noninvasive examination to detect the chemical substances in living organisms in vivo. MRI provides morphological information about the normal and abnormal tissues, while MRS provides the metabolic information of the tissues. It is well known that during the occurrence and development of many diseases, metabolic changes often occur earlier than the morphological changes. Therefore, the metabolic information of tissues provided by MRS is undoubtedly very helpful for the early diagnosis of diseases. However, its clinical application is still being studied and explored.

5.3.1.8 Magnetization Transfer (MT)

MT is one of the new imaging techniques launched in recent years, which enhances the image contrast or creates a new contrast by physical approach. Recently, MT is more frequently used to examine the nervous system, such as its application in TOF MRA or contrast enhanced scan.

5.3.1.9 Other MRI Related Important Techniques

In clinical MRI examinations, some other techniques are extremely important for the quality of MRI, including respiration compensation technique, respiratory gating technique, ECG-gating technique, and ECG triggering technique.

5.3.2 Clinical Application of MRI

5.3.2.1 Diagnosis of the Diseases in Central Nervous System

MRI application for the diagnosis of diseases in the central nervous system is fairly mature. By the 3-dimensional imaging and flow void effect, localization of the lesions is more accurate and the relationship between lesions and blood vessels can be observed. MRI is obviously superior to CT in examining the brainstem, infratentorial area, foramen magnum region, spinal cord and intervertebral disc. And it also has high value in the diagnosis of cerebral demyelinate diseases, multiple sclerosis, cerebral infarction, brain and spinal neoplasms, hematoma, congenital spinal abnormalities, and syringomyelia.

5.3.2.2 Diagnosis of Thoracic Diseases

On MR imaging, fat and blood vessels in mediastinum show favorable contrast, which facilitates the observation of mediastinal neoplasm and its anatomical relationship with its vessels. It is also very helpful for the diagnosis of hilar lymph node and lung carcinoma of the central type.

5.3.2.3 Diagnosis of Cardiac and Major Vascular Diseases

On MR imaging, the cardiac and major vascular lumen can be well shown. Therefore, the studies on morphology and hemodynamics of the heart and major vessels can be conducted by this noninvasive examination.

5.3.2.4 Diagnosis of Abdominal and Pelvic Diseases

MRI examination also has certain diagnostic value for the diseases at abdominal and pelvic organs, such as liver, kidney, bladder, prostate, uterus, neck and breast. MRI is also superior to CT in showing early stage of malignancy and vascular invasion as well as staging of neoplasm.

5.3.2.5 Diagnosis of Skeletal and Joint Diseases

The bone marrow is demonstrated as high signal on MR imaging, and thus the lesions involving the bone marrow, such as neoplasm, infection and metabolic disease, can be favorably revealed by MR imaging. MRI also has advantages in showing intra-articular lesions and lesions in soft tissue. However, MRI shows its limitations in revealing skeletal lesions.

5.3.2.6 Diagnosis of Other Diseases

MRI is expected to be applied in the studies concerning blood flow volume, biochemistry and metabolic functions. It also makes the early diagnosis of malignancy possible.

Within the range of magnetic field intensity to complete MR imaging, it shows no negative effect on human health and thus is a noninvasive examination. However, MRI equipment is expensive and thus the cost for MRI examination is relatively high. In addition, it requires long time for examination and shows limited value in detecting some organs and diseases. Therefore, its use should be appropriately indicated.

5.4 PET/CT

Positron emission tomography (PET) is a molecular imaging technology for functional metabolic imaging. PET examination is performed by using a positron emitting radionuclide as tracer, and the functional metabolic condition of lesions can be detected by understanding of their uptake of the tracer. Based on such a mechanism, PET is applied to diagnose diseases.

PET/CT is one of the fairly advanced diagnostic imaging technologies in the field of radiology, and the apparatus integrates two technologies, namely PET and CT. And the technology integrates diagnostic capabilities of PET and CT, which is not simple functional addition but rather complementation of their advantages.

5.4.1 Application of PET/CT in Diagnosis of Neoplasm

PET/CT can be applied for: (1) early diagnosis of neoplasms as well as differential diagnosis of benign neoplasms from malignancies; (2) staging and grading of malignancies; (3) assessment of therapeutic efficacy as well as prediction of prognosis; (4) early identification of recurrent neoplasm as well as restaging of recurrent neoplasms; (5) searching for the primary neoplasm; and (6) localization of biological target for radiotherapy.

5.4.2 Application of PET in the Nervous System

Brain metabolism imaging facilitates accurate understandings about the activities and metabolic changes of nerve cells under normal and diseased conditions. In addition, it also demonstrates the metabolic status of cerebral cortex after different physiological stimulus and during different thinking activities. By using PET, we can directly know the metabolic activities of human brain as well as its various physiological or pathological metabolic changes via reading images. Currently, PET is mainly used in locating epilepsy, early diagnosis of dementia, brain receptor study, diagnosis of cerebrovascular disease, pharmacological assessment and guidance of neurological or psychiatric medication.

Some key factors restrict the development and wider use of PET, such as complicated system, high cost for patients and in maintenance, high sensitivity, insufficient specificity, and certain false positive and false negative rates. However, it is still indispensable radiological apparatus in providing information about neural activity. For instances, in the diagnosis of hepatic echinococcosis, PET/CT can provide information about the location, shape, quantity, boundary, calcification and surrounding tissues of the lesion as well as its biological boundary and activities. In addition, it also has value in diagnosing surrounding metastasis as well as screening and diagnosing distant metastasis.

5.5 Ultrasonography

Ultrasonographic diagnosis emerges in the 1940s in the medical field. Due to its noninvasive operation and no ionizing radiation to patients, it is highly recommended by medical professionals. Now new techniques of ultrasonography are continually developed with emergence of 3-dimensional ultrasonography, harmonic imaging, and intraluminal ultrasound, which have been widely used in diagnosis, therapeutic assessment and prognosis prediction.

5.5.1 Development and Clinical Application of Ultrasonography

5.5.1.1 2-Dimensional Ultrasound

B-mode ultrasound is based on the theory of echo, in that impulse ultrasound is emitted into human body, and the echoes from all the tissue interfaces are received to serve as the diagnostic basis. The formed 2-dimensional real-time dynamic images have such merits as authenticity for direct observation, non-invasiveness and operational conveniency. And thus it is the most widely used ultrasonography in medical field, which is mainly applied in the diagnosis of cardiovascular and cerebrovascular diseases, abdominal organ injuries, neoplasm, pediatric diseases, obstetrical and gynecological diseases, and other diseases.

However, 2-dimensional ultrasound cannot well define the cavities containing air (such as stomach and intestines), the tissues containing air (such as lung), and the skeleton. In addition, due to its limitations in section range and scanning depth, it poorly defines the adjacent structures to the organ or tissue with lesion.

5.5.1.2 3-Dimensional Ultrasound

The 3-dimensional ultrasonography is based on the mechanism of 3-dimensional geometric composition, contour extraction, and body element model. The procedures of 3-dimensional ultrasonography include data acquisition, 3-dimensional image reconstruction and 3-dimensional image display, and the images can be grouped into static and dynamic.

5.5.1.3 Interventional Ultrasound

Interventional ultrasound is based on the ultrasonography for invasive diagnosis or treatment. Guided by real-time ultrasound, punctures for biopsy, X-ray radiography, aspiration, intubation, and local medication can be performed. Along with the improvement and development of various catheters, puncture needles, biopsy needles and biopsy techniques, interventional ultrasound has advanced the ultrasound-guided cytodiagnosis into histopathological diagnosis.

5.5.1.4 M-mode Ultrasound

This technique is based on M-mode anatomical imaging, instead of probe imaging, to generate more accurate information under the circumstances of abnormal heart shape and location. Compared with traditional M-mode ultrasound that the sampling is limited within a fan-shaped 90° angle, the M-mode anatomical ultrasound realizes arbitrary sampling within 360° angle and analysis of the M-mode echocardiogram on any point from any perspective. Therefore, it greatly strengthens the advantages of M-mode ultrasound in precise quantification time and spatial resolution.

M-mode ultrasound can be applied in studies on ventricular systolic and diastolic functions of heart and atrial functions of normal heart as well as in examining atrial function, atrioventricular accessory pathway, and pulmonary hypertension.

5.5.1.5 Color Doppler Blood Flow Imaging (CDFI)

CDFI is an abbreviation for real-time 2-dimensional color doppler blood flow imaging. It is based on multi-channel to gain the impulse doppler signal from cross sections of different depth. High-speed computer is used to conduct phase detection, autocorrelation processing, and color gray-scale encoding, and then to display the average blood flow velocity in colors. Therefore, the real time 2-dimensional overlapping display of the space and time of anatomical sections and blood flow is realized. This is the first stage of color doppler blood flow imaging technique and also the milestone in the development of doppler technique.

5.5.1.6 Quantitative Tissue Velocity Imaging

Quantitative tissue velocity imaging (QTVI) is based on speed sampling of the myocardial motion from all the segments of the left ventricular wall to harvest a hodograph of a complete cardiac cycle so as to quantitatively assess the left cardiac function.

5.5.1.7 Tissue Tracking Imaging

Tissue tracking imaging (TTI) is an echocardiography based on tissue Doppler imaging. It can provide information for rapid assessment about the movement distance of all the myocardial tissue in the left ventricle towards the cardiac apex during its systolic period, with 7 layers of colors for display. TTI provides a brand new way to rapidly assess the function of left cardiac ventricle. Especially for the patients shown by low-quality images, TTI is more sensitive than the conventional examinations.

5.5.1.8 Contrast Ultrasonography

The integration of newly developed ultrasound contrast agent into new technology of ultrasonography can efficiently enhance the 2-dimensional ultrasound imaging and Doppler blood flow signals from some parenchymal organs such as myocardium, liver, kidney and brain. In addition, such integration can reveal different blood perfusion in normal and diseased tissues and obviously increase the diagnostic sensitivity and specificity.

5.5.1.9 Harmonic Imaging

When ultrasonic wave is transmitted in medium, it shows both linear and nonlinear effects. Due to the different molecular arrangement in elastic medium, such density changes cause variant velocity of each dot of the ultrasonic wave, which thus leads to morphological malformation of the ultrasonic wave during transmission, namely the production of harmonic wave.

Harmonic imaging includes contrast harmonic imaging and tissue harmonic imaging. In clinical practice, tissue harmonic imaging is performed by using fairly low fundamental frequency emission to amplify the harmonic wave frequency for imaging.

5.6 Endoscopy

Endoscopy is to visually examine the organ via incubation of an instrument into a hollow organ or cavity within human body. In its early stage, it was used for diagnosis, but now it has been one of the essential tools for interventional therapy. It can be categorized into diagnostic use and therapeutic use, including gastrointestinal endoscopy (gastroscopy and colonoscopy), respiratory endoscopy (nasoscopy and bronchoscope), urinary endoscopy (cystoscopy, ureteroscopy, and nephroscopy), female reproductive endoscopy (hysteroscope and colposcopy), close cavity endoscopy (laparoscopy, arthroscopy, thoracoscopy, mediastinoscopy, and amnioscopy), and fetoscopy.

5.6.1 Electronic Gastroscopy

5.6.1.1 Indications and Contraindications

Electronic gastroscopy is indicated to long-term repeated upper abdominal pain, abdominal distension and upset, nausea, vomitting, sour regurgitation, and eructation. It is also indicated to suspected malignancy at esophagus, stomach and duodenum, lesions detected but cannot defined by X-ray barium meal examination, acute and chronic upper gastrointestinal hemorrhage, poor appetite with unknown causes, weight loss and anemia. In addition, electronic gastroscopy is applicable to the patients with definitively diagnosed upper gastrointestinal disease, for following up examinations in patients after esophageal or gastric surgery, and to the patients who need emergency electronic gastroscopic treatment for removal of foreign substance from upper gastrointestinal tract. However, it is contraindicated to serious heart disease, cardiopulmonary insufficiency, severe hypertension, cerebrovascular event, coagulation disorder, low level of hemoglobin or blood platelet, cough, bronchial asthma, and psychiatric disorders.

5.6.1.2 Clinical Application

It is clinically applied in electric coagulation and excision, microwave treatment, laser treatment, medication via injection, removal of foreign substance, percutaneous endoscopic gastrostomy, dilatation for esophageal or pyloric stenosis, and esophageal stent implantation.

5.6.2 Fibrobronchoscopy

5.6.2.1 Indications

Indications for Diagnosis

Fibrobronchoscopy is indicated to hemoptysis with unknown causes, inexplainable chronic cough, hoarse voice with unknown causes, and elevated diaphragm with unknown causes.

Fibrobronchoscopy is also applicable for the diagnosis of lung carcinoma and its staging as well as harvest of tissues for biopsy.

Fibrobronchoscopy can be also applied for the diagnosis of benign bronchial lesions, such as acute or chronic bronchitis, endobronchial tuberculosis, respiratory inhalation injury, tracheal or bronchial stenosis, and suspicious bronchoesophageal fistula.

Fibrobronchoscopy is clinically applied to diagnose diffuse lung diseases.

Indications for Treatment

Fibrobronchoscopy can be applied to fetch the foreign substance from trachea and to aspirate secretions and blood clot in the trachea. In addition, in combination with laser equipment, it can be applied to remove neoplasms or granuloma in the bronchi. And for patients with tracheostenosis, it can be used for dilatation or intratracheal stent inplacement.

5.6.2.2 Contraindications

Absolute Contraindications

Fibrobronchoscopy is absolutely contraindicated to uncontrollable mental confusion, hemorrhagic tendency, hypoxemia, acute respiratory acidosis, severe arrhythmia or poorly controlled high blood pressure and untreated open tuberculosis.

Relative Contraindications

Fibrobronchoscopy is relatively contraindicated to terminal phase of any disease, cardio-pulmonary dysfunction, pulmonary arterial hypertension, asthmatic attack or poorly controlled asthmatic attack, and massive hemoptysis.

5.6.2.3 Clinical Application

Fibrobronchoscopy is clinically applied to qualitatively define the lump in lungs, to trace the origin of suspected and positive sputum cells, and to diagnose refractory cough, wheezing of unknown causes, hemoptysis, bloody sputum and pulmonary atelectasis. For patients with long-term tracheotomy and intubation, fibrobronchoscopy is applied to aspirate secretions from trachea and bronchi. For patients with pulmonary infections, it is applied to diagnose diffuse pulmonary lesions and suspected pulmonary tuberculosis as well as to assist preoperative staging of lung carcinoma and determining the part for surgical removal. For patients with burn, it is applied to diagnose alveolar proteinosis, severe asthma, and pneumoconiosis. In addition, it is applied to fetch foreign substance and for thoracic trauma after thoracic surgery. It is also applicable for treatment of lung cancer and its follow-up examinations.

5.6.3 Fibrocholedochoscopy

5.6.3.1 Application of Fibrocholedochoscopy During Operation

Fibrocholedochoscopy directly visualizes the condition inside biliary tract, and the morphology of bile duct mocosa, its branches and the function of Oddi sphincter can be observed. But its major clinical significance is to accurately diagnose and treat biliary diseases. Its indications include:

1. Choledocholithiasis and hepatolithiasis;
2. Extrahepatic bile duct obstruction and cholangiocarcinoma;
3. Parasites, foreign substances, and other detectable lesions such as benign neoplasm, polyp, stress ulcerations, and granuloma;
4. Thickening of common bile duct wall that exceeds 1 cm; turbid bile; palpable hard nodules at the inferior common bile duct or pancreas;
5. Obstructive jaundice, severe pancreatitis or gallstone pancreatitis;
6. Postoperative biliary syndrome, biliary tract bleeding with unknown causes, abnormal biliary tract pressure;
7. Biliary tract stricture, sclerotic cholangitis;
8. Intravenous cholangiography, percutaneous transhepatic cholangiography, duodenoscopic retrograde cholangiopancreatography, abnormality of intrahepatic and extrahepatic bile duct by preoperative ultrasound;
9. verification of the false positive detected by intraoperative radiography, such as bubble;

Its contraindications include thin common bile duct with a diameter of less than 0.5 cm or thin and crisp common bile duct wall.

5.6.3.2 Trans-T Tube Sinus Tract Fibrocholedochoscopy (Postoperative Choledochoscopy)

The application of trans-t tube sinus tract fibrocholedochoscopy provides an alternative way to treat residue biliary stones, which exempts many patients from the second operation.

Indications

The patients carrying T tube drainage with suspected residue biliary stones should receive the procedure. The patients experiencing fever due to obstruction of biliary tract by stones should receive trans-T tube sinus tract fibrocholedochoscopy.

Contraindications

The procedure is contraindicated to severe heart failure and bleeding tendency. The procedure should be posponed for the patients with fever due to reasons other than biliary tract.

In the diagnosis of parasitic diseases, the examination of choice is etiological examination. But for some parasitic diseases, the application of clinical etiology is limited for their diagnosis. For instances, for the diagnosis of cerebral cysticercosis, the collection of specimen is constricted; for the diagnosis of toxoplasmosis gondii and trichinelliasis, the location for specimen collection is hardly correct, and the specimen collection is invasive with a low detection rate; and diagnostic puncture is contraindicated to echinococcosis. However, radiological modalities are simple, rapid, noninvasive, and capable of positioning, which are modern diagnostic examinations widely accepted by patients.

Different radiology modalities have their own advantages and disadvantages in diagnosis. For some diseases, only one modality can define the diagnosis; otherwise several modalities define the diagnosis. Therefore, we should fully know the capabilities and limitations of each radiological modality to optimize their selection for diagnosis. One radiological modality is impossible to indicate all diseases and one radiological modality cannot fully replace the other modality. The radiological modalities show mutual complementation and verification.

Laboratory Test for Diagnosis of Parasitic Diseases

6

Jinli Lou, Yanhua Yu, and Fangfang Dai

Parasitic diseases generally develop chronically, and most patients experience non-specific symptoms and show unobvious physical signs. Except medical history and physical examination, the diagnosis is mainly based on the laboratory tests. The commonly applied laboratory tests for the diagnosis of parasitic diseases include etiological examination, immunological assay and molecular biological examination.

Etiological examination is to detect the parasitic pathogen from such specimens as stool, blood, bone marrow, sputum, body excretions and secretions as well as tissue from patient. It can be categorized into staining test and non-staining test. The etiological examination by staining is commonly applied to detect larvae, eggs, and cysts of parasite, while the examination with no staining is selected to detect parasite from blood and tissue fluid. Etiological examination is the most reliable way for a definitive diagnose of parasitic infection or disease.

Immunological assay serves as a convenient diagnostic examination of parasitic infection. With the development and improvement of technology in immunological antigen detection, its clinical value in diagnosis has been increasing and the technique has been applied in epidemiological studies and surveillance of epidemics. The conventional examinations include intracutaneous test, circunoval precipitin test, cercarien hullen reaction and indirect haemagglutination test. Nowadays, fluorescence immunoassay, enzyme immunoassay and immunoblotting have been gradually applied to the laboratory diagnosis of parasitic diseases. Therefore, the diagnostic sensitivity and specificity of parasitic diseases have been greatly improved.

In recent years, molecular biology for diagnosis of parasitic diseases has gained rapid development. By using the technologies of DNA probe, PCR and DNA sequencing, DNA fragments and sequence of parasite can be detected in specimens, which have been applied in laboratory diagnosis of trypanosomosis, leishmaniasis, pneumocystosis and toxoplasmosis. Via high throughput and automated DNA hybridization or immunoassay, the biochip technology can be applied to simultaneously examine multiple specific target molecules in one biochip. This will bring about a revolution to high throughput combined detection of infectious diseases and genetic disorders.

In this chapter, we will briefly introduce the laboratory examinations and their clinical application. In combination to the medical history of patients and their clinical symptoms, these laboratory examinations facilitate the diagnosis of parasitic diseases. In addition, they are significance in assessing therapeutic efficacy and predicting prognosis.

6.1 Virological Diagnosis

The virus is the simplest and smallest microorganism whose infection is very common. About 70–80 % of infectious diseases are caused by viruses. Up to now, it has been proved that more than 500 species of virus are pathogenic to humans and some are extremely dangerous, such as viruses inducing avain influenza or SARS. Therefore, virological diagnosis is of great significance in controlling the spread of virus as well as diagnosing and preventing corresponding diseases.

The general principles in laboratory diagnosis of viral diseases are specific, sensitive, rapid, and simple. Based on the epidemiological and clinical features, the suspected pathogenic virus should be primarily judged. And the corresponding laboratory examination should be then chosen based on the biological properties of the suspected virus, immune responses of the patients, clinical course, and the condition of patients. Currently, the laboratory diagnosis of viral infection mainly relies on the conventional examinations and molecular biological examinations. The conventional examinations include virus isolation and culture, identification and serological test, while the molecular bio-

J. Lou (✉) • Y. Yu • F. Dai
Beijing Youan Hospital, Capital Medical University, Beijing, China
e-mail: 13718641405@163.com

H. Li (ed.), *Radiology of Parasitic Diseases*, DOI 10.1007/978-94-024-0911-6_6

logical examinations include nucleic acid hybridization, PCR and modern immunoassays.

6.1.1 Specimen Collection and Transport

6.1.1.1 Specimen Collection

Sampling Time

Sampling should be performed preferably in the early stage and acute stage after onset of the disease or on the first day of hospitalization. It should be performed as early as possible, preferably before treatment. In the later stages of disease, immunity may be induced in patient with less or no virus detected to adversely affect the diagnosis and treatment.

Selection of the Type of Specimen

According to the clinical symptoms and epidemiological data, the infected virus should be primarily judged and the site for specimen collection as well as the type of specimen should be correspondingly selected. The biological properties of the suspected virus should be considered when handling the specimen. For the following diseases, the virus should be isolated for laboratory diagnosis: heart disease, central nervous system infection, congenital or neonatal infection, gastrointestinal disease and respiratory infection.

Commonly Applied Ways for Specimen Collection

For possible respiratory tract infection, nasopharyngeal lavage liquid or sputum is commonly collected for laboratory diagnosis; faeces specimen for possible intestinal infection; aseptic collection of cerebrospinal fluid for possible brain infection; effusion in herpes for possible skin rash; and blood for the cases of viremia. The collected specimen should contain infected cells. The specimen for serological test, especially those for detection of IgG antibody against pathogen, should be collected at the early and convalescent stages for double sera, followed by analyzing dynamic changes of the double sera antibody titer. Only when the serum antibody titer at the convalescent stage is at least four times as high as that in the early stage, the diagnosis can be defined.

6.1.1.2 Transport and Preservation of Specimen

After the collection, the specimen should be maintained aseptic in freezing and moisturized environment and sent for examination immediately. It is more favorable to deliver the specimen into the laboratory within 1–2 h after its collection. Serum specimen should be labeled and preserved at a temperature of −20 °C. Virus tends to be inactivated at room temperature, and the specimen should be maintained at a low temperature and delivered for examination immediately. If transport is needed, the specimen should be stored in thermos flask filled with ice or low temperature materials. And the tissue specimen should be stored at a low temperature with 50 % glycerol buffered saline containing antibiotics or DMSO. If the tissue specimen cannot be delivered immediately, it should be stored at a temperature of −70 °C.

6.1.2 Virus Isolation and Identification

The ways for virus isolation, culture and identification are complex, which require specific conditions and a comparatively long period of time, therefore, fail to be widely applied in clinical diagnosis. For the following conditions, virus isolation, culture and identification can be applied: (1) The patient shows a long-term illness course with a suspected diagnosis of viral infection, all laboratory tests for virus show negative, and isolation of the virus will provide valuable guidance for the diagnosis and treatment. (2) Emergent virus infection is suspected, or re-emergence of eliminated virus is suspected. (3) The same symptoms are suspected to be caused by different viruses. The virus isolation can be applied to define the virus pathogen. (4) It should be defined whether attenuated live vaccine used for survallience shows pathogenic mutant strain. Virus has strict intracellular parasitism. Therefore, sensitive cells, chicken embryos and sensitive animals should be selected based on different virus receptor for isolation, culture and identification of virus.

6.1.2.1 Virus Isolation and Culture

Cell Culture

Cell culture is the most commonly applied way for virus isolation. Appropriate cells should be selected based on the cytotropism of virus. The commonly used cells for virus isolation include: (1) Primary cultural cells, such as simian nephrocytes or human embryonic nephrocytes, are highly sensitive but difficult to obtain. (2) Diploid cell strain, being limited within about 50 generations, can be applied for laboratory use. But the cells may be subjected to aging and death after several generations. (3) Continuous or infinite cell line or strain, such as Hela or Hep-2 cell, can be conveniently preserved in laboratory and shows stability to virus infection, which has gained widespread application. After inoculation, the cytolytic virus can induce cytopathic effect (CPE), while the cells stably infected by virus do not show obvious CPE. However, the membrane surface of a cell infected by virus may show markers of virus expressed protein, such as hemagglutinin and specific antigen of the virus. Hatest or immunological assays may be applied to examine proliferation of the virus. Negative findings in CPE or other laboratory tests may be induced by small quantity of viruses in the specimen that is undetectable even if proliferation of the virus does occur. After blind passage for 3 generations, negative finding definitively indicates no virus in the specimen.

Chicken Embryo Culture & Animal Inoculation of Influenza Virus

Although canine kidney cells can be selected, chicken embryonic inoculation is still the most widely applied way with sensitivity and specificity. Hemagglutination test and hemagglutination-inhibition test may be applied for further identification. Animal inoculation is the most primitive way for virus isolation, but rarely applied nowadays. Mice intracerebral inoculation is still applied for isolation and identification of rabies virus or encephalitis B virus.

6.1.2.2 Virus Identification

Indices for Virus Proliferation in Cultured Cells

Cytopathic Effect (CPE)

Most of the viruses are cytolytic, whose proliferation in sensitive cells causes CPE. CPE may show increased particles within infected cells that cluster or fuse, sometimes with formation of inclusion body. The infected cells are finally subject to cytolysis, detachment and death. Different viruses show different CPE. For instances, adenovirus can cause rounded shrinkage and clustering of the infected cells, which typically shows the appearance of grapes cluster; poliovirus can cause rounded shrinkage, scattering, necrosis and detachment of the infected cells; respiratory syncytial virus may cause fusion of the infected cells to form multinuclear giant cells. Therefore, the infected virus can preliminarily suspected based on the type of selected cell and CPE features. The enveloped virus, such as influenza virus, releases progeny viruses via budding and shows stable infection with no or only undetectable CPE. Such viruses should be identified by other examinations.

Hemadsorption

The envelope of influenza virus carries hemagglutinin. After infection of influenza virus, hemagglutinin can be found on the membrane surface of sensitive cells with infection to render their binding with erythrocytes, which is known as hemaadsorption phenomenon and is an indirect indicator in detecting orthomyxovirus and paramyxovirus.

Determination of Virus Infectivity and Virus Load

Hemagglutination Test

After the virus containing hemagglutinin is inoculated into chicken embryo or infects cells, chicken embryonic amniotic fluid and allantoic fluid or cell culture fluid should be collected, followed by addition of animal erythrocytes. For the cases with virus proliferation and extracellular release, erythrocytic agglutination occurs, which is an indicator of virus proliferation. The virus suspension can be diluted into different levels, the highest dilution degree of hemagglutination may act as a hemagglutination titer for semi-quantitative detection of virus load.

Neutralization Test (NT)

The serum containing specific antibody against a known virus should be firstly mixed with the detected virus suspension, followed by inoculating sensitive cells after a certain period at an appropriate temperature. After culture, the absence of CPE or hemadsorption indicates neutralization of the specific antibody with the corresponding virus infectivity, which is a more reliable diagnostic examination for virus infection. Different concentrations of serum containing specific antibody can be used for the neutralization test to semi-quantitatively examine the detected virus suspension based on the antibody titer.

Plaque Formation Test

Plaque formation test is a way to examine the quantity of virus in the specimen. After a certain amount of appropriately diluted virus suspension is inoculated into sensitive mono-layer cells for culture of a certain period, the cells are then covered by a layer of melting but un-concreting AGAR for continued culture. The proliferation of a single virus can dissolve and detach the mono-layer cells with infection to form a plaque that can be observed by naked eyes. One plaque is induced by proliferation of one virus. The quantity of plaques, therefore, indicates the virus load in the specimen. The final result of plaque formation test is usually indicated by plaque formation unit (PFU) per ml virus suspension, (PFU/ml).

Detection of 50 % Tissue Culture Infectious Does (TCID50)

The detected virus suspension is firstly diluted for a series of ten times, followed by inoculation of mono-layer cells, separately. After culture, the indicators such as CPE should be observed. With the virus load with the highest dilution level that can infect cells as the terminal point, the value of TCID50 can be calculated. TCID50 detection is applied to assess the virus infectivity and the virulence via assessment of CPE.

Detection of Multiplocity of Infection (MOI)

MOI primarily refers to the average of bacteriophages infected by a single bacterial cell in a specific test. But currently it is used as a quantitative indicator of the virus infectivity.

6.1.3 Rapid Diagnosis of Virus Infection

Rapid diagnosis mainly depends on non-culture identification of the virus instead of isolation and identification of the virus. The examinations for rapid diagnosis include direct observation of viral particles in the specimen under an electron microscope or direct detection of the viral components such as antigen and nucleic acid and specific antibody of IgM.

6.1.3.1 Morphological Examination

Electron Microscopy and Immuno-Electron Microscopy

The specimen containing high concentration of viral particles ($\geq 10^7$ particles/ml) can be directly observed under an electron microscope. The specimen containing low concentration of viral particles can be observed after the viral particles are gathered by immune-electron microscopy. Otherwise, the specimen sediments after ultracentrifugation can be observed under an electron microscope to improve the detection rate. Morphological properties of the virus as well as its size and quantity can be observed and determined by electron microscopy.

Light Microscopy

Pathological specimen or specimen containing cast-off cells and needle aspiration cells may show basophilic or eosinophilic inclusions at the sites of virus proliferation such as nucleus and cytoplasm. And the inclusions has a certain value for the diagnosis of virus infection. A diagnosis can also be made according to the pathological properties of the specimen in combination to immunohistochemical staining.

6.1.3.2 Viral Protein Antigen Detection

Immunoassays are commonly applied for direct detection of the virus antigen and thus early diagnosis. Currently, the commonly used immunoassays include enzyme linked immunosorbent assay (ELISA), immunofluorescence assay (IFA) and radioimmunoassay (RIA). These immunoassays are simple in operational procedures but show high specificity and sensitivity. By using high-quality labeling of specific antibody, especially monoclonal antibody labeling, the antigen or hapten at the level of ng (10^{-9} g) to pg (10^{-12} g) can be detected. Due to the possible radioactive pollution induced by radioactive nuclide, the radioactive immunological markers are gradually replaced by non-radioactive markers, such as digoxin. Positive finding of the virus antigen in Western-blot (WB) test can define the diagnosis of virus infection.

6.1.3.3 Early Antibody Detection

Detection of Specific Antibody IgM

To detect specific antibody IgM produced after virus infection can facilitate an early diagnosis of virus infection. For instance, an early diagnosis of fetal congenital infection induced by certain virus can be made when the specific antibody IgM is detected in amniotic fluid from the pregnant woman. HBc antibody is produced early after virus infection, and the detection of specific antibody IgM against HBc is thus used as an indicator of acute hepatitis B virus infection. Detection of specific antibody IgM facilitates the early diagnosis of virus infection, but the production of specific antibody IgM shows obvious individual differences.

Western-Blot Test (WB)

The diagnosis of certain virus infection should be cautious. For instances, the initial screening test positive cases of AIDS and adult leukemia should be confirmatively diagnosed by WB test. In WB test, purified HIV is processed, followed by separation of virus protein based on their molecular weight via polyacrylamide gel electrophoresis. The virus protein is then electro-transferred to nitrocellulose membrane for preparation of blot strip, whose positive reaction to the serum from patient can confirm the diagnosis. If the serum contains antibody against an antigen of HIV, the antibody binds to the corresponding site on the virus protein blotted strip.

6.1.3.4 Detection of Virus Nucleic Acid

Most of the viral genes have been successfully cloned for the whole genome sequencing, which laid a foundation for detection of virus nucleic acid. And the detection of virus nucleic acid is another rapid and specific examination for the diagnosis of virus infection.

Nucleic Acid Electrophoresis

The nucleic acid of orthomyxoviridae and reovirus genus is segmental. For instances, the influenza A and influenza B virues have 8 segments of nucleic acid; the influenza C virus 7 segments; reovirus 10 segments; rotavirus 11 segments. The nucleic acid of rotavirus can be directly extracted from the specimen, and the 11 segments can be shown without endonuclease hydrolysis. After polyacrylamide gel electrophoresis and silver staining, 11 strips can be observed clearly on the gel plate. In combination with clinical manifestations, the diagnosis can be made.

Nucleic Acid Hybridization

Nucleic acid hybridization is performed by using a single-stranded nucleic acid with known sequence as a probe, which should be labeled in advance by a radioactive nuclide (such as ^{32}P or ^{131}I) or biotin, digoxin, horse radish peroxidase. Under certain conditions, the probe binds to target sequence in the specimen according to the complementary base pairing. By examining the marker, the existence of specific virus nucleic acid sequence is proved for etiological diagnosis of virus infection. The commonly used nucleic acid hybridization techniques are as follows:

Dot Blot Hybridization

The DNA or RNA for examination is dotted on the hybridation membrane. After deformation, it hybrids with the labeled probe. Radionuclide or non-radioactive marker can be detected by autoradiography or enzyme reaction.

In Site Hybridization

It is a special examination combing nucleic acid hybridization with cytology. On the pathological section, the DNA or RNA released by cells in situ hybrids with labeled probe with specific nucleic acid. After staining, the intracellular distribution of examined virus nucleic acid and its relationship with cell chromosomes can be directly observed.

Southern Blot and Northern Blot Hybridization

The DNA or RNA extracted from the specimen is cut by using restriction endonuclease, followed by agarose electrophoresis to form endonuclease strips, which are then transferred onto nitrocellulose membrane to hybrid with the labeled probe sequence. In such a way, specific sequence of virus DNA or RNA can be detected.

Polymerase Chain Reaction (PCR)

Specific and conservative virus fragments were selected as target gene. Under the effects of polymerase (Taq enzyme), the designed specific primer sequence is used to amplify the specific sequence of the virus. In such a way, virus infection can be diagnosed. Otherwise, the susceptible area of virus is selected, in combination to restrictive fragment length polymorphism (RFLP) analysis or sequencing, for studies on virus typing and mutation. By using reverse transcription PCR (RT-PCR), the RNA virus is detected. And along with increased demands for laboratory diagnosis, the PCR technology has been gaining development, with emergence of ligase chain reaction (LCR) and real-time PCR in recent years. These new techniques integrate gene amplification, molecular hybridization and photochemistry, which can be applied for real-time dynamic quantitative detection of the PCR amplification products.

Gene Chip

The integration of harvested biological information about virus gene sequencing into automation technology produces gene chip technology. Specifically, gene chip technology derives from combination of single nucleotide polymorphisms labeling with automated chain trace analysis. Its principle is to arrange a biological molecular probe or gene probe with known sequence on a tiny silicon chip or other carrier in a large scale or orderly. The gene chip is then reacted to the biological molecular sequence or gene sequence in the examined specimen. Under sequential excitation by laser, the produced fluorescence spectrum signal is collected by a receiver. And the result can be obtained after data calculating, analyzing and processing. In such a way, a large quantity of specimens can be detected simultaneously for DNA sequencing, which has prospective widespread application in the etiological diagnosis and epidemiological studies of virus infection.

Gene Sequencing

Gene sequenceing includes whole gene sequencing of the virus and characteristic gene fragment sequencing of the virus. At present, the whole gene sequencing of the known pathogenic viruses has almost been completed. The virus gene sequences in the gene pool has laid a solid foundation for the genetic diagnosis of virus infection.

6.2 Bacteriological Diagnosis

6.2.1 Specimen Collecting and Processing

The quality of specimen directly affects the validity of the laboratory test, and inappropriate collection of specimen can lead to false-negative or false-positive result. Therefore, standardized operations for quality control are necessary in each procedure including collecting, transporting and preserving the specimen to guarantee the accuracy and reliability of the laboratory test results.

6.2.1.1 General Principles of Specimen Collection

Early Collection

The specimen should be collected at the early or acute stage of the disease, otherwise during the stage with typical symptoms. In addition, the specimen should be collected before the use of antibacterial agents.

Aseptic Collection

The specimen should be protected from exogenous contamination. During the collection of such specimens from sterile parts as blood, cerebrospinal fluid, pleural effusion and joint fluid, local skin and its surrounding area should be sterilized and the procedure should be performed according to aseptic operations. During the collection of specimens from cavity or canal with opening to the external environment, such as the sinus, the fudus should be chosen for collecting tissue specimen instead of the orifice to prevent contamination from the bacterial colonies at the skin surface. Otherwise, misdiagnosis may occur. During the collection of specimens from the sites where bacteria colonies commonly exist (e.g. oral cavity), the target bacteria should be firstly determined, and the culture mediums should be specifically chosen for pathogenic culture and isolation. The collected specimen should be placed in an aseptic container after it has been physically sterilized such as autoclaving, boiling and dry-heating. Otherwise, the specimen should be placed in a disposable sterile container which should not be processed with disinfectant or acid solution.

Choosing Different Ways Based on the Properties of the Target Bacteria

Different examinations should be applied for the collection of anaerobic bacteria, aerobic bacteria, facultative anaerobe and L-shaped bacteria. As for a urine specimen, when anaerobic infection is suspected, a sterile syringe can be used for suprapubic bladder puncture for specimen collection; when aerobic or facultative anaerobic infection is suspected, natural catheterization or clean-catch midstream urine can be chosen for its collection.

Specimen Collection for an Appropriate Quantity

The quantity of collected specimen should appropriate with representativeness. Meanwhile, some specimens should be collected from different sites at different time points. For instance, for patients with enteric fever, blood specimen should be collected in the first week after the onset; while stool specimen in the second week and urine specimen in the third week. Otherwise, the detection rate of the pathogenic bacteria can be affected.

Safe Collection

During collection, the specimens should be prevented from contamination by bacterial colonies on skin and mucosa. Meanwhile, the specimen collectors should pay attention to safety issues to prevent the spread of bacteria and to avoid being infected.

All the specimens should be collected according to the operational procedures for quality control. Clinical microbiology lab should refuse receiving specimens that fail to meet the requirements.

6.2.1.2 Specimen Processing

As for bacteria that are sensitive to the environment, such as neisseria meningitidis, neisseria gonorrhoeae and haemophilus influenzae, the specimens should be preserved at a certain temperature and immediately sent for examination. Other specimens should be sent to the lab within 2 h after collection. If immediate delivery is possible, the specimen should be preserved under a certain circumstance. For instances, such specimens as autopsy tissue, bronchial lavage, pericardial fluid, sputum and urine should be preserved at a temperature of 4 °C, while such specimens as cerebrospinal fluid and synovial fluid should be preserved at a temperature of 25 °C. Generally, the specimens for bacterial culture should not be preserved for above 24 h.

As the specimens from patients may contain a large quantity of pathogenic bacteria, safety and protection are always important regardless of the transporting distance. The opening and external surface of the container with specimen should not be contaminated by the specimen. The container should be well packed to avoid being tipped or spilt during the transport. As for the specimens of suspected fulminating infectious diseases, they must be packed according to the requirements and transported by specialized staff.

Anaerobic specimen should be placed in special bottles or tubes for transport, and sometimes it can be transported within the syringe that is used for tissue aspiration.

6.2.2 Morphological Examination of Bacteria

Morphological examination, as one of the most important bacteriological examinations, includes examinations of unstained specimens and stained specimens. Microscope is a necessary tool for the morphological observation of the bacteria.

Microscopy facilitates rapid understandings about presence of bacteria in the specimen and approximate quantity of the bacteria. In addition, the microscopic observation of the shape, structure and characteristic staining of bacteria is helpful for the preliminary identification and classification of the pathogenic bacteria, which also provides basis for further biochemical reaction and serological identification. As for certain types of bacteria, such as the acid-fast bacilli in sputum and neisseria meningitidis in cerebrospinal fluid, a preliminary diagnosis can be made based on the morphological examination, which is also an important reference for early clinical diagnosis and following treatment.

6.2.2.1 Unstained Specimen

Direct microscopic examination of unstained bacteria is mainly applied for studying the motility and mobility of the living bacteria, which includes pressure drop and hanging drop for observation under a common light microscope or dark-field microscope. Bacteria with motility can be observed moving from one site to another with obvious direction. However, bacteria without motility are shown in Brownian movement due to the push of water molecules, namely vibrating at the same location.

In clinical practice, certain pathogenic bacteria can be preliminarily identified via motility examination of the unstained specimen. For instance, for patient with a suspected diagnosis of cholera, the watery stool specimen should be collected and prepared into hanging drop or pressure drop for observation of the bacterial motility under a high power microscope or a dark field microscope. If meteorites like bacteria are observed moving back and forth, the same procedure should be repeated to prepare another hanging drop or pressure drop, followed by addition of serum containing O_1 group of vibrio cholerae for further diagnosis.

If the active motility of bacteria terminates (immobilization test positive), O_1 group of vibrio cholerae infection can be preliminarily diagnosed. In addition to the bacteria specimens, due to the difficulties in staining and the morphological characteristics, spirochetes can be prepared into unstained specimen for observation under a dark-field microscopes.

6.2.2.2 Stained Specimen

After being stained, the specimen for bacteriological examination can be observed for the shape, size and arrangement of the bacteria. In addition, the bacteria can also be classified based on their response to staining. Therefore, staining for examination is one of the most widely-used way for and plays a very important role in bacterial identification. The clinically common staining for bacterial examination includes Gram staining, acid-fast staining, fluorescence staining and special staining.

Gram Staining

Gram staining is the most classical and frequently applied staining for bacteriological examination. Except for extremely rare types of specimens such as blood, most of the specimens undergo Gram staining and microscopic examination before bacteriological culture and isolation. Based on Gram staining, the bacteria is categorized into Gram positive and Gram negative, which facilitates preliminary identification of the bacteria. In combination to characteristic morphology and arrangement of the bacteria, the pathogens can even be preliminarily identified in some cases. For instances, in patients with cerebrospinal meningitis, the cerebrospinal fluid can be collected for smear, Gram staining and microscopy. If intracellular or extracellular Gram negative diplococcocci with kidney liked shape and concave surface parallel to each other are detected, the finding of "Gram negative diplococcocci like neisseria meningitidis" can be reported. If Gram positive diplococcocci with clear surrounding capsules are detected, the finding of "Gram positive diplococcocci like streptococcus pneumoniae" can be reported. The finding after Gram staining and microscopy can provide basis for early clinical diagnosis and the following treatment.

In addition to bacterial identification, the finding by Gram staining can provide reference for clinical medication and facilitate therapeutic planning. This is because the Gram positive and Gram negative bacteria show different sensitivity to certain antibacterial agents, and their pathogenic substances (with the former generates exotoxin, while the latter generates endotoxin) and mechanisms are different.

Acid-Fast Staining

The acid-fast staining facilitates identifying acid-fast bacteria from non-acid-fast bacteria. Since most pathogens are clinically identified as non-acid-fast bacteria, acid-fast staining is not a routine bacteriological examination in clinical practice but only targets on suspected cases of such diseases as tuberculosis and lepriasis. The specimen from suspected cases of mycobacterium tuberculosis can be examined by acid-fast staining and following oil immersion lens for preliminary diagnosis. Based on the result of acid fast staining, "detected or un-detected acid-fast bacteria" can be reported, which provides important reference for clinical diagnosis and treatment.

Fluorescent Staining

With high sensitivity and efficiency, fluorescent staining provides results that can be conveniently observed and has great practical value in clinical bacteriological identification. It is mainly applied to diagnose infections of mycobacterium tuberculosis, mycobacterium leprae, corynebacterium diphtheria, shigella dysenteriae, and spirochete.

Special Staining

To observe the special bacterial structure for further identification, special staining is selected, which includes capsule staining, ink staining, spore staining, metachromatic staining and flagella staining. For instances, the specimen from suspected cases of cryptococcus neoformans infection should be performed ink negative staining. On the black background, white shining capsule of the bacteria can be observed if cryptococcus neoformans is infected. The specimen from suspected cases of corynebacterium diphtheria infection should be performed smear for examination. Based on the finding of Gram positive corynebacteria, the specimen should be further microscopically examined after metachromatic staining. If metachromatic granules are detected, the finding of "corynebacterium diphtheria like bacteria" can be reported, which provides basis for the early clinical diagnosis.

6.2.3 Bacteriological Culture and Identification

In most cases, the etiological diagnosis of bacterial infection can only be defined based on bacteriological culture and identification. Therefore, bacterial culture plays an important role in the diagnosis, prevention and treatment of bacterial infections.

6.2.3.1 Bacteriological Culture

After being transported to the bacteriological laboratory, the specimen should be immediately inoculated to an appropriate culture medium for bacterial isolation. Based on the source of specimen and the suspected pathogenic bacteria, an culture medium should be appropriately selected. For instances, blood plate, Chinese blue/MacConkey or chocolate plate is always selected for the bacteriological culture of sputum. Specifically, blood plate is selected for culture of streptococcus pneumoniae and corynebacterium diphtheria; the Chinese blue/MacConkey plate for screening of Gram negative bacillus; chocolate plate with bacitracin for screening of haemophilus. Appropriate use of culture medium can increase the positive rate of bacteriological culture.

For such liquid specimens as blood, pleural effusion, ascites, cerebrospinal fluid, bile and pus, blood culture bottle can be selected for automated blood culture.

6.2.3.2 Identification of Pathogenic Bacteria

The suspected pathogenic bacteria that shows positive to bacteriological culture should be further identified. The mechanism underlying the bacterial identification is that different bacteria produce different metabolites due to their respective unique enzyme system with variance in substrate decomposition. Based on their own unique biochemical properties of these metabolites, biochemical examinations are applied to identify the pathogenic bacteria. In clinical practice, a preliminary bacterial identification is performed based on bacteriological morphology, staining and culture. However, most isolated unknown bacteria are identified by manual or automated operations of biochemical examination. Currently, the isolated bacteria commonly undergo strain identification and drug sensitivity test based on automated operations of biochemical reaction test. However, nowadays, mass spectrometry identification system has been applied for bacterial identification via preparation, isolation and gas ion detection. The results have demonstrated favorable consistency with those by routine biochemical examinations, and high accuracy.

6.2.4 Non-culture Bacterial Identification

In addition to microscopic examination of smears and direct bacteriological culture, non-culture bacteriological examination is also applied in the etiological identification of bacterial infections. Such non-culture examinations include immunoassay, molecular biological examination, bacteriotoxin detection and animal experiment, which can result in the etiological diagnosis in combination of the clinical manifestations.

6.2.4.1 Immunoassay

Based on the immunological principles, a known antibody is used to detect an unknown antigen or a known antigen is used to detect an unknown antibody, which is an important way for clinical diagnosis of bacterial infection.

Antigen Detection

Agglutination reaction, immunofluorescence and enzyme immunoassay are the most widely used immunoassays for antigen detection.

Agglutination Reaction

Agglutination reaction is applied to directly detect possibly existent antigen of a small quantity in blood, cerebrospinal fluid and other secreted fluid in human body at the early stage of bacterial infection. For instance, neisseria meningitidis can be directly detected in the cerebrospinal fluid collected from a patient with epidemic encephalitis. Agglutination reaction facilitates rapid diagnosis of infectious disease.

Immunofluorescence

Immunofluorescence is a microscopic examination which combines immunological specific reaction and fluorescent trace. It maintains high serological specificity and greatly improves the sensitivity of detection, thereby playing an important role in bacterial detection.

Enzyme-Linked Immunosorbent Assay (ELIZA)

In addition to detection of both pathogen and antibody, ELIZA can also be applied to identify bacterial metabolites. Almost all the soluble antigen-antibody reaction systems are detectable by ELIZA, with the detectable minimum value reaching ng or even lg. And both specificity and sensitivity are high.

In addition to the above immunoassays, countercurrent immunoelectrophoresis, immunoblot assay, and luminescence immunoassay can also be applied in clinical practice to detect pathogenic bacteria.

Antibody Detection

After human is infected by pathogenic bacteria, the immune system is stimulated to produce immune responses, and specific antibodies are then produced. The quantity of antibodies commonly increases along with the progression of infection, with manifestation of increased titer. Therefore, a known bacteria or its specific antigen can be used to detect the existence of the corresponding antibody in the serum as well as the dynamic changes of the titre. This facilitates the diagnosis of certain infectious diseases. And antibody detection is mainly applied in the diagnosis of infectious diseases caused by bacteria with strong antigenicity and a long illness course.

That the antibody titre is obviously higher than the normal level or the titre value at the convalescent stage is at least 4 times as high as that at the acute stage indicates a serological diagnosis with clinical significance.

6.2.4.2 Molecular Biological Detection

Nucleic Acid Hybridization

Hybridization occurs when two individual single-stranded DNAs bind complementarily to form a double-stranded DNA. Based on the process of hybridization, a specifically sequenced DNA segment is prepared to serve as a labelled probe. Under certain conditions, it can hybridize with the denatured bacterial DNA in the specimen based on the base complementary pairing rule. Since the occurrence of hybridization can be detected via hybridization signals, the existence of the pathogenic bacterial genes in the specimen can be determined. Nucleic acid hybridization is a bacterial examination with high sensitivity, high specificity, simple operations and rapid result that can directly detect pathogenic bacteria in specimens. And the detection successfully avoids effects from non-pathogenic bacteria. Therefore, for those pathogenic bacteria that cannot be cultured or are difficult to be cultured, nucleic acid hybridization is of great significance for their detection.

Polymerase Chain Reaction (PCR)

PCR is applicable to detect pathogenic bacteria which cannot be precisely and immediately detected via conventional cultures, show low sensitivity to conventional cultures, or need a long culture time. For instances, the culture of mycobacterium tuberculosis consumes 2–3 months, which delays the diagnosis and immediate treatment. The infection of chlamydia trachomatis commonly shows no characteristic symptoms and the culture of pathogenic bacteria is challenging. Such pathogenic bacteria also include legionella, mycoplasma pneumoniae and rickettsia. And PCR is definitely a good choice for their detection. In addition, PCR is also widely used in bacteriotoxin detection. Since different bacteria produce different toxin, a specific primer can be designed and synthesized based on the specific toxin gene and the specific toxin gene segment can be amplified by PCR. Such a way shows high specificity and sensitivity. Currently, with the technological development of PCR, the fluorescent quantitative PCR overcomes the weakness of traditional PCR in relatively high rate of false positive and shows precise quantification.

Biochip

Biochip, as an emerging high-tech in the life science, has been rapidly developing in recent years. By establishing a mini biochemical analysis system on the surface of a solid chip via micromachining and microelectronics, biochip can process a large quantity of data precisely and rapidly to detect cell, protein, DNA and other biological components. The widely used biochips include gene chip and protein chip. The diagnostic biochip for pathogenic bacteria allows simultaneous detection of different pathogens in different specimens in one chip. Only with extremely small quantity of specimen, it offers a large quantity of diagnostic information within an extremely short period of time. Thereby, biochip technology provides a rapid, sensitive and high-flux way for clinical diagnosis of bacterial infection.

6.2.4.3 Bacteriotoxin Detection

Endotoxin Detection

Endotoxin detection is mainly applied for definitive diagnosis of Gram negative bacterial infection. Most of gram negative bacteria can produce endotoxin, which is released after death and lysis of the bacteria, with multiple biological effect. As an exogenous pyrogenic agent, the endotoxin can stimulate the white blood cells to release endogenous pyrogens, which then acts on the body temperature center to cause fever.

Limulus test is always applied for endotoxin detection, which shows high specificity to endotoxin generated by Gram negative bacteria but shows negative to non-endotoxin substances, Gram positive bacteria and virus toxin. The test also has high sensitivity, with (0.005–0.0005) μg/ml of endotoxin detectable. The operations are simple and the test result can be obtained within 2 h for typing of the pathogenic bacteria. The test result facilitates reasonable medication and early treatment.

Exotoxin Detection

Exotoxin detection can be used to identify an unknown bacteria and to distinguish toxigenic strain from non-toxigenic strain.

Internal Virulence Test

The toxic effect of bacterial exotoxin can be neutralized by the corresponding antitoxin. An animal fails to experience toxic symptoms after injection of extotoxin following antitoxin. Based on such a phenomenon, the production of corresponding exotoxin in the bacteria can be identified.

External Virulence Test

With strong antigenicity, exotoxin stimulates the organism to produce corresponding antibody. The serum with specific immunity to the bacterial exotoxin, selected as the antibody, should respond to the detected bacterial exotoxin (antigen) in vitro to verify the existence of the bacterial exotoxin.

In addition to the above examinations, the exotoxin produced by most bacteria can be detected by ELISA, such as staphylococcal enterotoxin as well as heat-labile toxin (LT) and heat-stable toxin (ST) produced by enterotoxigenic escherichia coli.

6.2.4.4 Animal Experiment

Animal experiment is an important part of the clinical bacteriological examination, and sometimes cannot be replaced by other examinations. It is mainly applied to isolate and identify the pathogenic microorganism, to determine the bacterial virulence, to prepare serum with specific immunity, to establish pathogenic animal model, to collect animal blood for bacteriological culture mediums, and to test the safety, toxicity and therapeutic efficacy of biological products and some drugs. In addition, the bacterial virulence and immunogenicity may change via the passage in susceptible and non-susceptible animals. Animal experiment requires not only understanding about the classification of experimental animals, but also appropriate selections of experimental animal and inoculation way based on the purpose and demands of experiment. The widely used experimental animals include mice, guinea pigs, rabbits and sheep, while the commonly used inoculation ways include subcutaneous injection, intradermal injection, intramuscular injection, intraperitoneal injection, intravenous injection and intracerebral injection.

6.3 Immunodiagnosis

6.3.1 Immunodiagnosis of Parasitic Diseases

Immunodiagnosis refers to diagnosis by *in vitro* detection of antigen or antibody based on the immune responses in human body triggered by invasion of parasites, including intradermal test serological test. The intradermal test is selected for initial screening of patients due to its low specificity. The serological test is applied to detect specific antigen or antibody via different examinations, with specific antigen positive indicating present infection while specific antibody positive indicating past or present infection. Therefore, serological test can be applied for diagnosis or to assist the diagnosis.

6.3.2 Classification of Immunodiagnostic Examinations for Parasitic Diseases

Based on immediately occurring allergy, the intradermal test is performed by injection of skin test antigen into the inner layer of epidermis. Positive or negative result can be obtained after observation of the skin mound. It is commonly applied for the diagnosis of helminthiasis or allergy induced by certain mites.

Serological diagnostic examinations include precipitation reaction, agglutination reaction, complement fixation test, immunofluorescence antibody assay, immunoenzymatic assay, radioimmunoassay, and immunoblotting. The serological diagnosis has evolved from serum sedimentation test and agglutination test to immunolabelling technique with efficiency requiring a trace amount of specimen and enzyme linked immunoblotting at the molecular level. These immunoassays can be applied to detect circulating antibody or antigen in infected human body, and are expected to be applicable for staging of an infection, identifying active stage of a new infection and assessing the therapeutic efficacy. The serological diagnosis is playing an increasingly important role in the clinical diagnosis that the etiological diagnosis fails to. Almost all the immunoassays are applicable for the diagnosis of parasitic diseases but not always effective. In China, several serological diagnostic examinations have been developed for the diagnosis of parasitic diseases that serve as assisting tools for the diagnosis and provide reference for medication. These examinations have been gradually promoted to clinical application, and herein we introduce several commonly used immunoassays in clinical practice.

6.3.3 Immunodiagnostic Examinations

6.3.3.1 Intradermal Test

Basic Principle

The skin test antigen is injected into the inner layer of epidermis. A positive or negative result can be obtained by assessing the skin mound. It is mainly applied to diagnose helminthiasis or allergy induced by certain mites.

Advantages and Disadvantages

The intradermal test is rapid and requires simple operations and a short period of time. Its positive detection rate generally reaches above 90 % but with low specificity due to cross reactions among different parasitic diseases. In some cases, the patients always showed positive after being treated for years. Therefore, the result of intradermal test fails to serve as the basis for definitive diagnosis and for assessing the therapeutic efficacy. It is only applicable to screen the suspected cases in affected region.

6.3.3.2 Precipitation Test

With appropriate quantity of electrolytes, soluble antigen (such as exotoxin, endotoxin, lysate of bacteria, soluble antigen of virus, serum, tissue exudates) binds with corresponding antibody to form whitish deposits that are observable by naked eyes. The procedure is known as precipitation test. For instance, circumoval precipitin test (COPT) is one of the most common and effective serological tests for the diagnosis of schistosomiasis. And the test is based on the mechanism of specific immune response in human body to the antigen, eggs of schistosoma. Mature miracidium in schistosoma eggs can secrete soluble egg antigen (SEA), which exudates from micropores in the eggshell to attach to the surface of eggshell and binds to the antibody in the to-be-detected serum. The antigen-antibody complex deposites are thus formed around the eggshell, which are shown as bubble like or finger shaped deposits on the surface of eggs under a microscope. Such a finding indicates positive. In the serum of a healthy human body, no specific deposits can be observed around eggs due to absence of corresponding antibody, which is defined as negative. Based on the perioval precipitation rate (the number of eggs with precipitation in per 100 eggs, which is calculated by the number of positive eggs/the number of observed eggs × 100 %), the COPT of examined serum can be assessed as positive or negative. Based on the size of deposits, the intensity of COPT can be understood. Currently, the test has been improving with favorable diagnostic efficacy and is applicable for population field application. The processing of eggs includes formaldehyde processed frozen dry egg antigen and heat processed ultrasonic dry egg antigen.

6.3.3.3 Indirect Heamagglutination Assay

Basic Principle

Indirect heamagglutination assay (IHA) is one of the agglutination tests, which is based on the biological mechanism that antigen binds to corresponding antibody to form complex whose agglutination, if electrolytes exist, can be observed as small pieces of deposits. Based on the produced agglutination, the existence of antigen or antibody can be detected. The procedure is known as the agglutination assay. The agglutination assay based on direct binding of granular antigen to antibody is known as direct heamagglutination assay. As for the indirect heamagglutination assay, a soluble antigen or antibody is firstly adsorbed to the surface of an immunity independent carrier particle with certain size, followed by its reaction with corresponding antibody or antigen. Under appropriate conditions agglutination of the carrier particles occurs along with the specific binding of antigen with antibody, showing observable agglutination. The commonly used carrier particles include red blood cells (O shaped human red blood cells or sheep red blood cells), polystyrene latex particles, kaolin, ion exchange resin, and collodions. The test with red blood cells selected as the carrier particles is known as indirect heamagglutination test.

Categorization

IHA has been applied for diagnosis and epidemiological studies of many parasitic diseases. It can be categorized into the following 4 types, all of which are for detection of antigen or antibody with high sensibility and certain specificity.

Positive Indirect Heamagglutination Test

An antigen is firstly adsorbed to the surface of red blood cells, followed by detection of the unknown serum antibody with a known heamagglutination antigen. Such a test is known as positive indirect heamagglutination test.

Reverse Indirect Heamagglutination Test

A specific antibody is used to sensitize red blood cells for detection of the antigen in the specimen.

Indirect Heamagglutination Inhibition Test

In the test, antigen sensitized red blood cells are used to detect the corresponding antibody. Specifically, the corresponding antibody is firstly added into the specimen, followed by addition of sensitized red blood cells after a certain period of time. If antigen exists in the specimen, the antigen binds to the antibody, with no agglutination after addition of the sensitized red blood cells. If no antigen in the specimen, agglutination is thus observable.

Reverse Indirect Heamagglutination Inhibition Test

In the test, antigen sentisized red blood cells are used to detect the corresponding antibody in the specimen. The corresponding antigen is firstly added into the specimen, followed by addition of the sensitized red blood cells. If antibody exists in the specimen, the antibody binds to the antigen with no agglutination. But if no antibody, agglutination is observable.

Advantages and Disadvantages

IHA has been widely used in the diagnosis of parasitic diseases and epidemiological studies, with high sensitivity and certain specificity. For instance, the positive rate for the diagnosis of schistosomiasis japonica has reached up to 91.9–100 %, with a false positive rate of only 0.7–3.2 %. The frozen and dried sensitized red blood cells can be preserved for 1–2 years at a temperature of 4 °C. The necessary equipment is not so complicated and has been manufactured in China. The test procedures are simple and applicable in community based clinics and hospitals. However, the red blood cells prepared in different laboratories or the red blood cells of different batches may show variance in sensitivity and specificity. Therefore, the preparation of the red blood cells should be standardized and commercialized. In addition, certain non-specific response may occur, which is related to the quality of sensitized red blood cells and hemagglutination inhibition test can be performed to exclude the possibility of non-specific response if necessary.

6.3.3.4 Immunofluorescent Assay

The assay can be applied to detect the antibody in serum for the diagnosis of malaria, filariasis, toxoplasmosis, schistosomiasis, paragonimiasis, clonorchiosis, echinococcosis, and amoebiasis. It can also be selected to detect parasites in tissues such as leishmania, amoeba, plasmodium and T. gondii.

Basic Principle

Immunofluorescence method is an immunoassay, which is also known as fluorescent antibody labeling. An antibody or antigen is firstly labeled with fluorescein, followed by observation of the labeled fluorescence under a fluorescence microscope to analyze the trace of the antigen or antibody.

Fluorochrome

Certain substances catch and absorb light energy to be excited, but emit light energy in the form of electromagnetic radiation when they regain their baseline state, which is known as photoluminescence. If exposed to light with short wave length, such as ultraviolet light, these substances emit light with longer wave length (such as visible light) within an extremely short period of time, which is known as fluorescence. The substances capable of emitting fluorescence are known as fluorochrome or fluorescein that can be categorized into various types. But the fluorescein capable of labelling antibody should: (1) be capable of covalent binding with immunoglobulin to form stable conjugate; (2) have no impact on immunoactivity of the immunoglobulin and no obvious effect on the fluorescence efficiency after the covalent binding; (3) be simple, fast, safe and non-poisonous in labeling. Currently, the most commonly used fluoresceins include fluorescein isothiocyanate, tetraethyl rhodemin, and tetraethyl rhodamine isothiocyanate. Nowadays, immunofluorescent assay has been applied in the diagnosis of many parasitic diseases, such as schistosomiasis, trypanosomosis, trichinosis, toxoplasmosis, and leishmaniasis.

6.3.3.5 Immunoenzymatic Techniques

Immunoenzymatic techniques are recently developed enzymatic immunoassays with both high specificity and high sensitivity, and have been widely applied in immunodiagnosis of parasitic diseases, such as schistosomiasis, paragonimiasis, black fever, echinococcosis, toxoplasmosis, cryptosporidiosis, pneumocystosis carinii, cysticercosis, and amoebiasis. In addition to traditional ELISA, K-ELISA, ABC-ELISA, DOT-ELISA, and film ELISA have been developed for the diagnosis of such diseases as schistosomiasis, paragonimiasis, and clonorchiasis. These immunoenzymatic assays can be applied to detect antibody, circulating antigen, stool specimens, pus and antigen in other body fluids. Currently, the procedures for antigen preparation and the procedures in laboratory test need to be clarified and standardized.

Basic Principle

Immunoenzymatic assays are another type of immune labelling technique following immunofluorescence method. They are based on the principles of specific binding of antigen with antibody as well as highly efficient catalytic effect of enzyme on substrate with the enzyme as marker. After enzyme labelled antibody or antigen binds with corresponding antigen or antibody, complex of enzyme-labelled antibody and antigen is formed. When the enzyme meets the corresponding substrate, it catalyzes the substrate to decompose, followed by oxidization of the hydrogen donor to form colored substance. The emergence of colored substance objectively demonstrates the existence of enzyme. Based on the presence and concentration of colored substance, the existence of suspected antigen or antibody as well as its quantity can be speculated so as to diagnose qualitatively or quantitatively.

Categorization

Methodologically, immunoenzymatic assays can be categorized into two types. One type is applied to detect and localize the antigen or antibody in tissues and cells, which is known as immunohistochemistry or immunoenzymatic staining. The other type is applied to detect soluble antigen or antibody in various body fluids, which is known as enzyme linked immunosorbent assay (ELIZA). After preparation of specimen, immunoenzymatic staining is performed with the following procedures: inhibition of the endogenous enzymes and the following immunoenzymatic staining for examination. Its basic principle and procedures are the same as those of fluorescence antibody method, only with enzyme to replace fluorescein as the marker and emergence of colored substances in the substrate as positive. Routine immunoenzymatic staining can be further categorized in to direct and indirect staining. Direct staining is performed to label specific antibody with enzyme for direct detection of parasite or its antigen. After fixation of the specimen containing parasite or its antigen, the inside endogenous enzymes are firstly inhibited. After that, enzyme labelled antibody is directly added for its binding to the antigen in the specimen, followed by addition of substrate for coloration and microscopic examination. Indirect staining is performed by processing of the tissues or cells containing parasite or its antigen with specific antibody for binding of the antigen to the antibody. The unbinding components are cleansed and eliminated, followed by addition of enzyme labelled antibody for formation of antigen-antibody-enzyme-labeled-antibody complex. Finally, the substrate is added for coloration and microscopic examination. Although with additional one step, the indirect staining shows higher specificity and has gained wider application. The second enzyme labelled antibody can be replaced by staphylococcal protein A (SPA) or biotin-avidin system, which has also be successfully applied to detect many antigens and antibodys. Meanwhile, both specificity and sensitivity of the staining have been increased by varying degrees. ELIZA can be further categorized into solid-phase ELISA and homogeneous ELISA. Solid-phase ELISA is to chemically or physically link antigen or antibody to a solid phase carrier to prepare an immunosorbent for immunoenzymatic assay. ELIZA is the most widely used solid phase immunoenzymatic assay. Homogeneous ELISA is applied to directly detect antigen or antibody in the solution with no isolation of the free enzyme marker from the binding enzyme marker and no use of solid phase carrier. It is mainly used to detect small molecular hapten such as hormone and antibiotics.

Enzyme and Substrate Used for Labelling

Horseradish peroxidase, glucose oxidase, acid phosphatase, alkaline phosphatase and β-galactosidase are commonly used enzymes for labeling, of which horseradish peroxidase has gained the widest application.

Enzyme Labeled Antibody

The activity and purity of enzyme play a crucial role in its labeling of antibody. The antibody is also required to be highly purified, more favorably by affinity chromatography. However, mice ascites can be directly labeled in labeling of monoclonal antibody. An ideal binding agent should be of high productivity, producing stable conjugate, showing no effect on activity of enzyme and antibody, producing no interfering substances, and simple operations. Currently, glutaraldehyde labeling and sodium periodate labeling are mainly adopted. The principle of glutaraldehyde labeling is the covalent binding of its aldehyde group to amidogen of immunoglobulin. Namely, one of the two active aldehyde groups of glutaraldehyde binds to amidogen of enzyme molecules, while the other binds to amidogen of immunoglobulin to form enzyme-glutaraldehyde-immunoglobulin complex. Sodium periodate labeling is mainly used to label HRP. Its advantages include high labeling rate and small quantity of unlabeled antibody. However, the conjugate with large molecular weight shows less favorable penetration through the cells than glutaraldehyde labeled antibody. Therefore, it should not be applied in immuno-electron microscopy.

6.3.3.6 Latex Agglutination Test (LAT)

With latex microparticle as the carrier, LAT is used to replace agglutination test using red blood cells. It is mainly applied to diagnose toxoplasmosis, cysticercosis, trichinosis, schistosomiasis, and echinococcosis, with favorable sensitivity and specificity. The procedures of LAT are simple and the test results can be rapidly obtained that is preferable for field application.

6.3.3.7 Immunoblotting Test (IBT)

IBT, also known as Western blotting, is a rapidly developed technique in recent years that integrates polyacrylamide gel electrophoresis, transfer electrophoresis and solid-phase ELISA. It has been applied for the diagnosis of schistosomiasis, teniasis, echinococcosis, paragonimiasis, pneumocystosis carinii, amoebiasis, and cysticercosis. With its ongoing rapid development, IBT is expected to be an effective diagnostic immunoassay with high sensitivity and specificity for the diagnosis of parasitic diseases and for identifying the stage of parasitic infections.

6.3.3.8 Immunochromatography (ICT)

ICT is a rapidly developed diagnostic technique in recent years, which is used to detect antibody or antigen. Antigen detection for plasmodium falciparum/plasmodium vivax, filaria bancrofti, and leishmania has gained clinical application. It is applicable for rapid clinical diagnosis and field studies due to its high sensitivity and specificity as well as its rapidity in obtaining results.

Generally speaking, with the rapid development of immunological technology, the immunodiagnosis of parasitic diseases has advantages of simple operations, micro quantity requirement of the specimen, rapidity, high accuracy and low cost. With the application of the high-tech, immunodiagnosis of parasitic diseases will be of greater clinical value.

6.4 Genetic Diagnosis

With the rapid development of modern life science and technology, the laboratory techniques in the diagnosis of parasitic diseases have been extended beyond etiological and immunological examinations into genetic diagnosis. Its application plays an important role in the laboratory diagnosis of parasitic diseases. The genetic diagnosis targets on the specific DNA fragment in genome of parasite, which shows higher specificity and sensitivity compared to etiological and immunological diagnosis. In this section, we briefly introduce the genetic diagnosis of parasitic diseases.

6.4.1 Nucleic Acid Probe

6.4.1.1 Introduction

Nucleic acid probe is a widely used technique in genetic diagnosis with high sensitivity and specificity that gains rapid development in recent years. It is based on the principle that two single-strand nucleic acids with certain homogeneity anneal to form double strands by pairing of complementary bases. Therefore, the under-detected sequence of nucleic acid can be detected with a known probe. Nucleic acid probe refers to a DNA or RNA fragment that is labeled by radionuclide or other markers and is thus capable of binding specifically with specific target molecule.

6.4.1.2 Procedure

Generally speaking, four operational procedures are performed, including selection of nucleic acid fragment as probe, labeling, hybridization, demonstration or detection of hybridization signal.

Types of Probe and Its Selection

Based on its origin and property, nucleic acid probe can be categorized into genomic DNA probe, cDNA probe, RNA probe and artificially synthesized oligonucleotides probe. It also can be categorized into radiolabeled probe and non-radiolabeled probe based on its markers. The probe should be selected based on the following principles:

1. High specificity: For instance, exon is commonly selected as probe for human genome.
2. Simplicity in preparation: In simplicity of preparation, the single strand is superior to double strands with a more efficient hybridation.
3. Conveniency in detection and high sensitivity: When a short fragment is selected, it carries less marker with low sensitivity.
4. A general length of 17–50 bp: A longer fragment generally consumes longer period time for hybridization, while a shorter fragment shows lower specificity.
5. A probe containing 40–60 % G-C shows higher rate of specific binding.
6. Less than 4 repetitive bases: Intron is generally not selected as probe for eukaryocyte genome.
7. The homogeneity should be less than 70 % or less than 8 consecutive base in the non-target molecular area.

Labeling

An ideal probe marker should embed the following properties: (1) highly sensitive but low false positive rate; (2) no impact of the binding between marker and probe on specificity of base pairing; (3) no impact on the main properties of the probe, such as physicochemical properties, hybridization specificity and stability, as well as Tm value. In addition, the procedure should show high sensitivity and specificity, which requires simple procedures for labelling and detection and the labeled marker can be preserved for a long period of time. It also should not contaminate the environment and do no harm to human health, with a low cost. The applicable markers include radioactive markers such as ^{32}P, ^{3}H, ^{35}S and non-radioactive markers such as biotin, digoxin and enzymes. The radioactive markers are highly sensitive and specific, but their use requires special safety procedures in addition to their contamination to the environment and a short half-life period. Non-radioactive markers are free of contamination and special safety procedures, with favorable stability for prolonged preservation. However, they are inferior to radioactive isotopes in term of sensitivity and specificity. The labeling of probe can be categorized into in vivo labeling and in vitro labeling. In vitro labeling is more commonly applied, including chemical labeling and enzymatic labeling.

Hybridization

The molecular hybridization of nucleic acid is actually denaturation of double-stranded DNA and renaturation of two homogeneous single-strands. The molecular hybridization of nucleic acid can be categorized into solid-phase hybridization and liquid-phase hybridization based on the medium and responsive environment. Solid-phase hybridization is achieved by first fixation of one nucleic acid chain on a solid support while the other chain is free in liquid. It has gained wide application due to the simplicity in operations. Liquid-phase hybridization, however, refers to the test procedure that two nucleic acid chains participating in hybridation are free in liquid, with rapid hybridation. Liquid-phase hybridation is commonly applied together with nucleic acid electron microscopy to study the homogeneity of different DNA and the relationship between mRNA and chromosomal DNA. Liquid-phase hybridization can be classified into Southern blotting, Northern blotting and Western bloting based on different target of identification. Southern blotting is employed to analyze recombinant DNA, recombinant plasmid and bacteriophage, while Northern blotting is used for qualitative and quantitative analysis of RNA. Western blotting is applied for qualitative and quantitative analysis of proteins and their interactive effect. According to the type of nucleic acid and approaches for detection, liquid-phase hybridization can also be categorized into Southern blotting, dot blot hybridization, in situ hybridization and Northern blotting.

Detection of Hybridization Signal

Approaches to detection of hybridization signal vary with the markers labeled on the probe. The probes labeled by radioactive isotope are shown on X-ray film via radiography, while the probes labeled by biotin are indicated by enzyme-chromogenic substrate via ABC system. The probe labeled by digoxin with enzyme labeling digoxin antibody is also indicated by chromogenic substrate.

6.4.1.3 Application of Nucleic Acid Probe in Diagnosis of Parasitic Diseases

In the diagnosis of parasitic diseases, probe is a specific nucleotide sequence of pathogen, which can be used to detect the existence of pathogenic parasite. And preparation of highly specific probe is the key procedure. Currently, parasite-based nucleic acid probes available include complete genomic DNA probe, kinetoplast DNA probe, recombinant plasmid DNA probe or bacteriophagic recombinant DNA probe, artificially synthesized oligonucleotide probe.

To date, concerning the detection of parasite, nucleic acid probe is mainly used to detect parasite in blood such as plasmodium, leishmania, trypanosome and toxoplasma gondii. Microscopy fails to detect these protozoa when the protozoal density in blood is low and, therefore, is not applicable in large-scale epidemiological studies. Although detection of antibody in serum can provide important information about protozoa infection, it fails to define the existence of viable protozoa due to detectable antibody long after the disappearance of protozoa. With the development of molecular biology, the identification of pathogen via detection of nucleic acid emerges. Generally speaking, heterologous nucleic acid can be resolved rapidly in blood. Therefore, the finding of pathogenic nucleic acid in blood demonstrates the existence of viable pathogen.

6.4.2 Polymerase Chain Reaction (PCR)

6.4.2.1 Introduction

PCR is a technique for nucleic acid amplification in vitro which is developed in mid 1980s, with advantages of high sensitivity and specificity, being rapid, convenient and automatic as well as favorable repeatability. The target gene or certain DNA segment can be amplified into 100 thousands to one million times within several hours by PCR, which renders direct observation with naked eyes possible. PCR can amplify the target gene or certain DNA segment into enough quantity of DNA from a single hair, a drop of blood or even a single cell for analysis and identification. PCR technology is a revolutionary invention and milestone breakthrough in biomedical field.

6.4.2.2 Principle and Procedure

The fundamentals of PCR resemble to the natural replication of DNA, and its specificity depends on the oligonucleotide primers which complement with both terminals of target sequence. The procedures of PCR include denaturing, annealing and extending.

Template DNA Denaturation

After heated to the temperature of 93 °C for a certain period of time, the double-strands template DNA or double-stranded DNA formed by PCR amplification break down into single strand. The single strand then binds to the primer for another reaction.

Template DNA

The template DNA denatures into single strand after heating along with annealing of primer. When temperature reduces to about 55 °C, the primer binds to the complementary sequence of the single strand template DNA.

Extension of Primer

Under the effect of Taq DNA polymerase, the conjugate of template DNA with primer produces a new semiconservative replication strand by using dNTP as raw material and target sequence as template. The production is achieved via base pairing and semiconservative replication.

More semiconservative replication strands can be obtained when denaturing, annealing and extending, the 3 processes in PCR, are repeated. And these new strands act as template for next cycle of replication. Each cycle of these 3 processes requires about 2–4 min and the target gene can be amplified to millions of times within 2–3 h. The quantity of cycles to reach the plateau level depends on template copy in the specimen.

6.4.2.3 Reaction Component

The basic system of PCR is composed of DNA template, oligonucleotide primer, dNTP, DNA polymerase and reaction buffer with necessary ions, all of which affect PCR reaction.

DNA Template

DNA template is also referred to as target sequence, which can be either single-stranded DNA or double-stranded DNA. The amplification efficiency of ring-closed DNA template is relatively lower than that of linear DNA. Therefore, it is rather better to choose linear plasmid as template. Protease, nuclease, DNA polymerase and proteins that can bind with DNA should not mix with template DNA. To some extent, PCR production increases markedly along with concentration of template DNA. However, steep concentration of template DNA can increase non-specific reaction. To guarantee specificity of the reaction, the concentration of genomic DNA template should be about 1 μg; and plasmid DNA template, about 10 ng.

Primer

Primer is the key material in specificity of PCR, which depends on the degree of complementarity between primer and template. Theoretically, based on any template DNA with known sequence, complementary oligonucleotide strand can be designed as primer. By PCR, template DNA can be amplified considerably in vitro.

Thermostable DNA Polymerase

In addition to classic Taq DNA polymerase, many thermostable DNA polymerases are found, such as Tth DNA polymerase, Vent DNA polymerase and Pfu DNA polymerase.

Taq DNA Polymerase

Natural Taq DNA polymerase is directly isolated from Thermus aquatic YT-1 strain, which has favorable thermal stability. Its half life period is 130 min, 40 min, and 6 min at the temperature of 92.5 °C, 95 °C, 97.5 °C, respectively, and the optimum temperature for catalyzing DNA synthesis is 72–80 °C. Taq DNA polymerase shows the following properties: directional polymerization of 3′–5′, directional exonclease activity of 5′–3′, reverse transcriptase activity, relatively weak non-template dependency, absence of directional 3′–5′ exonclease activity.

In addition to natural Taq DNA polymerase, there is also modified Taq DNA polymerase. One is recombinational Taq

DNA polymerase which is acquired from the expression of colibacillus via genetic technology. The other is known as Stoffel segment that shows favorable thermal stability (Vainshtein et al. 1996). It was firstly obtained by Stoffel et al. via cutting off the N-terminal of natural Taq DNA polymerase.

Tth DNA Polymerase

Tth DNA polymerase is a thermostable DNA polymerase isolated from Thermus thermophilus HB8. It can effectively reverse transcribe RNA at a high temperature with the presence of $MnCl_2$. When Mg^{2+} is added in chelated Mn^{2+}, polymerization of Tth DNA polymerase increases which can enable synthesis and amplification of cDNA with one catalyzer.

Vent DNA Polymerase

Vent DNA polymerase is also known as Tli DNA polymerase, which is isolated from pyrococcus horikoshii at the vocalnic vent in deep sea. In addition to favorable thermal stability, it also has directional exonclease activity of 5′–3′ and the function of correcting. Additionally, it can amplify template NDA that is larger than 12 kb.

Pfu DNA Polymerase

Pfu DNA polymerase is purified from pyrococusfurisus strain, with directional polymerization of 5′–3′ and directional exonclease activity of 3′–5′. Its fidelity in DNA catalysis is 12 times as high as that of Taq DNA polymerase and it also has favorable thermal stability. Its half life period at the temperature of 97.5 °C is more than 3 h.

The concentration of DNA polymerase is a key factor affecting PCR reaction, with an optimal concentration level for each PCR reaction. Excessive polymerase can reduce specificity of the reaction, while insufficient polymerase may affect the production of reaction. For instance, the optimal quantity of Taq DNA polymerase is between 0.2 and 2.5 units in the 50 μl PCR system.

Deoxy-Ribonucleoside Triphosphate (dNTP)

dNTP in PCR can be grouped into 4 types, including dATP, dGTP, dCTP and dTTP. The molar concentration of these 4 dNTPs should be equal, otherwise, the variance may induce misincorporation of polymerase to reduce the velocity of new strand synthesis. The concentration of dNTP depends on the length of the amplified segment, concentration of MgCl and concentration of primer. The final concentration is commonly ranges from 50 to 200 μmol/L. A high concentration of dNTP may produce error base incorporation, while a low concentration of dNTP may reduce the production of reaction. The concentration of dNTP should be adjusted to a pH value of 7.0 via addition of NaOH to guarantee a pH value of being not lower than 7.1 during reaction.

Clinically, dTTP is usually replaced by dUTP to prevent contamination from amplified product and control the false positive rate.

Buffer

The standard buffer in PCR is usually consisted of Tris–HCl, KCl, and MgCl with a Tris–HCl concentration of (10–50) nmol/L. Mg^{2+} is of great importance because its concentration may affect the activity of DNA polymerase and melting temperature of double-stranded DNA. As a result, Mg2+ has significant effect on the specificity and the production of PCR. Its lower concentration may reduce both the activity of DNA polymerase and the production of PCR, while its superfluous concentration may affect the specificity of PCR. The optimal concentration of Mg2+ is between 1.5 and 2.0 mmol/L. All of the DNA template and primer in PCR compound, phosphate group of dNTP as well as chelating agent such as EDTA can bind with Mg2+ to reduce the actual concentration of Mg2+.

6.4.2.4 Conditions for Reaction

Denaturation Temperature and Time

The denaturation of template DNA or PCR product in PCR is of great importance in its complete melting into single strands. Generally speaking, the higher temperature and longer time guarantee complete denaturation. However, too high temperature and prolonged time may affect the activity of Taq DNA polymerase. Therefore, denaturation temperature and time are commonly selected to be 95 °n for 30 s. The first circular denaturation of PCR may require more time due to the relatively longer template DNA.

Annealing Temperature and Time

Optimal denaturation temperature and time are the key points for the success of PCR, while annealing temperature and time affect the specificity of PCR. Lower temperature causes better renaturation, but with possible mismatch of primer and target DNA to increase non-specific binding, while higher temperature is disadvantageous for renaturation. Generally, the temperature for renaturation is about 55 °C.

Extension Temperature and Time

The extension temperature of primer is generally 72 °C due to the considerations about both the activity of Taq DNA polymerase and the binding of primer and targeted gene. Unfavorable extension temperature affects both the specificity and quantity of amplified product.

Cycle Number

The quantity of PCR amplified product is determined by the number of reaction cycles. The optimal cycle number is determined by the original concentration of target sequence when the other parameters have been optimized. A low original concentration of target sequence requires more cycles of reaction. In the case of reduced enzyme activity or insufficient enzyme, more cycles are also necessary for effective quantity of amplification.

6.4.2.5 Detection and Analysis of Amplified Product

Gel Electrophoresis

After electrophoresis and ethidium bromide staining, the PCR product can be observed under a UV detector for preliminary judgement of its specificity. The size of PCR product segment should be in consistency with the predicted size.

Agarose Gel Electrophoresis

One to two percent of agarose gel is commonly employed for detection.

Polyacrylamide Gel Electrophoresis

Six to ten percent of polyacrylamide is better than agarose in gel electrophoresis, which has more concentrated strips and is applicable in scientific research and surveillance.

Enzyme Digestion Analysis

Theoretically consistent segment is obtained via corresponding enzyme digestion and gel electrophoresis based on restriction enzyme site. Enzyme digestion analysis can be applied to identify the PCR product, to type the target cells and to analyze their variability.

Molecular Hybridization

Molecular hybridization can provide sufficient evidence for the specificity of PCR product, and is an effective approach in detecting the base mutation of PCR product.

Southern Blotting

By using Southern blotting, a oligonucleotide strand is synthesized between two primers, which can be used as probe after being labelled to hybridize with PCR product. Southern blotting can be used to identify the specificity of PCR product, and can increase the sensitivity in detecting PCR product. In addition, it can also be used to understand its molecular weight and the shape of strip. Southern blotting is mainly used in scientific research.

Dot Blot Hybridization

The PCR amplified product is firstly dropped on nitrocellulose filter membrane or thin nylon membrane, followed by hybridization with inner oligonucleotide probe to observe the presence of stained spots. Dot blot hybridization is mainly used to identify the specificity of PCR product and to analyze its variance.

6.4.2.6 Application of PCR in the Diagnosis of Parasitic Diseases

During recent years, PCR has been applied to detect infection of plasmodium, amebic protozoa, tyoxoplasma gondii, trypanosome, leishmania, cryptozoite and giardia lambli. For some protozoal diseases, the quantity of pathogen is extremely rare and thus undetectable with conventional diagnostic examinations. PCR amplification is an approach to define the diagnosis, which provides findings that conventional etiological examinations fail to, especially in detecting parasite in tissues. For instance, in the diagnosis of trypanosomiasis disease, single polypide in blood specimen can be detected by PCR amplified and purified DNA probe. Currently, the diagnostic procedures of toxoplasmosis and falciparum malaria by PCR have been established in China. PCR is a promising and prospected molecular biological technology in the diagnosis of parasitic diseases.

6.4.3 Gene Chip Technology

6.4.3.1 Introduction

Gene chip, also known as DNA chip and DNA microarray, is a technique for the diagnosis of diseases based on analyzing the intensity and distribution of hybridization signals. A large quantity of DNA probe molecules should be firstly fixed on the support, followed by their hybridization with labelled sample. The intensity and distribution of hybridization signals are then detect for analysis. In recent years, gene chip technology has been continually improved and shows its application value in genetic diagnosis, gene expression research, genome research, discovery of new genes and detection of various pathogens.

6.4.3.2 Principle

Gene chip technology is a rapid, effective way in analyzing nucleotide sequence based on gene recombination technology. Firstly, a large quantity of DNA probe molecules are fixed on the support and subsequently hybridize with labelled sample. Via detecting the intensity and distribution of hybridization signals, gene analysis is conducted. On a gene chip of 1 cm^2 in size, necessary thousands of, even millions of genes can be fixed to form a gene array for synchronous detection of genes. In the gene chip technology, 4 major steps should be performed, including chip preparation, sample preparation, hybridization and detection of hybridization signal, and analysis.

6.4.3.3 Operational Procedures

Chip Preparation

Probe Synthesis

Probe synthesis is the first important step in gene chip preparation. The probes of gene chip include genomic probe, cDNA probe and oligonucleotide probe. As for the genomic probe, a sequence which completely matches with the target sequence for detection or carries mutant site is firstly selected from genomic bank. After PCR amplification, a genomic probe is prepared. As for short oligonucleotide probe, it can be directly synthesized by DNA synthesizer. However, cDNA probe can be prepared by selection in cDNA bank with following PCR amplification based on the expressed mRNA of genes for detection. The selected sequence for PCR amplification should be close to 3′-terminal of cDNA, with a length of about 1 kb.

Fixation of Probe on the Surface of Carrier

Many approaches can be chosen to fix the probe on a solid support, which can be categorized into in situ synthesis and off-chip synthesis. In situ synthesis refers to synthesis of oligonucleotide probe on the surface of support, including in situ lithography synthesis, piezoelectric printing and molecular stamp in situ synthesis. By off-chip synthesis, pre-synthesized probe, cDNA or genomic DNA are directly placed on chip by a specific rapid off-chip robot, which is more commonly applied for large segment of DNA and sometimes oligonucleotide and even mRNA. The solid carrier used as gene chip mainly include solid flakes (such as glass, silicon chip, and ceramic chip) and thin films (such as nitrocellulose filter membrane, nylon membrane and polypropylene film). Glass slide or nylon membrane are usually employed as carrier of the gene chip.

Sample Preparation

The sample to be detected is mainly DNA/mRNA obtained from blood cells or tissues. Many sample molecules are needed in detection, and the sample is commonly amplified before labelling and analysis in order to improve the sensitivity of reading. As for the isolated genomic DNA, direct amplification by PCR can be performed, while as for mRNA, cDNA should be firstly prepared by using RT-PCR.

Many materials can be used to label the sample to be detected, such as Cy3, Cy5, biotin, and radioactive isotope ^{32}P. The labelling of sample refers to the process that dNTP carrying marker, such as Cy3 or Cy5-dNTP, biotin-dNTP, ^{32}P-dNTP, is interfused into target molecule sequence via amplification of DNA or cDNA. Otherwise, the dNTP carrying marker is interfused into the target molecule sequence during synthesis of primer, which is then used to amplify the target sequence, with the terminal of amplified product labelled. Fluorescence labeling is more commonly employed, which is based on the fact that the labeled molecules can be activated by laser to emit fluorescence within specific wave length. Therefore, the sample containing labeled molecules can be detected. This approach does not require labeling by isotope but with extremely high sensitivity and is applicable for quantitative detection. Therefore, it is widely applied in labeling of samples on gene chip.

Hybridization

Hybridization refers to the reaction between fluorescence labeled sample with the probe on the chip to produce a series of information. Appropriate conditions can optimize reaction among biological molecules to reduce the mismatch rate. After hybridization, the matrix is inserted in a scanner to detect the mode of hybridization. The data of hybridization signal can be obtained when fluorescence signal labeled in the target is stimulated. The signal produced by matched probe and target is often stronger than mismatched probe and target. As the sequence and site of each probe in the matrix are known, the properties of target nucleotide can be identified by using the probe matrix.

Detection and Analysis of Hybridization Signal

Many approaches can be used to detect hybridization signal, such as fluorescence microscopy, mass spectrometry, chemiluminescence and optical fiber, among which fluorescence microscopy has gained the widest application.

As for detection technique based on nucleic acid hybridization, the main process of fluorescence detection is described as follows: The target sequence or sample that has been amplified is firstly labeled by fluorescence, followed by hybridization with the large quantity of probes on the chip. After the non-hybridized molecules are flushed away (the procedure can be omitted when real-time fluorescence is used), a fluorescence microscope is used to scan the gene chip to collect fluorescence intensity of each site for comparative analysis. As double strands with normal base pairing show more favorable thermal stability than double-strands with mismatched base pairing, fluorescence signal intensity of the double strands with well matched base pairing is 5–35 times as high as that double strands probe with one or two mismatched base pairing. Additionally, the intensity of fluorescence signal shows certain linear relationship with the quantity of target molecule in the sample. Therefore, precise detection of fluorescence signal intensity can demonstrate the specificity and intensity of hybridization. The approaches for quantitative analysis of fluorescence intensity from each site of highly dense probe matrix mainly include confocal laser scanning microscopy and charge-coupled camera.

6.4.3.4 Application of Gene Chip in the Diagnosis of Parasitic Diseases

Due to its simple operations and simultaneous sensitive rapid detection of thousands of genes, in addition to its highly specific stable automatic information, DNA gene chip technology is believed to be a revolutionary milestone in genetic information analysis. Furthermore, it provides a prospected future for academic research and clinical application in parasitology. Its present or potential application fields include categorization and evolution of parasites at the genetic level, relationship between parasites and their environment, research in vaccines for protection against parasites, molecular diagnosis of parasitic diseases, research in drug resistance of parasites and new drug development.

6.4.4 DNA Sequencing

6.4.4.1 Introduction

DNA sequencing is to map out the sequence of 4 chemical bases that comprise a stand of DNA, namely first grade detection of nucleic acid sequence, which is an important technology in modern molecular biology. The basic procedures in routine sequencing include: (1) to process the DNA molecules to be detected to obtain a series of one nucleotide curtate DNA molecular complex; (2) to isolated these DNA molecules by gel electrophoresis to form a stripe in cascade shape arrangement for readings of each DNA base.

6.4.4.2 Principle

Approaches to DNA sequencing include the dideoxy chain termination proposed in 1977 by Sanger et al. (1981) and chemical degradation proposed in 1977 by Maxam and Gilbert. The principle of Sanger DNA sequencing is to use a DNA polymerase to extend primer that binds to the template with sequence to be detected until termination of nucleotide by interfusing one strand. Each sequencing is consisted of 4 independent reactions in a series. And each reaction contains all 4 dNTPs with addition of limited quantity of different ddNTP. Due to the absence of necessary 3-OH group in ddNTP extension, the extending oligonucleotides are terminated selectively at site G, A, T or C, which is determined by corresponding dideoxy in reaction. The relative concentration of each dNTP and ddNTP can be adjusted to obtain a chain termination product with bases ranging from hundreds to thousands. They share common starting point, but terminate at different nucleotides, and can be separated into fragments in different sizes via high resolution denaturing gel electrophoresis. After that, X-ray film autoradiography or non-isotope labeling can be applied for their detection. The chemical modification sequencing proposed by Maxam and Gilbert is based on the fact that certain chemical reagent can induce base modification at specific site (or specific type of base) in DNA strand, with following base shedding or replacement and finally occurrence of specific rupture. The rupture of different molecules at different sites can produce a series of DNA fragments in different sizes, which are then separated by gel electrophoresis. Before analysis, 5′-terminal of DNA is labeled by isotope for readings of DNA nucleotide sequence after autoradiography.

6.4.4.3 Next-Generation and Third-Generation of Sequencing Technology

With the development of science and technology, traditional Sanger sequencing is not sufficient for academic research. Both genome resequencing of model organism and genome sequencing of non-model organism require a sequencing technology with lower cost, higher flux and rapid sequencing. As a result, the next-generation sequencing emerged. The core of next-generation sequencing is sequencing by synthesis, namely capturing newly synthesized terminal labeling to determine DNA sequence. Currently, the technological platforms available include Roche/454 FLX, IIlumina/SolexaGenome Analyzer and Applied BiosystemSOLID system.

Researchers from University of Washington and other institutions have made a breakthrough in new-generation sequencing using nano-biotechnology. This new approach can provide individualized gene sequencing blueprint for more efficient individualized medical care for patients with cancer, diabetes or certain substance abuse. This achievement has been published on PNAS, which is known as the next-next-generation sequencing or the third-generation sequencing. The new approach is based on singular molecule reading technology via nanopore, which is different from the next-generation sequencing and can read information directly, rapidly and simply. The next-generation sequencing requires the aid of fluorescence or chemiluminescent material to read DNA polymerase or DNA ligase for indirect sequencing via detecting optical signal released during linkage of base into DNA strand.

6.4.4.4 Application of DNA Sequencing in the Diagnosis of Parasitic Diseases

In recent years, with maturation of DNA sequencing and improvement of automatic sequencer, the sequencing technology has been greatly improved in speed and accuracy. Different gene code sequences in the living organisms are the basis to distinguish their different species. Studies on DNA sequencing of paragonimus and other parasites have been conducted in China and DNA sequencing will be widely employed in the detection of parasites in the future.

6.4.5 Ribosome Display Technology

6.4.5.1 Introduction

Ribosome display technology (RDT) is a new technology using functional protein for mutal screening that was improved in Plückthun laboratory based on polyribosome display technology. Via the procedures, the correctly folding protein and its mRNA can simultaneously bind on ribosome to form mRNA-ribosome-protein tripolymer. Therefore, the genotype and phenotype of target protein are integrated for selection of antibody and protein bank as well as in vitro modification of protein. It has been successfully applied in screening some proteins with high affinity and specificity in binding to target molecule, such as antibody, polypeptide and enzyme, which is an important tool in screening of protein.

6.4.5.2 Principle

By amplification of DNA bank by PCR and simultaneous introduction of T7 initiator, binding site of ribosome and stemring structure, RDT is applied to transcribe DNA into mRNA, which is then translated in the cell-free translation system to display the translated product of target gene on the surface of ribosome. The formed mRNA-ribosome-protein compound composes the protein bank for ribosome display. After that, the corresponding antigens are chosen to screen the translated compound, with edetic acid to dissociate binded ribosome compound or with specific antigen to elute the whole compound, and to dissociate mRNA from the compound. Via RT-PCR, the next-round template for display is provided and the obtained DNA enters into the next round of accumulation. Some DNAs can be sequenced for analysis via cloning.

6.4.5.3 Development History of RDT

Kawasaki proposed to use similar approach to screen peptide ligand from peptide library. Previously early-stage peptide antibody had been used for immunoprecipitation and mRNA conjugating with ribosome and peptide had been successfully isolated. Mattheakis et al. (1994) put these forerunning ideas into practice for the first time and established polyribosome display technology for screening peptide ligands. By using this technology, they successfully screened peptide ligands of immobilized monoclonal antibody with an affinity constant of up to 10^9 (at the level of nmol) from a peptide library with storage capacity of 10^{12}. Gersuk et al. (1997) also successfully applied this technology to screen peptide ligands labeled by prostate neoplasm marker. During in vitro translation, the folding and translation of protein or peptide work synchronously. The natural peptide binding with ribosome also has enzymatic activity. These studies have demonstrated that folding of certain protein is not affected by ribosome-channel protein. Therefore, isolation of ribosome is not indispensable for protein to obtain native conformation. Based on studies mentioned above, Mattheakis's polyribosome display technology was greatly improved in Plückthun laboratory in 1997 and a new technology to screen complete functional protein, ribosome display technology, was established.

6.4.5.4 Operational Procedure

Transformation of Gene Fragment

In order to guarantee the effective transcription and translation of gene fragments, 5′ terminal should connect with T7 initiator sequence and SD sequence and the termination codon of 3′ terminal should be removed. Thus the ribosome can successfully remain on 3′ terminal of mRNA to connect protein and mRNA together. PCR extension is commonly performed for several times during transformation of gene fragment.

1. The necessary gene fragments and intervening sequences are amplified respectively.
2. Primer 4 and Primer 5 are used for connecting PCR to connect target gene and introns. Primer 5 contains SD sequence and Primer 4 contains 3′ terminal stem-loop structure sequence.
3. The PCR product undergoes PCR extension again, with Primer 6 containing T7 initiator sequence and 5′ terminal stem-loop structure sequence as well as the same downstream product. The 5′ terminal is connected with T7 initiator, stem-loop structure and SD sequence, while 3′ terminal integrates with intervening sequences and contains 3′ terminal stem-loop structure.

In Vitro Transcription and Translation

1. In vitro expression can be achieved with pronucleus originated E. coli S30 cell-free protein synthesis system, eukaryon-based downbreak fluid of rabbit reticulocyte and cell-free protein synthesis system for wheat embryo extraction. However, disputes exist concerning the question that which one is the most appropriate. In vitro transcription and translation can be conducted conjugately or respectively. Currently, in vitro protein translation system based on DNA template and in vitro coupled transcription and translation system based on RNA template are commercially available.

Transcription and translation should be conducted respectively in ScFv antibodies and other proteins containing disulfide bridges, because protein containing disulfide bridges can be only correctly folded under oxidizing condition. However, during transcription, T7 RNA polymerase requires β-mercaptoethanol with reducibility to maintain its stability. If target protein can be correctly folded under reducing condition, in vitro coupled transcription and translation system is recommended.

2. It is necessary to control the effect of RNase when in vitro transcription and translation is conducted respectively. VRC, transition state analog as inhibitor of RNase, can effectively inhibit nuclease and improve the display rate of E. coil ribosome. In addition, stem-loop structure of 3′ terminal and 5′ terminal can help mRNA to avoid the effect of exonuclease RNase II and endonuclease RNase E.

Affinity Screening

The mixture produced by translation is diluted by 4 times to terminate the reaction after the completion of in vitro translation. The concentration of Mg^{2+} in reagent remains consistently at 5 mmol/L to stabilize mRNA-ribosome-protein ternary complex for either direct application in screening experiment or storage at the temperature of 4 °C. The ribosome complex can be stabilized for at least 10 days at the temperature of 4 °C, but with gradually decreased productivity. During in vitro screening, Nonspecific reaction between antigen and antibody can be reduced when 1–2 % skim milk powder and 0.2 % heparin are added during in vitro screening. In addition, heparin can also inhibit nuclease.

Affinity Screening

The affinity screening can be categorized into solid-phase screening and liquid-phase screening, specifically ELISA and paramagnetic particle method. Hanes et al. (2001) believed that the antigen coats the plastic surface in ELISA, whose hydrophobic effect may affect spatial conformation of attachment protein and the screened antibody molecules are unable to identify the natural epitope of antigen. They also argued that paramagnetic particle method is to connect the catching label such as biotin with the antigen, with capturing of the label by using streptavidin beads after the formation of antigen-antibody complex for affinity screening.

Separation of mRNA

After screening, icy buffer containing 20 mmol LEDTA is added to elute mRNA, and the eluted mRNA is then processed by Dnase I to remove the residual DNA template. T7 initior is introduced again via RT-PCR and the necessary components for ribosome display such as SD sequence are used in next-round display. Otherwise, the components are directly used in Northern hybridization to assess the screening efficiency. The target genes acquired from the final-round screening are connected with plasmid, followed by transfer to E. Coli to obtain singular target clone. The single-strand antibody molecules are further expressed in vitro or in vivo (secreting type or inclusion body type) to identify their activity.

Oriented Molecular Evolution in Vitro

When ribosome display technology is used to screen non-variant library, mutation and recombination technology can be introduced via error PCR or DNA shuffering to increase the molecular diversity. Therefore, the possibility in obtaining target molecules with favorable affinity and stability as well as being capable of increasing enzyme activity can be greatly improved. Plückthumx's team has successfully screened 1.1 nmol ScFV with high affinity by using ribosome display technology. Subsequently, they increased its affinity by introducing DNA rearrange technique.

6.4.5.5 Application of RDT in the Diagnosis of Parasitic Diseases

RDT is performed completely in vitro, with simplicity in library establishment, large storage capacity, more molecular diversity, simplicity in screening, and no pressure in selection, compared with phage display technology or saccharomycetes display technology. In addition, the affinity of target protein can be improved by introducing mutation and recombination technology. Therefore, we believe it is a powerful approach in large-scale library screening and acquisition of molecular evolution and prospect its application in the diagnosis of parasitic diseases.

6.5 Pathological Diagnosis

The etiological examination of parasitic diseases is performed to morphologically identify the infected parasite in different stages of its development for clinical diagnosis. Different parasites gain their access into human body via diverse routes, reside at different parts within human body, and affect different organs and tissues. Therefore, all the procedures including collection of suspected specimen, identification of polypide and understandings about the possible parasitized site are of great importance for the etiological examination of parasitic diseases.

6.5.1 Site with Possibly Detectable Parasite

Generally speaking, each species of parasite has relatively fixed and specific parasitizing site, for instances, plasmodium parasitizes successively in hepatocytes and erythrocytes; and Paragonimus westermani parasitizes in lungs. However, many parasites may also be present at sites other than the common parasitizing body parts to cause ectopic parasitism. For instances, in addition to the mesenteric vein, schistosome may also grow in lungs, brain and other body parts to cause ectopic damages; protozoa parasitizing in blood or blood cells may spread to any part of human body along with blood flow to cause multiple organ damages or systematic extensive lesions, such protozoa as plasmodium, Leishmania donovani and Trypanosoma; parasites residing in the intestinal tract generally are confined to the digestive system to cause gastrointestinal symptoms, but the larvae of a notable proportion of intestinal nematodes may migrate inside the body via the blood flow, such as roundworm, hookworm and strongyloides stercoralis. The parasites residing in tissue may parasitize in different organs and tissues and incur inflammation and space occupying symptoms at the corresponding sites.

6.5.2 Selection of Methods and Sites for Specimen Collection as Well as Ways for Parasite Detection

Standardized procedures for specimen collection and processing are crucial to the accuracy in identifying parasitic infection. One organ may be infected by different species of parasites, characterized by variance in clinically manifestations. Therefore, the pathogen should be detected to obtain definite diagnosis for effective treatment of parasitic disease. The successful detection of parasite is based on knowledge about possible parasite at different sites, appropriate approaches in specimen collecting, and appropriate processing of the specimen.

The specimens collected for detection of parasite are diverse, ranging from excretions, such as stool, intestinal drainage and urine, and secretions, such as sputum, cerebrospinal fluid, bile fluid, tears, prostatic fluid and vaginal discharges; to tissues from different organs and different body parts for biopsy, such as specimens collected during surgery, histocytes harvested from puncture, skin scrapings and pus from necrotic and liquefied tissues. In addition, approaches for collecting different specimens via different routes may present varying difficulties. For instances, the collection of stool and urine specimens as well as their volume control are comparatively easy, while by needle aspiration of tissues for biopsy, only limited quantity of tissues can be obtained and the patients commonly experience pain. In addition to skillful practice of the procedures for specimen collection, timely processing of the specimen is also of great importance. The factors, such as selection of fresh or fixed specimen, selection of fixation fluid, approach for staining and the staining agent, selection of refrigerated or frozen specimen, procedures in specimen processing and the duration of preservation, all show effect on the result of etiological detection in different degrees. Some of these factors even directly affect the success of polypide detection or correct identification of the parasite if detected.

An appropriate selection of examination methods guarantees accurate diagnosis of parasitic disease. The parasite examinations are generally categorized into etiological examination, immunoassay and molecular biological examination, all of them have diverse specific methods and are of different values in identifying different species of parasite. For helminthic infections, successful detection of eggs, larva and/or imago in various clinical specimens, especially intestinal specimens, can define the diagnosis. For instances, by direct stool smearing or microscopic observation, the finding of roundworm, whipworm and hookworm eggs can define the diagnosis. As for the protozoa, because their polypides are minute ranging from only 1.5 μm (microsporidia) to 80 μm (balantidium coli), and some even grow within cells, the definitive diagnosis requires smearing and staining, or even special staining. For some parasites, the species can be defined by biochemical and molecular biological techniques. For instances, plasmodium infection can be defined by microscopic examination after blood collection, blood smearing and staining. As for the microsporidia, a definitive diagnosis requires electronic microscopy or molecular biological examination such as PCR. Different approaches in examining the same specimen show different sensitivity and specificity to the polypide. Therefore, the selection of appropriate examination is paramount to successful detection of polypide and accurate identification of parasite from the suspected specimen.

References

Gersuk GM, Corey MJ, Corey E, et al. High-affinity peptide ligands to prostate-specific antigen identified by polysome selection. Biochem Biophys Res Commun. 1997;232(2):578–82.

Hanes SD, Herring VL. Gentamicin enzyme-linked immunosorbent assay for microdialysis samples. Ther Drug Monit. 2001;23(6):689–93.

Mattheakis LC, Bhatt RR, Dower WJ. An in vitro polysome display system for identifying ligands from very large peptide libraries. Proc Natl Acad Sci USA. 1994;91(19):9022–26.

Sanger BKF. Pioneer of primary sequences. Biosci Rep. 1981;1(1):1–2.

Vainshtein I, Atrazhev A, Eom SH, et al. Peptide rescue of an N-terminal truncation of the Stoffel fragment of taq DNA polymerase. Protein Sci. 1996;5(9):1785–92.

Part II

Clinically Detected Parasitic Diseases

7 Protozoiasis

Bailu Liu, Hanqiu Liu, Yonghua Tang, Xiaochun Zhang, and Yuxin Yang

Protozoon is single cell eukaryote, which is made up of only one cell. Despite its simple structure, it can achieve all the physiological activities by itself. A variety of protozoa distributes widely in natural world, and about 200,000 species of protozoa have been identified and nominated. Most of the identified protozoa are autotrophic or saprophytic, living in the sea, soil, water or putrid substances, and nearly 10,000 species of the identified protozoa are parasitic, living in or on the surface of animals.

Medical protozoa refer to pathogenic or non-pathogenic protozoa that parasitize in cavities, canals, body fluids, tissues and cells of human body. And more than 40 species of protozoa have been identified and categorized into this group of protozoa. The diseases induced by these protozoa are known as protozoiasis, showing various symptoms and transmission routes due to different parasitic sites. Protozoiasis can be transmitted via mouth or media organisms with varying degrees of damages to human body due to the variances in species of protozoa, parasitic sites, and immunity of the host. The protozoa parasitizing in tissues are generally more harmful to human body than those parasitizing in cavities and canals. Because of the characteristic biological properties and transmitting routes, some protozoa threat the health of human and farm animals to cause wide regional prevalence.

B. Liu
The Second Affiliated Hospital, Harbin Medical University, Harbin, Heilongjiang, China

H. Liu
Affiliated Huashan Hospital, Fudan University, Shanghai, China

Y. Tang
Ruijin Hospital, Affiliated to School of Medicine, Shanghai JiaoTong University, Shanghai, China

X. Zhang (✉)
The First Affiliated Hospital (Xinan Hospital), The Third Military Medical University, Chongqing, China
e-mail: zxcylxyr@163.com

Y. Yang
The People's Hospital, Urumqi, Xinjiang Uygur Autonomous Region, China

7.1 Biological and Physiological Properties of Protozoa

Protozoon is basically composed of cytomembrane, cytoplasm and nucleus.

7.1.1 Cytomembrane

Cytomembrane is also known as pellicle or plasmalemma, which encloses the body of protozoon. Under an electron microscope, it is one layer or more than one layer of unit membrane whose outer layer is a cell coat (or known as glycocalyx) composed of lipid and protein molecules binding to polysaccharide molecules, In addition to the roles in separating and transporting, the cytomembrane also participates in protozoal physiological activities such as feeding, excretion, motion, sensation, invasion and elusion with its dynamic structure.

7.1.2 Cytoplasm

The cytoplasm is composed of stroma, organelles and internal substances.

7.1.3 Nucleus

The nucleus is the most important component of protozoon to maintain its life and reproduce its offsprings. It is composed of nuclear envelope, nucleoplasm, nucleolus and chromatin.

Facilitated by motion organelles, most protozoa perform activities of moving, ingesting and defense and their motions can be categorized into pseudopodial movement, flagellar movement, and ciliary movement. The protozoa with no motion organelles can move by sliding, twisting and bending. The parasitic protozoa live in a nutrient-rich environment,

H. Li (ed.), *Radiology of Parasitic Diseases*, DOI 10.1007/978-94-024-0911-6_7

which are supplied via osmosis, pinocytosis and phagocytosis. The parasitic protozoa reproduce asexually (by binary fission, multiple fission or budding), sexually (by zygogamy or gamogony) or in both ways (metagenesis). Meanwhile, they leave the primary host to invade other host to maintain the colony.

7.2 Types of Protozoal Life Cycle

The reproduction of protozoa is actually an adaptive survival of their colony, which is necessarily accompanied by alternation of hosts. Therefore, the life cycle of medical protozoa includes the stages of growth, development and reproduction as well as their transfer from host to host, which are characterized by diversity. Medically, the life cycle of protozoa is of great epidemiological significance. Based on their transmission routes, the life cycle of protozoa can be categorized into the following three types:

7.2.1 Interpersonal Transmission Type

The protozoa of this type have a simple life cycle, which parasitize only one species of host and transfer from one directly to another via contacts or intermediate media. Interpersonal transmission can be further divided into 2 sub-types.

7.2.1.1 Only Developmental Stage of Trophozoite

The life cycle of this sub-type only has the developmental stage of trophozoite and the protozoa commonly transfer their host via direct contacts, such as trichomonas vaginalis, trichomonas tenax, and entamoeba gingivalis.

7.2.1.2 Two Developmental Stages of Trophozoite and Cyst

The life cycle of this type includes two developmental stages, trophozoite and cyst. Trophozoite proliferates by binary fission while cyst remains static, which is also the infective (transmitting) stage of medical protozoa. The examples of protozoa with this type of life cycle include entamoeba histolytica and giardia lamblia.

7.2.2 Circular Transmission Type

One life cycle is completed in one or more vertebrates as the definitive or intermediate host. For instances, toxoplasma gondii parasitizes in cats as the definitive host and parasitizes in humans, rats or pigs as the intermediate host.

7.2.3 Insect Intermediated Transmission Type

One life cycle of protozoa includes a stage of sexual or asexual reproduction in blood sucking insect and the other stage of parasitizing in human or animal, such as leishmania and plasmodium.

7.3 Categories of Protozoiasis

Based on the parasitic sites, protozoiasis can be categorized into the following groups:

7.3.1 Protozoal Intestinal Infection

Protozoal intestinal infection, such as intestinal amebiasis, shows clinical symptoms of abdominal pain, diarrhea, loose stools or bloody purulent stool. Trophozoites or cysts can be detected by laboratory test in stool specimen.

7.3.2 Protozoal Hepatobiliary Infection

Protozoal hepatobiliary infection, such as amebic liver abscess, is clinically manifested with fever, hepatalgia, and hepatomegaly with tenderness.

7.3.3 Protozoal Blood and Lymphatic Infection

Protozoal blood and lymphatic infection, such as malaria and babesiasis, commonly shows local or systemic lymphadenectasis, anemia, chills, and fever.

7.3.4 Protozoal Infection of the Nervous System

Such infection cause diseases of primary amebic meningoencephalitis, African trypanosomiasis and cerebral malaria. These diseases are clinically manifested as fever, headache, increased intracranial pressure, epilepsy and coma.

7.3.5 Protozoal Infection of Skin and Muscle

Such infections cause the diseases of amebiasis cutis and cutaneous leishmaniasis with symptoms of red and swollen skin and ulceration.

7.3.6 Pulmonary Protozoal Infection

Pulmonary protozoal infection causes the diseases of amoebic lung abscess, with symptoms of fever, chest pain, and dry cough or cough with thick phlegm. In some cases, protozoa can be detected in the phlegm.

7.3.7 Ocular Protozoal Infection

Such infection causes the disease of ocular toxoplasmosis, and the patients experience iridocyclitis and retinochoroiditis.

7.3.8 Urogenital Protozoal Infection

Such infection causes diseases like trichomonas vaginitis, which clinically resembles urethritis or vaginitis. It should be noted that the lesion caused by protozoal infection are commonly not confined to only one organ but involves several organs.

7.4 Characteristics of Protozoiasis

7.4.1 Proliferation

7.4.1.1 Damaging Cells

Only when protozoon proliferates to a considerable quantity in certain stage of its life cycle, it can cause damages or clinical symptoms. Without repeated infection, such tremendous increase of pathogen in quantity is different from other common worm infections and constitutes a biological essential for tiny protozoa to threaten human health. The protozoal density in one unit of their parasitizing blood or hematocytes is known as parasitemia. As an indicator of the infection severity, it can be measured by enumeration. Proliferation of different protozoa induces different characteristic pathogenicity, which provides reliable information for clinical diagnosis of protozoiasis.

7.4.1.2 Spreading

After the protozoon proliferates to a considerable quantity, they are capable of spreading to neighbouring or distant tissues and organs to induce histopathological changes.

7.4.2 Opportunistic Pathogenicty

Clinically, it has been found that the conditions of patients with extreme malnutrition, advanced malignancy, long-term use of hormone therapy, immunodeficiency, compromised immunity or acquired immune deficiency syndrome (AIDS) are commonly complicated by fatal protozoal infection. Such patients, characterized by activated infection due to compromised immune mechanism that is induced by natural or interventional factors such as diseases and treatments, are known as immune compromised hosts. And the protozoa inducing such activated infection are known as opportunistic protozoa.

7.4.3 Toxic Effect

The toxic substances produced by pathogenic protozoa can damage the cells, tissues and organs of the host via various approaches. The secretions (including multiple enzymes) and excretions of protozoa as well as debris of dead protozoa show toxicity to their hosts.

7.5 Amebiasis

Amebiasis is a disease caused by entamoeba that has been categorized into the genus of Entamoeba and the family of Entamoebidae. The disease distributes worldwide, with a high incidence rate in tropical and subtropical areas and a high prevalence in Mexico, Eastern South America, and Southeast Asia. And the global average infection rate of amebiasis is above 20 %. In China, it commonly occurs in the northwestern, southwestern and North China, with an infection rate of above 2 % in the provinces of Yunnan, Guizhou and Gansu as well as Xinjiang Uygur Autonomous Region. The infection rate peaks in adolescents and children under the age of 14 years as well as adults aged above 40 years. Amebiasis is an intestinal protozoal zoonosis, with susceptibility in humans and animals, such as dogs, cats, pigs, rats and primates. The pathogen, entamoeba, parasitizes in colon and cecum of animals to cause mucosal erosion and ulceration of large intestines.

The sources of its transmission are mainly the entamoeba carriers who continually excrete entamoeba cysts along with stools or animals carrying entamoeba. The routes of its transmission include oral intake of contaminated food or water by mature entamoeba cysts in humans, and oral intake of entamoeba cysts contained stools or contaminated feedingstuff and water by entamoeba cysts contained stools in animals. The life cycle of entamoeba histolytica includes the stages of trophozoite, precyst, cyst and postcystic trophozoite, and its shape varies at different stages. Entamoeba histolytica completes its whole life cycle in only one mammal host and human is its optimal choice. However, animals such as dogs, cats and pigs also show epidemiological significance.

Amebiasis, characterized by chronic onset and obstinate diarrhea, may also involve other tissues and organs via blood flow. In some rare cases, it directly invades the liver, lung, brain and skin to cause secondary amebiasis, with amebic liver abscess as the most common. After humans or animals are infected, they are firstly asymptomatic carriers, followed by developing into various clinical types of amebiasis, including acute diarrhea type or abscess type. According to the clinical typing recommended by WHO, amebiasis can be categorized into asymptomatic infection type (accounting for above 90 % and mostly non-invasive infection of complexus) and symptomatic invasive infection type. The symptomatic invasive infection type can be further divided into intestinal amebiasis (including amebic dysentery, enteritis, amebic abscess, and amebic appendicitis) and parenteral amebiasis (including amebic liver abscess, amebic lung abscess, amebic cerebral abscess and skin amebiasis).

Case Study 1

[Summary of Medical History]

A 20-year-old man complained of black stool in Oct. 2010 and his stool changed into mucous bloody in late May, 2011. In a local hospital, his condition was diagnosed as hemorrhoids and the administered therapy showed no therapeutic response. On Aug. 14, 2011, his bowel movement was fresh bloody in a large quantity, and colonoscopy in a local hospital indicated ulcerative colitis. Oral intake of Etiasa in a dose of 2 packs every 6 h was prescribed. On Aug. 29, 2011, colonoscopy showed continuous diffuse mucosal erosion and ulceration at the observable colon 40–120 cm from the anus. The ulceration was covered by pus and showed bleeding after palpation. Local hyperplasia of granulation tissue was detected. Based on these findings, the diagnosis was suspected to be severe active ulcerative colitis. Intravenous administration of methylprednisolone in a dose of 40 mg per day was ordered, which was then changed into oral intake of Prednisone in a dose of 60 mg per day. Such anti-infection therapies showed good therapeutic responses and the symptoms were relieved. The patient was then discharged from the hospital. On Nov. 4, 2011, he was admitted to the hospital again due to aggravated bloody stool and was treated with therapies of fasting, parenteral nutrition, hemostasia, acid suppression, anti-infection, Etiasa, and Prednisone. After the symptoms were relieved, he was discharged from the hospital. At that time, his bowel movement was once per day, semi-solid. Long-term following up examinations showed occult blood test of stool positive, but no observable blood in his stools. The use of hormone then decreased gradually in dosage and was withdrawn on Mar. 2012. However, mesalazine was still given in a dose of 3 g per day. Meanwhile, he had been receiving therapy of traditional Chinese medicine till now. On Sep. 17, 2012, his bowel movements began to show observable bloody stool, twice a day, soft and dark red. On Oct. 9, 2012, the patient caught a cold and experienced fever with a peak body temperature of 38 °C. In a local hospital, intravenous administration of antibiotics was given and the patient showed decreased body temperature after 4 days. One week ago, his condition began to aggravate, with 5–6 bowel movements per day, which was bloody with pus in a volume of 100 ml. He also experienced paroxysmal periumbilical pain, dizziness, normal body temperature, nausea and vomiting, and increased the dose of mesalazine to 4 g per day by himself. After paying emergency clinic visit, he was treated by fasting, acid suppression, hemostasis, anti-infection and fluid infusion. In recent 2 days, his stools changed into watery bloody, 3–4 times a day in a volume of about 50 ml. He was admitted to our hospital for further diagnosis and treatment. Since the onset, the patient had a poor appetite and experienced a weight loss of nearly 10 kg in recent half a year. The laboratory test of stool specimen on admission showed that the stool was brown in color, loose, RBC positive (+++), WBC positive (++++), parasitic eggs negative (−), and occult blood test (OB) positive (+). Examination of parasites in the stool specimen showed amebic cyst positive (+) and Entamoeba histolytica trophozoite positive (+).

[Radiological Demonstrations] (See Fig. 7.1)

[Diagnosis]

Ulcerative colitis (moderate to severe in severity), colonic amebiasis.

[Discussion]

The onset of ulcerative colitis firstly involves the rectum. The inflammation is characterized by superficial diffuse mucosal and submucosal inflammatory lesions at the rectum and colon. During its active stage, it is histologically manifested as intestinal mucosal congestion and edema, infiltrated lymphocytes, neutropils and mononuclear cells as well as ulceration. The chronic lesions show infiltrated karyocytes, crypt swelling, deranged and atrophic gland structure, and villi like mucosa. Radiologically, the lesions are demonstrated as slight or moderate thickening of the colonic wall, which is symmetrical and continuous and is milder than the condition of Crohn's disease (CD). The thickening is commonly detected at the rectum and left colon. Otherwise, it may diffuse across the whole colon, showing lead-pipe like sign. By contrast scanning, the thickening shows laminal enhancement. Compared to CD, the thickening has a lower degree of enhancement, with no typical comb sign; and rarer lymphadenectasis, peri-intestinal leakage, and ileocecal valve involvement as well as rarer intestinal fistulation. Toxic megacolon is the most common and characteristic complication of ulcerative colitis, with rarer occurrence of intestinal abscess and intestinal lumen narrowing than CD. The possibility of various infectious diseases should be excluded for the clinical diagnosis, especially bacillary dysentery, amebic dysentery

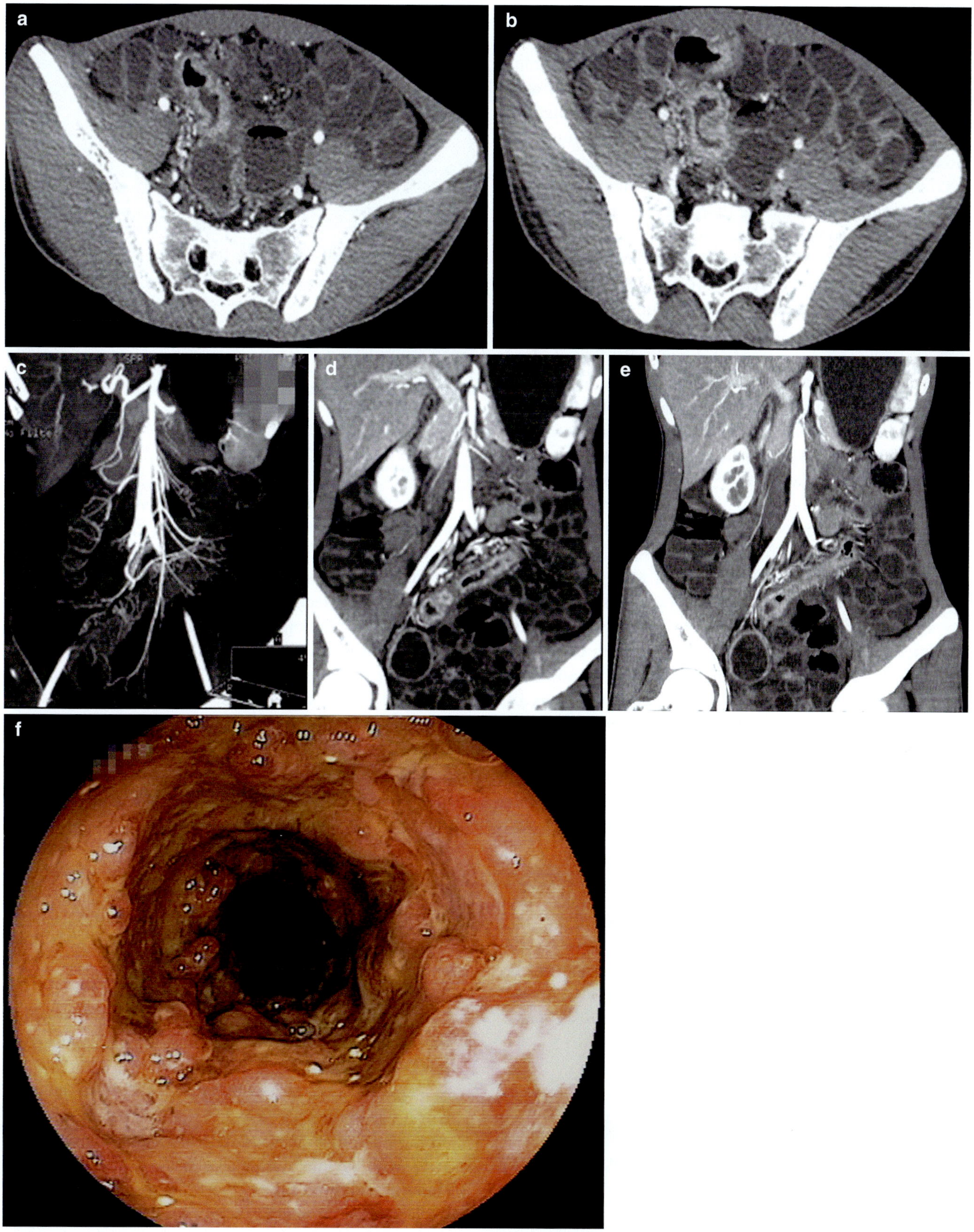

Fig. 7.1 (**a**–**e**) Small intestinal CT scanning and reconstruction showed thickening of parts of transverse colonic wall, descending colonic wall and sigmoid colonic wall with narrowed intestinal lumen. And such changes were shown to be the most obvious at the sigmoid colon, The neighboring intestines were demonstrated with slight dilation. By enhanced scans, the intestinal walls were shown with stratified enhancement and obvious mucosal enhancement. The submucosal tissues were visualized with edema, showing attenuated enhancement. The mucosal surface was unsmooth, with ulcerations. (**f**) Colposcopy showed irregular ulcer like lesions that densely distributed in the diseased intestinal lumen and were covered by white coating. The surrounding mucosa was shown with obvious congestion and edema, with observable particle like hyperplasia

and intestinal tuberculosis. Amebiasis is a group of protozoal diseases caused by entamoeba histolytica infection. Based on the infected site, it can be categorized into intestinal amebiasis and ab-enteric amebiasis. Ab-enteric amebiasis is caused by access of trophozoites in intestinal submucosa and muscularis into the vein and their migration along with blood flow to invade other tissues and organs, the most commonly amebic liver abscess. Intestinal amebiasis is an intestinal parasitic diseases caused by intake of contaminated foods and water by amebic cysts, which is characterized by symbiosis and potential pathogenicity. Its diagnostic criteria is detected trophozoites of entamoeba histolytica in stool specimen. Co-infection of the two diseases shows high possibility of invasive disease. In this case, enteroscopy demonstrated densely distributed ulcers at the colon, particle like hyperplasia of the lesions and accompanying luminal narrowing. Small intestinal CT scanning indicated that the range, degree and enhancement of the lesions were different from common ulcerative colitis, which is characterized by superficial and scattering ulcerations. In combination with the laboratory fecal examination, the diagnosis was defined. In addition, the disease should be differentiated from Crohn's disease. By CT scanning, Crohn's disease is demonstrated as segmental wall thickening of the small intestine or colon, which is asymmetric with enhancement by contrast scanning. Ab-enterically, Crohn's disease is shown with comb sign and penetration of inflammation into the serosa and other mesentery to cause exudation. Contrast scanning shows different degrees of enhancement. In addition, endoscopy shows longitudinal full-thickness ulceration, which is also the key points for the differential diagnosis from ulcerative colitis (characteristically superficial ulceration) and intestinal amebiasis (characteristically map like ulceration).

Case Study 2

[A summary of Medical history]

A 59-year-old man complained of liver swelling for 6 days and heated urine for half and a month, and was reported with mental disorder. The body temperature was 36.8 °C. By laboratory tests, the WBC count was 8.9×10^9/L, and NEUT% 63.9 %. By ultrasound guided liver abscess puncture, the diagnosis was defined to be liver abscess.

[Radiological Demonstrations] (See Fig. 7.2)

[Diagnosis]

Amebic liver abscess

[Discussion]

The onset of amebic liver abscess is chronic, with a long illness course and the lesion commonly at the right liver lobe. The lesion is commonly singular large abscess with smooth margin. And the ultrasonogram is characterized by no-echo area at the liver due to formation of abscess cavity. During the pathological progress of the abscess, it shows different ultrasonographic signs.

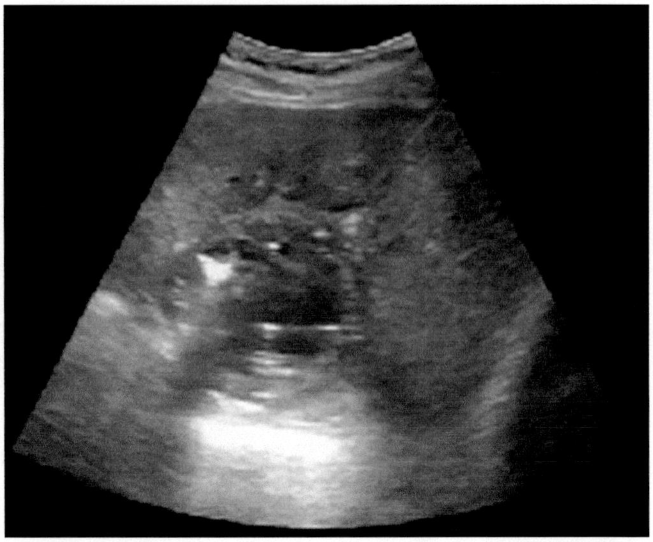

Fig. 7.2 Ultrasound demonstrated cystic-solid mass at the right liver lobe in a size of 6.3 × 7.2 cm, which was poorly defined with heterogeneous echo inside

During the early stage of abscess (pre-abscess stage), the liver tissue can be demonstrated with congestion and edema, with confined low echo at the liver area that is poorly defined.

During the abscess stage, the lesion progresses into tissue necrosis and liquefaction, which commonly occurs during 10–30 days after the onset. The abscess cavity shows as a no-echo liquid like dark area, which is commonly round or oval in shape but poorly defined. The echo from its posterior wall and deep liver tissue are commonly enhanced, and irregular sonolucent area can be visualized due to thick pus, honeycomb like abscess cavity, and incomplete liquefaction.

During the convalescence stage, the abscess lesion should be treated by puncture to discharge the pus. Along with the shrinkage of the abscess cavity, the pressure is decreased to show collapse of the abscess cavity and newly emerging liver tissues. CDFI may demonstrate no obvious surrounding blood flow signals.

Amebic liver abscess should be distinguished from the following diseases:

1. **Bacterial liver abscess**
   The lesions are commonly multiple abscesses, showing no-echo areas in different sizes with unsmooth edges. In the cases with singular abscess, the lesion is demonstrated as large no-echo area with thick wall. Bacterial liver abscess is commonly secondary to biliary infection or other suppurative diseases. The condition rapidly aggravates, with obvious systemic sepsis.
2. **Hepatocarcinoma cavity**
   The lesions are commonly singular with uneven thickness of the wall. CDFI demonstrates the lesion with abundant surrounding blood flow. And the patients experience hepatalgia and weight loss. By laboratory tests, AFP is positive, with ascites in some cases.

7.6 Toxoplasmosis

Toxoplasmosis, also known as Toxoplasmic disease, is a zoonosis caused by toxoplasma gondii. It parasitizes the karyocytes of humans and animals whose infection is commonly asymptomatic. Those patients experience the onset show complex clinical manifestations with no specific symptoms and signs, and misdiagnosis is quite common. The disease usually provokes lesions at the eyes, brain, heart, liver, and lymph nodes, and is also closely associated with HIV/AIDS. Toxoplasma gondii is one of primary pathogens inducing embryonic malformation due to intrauterine infection during pregnancy. Toxoplasma gondii is an obligate intracellular parasite, which is categorized into the genus of Toxoplasma, the family of Isospora, the order of Eucoccidia, and the sub-class of Coccidia.

7.6.1 Epidemiology

7.6.1.1 Source of Infection

Animals infected by toxoplasma gondii are the source of infection of toxoplasmosis, among which felines infected by toxoplasma gondii are the most important because their feces containing oocyst of toxoplasma gondii are the most important source of human infection.

7.6.1.2 Route of Transmission

Congenital Infection

The fetus can be infected from mother due to transplacental transmission.

Acquired Infection

Acquired infection is commonly transmitted via oral intake of not thoroughly cooked meat products, eggs, and milk contaminated by toxoplasma gondii. Other important route of transmission includes contacts to soil and water contaminated by oocysts of toxoplasma gondii. Toxoplasmosis can spread via blood transfusion and organ transplantation, and the pathogen can also invade its host via wounded skin and mucosa.

7.6.1.3 Susceptible Population

Populations are generally susceptible to toxoplasmosis, especially fetus, infants, as well as patients with neoplasm and/or HIV/AIDS.

7.6.2 Pathogenesis and Pathological Changes

During the stage of tachyzoite, toxoplasma gondii shows the strongest pathogenicity. After its invasion into karyocytes of the host, it rapidly develops and reproduces to induce rupture of the cells. The released toxoplasma gondii from ruptured cells gain their access into other cells to stimulate infiltration of lymphocytes and macrophages, which further causes acute inflammation and tissue necrosis. Bradyzoites in cysts of toxoplasma gondii can provoke chronic infection, whose reproduction in the cysts compresses the neighboring organs and induces dysfunction of the compressed organs. When cysts enlarge along with reproduction of bradyzoites to a certain degree or finally rupture, the released toxoplasma gondii triggers delayed allergy to induce formation of granuloma and occurrence of fiber calcification. Such lesions are commonly located in the brain and eyes.

Regardless of its route of infection, toxoplasma gondii gains its access into human blood flow directly or along the lymphatic system to cause parasitemia, followed by dissemination into other tissues and organs. The pathological changes of toxoplasmosis can be classified into the following 3 types: (1) The tachyzoites multiply in the host cells to induce necrosis, which can be replaced by new cells. Otherwise, fibrous scars may develop, which are surrounded by cysts with no inflammatory responses. (2) Pathological changes are provoked by ruptured cysts. (3) Local lesions are caused by toxoplasma gondii.

7.6.3 Clinical Manifestation

Clinically, toxoplasmosis falls into 2 groups, the congenital type and the acquired type, both of which are commonly asymptomatic. In some patients with severe conditions, serious symptoms may occur due to damages to multiple organs, which are commonly induced by newly developed acute infection or activated latent lesions.

7.6.3.1 Congenital Toxoplasmosis

Congenital toxoplasmosis usually shows neurological lesions, such as anencephaly, hydrocephalus, microcephaly and cranial fracture. Cerebral calcification is also a common sign of toxoplasmosis. Infants may experience mental retardation of different degrees, low intelligence quote, convulsions, spasm and paralysis due to brain damages. The disease has a poor prognosis, and the survivors commonly sustain such sequela as convulsion, mental retardation, chorioidoretinitis, strabism and blindness.

7.6.3.2 Acquired Toxoplasmosis

In the cases with mild conditions, the patients are asymptomatic, with clinical manifestations of enlarged head&neck lymph nodes. In some serious cases, the patients experience myocarditis, pneumonia, encephalitis, and choroidoretinitis.

7.6.4 Diagnostic Examination

7.6.4.1 Laboratory Tests

Etiological Test

The smear section and histological section can be prepared with body fluid or tissues for examination. The finding of tachyzoites can define the diagnosis of its acute infection.

Acute infection can also be diagnosed after toxoplasma gondii is detected in the peritoneal fluid of mice inoculated by blood, body fluid, cerebrospinal fluid and placenta.

Immunological Assay

Assay of specific antibody has been widely applied for the clinical diagnosis and epidemiological study of toxoplasmosis.

Assay of specific antigen, such as CAg of toxoplasma gondii, can be used for the early diagnosis or definitive diagnosis of toxoplasma gondii infection.

Assay of circulating immune complex can be used as a reference index for the diagnosis of acute toxoplasma gondii infection.

Molecular Biological Examination

Polymerase Chain Reaction (PCR) can be used to detect DNA of toxoplasma gondii in clinical speciments, which possesses such advantages as real-time, accuracy, high sensitivity, and high spcificity.

7.6.4.2 Diagnostic Imaging

Head CT scanning or MR imaging is capable of assisting the diagnosis of cerebral toxoplasmosis, indicating the site and range of lesions as well as their development in follow-ups.

Case Study 1

[Brief medical History]

A 42-year-old man complained of dizziness and headache for above 1 month that aggravated with accompanying vomiting in recent 2 days. By laboratory tests, he showed HIV positive (+). And by serological ELIZA, toxoplasma gondii IgM positive (+) and IgG positive (+) were detected.

[Radiological Demonstrations] (See Fig. 7.3)

[Diagnosis]

Cerebral toxoplasmosis (right frontal-parietal lobe)

[Discussion]

If patients with HIV/AIDS show neuropsychiatric symptoms, the condition should be highly suspected to be AIDS complicated by cerebral toxoplasmoisis and CT scanning and MR imaging should be ordered as early as possible. The radiological finding of characteristic eccentric target sign, especially at the basal ganglia, cerebral toxoplasmosis can be primarily diagnosed. In combination to serological toxoplasma gondii complement fixation test positive or good therapeutic responses by trial therapy for toxoplasmosis, the diagnosis can be defined.

In this cases, the diagnosis can be defined based on differentiation from the following diseases:

1. Primary brain lymphoma (PBL)
   The disease usually shows singular lesion. Compared to cerebral toxoplasmosis, peri-focal edema is mild but the lesion shows obvious space occupying effect, commonly invading ependema and callosum. By contrast scanning, the contour of lesion can be visualized with map liked or serration liked enhancement. And the condition fails to be improved by anti-toxoplasmosis therapy, but can be improved by radiation therapy. Diffusion MR imaging is of great significance in differentiating toxoplasmosis abscess from lymphoma.
2. Cerebral cysticercosis
   Radiologically, the lesion of cerebral cysticercosis can be demonstrated with small target sign, thin and smooth wall, and eccentric scolex inside. By contrast scanning, ring shaped enhancement can be shown. Immunological assay can assist its diagnosis.
3. Metastatic encephaloma
   Metastatic encephaloma always has its primary lesion. The metastatic neoplasm has a wall of uneven thickness, with attached nodular small lesions that are susceptible to hemorrhage and necrosis. Peri-focal edema widely exists, which facilitates the diagnosis of metastatic encephaloma.
4. Brain tuberculoma
   The lesion of brain tuberculoma is usually located close to the basal cistern, in the cortex and medulla of frontal, parietal and temporal lobes, and in the cerebellar hemisphere and vermis. Contrast scanning usually demonstrates multiple small nodular enhancement or ring shaped enhancement with typical target sign, obvious brain edema, and accompanying enhancement of basal cistern and meninges at the brain surface, hydrocephalus, cerebral infarction and cerebral vasculitis. The presence of pulmonary tuberculosis supports its diagnosis.

Case Study 2

[Brief Medical History]

A 60-year-old man experienced successful allogenic cadaveric kidney transplantation in our hospital for chronic renal failure (uremic stage) due to polycystic kidney on Sep. 2, 2010. After the operation, he received triple therapy of Ciclosporin, Myfortic and Hormone to suppress his immunity. On discharge from the hospital, his Cr level was about 80 μmol/L. After that, he was hospitalized due to pulmonary infection in April, 2011, Feb. 2012 and July, 2013, respectively, and discharged after healing. In a clinic follow-up examination in Feb. 2013, his Cr level was 170 μmol/L. The

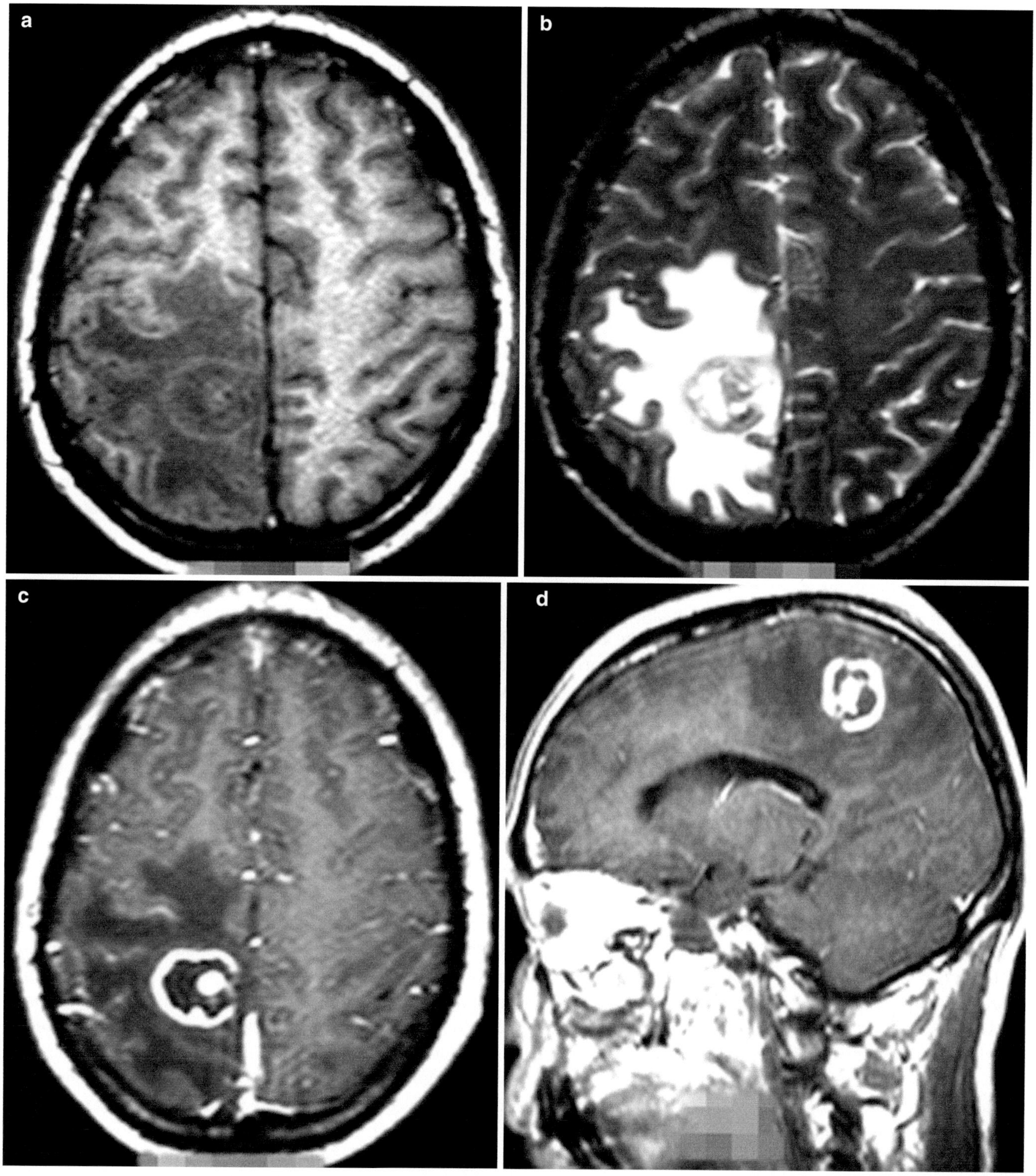

Fig. 7.3 (**a**, **b**) MR imaging demonstrated the lesions at the right frontal-parietal lobe with low T1WI signal but high W2WI signal and surrounding edema. (**c**, **d**) Contrast imaging showed the lesions with 3 different parts: inner part with enhancement, middle part with low signal, and peripheral part with ring shaped high signal enhancement. In addition, the inner part was visualized to be eccentric, which is known as the eccentric target sign

patient experienced cough with no expectoration, no fever, no chest distress, no shortness of breath, and no chest pain 6 days ago, with no known causes. But he paid no special attention to the condition. And 2 days ago, cough aggravated with chest pain and abdominal distension but no fever, chest distress and shortness of breath. He then paid clinical visit in a local hospital. By routine blood test, the WBC count was 8.0×10^9/L, and NEUT% 86 %; and by chest X-ray, negative. Anti-infection therapy was then prescribed (details unavailable). Today, the patient paid his clinical visit to the Emergency Department of our hospital for further treatment. CT scanning indicated bilateral pneumonia with slight pericardial thickening and slight bilateral pleural thickening. Routine blood test indicated WBC count 13.89×10^9/L, NEUT% 86.7 %, and CRP 144 mg/L. Renal function examination indicated BUN 11.3 mmol/L, and Cr 254 μmol/L. The patient complained of cough accompanied by chest pain, no expectoration, no fever, no chest distress and shortness of breath. He was then hospitalized with definitive diagnosis of pulmonary infection. The present immunosuppressive therapy include Ciclosporin 50 mg/50 mg, Myfortic 540 mg/540 mg, and Hormone 10 mg. After hospitalization, etiological examinations were ordered and anti-infection therapy was administered, including Rocephin, Fortum and Diflucan, but with poor therapeutic efficacy. The immunosuppressants were then retrieved and Methylprednisolone was administered to maintain immunosuppression. The blood specimen was sent to National CDC of China for further examination. Toxoplasma gondii IgG was detected positive, being 21.20 on Jan. 16th, 2014, and IgM 0.11, indicating toxoplasma gondii pneumonia. The anti-toxoplasmic therapy, including SMZco and Azithromycin, was then added. After treatment, the patient showed stable body temperature, relieved cough and expectoration as well as improved renal function.

[Radiological Demonstrations] (See Fig. 7.4)

[Diagnosis]

Toxoplasmosis, pulmonary infection, hypertension, renal dysfunction after kidney transplantation.

[Discussion]

Toxoplasma gondii parasitizes only in karyocytes of its host to induce damages to many organs and tissues, which further develop into toxoplasmosis. The disease can invade any organ, in some serious cases, to cause disabilities or death. Its clinical manifestations are quite complex with low specificity, and misdiagnosis is quite common. Mammals and birds are the most common host of toxoplasma gondii, and both can act as the source of its infection. The routes of its transmission include: (1) gastrointestinal infection after intake of meat that is uncooked or unthoroughly cooked; (2) infection via mucosa or wounded skin after contacts to domestic pets such as cats and dogs; (3) transmission via blood transfusion or organ transplantation; (4) fectus infection via placenta blood flow from mother with primary acute infection during pregnancy. In healthy population with good immunity, most of toxoplasma gondii infections are asymptomatic. In individuals with suppressed or compromised immunity, such as patients with malignancy, transplanted organ or AIDS, the infection may change from latent into active, with common symptom of enlarged lymph nodes. The patients may also experience neuropsychiatric symptoms, such as encephalitis, meningocephalitis, epilepsia and mental disorders. The ocular symptoms include choroidoretinitis and iridocyclitis. In addition, myocarditis,pericarditi s,pneumonia and hepatitis may also occur, with accompanying systemic symptoms of fever, headache, muscle pain, and skin rash. The definitive diagnosis should be based on findings by etiological examination and immunological assay. In this case, the patient had a past medical history of renal polycyst, renal failure and kidney transplantation, followed by triple maintenance therapy for immunosuppression with Cyclosporine, Myfortic, and Hormone. The patient presented an onset of pulmonary infection, with cough and chest pain. The routine blood test showed WBC count 13.89×10^9/L, and NEUT% 86.7 %. CT scans demonstrated widely scattering flakes and cotton like high-density blurry opacity in both lungs. Routine anti-inflammation therapy showed poor therapeutic efficacy. His blood specimen was sent to National CDC of China for further examination, with the finding of toxoplasma gondii IgG positive (+), with obviously increased levels of IgG and IgM against toxoplasma gondii. Then toxoplasmosis pneumonia was highly suspected. A trial anti-toxoplasmic therapy with SMZco and Azithromycin was prescribed, with good therapeutic responses. Based on all of these findings, the diagnosis was finally defined to be toxoplasmic pneumonia. For differential diagnosis, due to his past medical history of kidney transplantation, the possibilities of other opportunistic infections should be excluded. In China, clinicians and radiologists should raise their awareness of toxoplasmosis in patients with suppressed or compromised immunity. Its early diagnosis is of great significance due to the favorable clinical prognosis after immediate administration of anti-toxoplasmic therapy. Literature reports show that toxoplasma gondii infection is one of the most important opportunistic infections in AIDS patients, with obviously higher incidence of toxoplasmosis in patients with AIDS than in immunosuppressed population. The radiologists should gain more knowledge about the neuroimaging signs in such patients. In addition, literature reports also showed that patients receiving organ transplantation might be infected with toxoplasma gondii due to infection of donor, blood transfusion and activated latent infection in recipient. Among these approaches, infection of donor is the most common. Therefore, timely diagnosis of latent infection plus

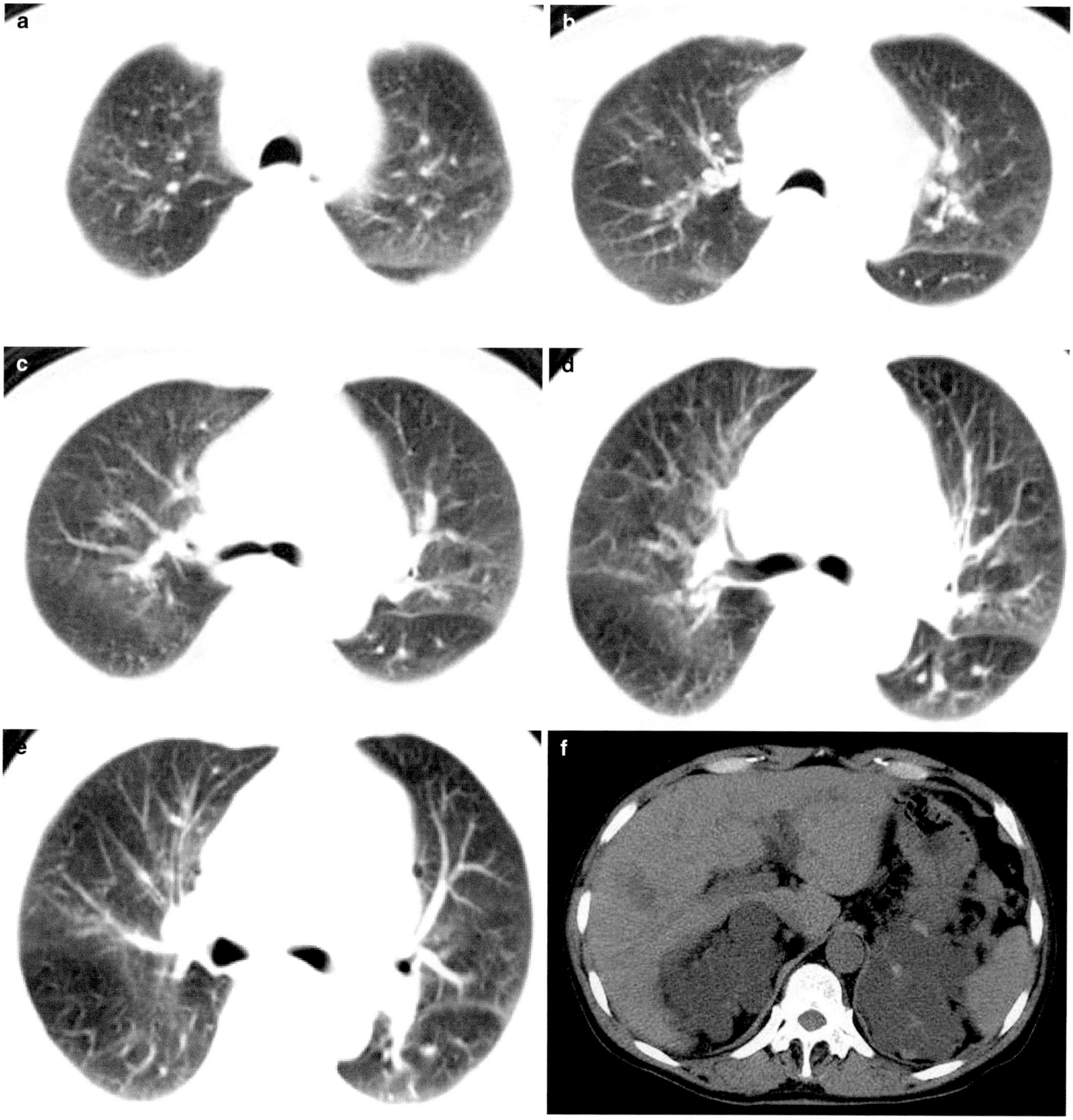

Fig. 7.4 (**a**–**e**) CT scanning displayed widely scattering flakes and cotton like high-density blurry opacity in both lungs as well as increased and thickened pulmonary markings. (**f**) CT scanning showed bilateral renal polycyst

following appropriate treatment can reduce the chances of its occurrence and transmission. Routine examination of toxoplasma gondii before blood donation has its clinical significance in eliminating its hematogenous transmission. From the perspective of eugenics, routine serum test before pregnancy in women at the childbearing age can reduce the risks of its infection in both pregnant women and fetuses.

7.7 Malaria

Malaria is a mosquito-borne infectious disease transmitted via stings and bites of anopheles (commonly female anopheles), and can be categorized into the following four groups: plasmodium vivax, plasmodium ovale, plasmodium malariae and plasmodium falciparum.

7.7.1 Epidemiology

7.7.1.1 The Source of Infection

The patients with malaria and asymptomatic plasmodium carrier are the source of its infection, but only existence of gametophytes in peripheral blood has epidemiological significance. The ratio of male to female gametophytes and their maturity also affect its transmission and infection. Generally, the communicable period of plasmodium vivax lasts for about 1–3 years; plasmodium malariae, 3–10 years; and plasmodium falciparum, less than 1 year.

7.7.1.2 Route of Transmission

Malaria is transmitted via stings and bites of female anopheles. In some rare cases, the disease is transmitted via blood transfusion or vertical spread from mother to fetus.

7.7.1.3 Susceptible Population

Populations living in non-affected regions by malaria are generally susceptible to the disease. Populations living in highly prevalent regions of malaria above the age of 25 years show certain immunity against malaria, but those at a younger age are still vulnerable, especially young children aged under 2 years. Although certain immunity can be acquired after its infection, the immunity only lasts for a short period of time. In addition, no cross immunity is found between different types of malaria.

7.7.1.4 Prevalence

Malaria prevails mainly in the tropical and subtropical regions, followed by the temperate region, which is closely related to the ecological environment that the transmitting vector lives. In the prevailing regions of malaria, plasmodium vivax spreads the most widely, while plasmodium falciparum mainly prevails in the tropical region. Plasmodium malariae and plasmodium ovale are relatively rare. In China, mixed prevalence of plasmodium vivax and plasmodium falciparum is found in the provinces of Yunnan and Hainan, and plasmodium vivax prevails in other provinces and regions. The onset often occurs in summers and autumns, but with no seasonal effects in the tropical region. In addition, many imported cases of malaria have been clinically detected in China.

Among all the parasitic diseases worldwide, malaria is the deadliest, followed by schistosomiasis and amoebiasis. Currently, about 300–500 million new cases of malaria are reported per year, with death cases of about 3 million including 1 million children. The death rate in children patients is the highest among young children aged under 5 years. In some malaria affected regions, about 10 % deaths of children occur due to malaria.

7.7.2 Clinical Manifestation

7.7.2.1 Common Clinical Manifestation

Clinically, malaria is characterized by repeated episodes of chills as well as high fever, headache, sweating, anemia and splenomegaly.

7.7.2.2 Complication

Cerebral Malaria

Cerebral malaria is the most common, with critical condition and high mortality. The disease onsets with symptoms of high fever, chills and accompanying severe headache, nausea, and vomiting. Subsequently, the patients may experience sleepiness, delirium, generalized convulsions, and gradually fall into coma. Splenomegaly and anemia are common signs of malaria, while hepatomegaly is rare.

Pulmonary Malaria

Concurrent to typical and atypical systemic symptoms of malaria, the patients show clinical syndrome, with predominant respiratory symptoms including malarial bronchitis, malarial pneumonitis, malarial asthma, malarial interstitial pneumonia, pulmonary edema, acute respiratory distress syndrome.

Malarial Nephropathy

Acute Renal Failure

Due to high fever and profuse sweating induced by malaria as well as insufficient intake of water, the patients may develop renal failure.

Nephrotic Syndrome

Nephrotic syndrome occurs commonly after repeated episodes of plasmodium malariae, while it rarely occurs in the cases of plasmodium falciparum. It is caused by deposition of malarial antigen-antibody complexes in glomerular capillary basement membrane and glomerular vascular interstitium.

Blackwater Fever

Blackwater fever, also known as hemolytic uremic syndrome (HUS), is clinically characterized by microvascular hemolytic anemia, acute renal failure and thrombopenia. Its clinical manifestations include acute onset of chills, high fever, accompanying lumbar and abdominal pain, and sharply decreased urine. Several hours later, hemoglobinuria occurs with bloody red urine.

7.7.3 Diagnostic Examination and Diagnostic Basis

7.7.3.1 Epidemiological History

The patients may report visits to the malaria affected regions in recent 2 weeks. Otherwise, the patients may report a history of blood transfusion in past recent 2 weeks.

7.7.3.2 Typical Clinical Manifestation

The typical symptoms include repeated episodes of chills, high fever, and sweating, which are absent during intervals of the episodes. In some cases, the patients may also experience progressive anemia and splenomegaly.

7.7.3.3 Laboratory Test

Blood Smear

The diagnosis can be defined based on the finding of plasmodium by blood smear.

Immunologic Assay

Immunologically, the diagnosis can be established based on the finding of plasmodium antigen or its specific antibody.

Genetic Diagnosis

By PCR, DNA of plasmodium can be directly detected for diagnosis, which has high sensitivity and specificity.

Diagnostic Therapy

For the cases showing similar clinical manifestations to malaria but negative by blood smear, a trial anti-malarial therapy with chloroquine for 3 days can be prescribed. Those patients showing good therapeutic responses can be suspectively diagnosed with malaria.

Case Study 1

[Brief Medical History]

A 26–year-old woman experienced cardiac arrest and anemia following convulsion for about 20 min 3 days after delivery. Blood smear showed plasmodium positive, with a protozoal density of 300 per μl blood.

[Radiological Demonstrations] (See Fig. 7.5)

[Diagnosis]

Malaria complicated by encephaledema.

[Discussion]

Pathologically, cerebral malaria shows diffusely distributed erythrocytes containing plasmodium in cerebrovascular blood, which adhere to endothelial cells of cerebral capillaries to

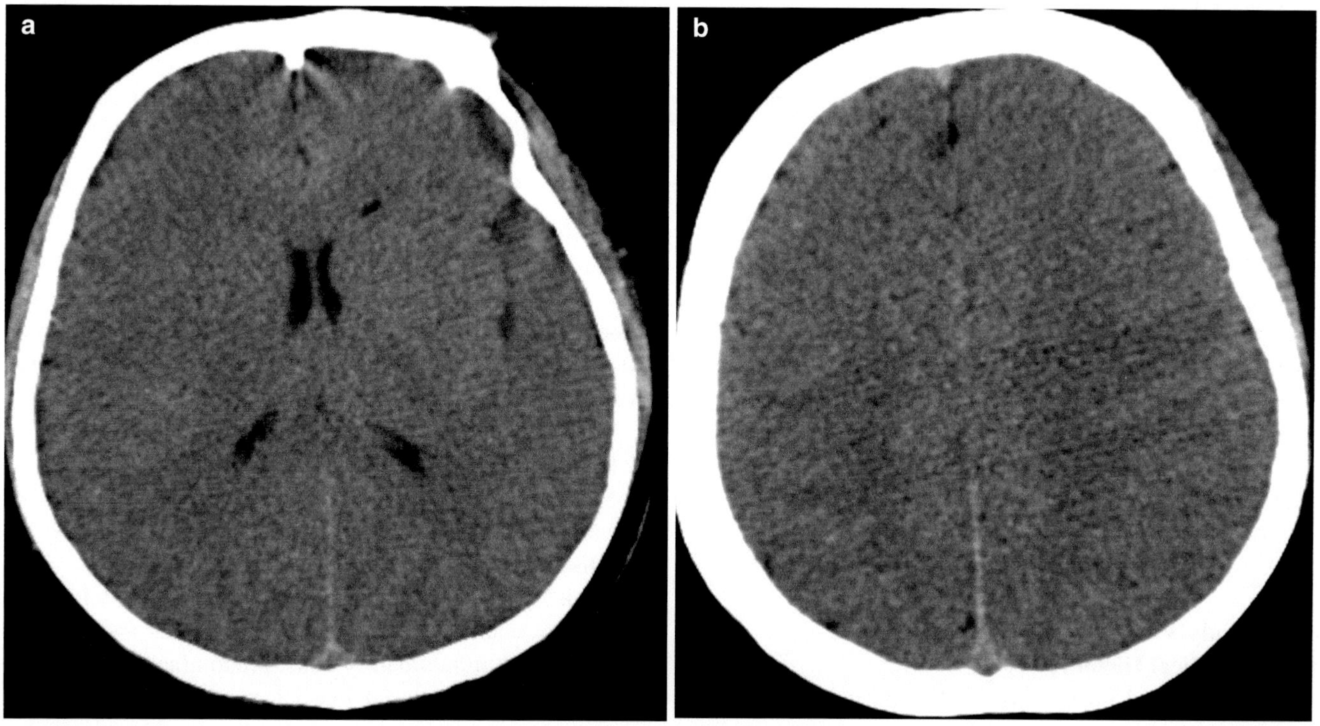

Fig. 7.5 (**a**, **b**) Plain CT scanning demonstrated diffusely decreased density of the brain parenchyma, poorly defined interface between gray and white matters as well as narrowed ventricles and cisterns

block the blood flow. Therefore, gas exchange in brain tissue is impaired to result in cerebral anoxia, which further causes metabolic disorders. In addition to toxic effects of plasmodium, serious brain lesions emerge, with encephaledema as the most common. CT scanning demonstrates diffusely decreased density of brain parenchyma, and poorly defined interface between gray and white matters. MR imaging demonstrates slightly long T1and T2 signals, and shallow sulus and gyrus. Clinically, encephaledema should be differentiated from ischemic lesions, and postpartum eclampsia. In addition, embolism of focal small vessel or post embolism hemorrhage should be differentiated from multiple lacunar infarction or rupture and hemorrhage of malformed small vessel.

Case Study 2

[History]

A 55–year-old man complained of fever for 5 days with convulsion and disturbance of consciousness for 3 days after being stinged by mosquito. Blood smear showed plasmodium positive, with a protozoal density of 2000 per μl blood.

[Radiological Demonstrations] (See Fig. 7.6)

[Diagnosis]

Malaria complicated by encephaledema.

[Discussion]

Radiologically, it should be differentiated from viral encephalitis and cerebral infarction. Encephaledema occurs commonly in patient with cerebral malaria, with intracerebral multiple low-density lesions and spots of hemorrhagic foci. Patankar et al categorized cerebral malaria into four types based on CT demonstrations: (1) negative type; (2) diffuse encephaledema; (3) diffuse encephaledema with low density lesions at bilateral thalamus; (4) diffuse encephaledema with low density lesions at bilateral thalamus and cerebellum. For the later two types, the low-density lesions are well defined, with no hemorrhage. Patankar et al reported that the patients of negative type by CT scanning show good prognosis, while the patients with their lesions involving cerebellum show poor prognosis. Death occurred in all the five cases in their report, with lesions involving cerebellum. This case was categorized into the type 2, with diffuse encephaledema.

Case Study 3

[Brief Medical History]

A 38–year-old man complained of dizziness, headache, dysphagia, and slurred speech for 2 weeks. He was diagnosed with cerebral malaria due to plasmodium infection. Blood smear showed plasmodium positive, with a protozoal density of 670 per μl blood.

[Radiological Demonstrations] (See Fig. 7.7)

[Diagnosis]

Malaria complicated by encephaledema, with accompanying cerebral ischemia.

[Discussion]

Pathologically, malaria is caused by hyperplasia of mononuclear phagocyte system. Proliferation of plasmodium in human body induces intense phagocytic response. Intraerythrocytic schizogony of plasmodium vivax and plasmodium malariae proliferates in peripheral blood, which causes obvious hyperplasia of systemic mononuclear phagocyte system, showing hepatosplenomegaly, hyperplasia of bone marrow, increased monocytes in peripheral blood, and elevated plasma globulin. Intraerythrocytic schizogony of plasmodium falciparum commonly occurs in microvessels of organs. The infected erythrocytes often adhere to capillary endothelial cells to cause damages to the organ. The damages to the brain are the most obvious, and kidney and other organs are also affected. Within a short period of time, a large quantity of erythrocytes containing plasmodium adhere to cerebral capillaries or microvascular endothelial cells to cause microvascular blockage, which further leads to anorexia of the subcortical white matter. The collaborative effect with toxic factors of plasmodium induces extensive hypoxic-ischemic changes and diffuse edema of the white matter. CT scanning demonstrates white matter with diffuse low-density opacity and narrowed lateral ventricles due to compression. MR imaging demonstrates white mater with extensive long T_1 and T_2 single shadow and high FLAIR signal shadow. The disease should be differentiated from leukoencephalopathy induced by other etiological factors.

Case Study 4

[Brief Medical History]

A 18–year-old young woman complained of repeated upper abdominal pain for 15 days and gingival swelling for 10 days with recurrent fever. Blood smear showed plasmodium positive with a protozoal density of 150 per μl blood.

[Radiologial Demonstrations] (See Fig. 7.8)

[Diagnosis]

Malaria complicated by pulmonary edema and infection as well as hepatosplenomegaly.

[Discussion]

By autopsy, the death cases from malaria definitely show pulmonary edema. The pulmonary capillaries and veinlets are filled with inflammatory cells, predominantly including neutrophils, hemocytes in plasma, macrophages with pigmentation, and malaria infected erythrocytes. Vascular endothelial edema can cause narrowed lumen of the capillaries, pulmonary interstitial edema, and formation of glass membrane. Secondary bronchitis is also common.

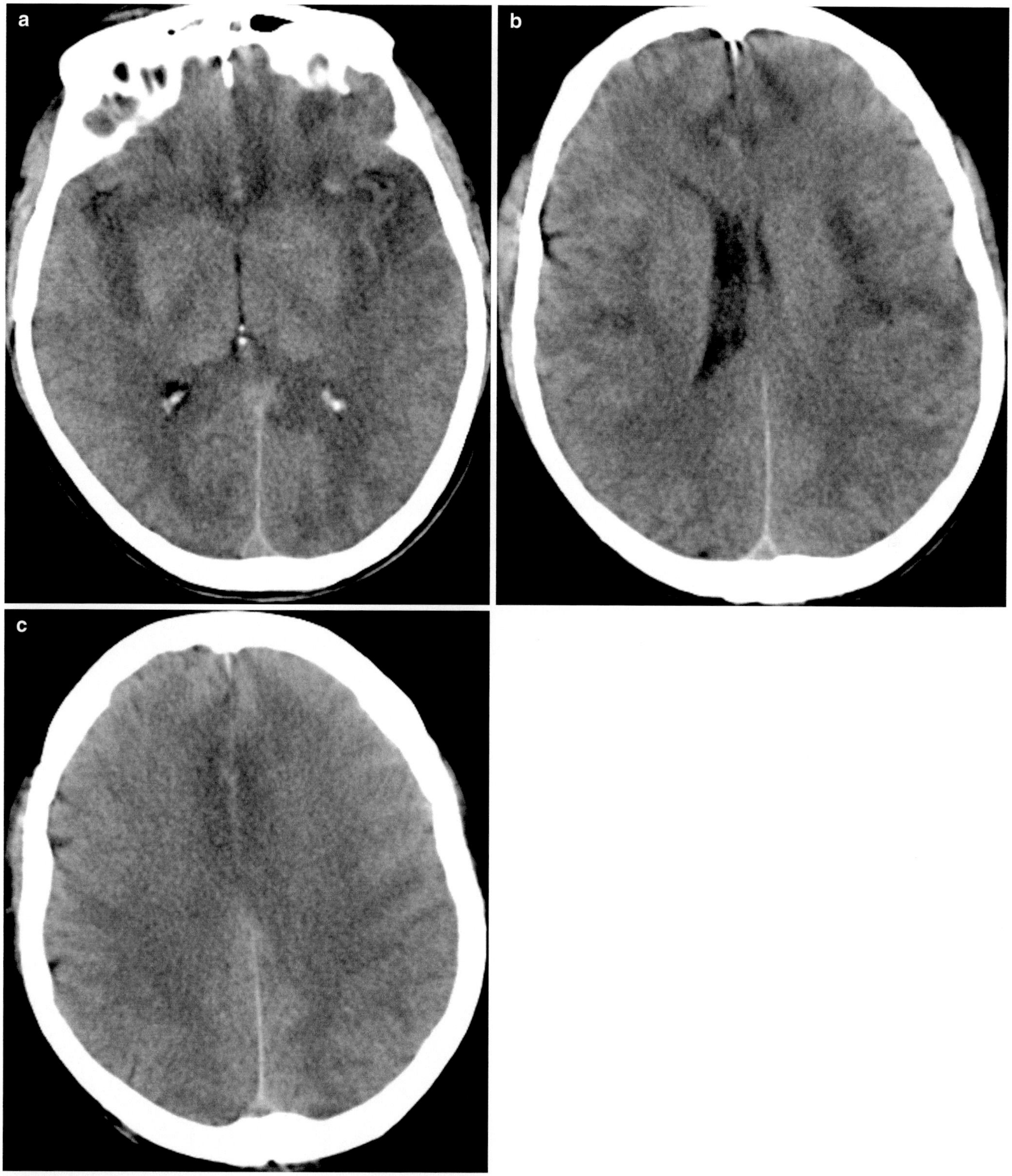

Fig. 7.6 (**a**–**c**) Plain CT scanning demonstrated flakes of low density opacity at bilateral temporal and frontal lobes, and narrowed left lateral ventricle due to compression

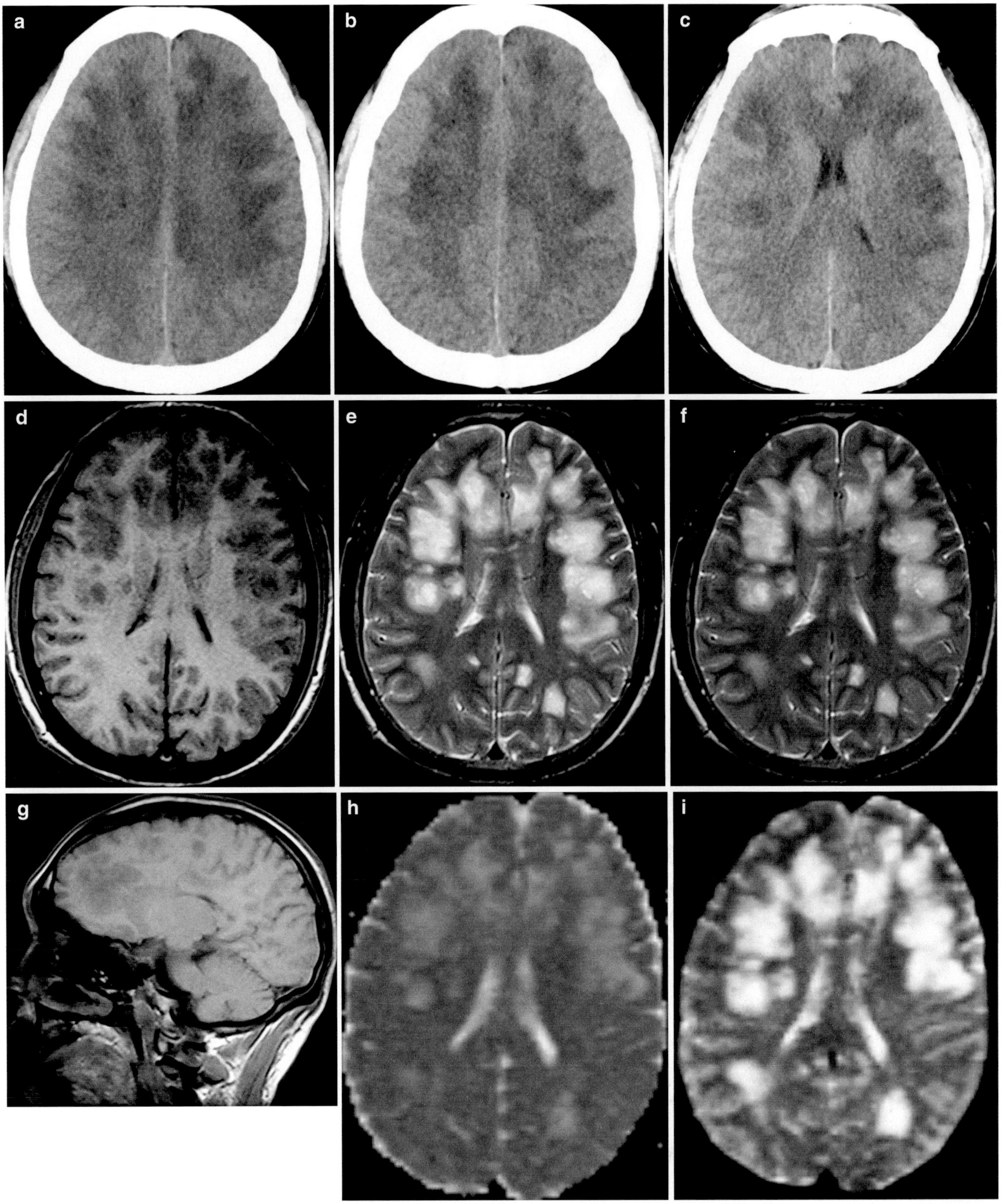

Fig. 7.7 (**a**–**c**) Plain CT scanning demonstrated flakes of low density opacity at bilateral temporal and frontal lobes, and narrowed left lateral ventricle due to compression. (**d**–**g**) Plain MR imaging demonstrated patches of long T1 long T2 signal opacity at bilateral cerebral hemispheres. (**h**–**k**) FLAIR demonstrated hyper-intense signal of the above lesions, with limited diffusion

Fig. 7.7 (continued)

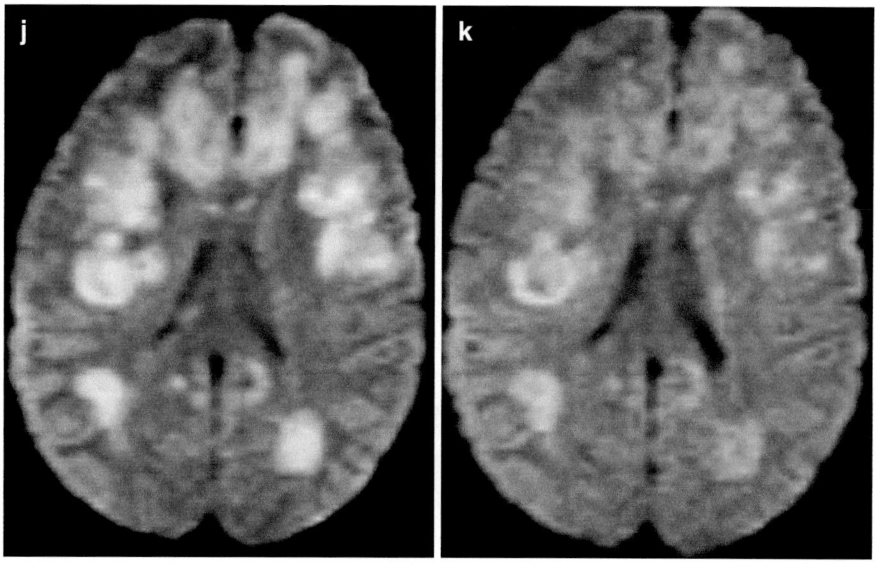

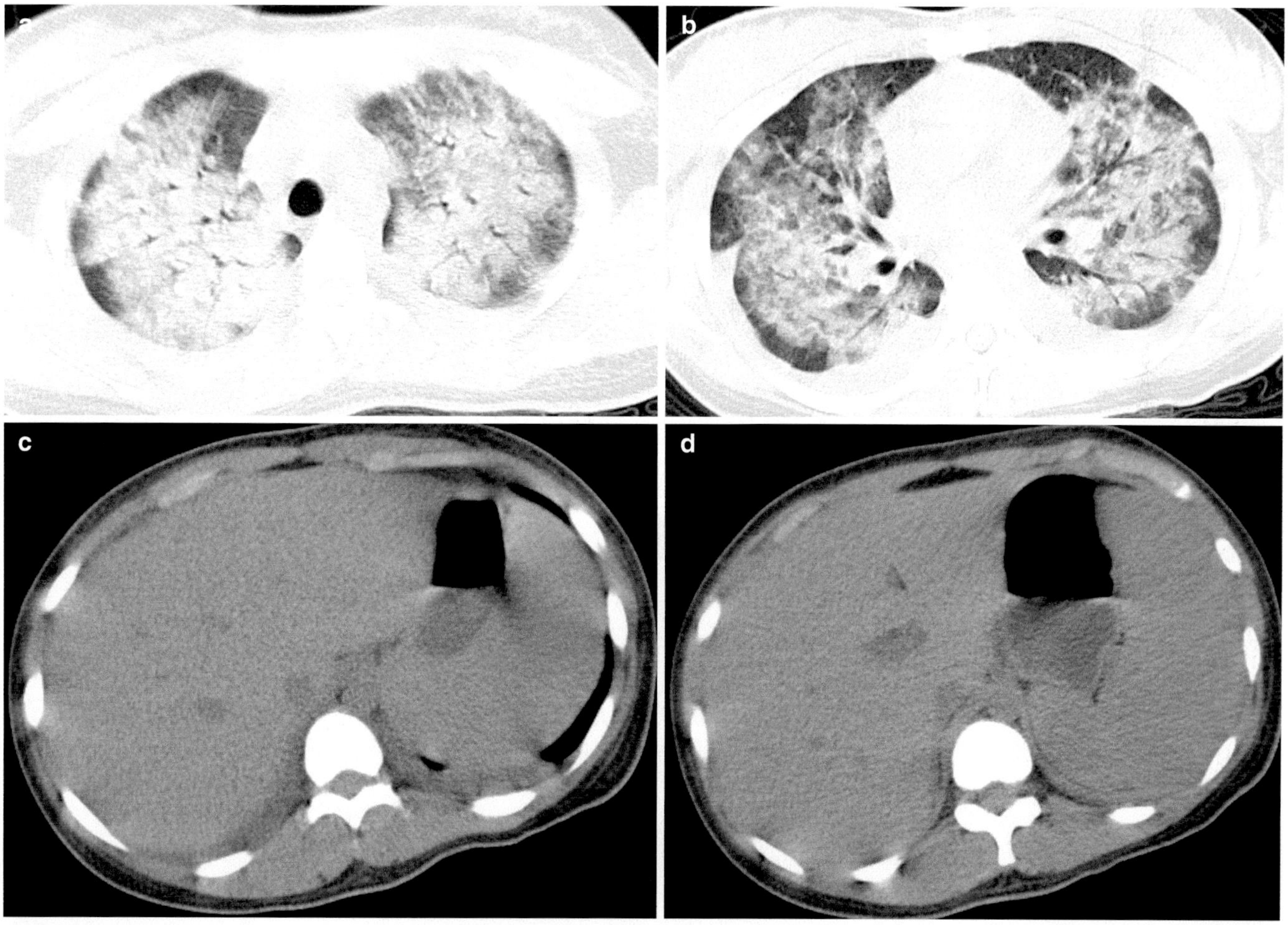

Fig. 7.8 (**a**, **b**) CT scanning demonstrated reduced transparency of both lungs, poorly defined bronchovascular bundle in both lungs, scattering cotton like opacity with high density in both lungs that is more obvious around the hilum. A small quantity of pleural effusion was shown. (**c**, **d**) Hepatosplenomegaly was demonstrated

Case Study 5

[Brief Medical History]

A 44–year-old woman complained of fever with with fatigue and shortness of breath after physical activities for about 1 month. She experienced recurrent fever and night sweat. Blood smear showed plasmodium positive with a protozoal density of 100 per μl blood.

[Radiological Demonstrations] (See Fig. 7.9)

[Diagnosis]

Malaria complicated by pulmonary edema and hepatosplenomegaly.

[Discussion]

By autopsy, the death cases from malaria definitely show pulmonary edema. The pulmonary capillaries and veinlets are filled with inflammatory cells, predominantly including neutrophils, hemocytes in plasma, macrophages with pigmentation, and malaria infected erythrocytes. A small quantity of erythrocytes containing plasmodium are filled in the pulmonary capillaries and veinlets to adhere to vascular endothelium, causing narrow lumen of the capillaries. Due to effect of a few toxins, the exudates are in a small quantity and pulmonary edema is mild, with radiological demonstra-

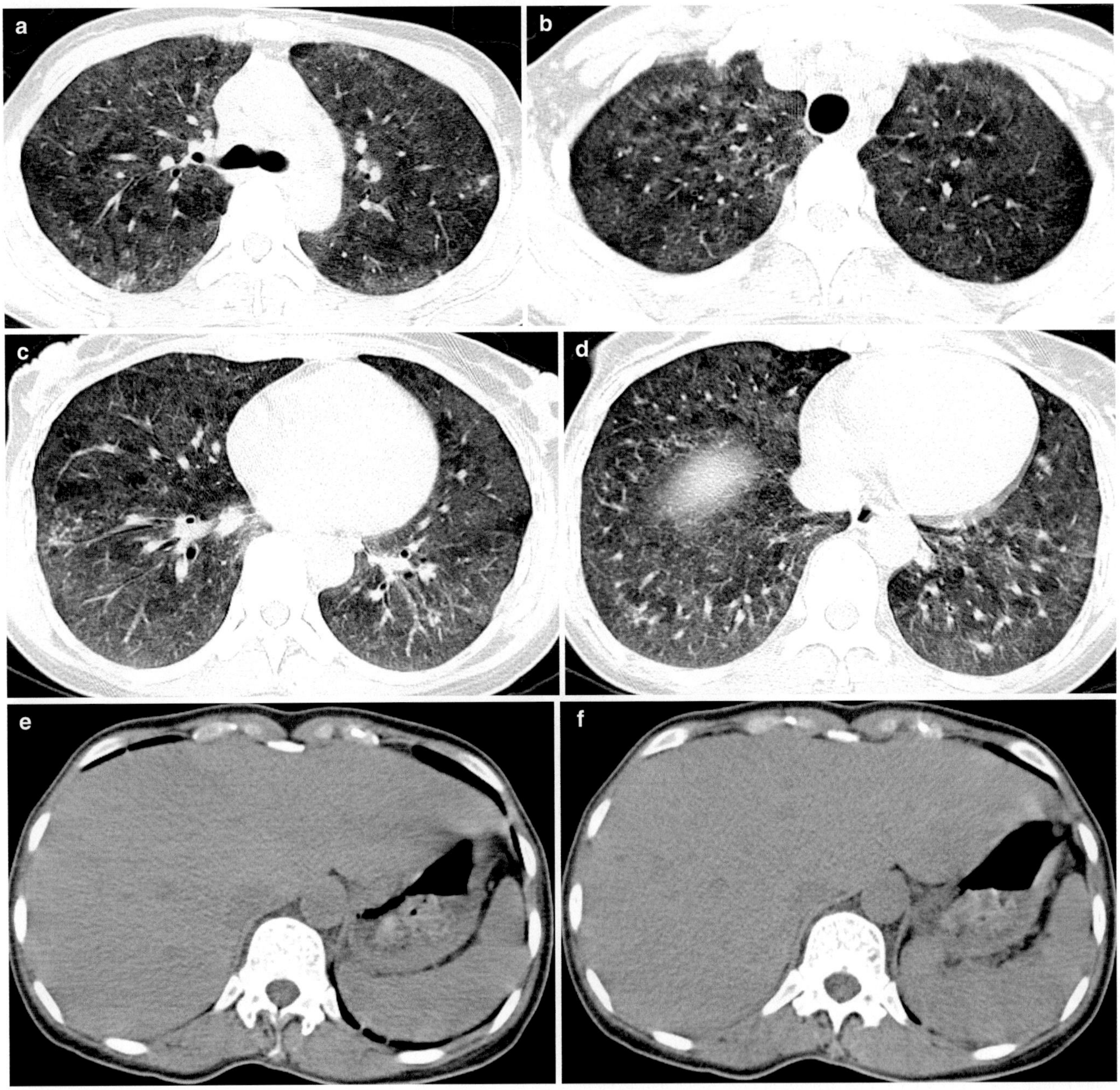

Fig. 7.9 (**a–d**) CT scanning demonstrated diffuse ground-glass opacity in both lungs, with sporadic small patches of opacity. (**e**, **f**) Hepatosplenomegaly was demonstrated

tion of diffuse opacity with ground-glass like density. These radiological findings should be differentiated from common inflammation or hemorrhage of pulmonary alveoli in a small quantity.

Case Study 6

[Brief Medical History]

A 3.6-year-old boy complained of high fever and shortness of breath for 6 days. Blood smear showed plasmodium positive with a protozoal density of 50 per µl blood. For case detail and figures, please refer to Fig. 29.3 of the book (Hongjun Li, Radiology of Infectious diseases, Volume 2, 2015)

[Diagnosis]

Pulmonary malaria (lobar pneumonia type).

[Discussion]

According to severity of the condition, chest X-ray demonstrations of lungs in the cases of pulmonary malaria can be categorized into the following five types.

1. **Bronchitis type**
   Chest X-ray demonstrated obviously increased and thickened lung markings, which are mainly distributed in the inner and middle areas of both lower lung fields. The outer area commonly shows interstitial changes, which are nonspecific.
2. **Interstitial pneumonia type**
   In addition to chest X-ray demonstrations of increased and thickened lung markings, there are accompanying network like opacity and small spots of opacity at the interlobular septum, which are predominantly distributed at the middle and outer areas, and, otherwise across the whole lung.
3. **Bronchial pneumonia type**
   Chest X-ray demonstrated patches of blurry opacity in the thickened lung markings at both lower lung fields.
4. **Lobar pneumonia type**
   Chest X-ray demonstrated large flakes of cloudy high density opacity with lobar or segmental distribution. In addition, consolidation of pulmonary segment is mainly presented as basal lesion.
5. **Pulmonary edema**
   Chest X-ray also demonstrated pleural thickening, pleural effusion and other changes.

The studied case No. 6 can be categorized into the type 4.

Case Study 7

[Brief Medical History]

A 33–year-old woman complained of fever, cough and chest pain for half a month. Blood smear showed plasmodium positive with a protozoal density of 300 per µl blood. For case detail and figures, please refer to Fig. 29.5 of the book (Hongjun Li, Radiology of Infectious diseases, Volume 2, 2015)

[Diagnosis]

Malaria complicated by pulmonary edema, cardiac enlargement, pericardical effusion, right pleural effusion.

[Discussion]

In addition to symptoms of pulmonary malaria, the patient may experience dry and sore throat, chest pain, cough, expectoration or bloody phlegm. In severe cases, the patient may even experience shortness of breath, cyanosis, dry and moist rale at the lung. CT scanning mainly reveals pulmonary edema.

Case Study 8

[Brief Medical History]

A 45-year-old man was hospitalized due to recurrent fever for 2 months with systemic soreness. On May 20th, the patient experienced fever, chills and slight systemic muscle soreness, with the highest body temperature of 39.4 °C. He was diagnosed with catching a cold and received treatment in a local hospital. After profuse sweating, his body temperature dropped. However, on the following day, he experienced syncope and was hospitalized in Nanchang Ninth Hospital with a suspected diagnosis of malaria. The therapy of Artequick and Primaquine was administrated, followed by a decreased body temperature to the normal level and renal dysfunction. Laboratory tests revealed BUN 35.38 mmol/L, CREA 944 µmol/L, UA 713 µmol/L, and HGB 5.2 g/L. Following blood transfusion and diuretic treatment, the renal dysfunction was gradually improved. But fever, bloody pleural effusion and rash developed. CT scanning demonstrated right pleural effusion in a large quantity, hepatomegaly, subcapsular hemorrhage of the spleen. He was then transferred to the First Affiliated Hospital of Nanchang University and received treatment after a suspected diagnosis of epidemic hemorrhagic fever with pleural effusion. After the medications of Meropenem, Sulperazone, Amoxyxillin, Haikunshenxi capsules, the patient showed improved condition of pleural effusion and smaller subcapsular hematoma of the spleen. After being discharged, the patient experienced fever again with a body temperature of 38.1 °C. This time, he reported a history of working in Zambia, Africa 1 year ago and returning to China in early May. About 90 % of his colleagues were diagnosed with malaria in Africa. For further diagnosis and treatment, the patient was admitted to our hospital due to fever with uncertain cause. Ultrasound revealed impaired liver and splenomegaly. A laboratory test report by National CDC on July 9th, 2015 showed plasmodium falciparum by microscopic examination of plasmodium with a finding of plasmodium in a small quantity. Concurrent administrations of anti-malarial therapy with other symptomatic therapies, the body temperature stabilized with improved conditions.

[Radiological Demonstrations] (See Fig. 7.10)

[Diagnosis]

Plasmodium falciparum, pleural effusion, pulmonary infection, splenomegaly and hepatomegaly.

[Discussion]

Malaria is one of the major infectious diseases that seriously threatens human health and life safety. WHO has listed malaria, HIV/AIDS and tuberculosis, as the top 3 public health problems in the world. Malaria used to be an important and major infectious disease in China. However, due to focused attention from government, the prevention and treatment of malaria in China has achieved success, which has been praised worldwide. In 2010, malaria elimination program was launched in China. However, gaps still existed in terms of awareness, knowledge and understandings about the disease. Technically, difficulties were found in terms of detecting cases with low protozoal density and carriers of plasmodium. In addition, challenges existed in aspects of complete cure of plasmodium vivax, emergence of drug resistant malaria, and increased cases of imported plasmodium falciparum.

Plasmodium falciparum prevails in African countries southern to the Saharah desert. In this case, the patient has a history of working in Zambia, Africa, 1 year prior to the onset of symptoms. Imported plasmodium falciparum from Africa showed non-specific clinical symptoms, commonly with fever, aversion to cold, headache, muscular soreness, diarrhea, and splenomegaly. Clinically, the condition is commonly misdiagnosed as upper respiratory tract infection.

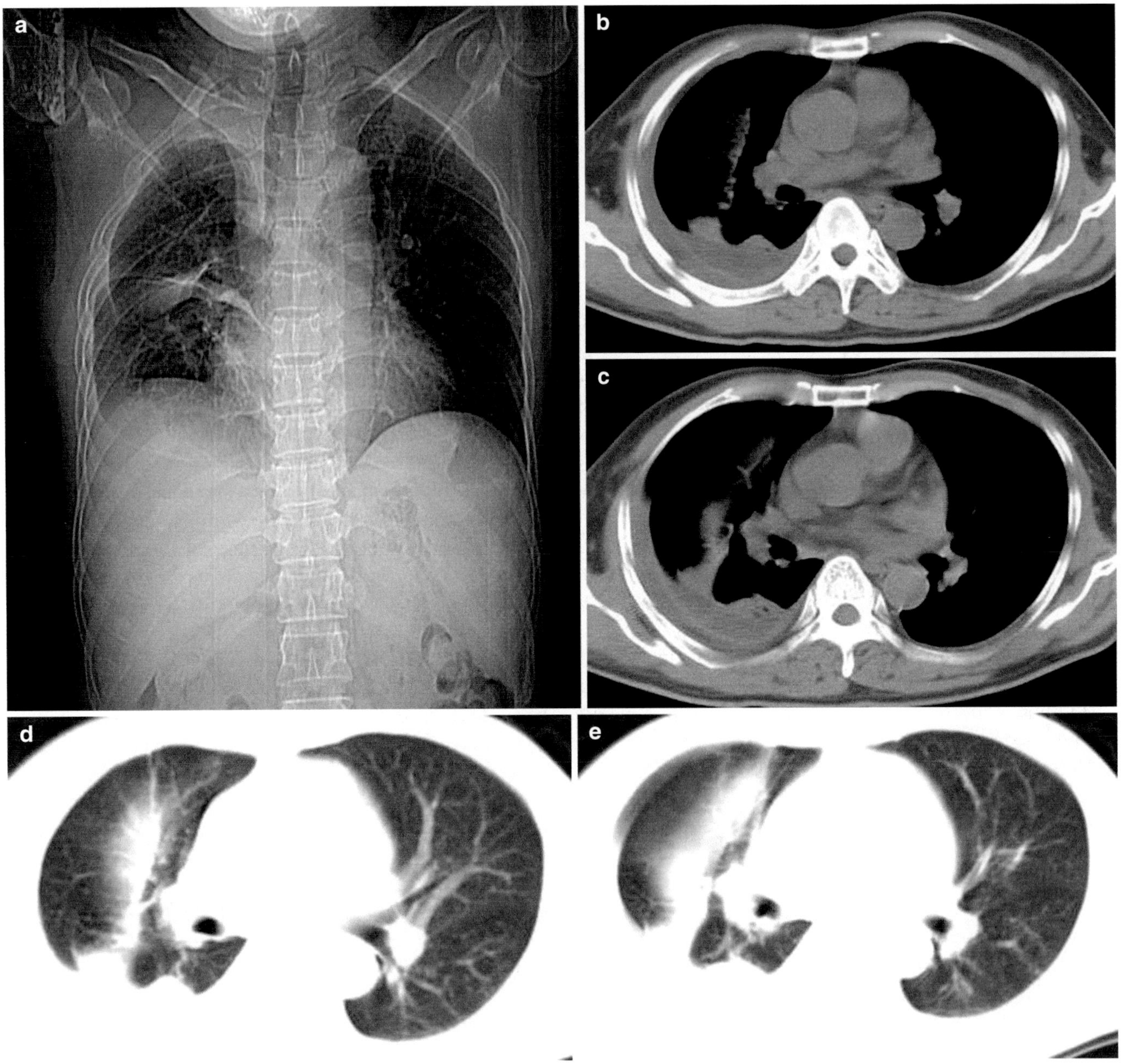

Fig. 7.10 (**a**) Fixed location CT scanning demonstrated fibrous cords of opacity at the right middle lung field as well as obviously enlarged liver and spleen. (**b–e**) Plain CT scanning demonstrated right pleural effusion, interlobar effusion, atelectasis of the right lower lung, and multiple cords of opacity at both lower lungs

However, plasmodium vivax is more commonly characterized by aversion to cold with splenomegaly and markedly reduced platelets, which are consistent with the case we reported. In addition, subcapsular hematoma of the spleen is found in this case and the patient experienced acute renal failure during hosptalization. Literature reports have demonstrated that the pathogenesis contributing to malaria complicated by acute renal failure is as the following:

1. Hemoglobin casts in a large quantity obstruct the collecting tube and renal tubule to cause oliguria and anuria.
2. In the case of serious malarial infection, the capillary endothelium is impaired to increase the vascular permeability. Therefore, protein and water infiltrate into tissue space to cause microcirculation disturbance.
3. Erythrocytes infected by plasmodium adhere and obstruct the capillary endothelium. Meanwhile, inflammation and hemolysis cause local intravascular coagulation. In our reported case, CT scanning demonstrated obviously enlarged liver and spleen, right pleural effusion, interlobar effusion, atelectasis of the right lower lung and fibrous lesions in both lower lungs, which are in accordance with chest X-ray findings in patient with plasmodium falciparum that Song, XB et al have summarized in their medical practice in Africa. These pulmonary signs are related to hemodynamic abnormalities caused by plasmodium as well as pathological changes of involved tissues. The whole treatment course of this imported plasmodium falciparum from Africa indicated that the patients with fever and a recent visit to Africa should be firstly suspected with plasmodium falciparum. The clinicians should inquire the epidemiological history and microscopic examination of plasmodium should be ordered as early as possible for early treatment with medications of artemisinin and its derivatives.

Case Study 9

[Brief Medical History]

A 44-year-old woman complained of fever for more than 1 month, with abdominal pain and distention. Blood smear showed plasmodium positive with a protozoal density of 650 per μl blood.

[Radiological Demonstrations] (See Fig. 7.11)

[Diagnosis]

Malaria complicated by splenomegaly and anemia.

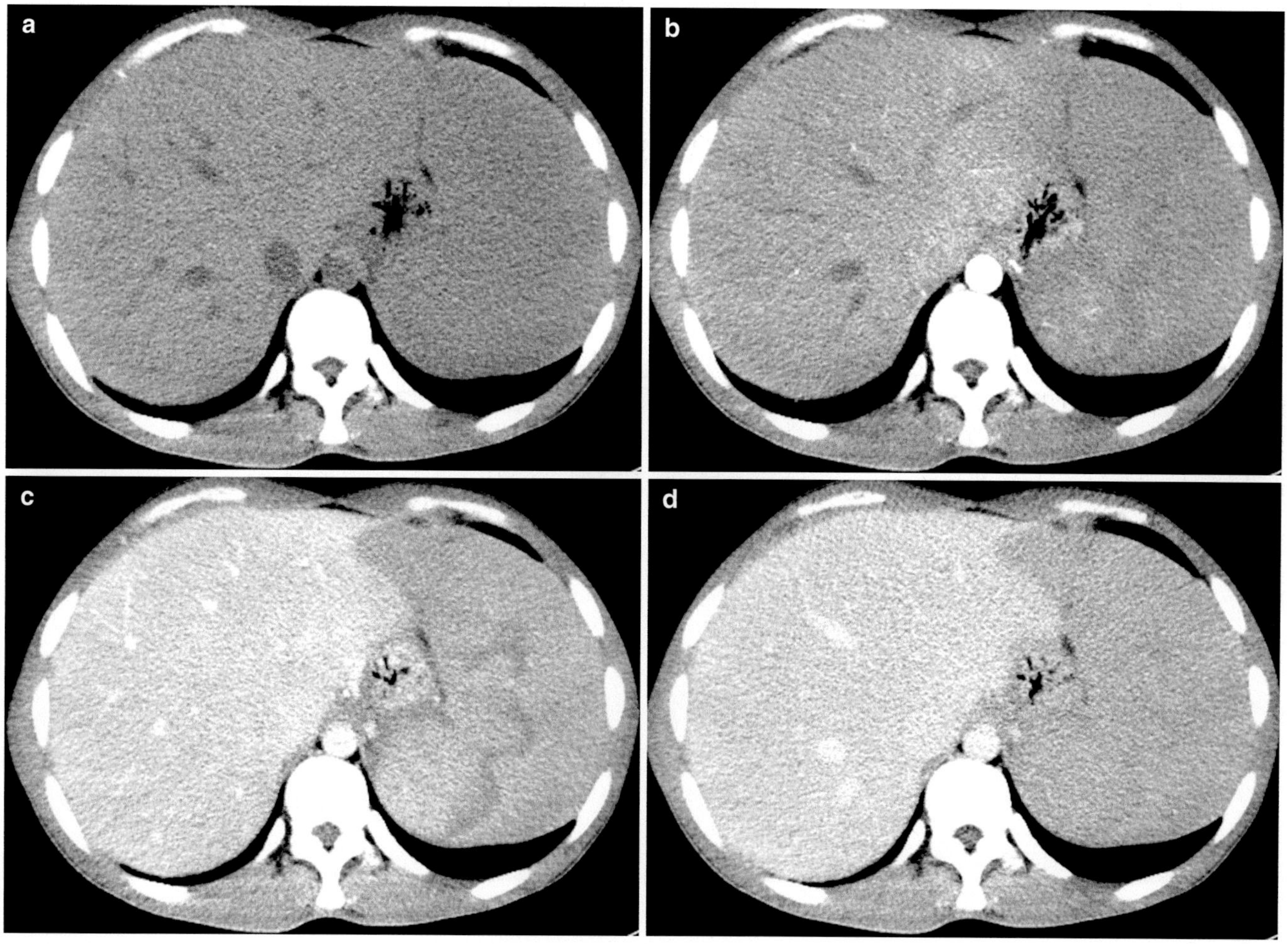

Fig. 7.11 (**a**) CT scanning demonstrated obviously enlarged liver, which crosses the middleline of the body, as well as migration of the stomach due to compression. (**b–d**) Contrast scanning demonstrated small mass and flakes of slight low-density opacity in the spleen at the arterial phase, which is iso-intense at the equilibrium phase

[Discussion]

Splenomegaly can be attributed to congestion, lymphoid-macrophage hyperplasia, as well as dilated venous sinus due to filling of erythrocytes and macrophags. Macrophages containing a large quantity of malarial pigment cause dark coloration of the spleen. About 3–4 days after onset, swelling of the spleen occurs. after repeated onset or repeated infection by plasmodium, the spleen may enlarge down to navel, with a weight up to 100 g in some serious cases. In chronic cases, splenic fibrosis occurs with capsular thickening, inducing increased hardness of the spleen. Antimalarial drug therapy cannot improve the spleen into its normal size.

Case Study 10

[Brief Medical History]

A 36–year-old man complained of chills, aversion to cold, hot flash and night sweat. Blood smear showed plasmodium positive with a protozoal density of 170 per μl blood.

[Radiological Demonstrations] (See Fig. 7.12)

[Diagnosis]

Malaria complicated by megalosplenia.

Case Study 11

[Brief Medical History]

A-72-year-old man complained of limbs weakness after shower. Blood smear showed plasmodium positive with a protozoal density of 90 per μl blood.

[Radiological Demonstrations] (See Fig. 7.13)

[Diagnosis]

Malaria complicated by megalosplenia and splenic infarction.

[Discussion]

The abdominal lesion complicating malaria is the most commonly hepatosplenomegaly. By plain scanning, the spleen can be revealed with with multiple wedge shaped long hypointense opacity, with no enhancement by contrast scanning. The wedge shaped lesion is believe to be typical sign of splenic infarction. The pathogenesis of splenic infarction may be related with hyperplasia of reticuloendothelial system induced by increased clearance of the spleen. Studies in animal models have shown that the long hypointense lesion is the organized thrombus caused by infected erythrocytes in the dilated splenic veins. Kin et al have reported that the lesions in the spleen are reversible. By follow-up examinations, the long hypointense lesion can be cured, with concurrently normalized spleen size.

Case Study 12

[Brief Medical History]

A 17–year-old young man complained of fatigue, poor appetite, and upper abdominal upset for above 7 days. Blood smear showed plasmodium positive with a protozoal density of 35 per μl blood.

For case detail and figures, please refer to Fig. 29.10 of the book (Hongjun Li, Radiology of Infectious diseases, Volume 2, 2015)

[Diagnosis]

Hepatomegaly (intrahepatic lymphatic stasis) and splenomegaly.

[Discussion]

Liver can be infected by all four types of plasmodium, among which plasmodium falciparum shows the most serious damages to the liver. After the hepatocytes are infected by plasmodium, the spores can gain their access into the

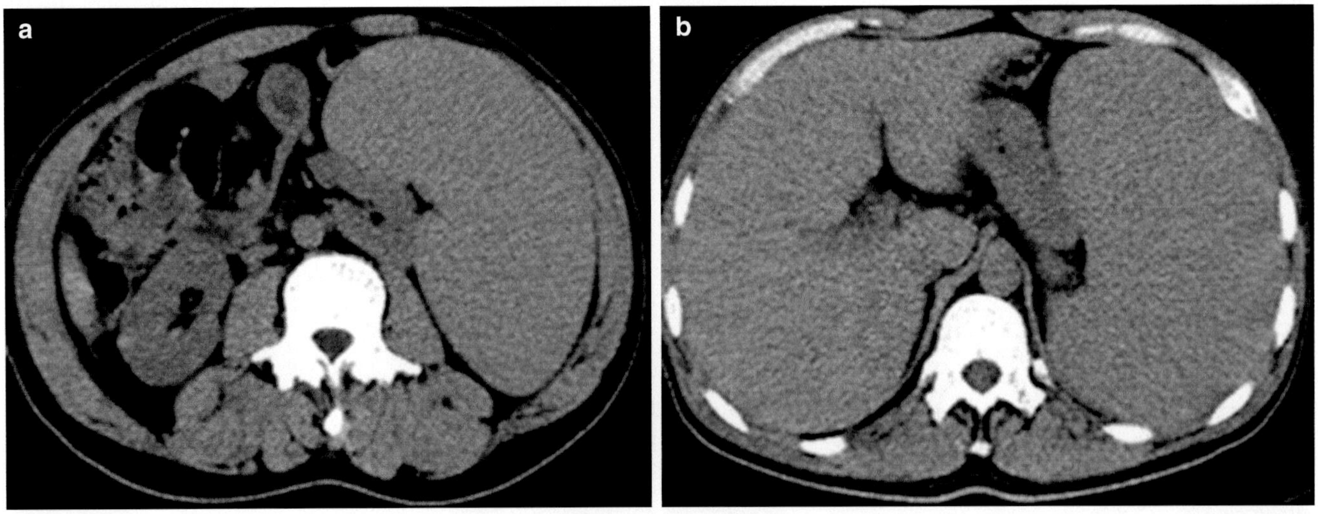

Fig. 7.12 (**a**, **b**) CT scanning demonstrated obviously enlarged spleen and migration of the left kidney toward the middleline due to compression

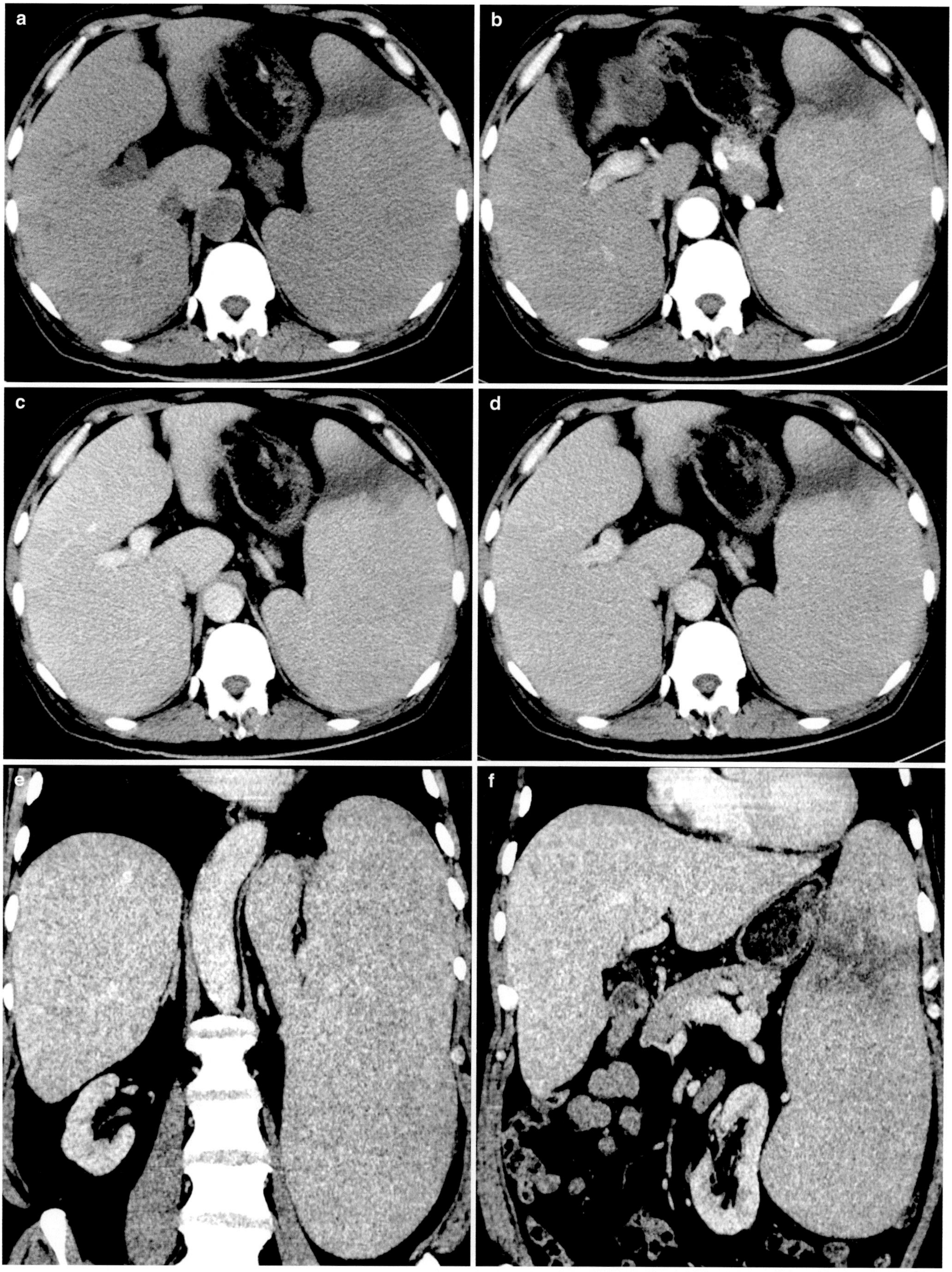

Fig. 7.13 (**a**–**f**) CT scanning demonstrated obviously enlarged spleen, with its lowest point reaching the pelvic cavity, rightward shift of the left kidney due to compression, and multiple strips and flakes of hypointense opacity in the spleen

hepatocytes along with blood flow for further development. Subsequently, the infected hepatocytes are subject to turbidity, swelling, degeneration and rupture, with mature merozoites released into blood flow again, which can be swollen by macrophagocytes or bind to erythrocytes. The erythrocytes and macrophagocytes containing plasmodium enter into the blood flow to fill in the hepatic sinus and the central vein. Therefore, congestion of hepatic sinus and central vein as well as swelling and edema of the liver occur. The liver is then grayish blue or black in color due to hyperplasia of Kupffer cells, malarial pigment in the blood, fragments of erythrocyte with or without plasmodium, and a small quantity of xanthematins with iron. The patients are clinically characterized by jaundice hepatitis with aversion to cold, fever, poor appetite, vomiting, icteric sclera, yellowish urine, right upper abdominal distending pain, hepatomegaly and hepatic dysfunction. In serious cases, acute hepatic failure may develop. Radiologically, the patients commonly show hepatomegaly and hepatic dysfunction but rarely show hepatic abscess.

Case Study 13

[Brief Medical History]

A 60–year-old man complained of abdominal distension and edema of both lower limbs for about 8 days and recurrent fever. Blood smear showed plasmodium positive with a protozoal density of 2100 per μl blood.

[Radiological Demonstrations] (See Fig. 7.14)

[Diagnosis]

Malaria complicated by edema of colonic wall and ascites.

[Discussion]

Gastrointestinal malaria is mainly manifested as extensive edema of the gastrointestinal wall and ascites. Radiologically, the patients mainly show obvious thickening of the gastrointestinal wall, quite smooth intestinal wall, enhancement of the mucosa, and the complication of abdominal effusion. By radiology, the condition should be differentiated from tubercular enteritis and ulcerative enteritis. Tubercular enteritis is mainly demonstrated with uneven thickening of the ileocecal wall, which can be unevenly

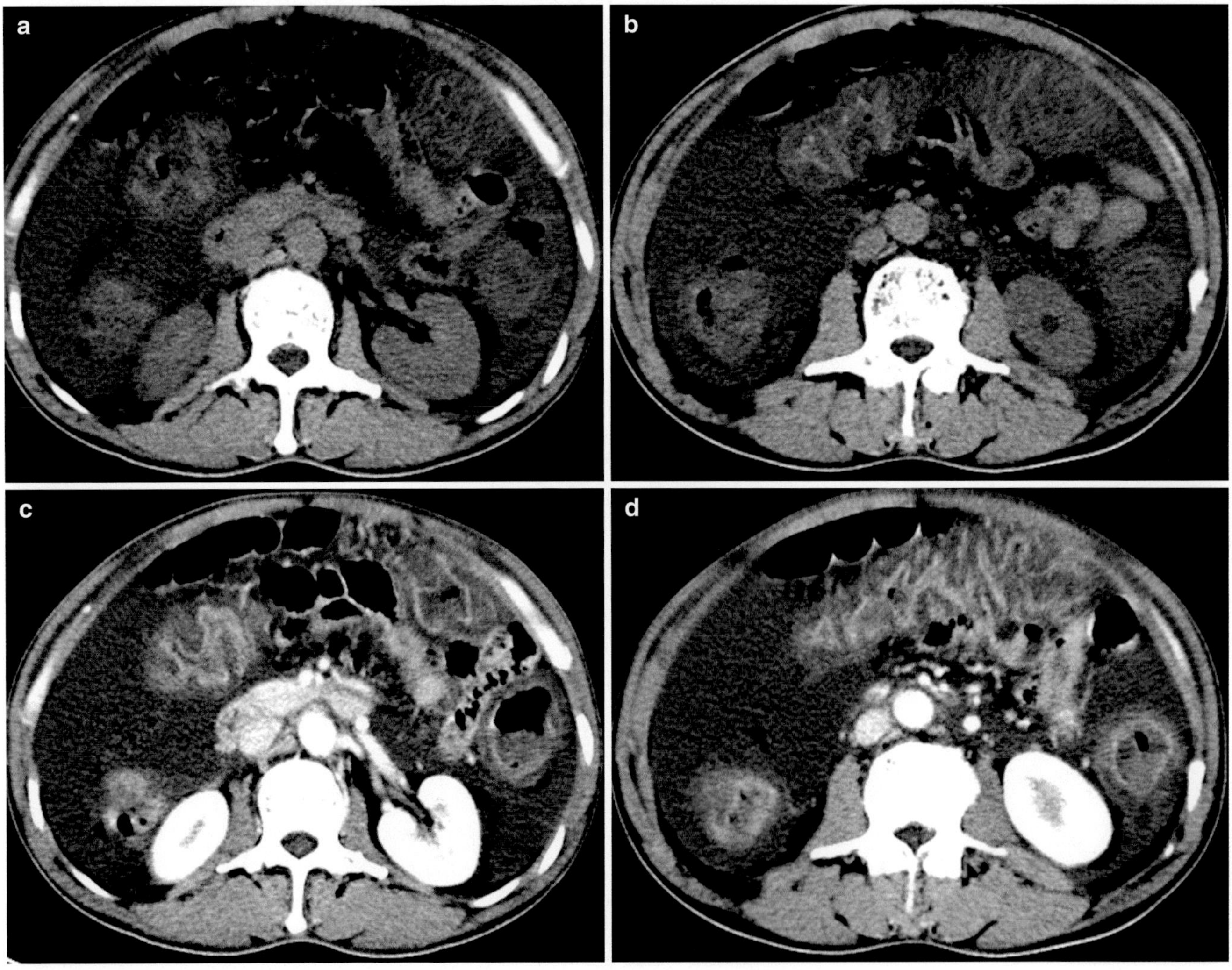

Fig. 7.14 (**a**–**d**) CT scanning demonstrated obvious thickening of the whole colonic wall, with enhancement of mucosa by contrast scanning, as well as liquid-density opacity around the intestinal tube

enhanced by contrast scanning. Ulcerative enteritis is mainly demonstrated with extensive or segmental thickening of the intestinal wall, with no luminal stenosis, but the intestinal wall can be obviously enhanced by contrast scanning.

Case Study 14

[Brief Medical History]

A 47–year-old man complained of abdominal distension and pain for more than 1 month. Blood smear showed plasmodium positive with a protozoal density of 400 per μl blood.

[Imaging Demonstrations] (See Fig. 7.15)

[Diagnosis]

Malaria complicated by extensive edema and ascites.

[Discussion]

Gastrointestinal malaria is mainly manifested as extensive edema of the gastrointestinal wall and ascites. Radiologically, it should be differentiated from ischemic bowel disease and immunity related enteritis.

Case Study 15

[Brief Medical History]

A 61–year-old man complained of recurrent edema in both lower limbs for about 8 months, which aggravated with facial edema for about 1 month. Blood smear showed plasmodium positive with a protozoal density of 700 per μl blood.

[Radiological Demonstrations] (See Fig. 7.16)

[Diagnosis]

Reduced function of both kidneys, ascites, inflammation and atelectasis of both lower lungs, bilateral pleural effusion, edema of the abdominal wall.

[Discussion]

The radiological demonstrations of the patient include shrinkage of both kidneys in volume, decreased density of both kidneys with poorly defined cortex–medulla interface, ascites, inflammation and atelectasis of both lower lungs, bilateral pleural effusion, and edema of the abdominal wall. Based on the radiological findings and clinical history, the diagnosis of renal malaria can be made. Renal malaria is mainly manifested as shrinkage of the kidneys in volvume and their decreased density induced by renal dysfunction, with secondary ascites. The kidneys show attenuated enhancement by contrast scanning. Plasmodium malariae is the most common malarial nephropathy. When antibody-antigen complex deposits in the capillary basement membrane of glomerulus along with blood flow to activate complements. The leukocytic chemokines can be induced to cause local aggregation of neutrophils. The proteolytic enzyme is then released to cause vascular thrombosis and local necrosis, mostly IgM and a small quantidy of IgG. Some patients with long illness course may develop nephrotic syndrome, with clinical manifestations of systemic edema, ascites, proteinuria and hypertension, which may finally develop into renal failure.

Case Study 16

[Brief Medical History]

A 17–year-old young man complained of intermittent fever for 5 days, vomiting and abdominal distension for 4 days, lumbar distension for 3 days. Blood smear showed plasmodium positive with a protozoal density of 70 per μl blood.

For case detail and figures, please refer to Fig. 29.12 of the book (Hongjun Li, Radiology of Infectious diseases, Volume 2, 2015)

[Diagnosis]

Malaria complicated by renal dysfunction.

[Discussion]

In this case, upper abdominal CT scanning demonstrated both kidneys with decreased density, thinner cortex, poorly defined cortex-medulla interface, slight thickening of fascia around kidneys, and slightly attenuated and delayed enhancement. In combination to the clinical case history and radiological finings, the disease should be differentiated from renal insufficiency due to lithangiuria and nephritis. In patients with falciparum malaria, nephropathy is mainly characterized by acute hyperplastic nephritis and nephrotic syndrome that are reversible. Immune complex nephropathy complicating plasmodium malariae is a chronic progressive membranous glomerulonephritis. By kidney tissue biopsy, antigen and specific antibody of plasmodium malariae can be detected. The disease is an immune reactive nephropathy, which is irreversible after sufficient anti-malarial therapy. In some serious, renal failure may occur.

Case Study 17

[Brief Medical History]

A 39–year-old man complained of fever, cough and headache for 3 days. Blood smear showed plasmodium positive with a protozoal density of 600 per μl blood.

[Imaging Demonstrations] (See Fig. 7.17)

[Diagnosis]

Malaria with multiple organs involved including brain, liver and lungs.

[Discussion]

In this case, brain CT scanning demonstrated multiple flakes of hypo-intense opacity at bilateral cerebral hemispheres with local slightly hyper-intense opacity as well as shallow cistern and sulus. Chest and abdominal scanning revealed increased vascular and bronchial bundles at both lungs with multiple cotton-like flakes of opacity, arch shaped hypo-intense opacity in bilateral pleural cavities, and enlarged liver with plump capsule. Based on its clinical case history and radiological demonstrations, it should be differentiated from ischemia of deep white matter and cardiogenic pulmonary edema. Malaria is transmitted via stings and bites of mosquito and has been listed the top in the fatal parasitic diseases. Plasmodium falciparum prevails mainly in the tropical region,

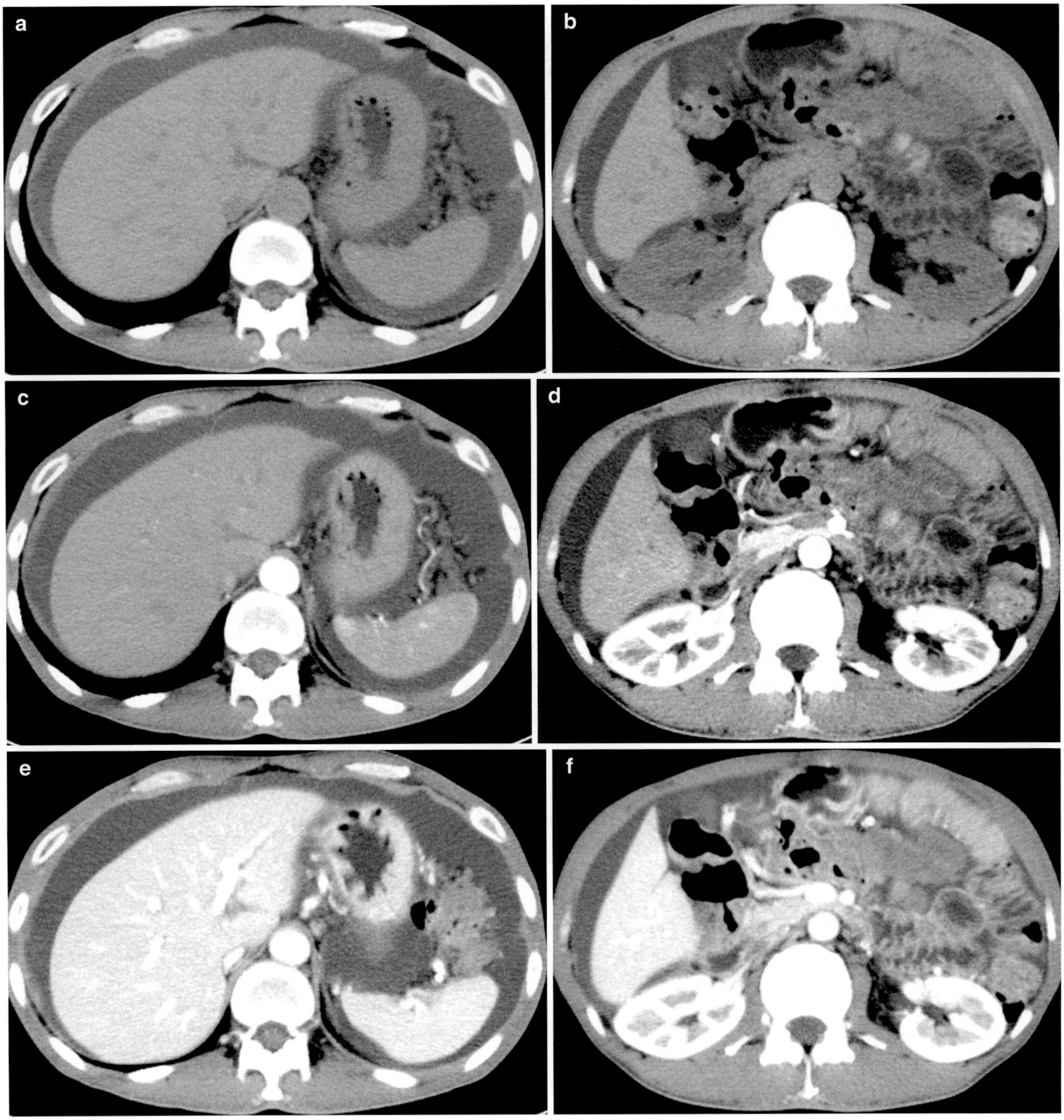

Fig. 7.15 (**a**–**h**) Plain CT scanning demonstrated swelling of small intestinal wall. Contrast scanning demonstrated absence of the 3-layer structure of intestinal wall, increased density of adipose mesenchyme between intestines, and liquid-density opacity around liver, spleen and intestinal tube

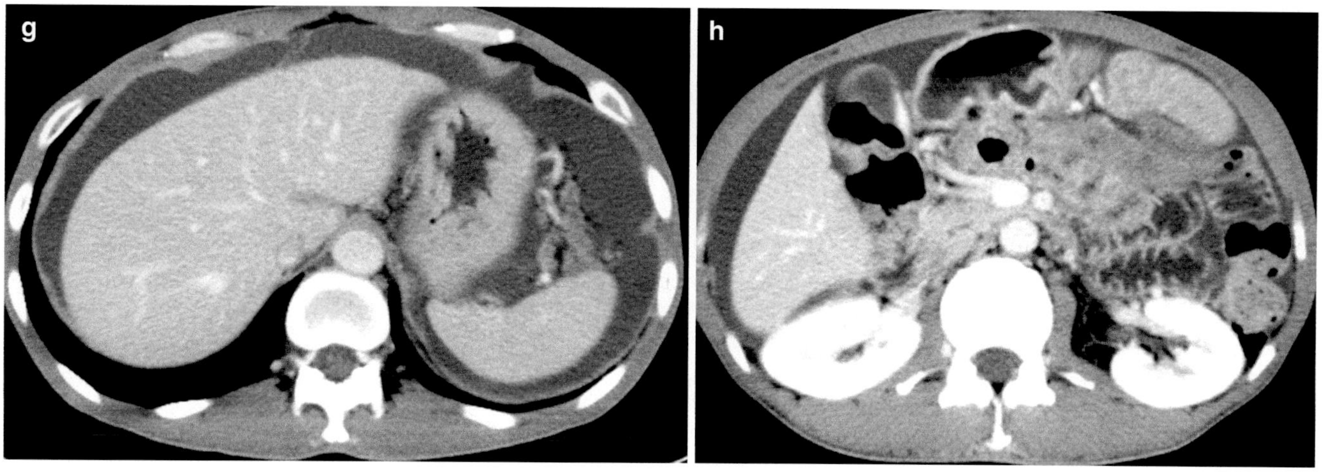

Fig. 7.15 (continued)

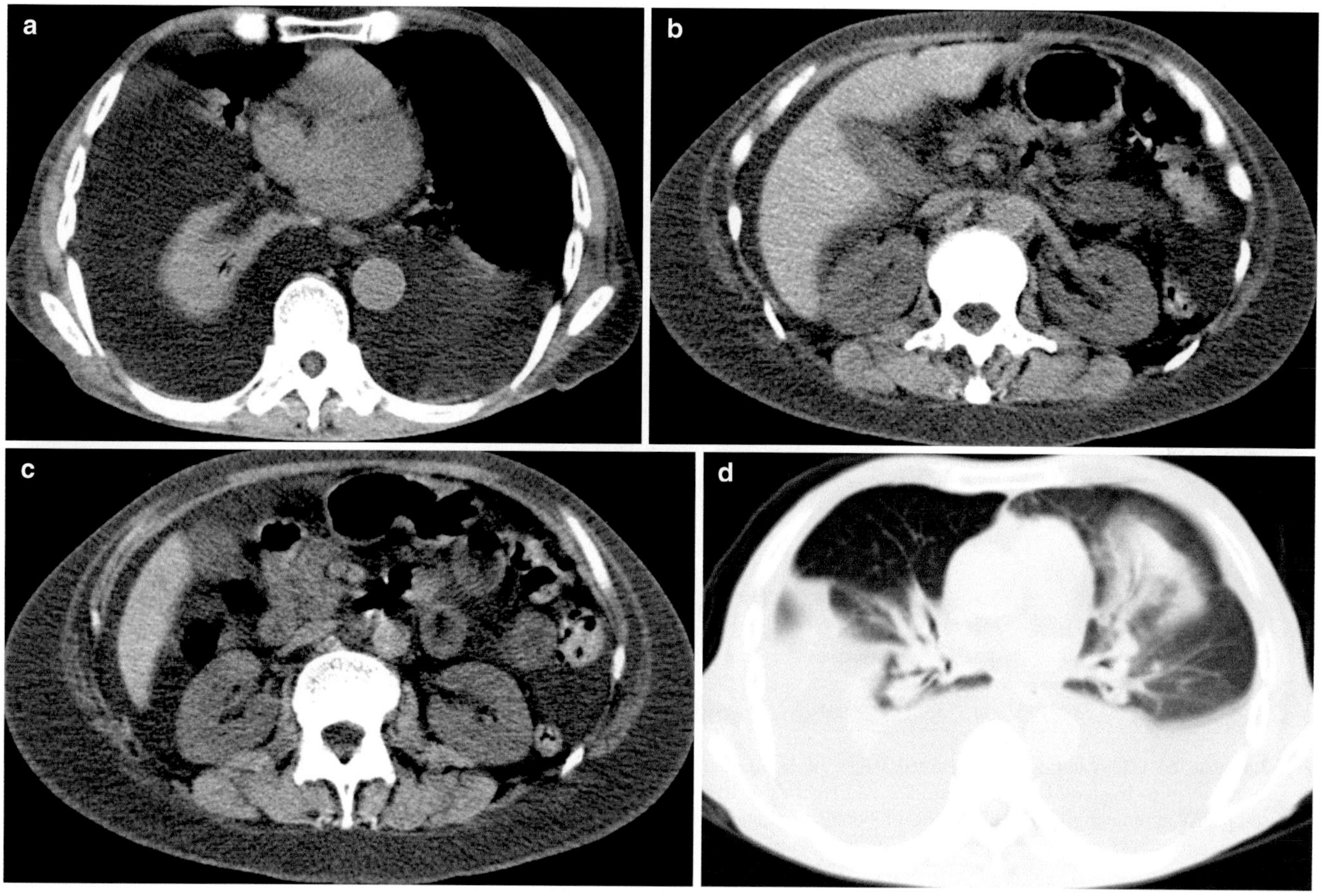

Fig. 7.16 (**a**–**c**) CT scanning demonstrated both kidneys with shrinkage in volume, decreased density, and poorly defined cortex-medulla interface as well as ascites. (**d**) Both lower lungs were revealed with strips and flakes of hyper-intense opacity, and atelectasis of both lower lungs; the bilateral thoracic cavity with arch shaped liquid-density opacity; the abdominal wall with edema

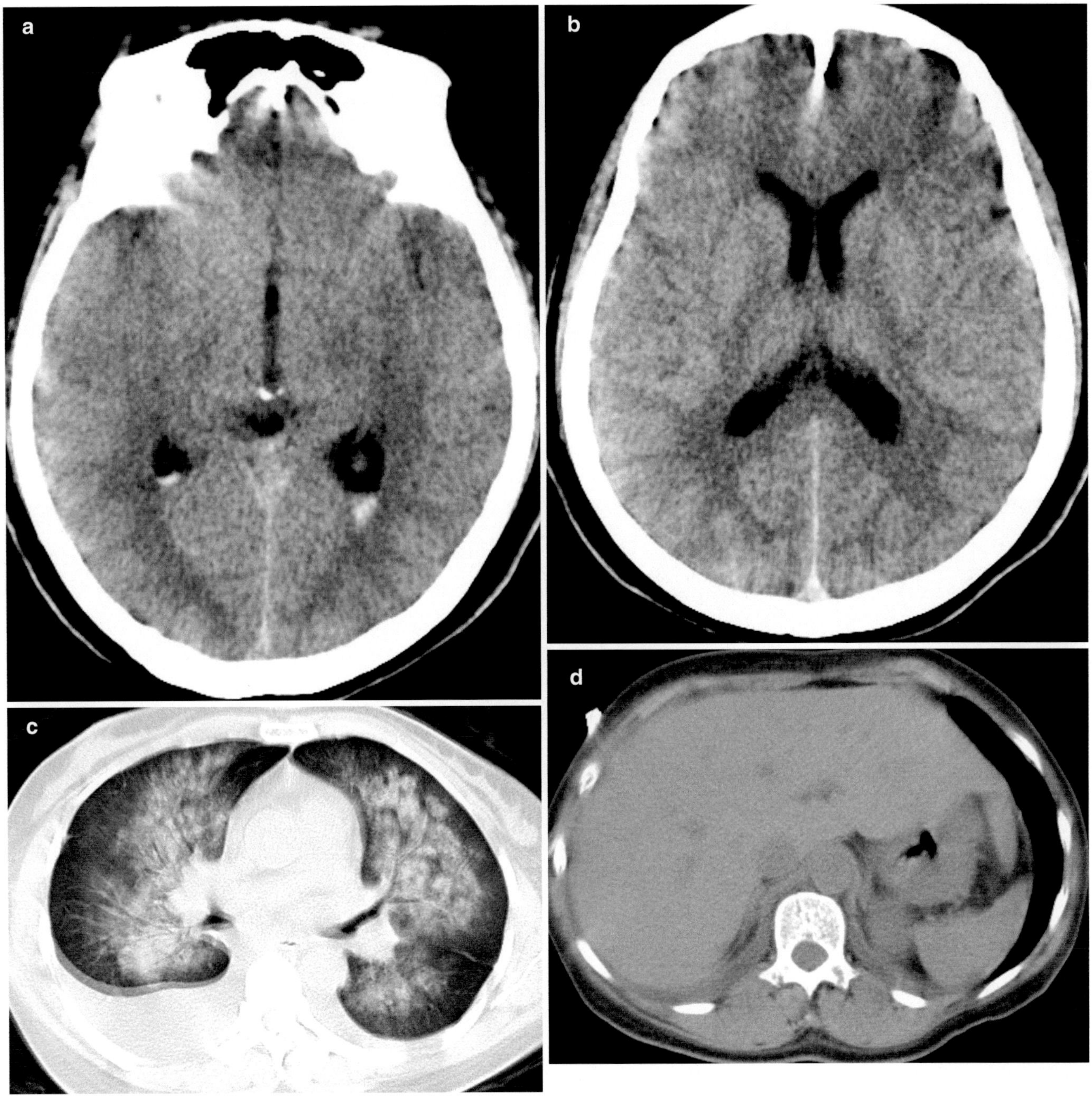

Fig. 7.17 (**a**, **b**) CT scanning demonstrated flakes of hypo-intense opacity adjacent to bilateral cerebral ventricles and patches of hyper-intense opacity at the margin of right temporal lobe, diffuse cotton like flakes of hyper-intense opacity at both lungs. (**c**) The bilateral thoracic cavities were demonstrated with arch shaped liquid-density opacity. (**d**) The liver was demonstrated to be enlarged (Reprint with permission from Hongjun Li, Radiology of Infectious diseases, Volume 2, 2015)

which can induce fatal episode of cerebral malaria that has a high death rate. When mosquitos suck blood from human, plasmodium gains its access into human body along with its secretions of salivary gland. The plasmodium rapidly reaches the liver along with blood flow. The infected hepatocytes may be subject to rupture to release a large quantity of merozoites, which gain their access into blood flow again to invade erythrocytes. Imported malaria may induce production of some adhesive factors which can bind to endothelial cells, phagocytes and monocytes via intercellular adhesion molecule-1 (ICAM-1, CD_{54}). In such a way, these adhesive factors adhere to hepatocytes in human body to cause local disturbance of blood circulation and subsequent hepatosplenomegaly, with manifestation of hepatic dysfunction. By microscopic examination of the liver, congestion of hepatic sinus and central vein as well as hyperplasia of Kupffer cells can be observed. Plasmodium develops and reproduces in erythrocytes, which destroys a large quantity of erythrocytes to cause jaundice. In patients with plasmodium falciparum, a large quantity of infected erythrocytes by plasmodium can be destroyed within a short period of time to induce hemoglobinuria, which may further develop into renal failure.

7.8 Kala Azar

Kala azar, also known as visceral leishmaniasis, is a chronic endemic infectious disease caused by leishmania donovani and transmitted by sandfly. It is clinically characterized by irregular fever, progressive hepatosplenomegaly, weight loss, anemia, and pancytopenia with increased plasma globulin level. In its advanced stage, edema, jaundice, varicosity in abdominal wall and ascites may develop.

7.8.1 Epidemiology

7.8.1.1 Source of Infection

Patients and infected dogs are major sources of its infection. In China, the major sources of infection are patients in plain areas of north Anhui province and east Henan province, while infected dogs in northwest plateau region.

7.8.1.2 Route of Transmission

Chinese sandfly is the major media for transmission of kala azar in China, and the disease is transmitted via stings and bites by sandfly.

7.8.1.3 Susceptible Population

Populations are generally susceptible to kala azar, and the susceptibility decreases along with aging. Human develops persistent immunity after its infection.

7.8.2 Pathological Changes

The basic pathological changes of kala azar include obvious hyperplasia of macrophages and plasmocytes. The lesions are mainly found in spleen, liver, bone marrow and lymph nodes, which contain abundant macrophages.

7.8.3 Clinical Manifestation

Fever is the most common symptom of kala azar, which is typically double quotidian fever. The patients usually experience perspiration, fatigue, weakness and general malaise. Symptoms at the advanced stage include anemia, malnutrition, nasal bleeding, gum bleeding and hemorrahic spots on skin. In some serious cases, the patients may gradually develop dark black skin on face and limbs, which is the reason for its nomination as kala azar denoting black fever.

Clinical manifestations also include mild lymphadenectasis, hepatomegaly and splenomegaly. The spleen shows progressive enlargement, which is palpable 2–3 weeks after onset, can reach the navel 6 months after onset, and reach pelvic cavity about 1 year after onset. The spleen can be palpated to be soft in the early stage, but hard in the advanced stage.

During the illness course, alleviation and deterioration alternate, and the disease develops into the convalescent stage about 1 month after onset, with decreased body temperature, relieved symptoms, shrinkage of the enlarged spleen and improved hemogram. However, the improved condition may only last for several weeks before the condition aggravates again, and the course may last for several months.

7.8.4 Diagnostic Examination

7.8.4.1 Laboratory Test

Etiological Examination

The etiological examinations mainly include smear and culture.

Serum Immunological Assay

The serum immunological assays mainly include specific antibody detection specific antigen detection.

Molecular Biological Examination

PCR and DNA probe technique can be applied to detect kDNA or DNA of leishmania donovani, with high sensitivity and specificity.

7.8.4.2 Radiological Examination

Ultrasound, CT scanning and MR imaging are commonly applied to detect abdominal lesions and complications. However, qualitative diagnosis cannot be made based on the radiological findings.

Case Study 1

[Brief medical History]

A 25-year-old young man had a history of military service in Gansu province in the year of 2009, followed by a 4-month military service in Tibet, China since Apr. 2011. He was hospitalized in Aug. 2011 due to fever for 12 days and spiritual indifference for 4 days. Laboratory tests showed normal level of WBC count, HBG 104 g/L, ALT 154 IU/L, AST 223 IU/L. Therapy for protection of the liver was prescribed. Reexamination of the bone marrow on Oct. 3 showed PK39 antibody positive, but shistosome, liver fluke and bronchial fluke negative. The disease was suspectively diagnosed to be kala azar, which was then treated by intramuscular injection of sodium stibogluconate monitored with ECG. Since Oct. 31, the second treatment course was started with intramuscular injection of sodium stibogluconate, oral intake of Cicolosporin and a therapy for liver protection. The condition was improved after treatment. For further definitive diagnosis, the patient paid his clinic visit to our hospital on Jan. 2012. The laboratory tests on Mar. 3, 2012 showed WBC 2.30×10^9/L, NEUT% 15.9%, NEUT 0.37×10^9/L, RBC 3.09×10^{12}/L, HBG 79 g/L, HCT 0.237, MCH 25.5 pg, PTL 121×10^9/L, and IgG 4320 mg/dl. Bone marrow puncture for biopsy revealed obvious and active hyperplasia of the bone marrow, decreased myeloid erythroid ratio, active hyperplasia of myeloid, erythroid and macrophagic cells, leftward shift of granulocytic nucleus, and visible scattering or clustering platelets. Bone marrow sections showed macrophagic L-D corpuscloid pathogen in cytoplasm of bone marrow, which may also scatter or cluster out of the macrophages. Abdominal B-ultrasound revealed hepatosplenomegaly. Since the onset of symptoms, the patient experienced poor appetite, quite good sleep, normal urination and bowel movement, and slight weight loss. By physical examination, the spleen was shown with splenomegaly of degree III. B-ultrasound revealed: (1) no abnormalities in gallbladder, pancreas and kidney; and (2) enlarged liver and spleen. After a whole series of physical examinations and laboratory tests, the diagnosis was defined to be kala azar. One course of sodium stibogluconate was prescribed then for disease control and a good rest was recommended. And the patient showed improved condition at present.

[Radiological Demonstrations] (See Fig. 7.18)

[Diagnosis]

Kala azar (visceral leishmaniasis) with hepatosplenomegaly.

[Discussion]

Kala azar, also known as visceral leishmaniasis, is a chronic endemic infectious disease caused by leishmania donovani. The life cycle of leishmania donovani includes 2 phases, amastigote (leishmaniform) and promastigote. Amastigote parasitizes in macrophages of endothelial system for reproduction in a large quantity to destroy numerous macrophages, which commonly occurs in spleen, liver, lymph nodes and bone marrow. Human is generally susceptible to the disease, which is a chronic endemic communicable disease transmitted via stings and bites by sandfly. Clinically, it is characterized by long-term irregular fever, enlarged spleen, liver and lymph nodes, peripheral pancytopenia and obviously increased plasma globulin, which are non-specific. Therefore, the disease tends to be misdiagnosed as hematologic neoplasm, cholera, TB, and liver cirrhosis. For this case, the patient was performing his military service in the disease affected area, with clinical symptoms of fever, hepatosplenomegaly, and lympadenectasis. Routine blood test revealed decreased myeloid and erythroid cells, IgG 4320 mg/dl. By bone marrow examination, the myeloid erythroid ratio was shown with a decrease. In addition, there were active hyperplasia of myeloid, erythroid and macrophagic cells, leftward shift of granulocytic nucleus. On bone marrow section, Leishman-Donovani body was observed in cytoplasm of some cells, which serves as etiological evidence. After treated by sodium stibogluconate, the patient improved, indicating a good therapeutic response. Based on these findings, the diagnosis was defined. For the diagnosis of kala azar, a report of epidemiological history about living in the disease affected area is very important. For differential diagnosis, the condition induced by liver cirrhosis decompensation after hepatitis should be considered and liver ultrasound, CT scanning and liver biopsy facilitate the differential diagnosis. The symptoms of kala azar resemble to those of Non-Hodgkin's lymphoma and primary macroglobulinemia. However, increase of immunoglobulin in kala azar is multi-clonal with no bone destruction and kidney impairment. In addition, bone marrow examination commonly reveals Leishman-Donovani body in cytoplasma, which serves as the etiological evidence for differential diagnosis. However, literature reports have shown that the positive rate of Leishman-Donovani body by bone marrow smear is relatively low and, therefore, PCR and strip for rk39 antibody detection are clinically recommended.

Case Study 2

[Brief Medical History]

A 35-year-old man complained of intermittent fever for 6 months. Laboratory tests showed serologically positive (+) for kala azar.

[Radiological Demonstrations] (See Fig. 7.19)

[Diagnosis]

Kala azar with splenomegaly.

Case Study 3

[Brief Medical History]

A 38-year-old man complained of persistent fever for 2 months with a body temperature of 39 °C. The patient was working as a construction worker in the disease affected area of kala azar and had a history of being biten by mosquito. The serological test showed kala azar positive.

[Radiological Demonstrations] (See Fig. 7.20)

[Diagnosis]

Kala azar with splenomegaly.

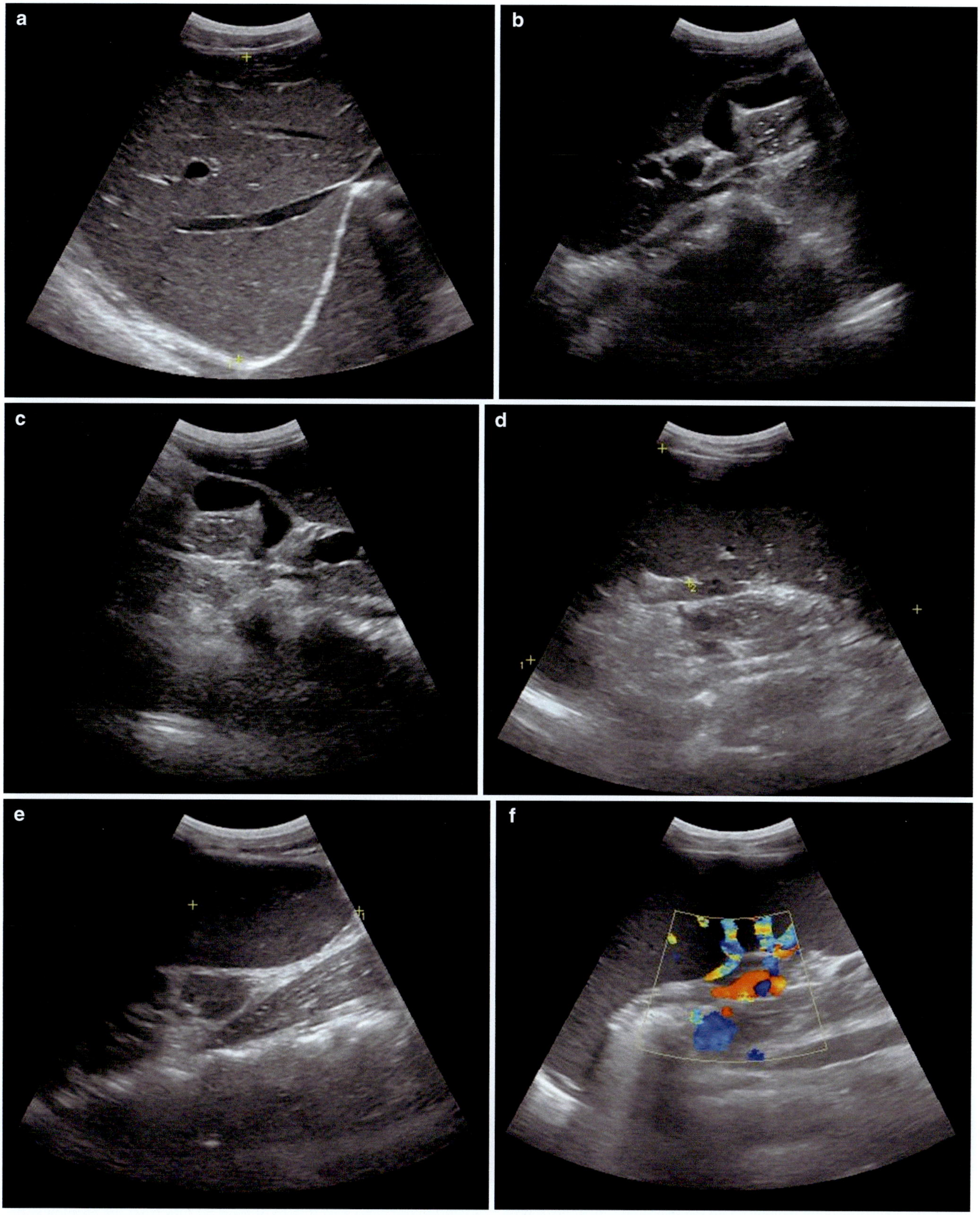

Fig. 7.18 (**a**–**f**) Ultrasound showed anlarged spleen and liver, as well as portal vein widened slightly

Case Study 4

[Brief Medical History]

A 43-year-old man complained of intermittent fever for 5 months and was hospitalized. He lived in the affected area of kala azar with a history of being biten by mosquito. The serological test showed kala azar positive.

[Radiological Demonstrations] (See Fig. 7.21)

[Diagnosis]

Kala azar complicated by liver abscess.

Case Study 5

[Brief Medical History]

A 38-year-old man complained of persistent fever for 3 months with a body temperature of 39 °C. He reported a history of being biten by mosquito in affected area of kala azar. The serological test showed kala azar positive.

[Radiological Demonstrations]

[Diagnosis]

Kala azar complicated by spleen infarction.

[Discussion]

Kala azar is caused by leishmania donovani, which commonly parasitizes in reticular-endothelial system of organs and whose antigen can be expressed at the surface of macrophages. Therefore, laboratory test findings are the important criteria for its qualitative diagnosis. The commonly applied laboratory tests for its qualitative diagnosis include serological antibody test, serological circulatory antigen test and bone marrow puncture for biopsy. Radiological modalities can be applied for accurate diagnosis of the range with lesions and related complications.

Ultrasound, as a routine examination, can detect hepatosplenomegaly at the early stage. Megalosplenia can also be identified in some patients, but insufficient for the qualitative diagnosis. Contrast CT scanning can facilitate to detect abdominal lesions and related complications. Especially, reconstruction of images after multislice CT scanning can show the lesion from multiple perspectives but still insufficient for the qualitative diagnosis. MR imaging has a high resolution for tissues and can help to visualize the target region from any perspective. Therefore, the lesion of abscess and involvement of the surrounding tissues can be more clearly demonstrated by MR imaging, which is more favorable than CT scanning in detecting involved tissues and the range of involvement at the early stage. MR imaging can serve as an important supplement for radiological examination of kala azar.

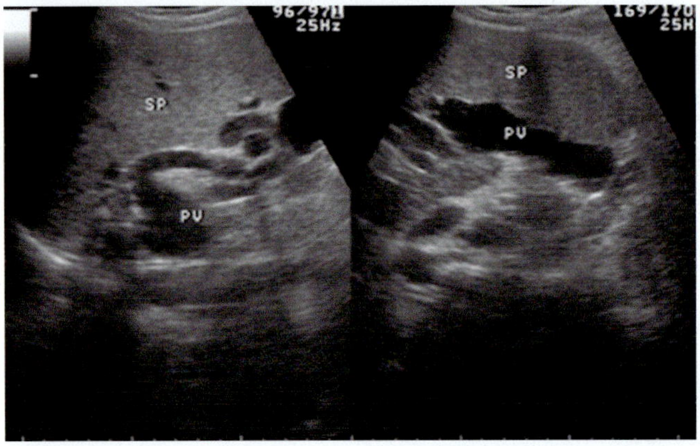

Fig. 7.19 Ultrasound showed enlarged spleen with its lowest point under the navel, indicating megalosplenia, as well as twist and dilated splenic vein

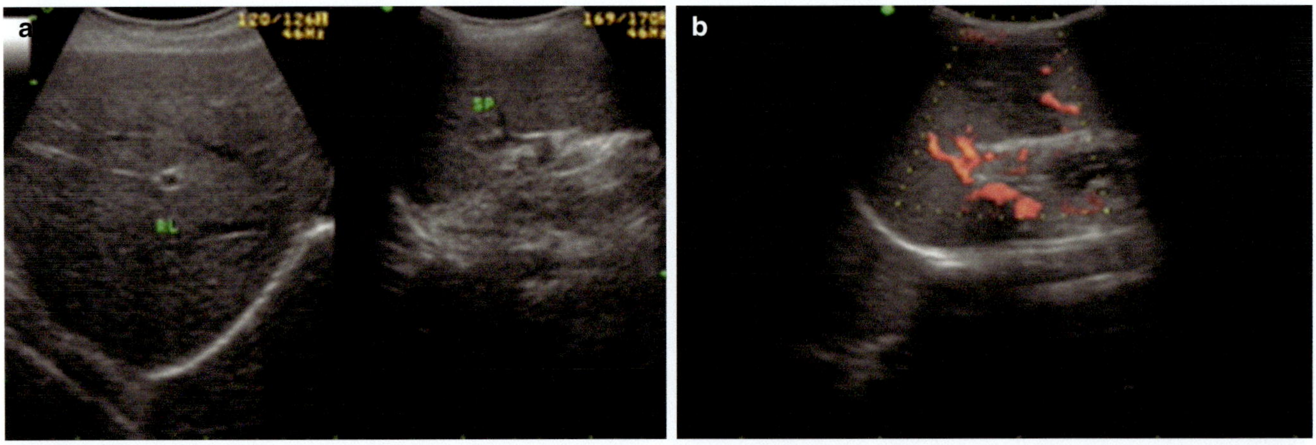

Fig. 7.20 (**a**) Ultrasound demonstrated the liver to be enlarged, with a maximal oblique diameter of the right liver lobe being 151 mm, slightly plump in appearance, and quite homogeneous echo from liver parenchyma. (**b**) Ultrasound demonstrated the spleen to be enlarged, with heterogeneous echo from spleen parenchyma, and a wedge shaped hypo-intense echo area in size of 49 × 39 mm from the envelop at the middle spleen to splenic hilus with rare spots like or rod like hyperintense echo inside. CDFI showed no obvious blood flow signal in the hypo-intense echo area, indicating spleen infarction

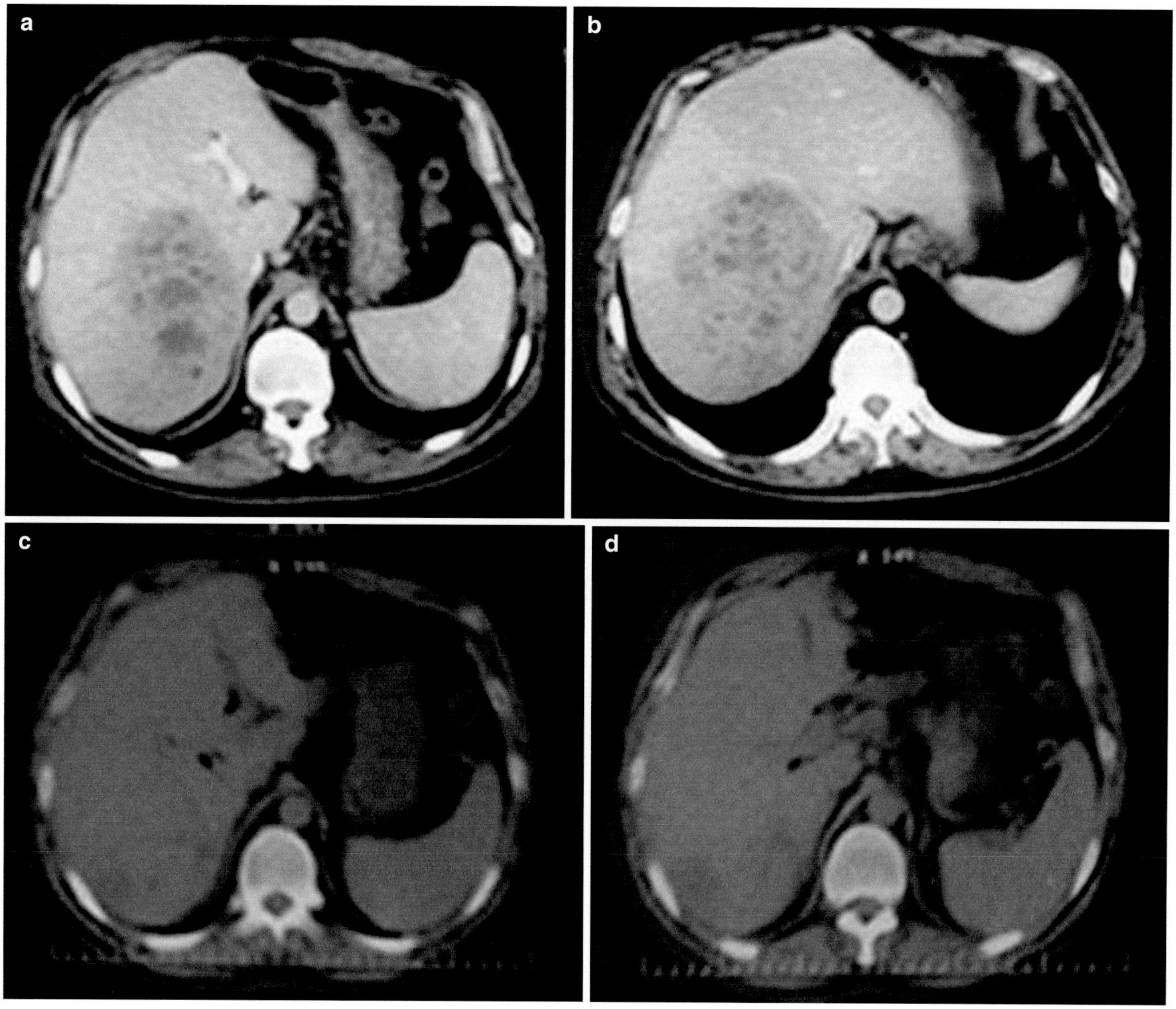

Fig. 7.21 (**a**, **b**) CT scanning showed large flakes of honeycomb like hypo-intense opacity at the right liver lobe, with poorly defined boundary and uneven inner density. Contrast scanning showed unobvious enhancement of the lesion, which was suspected to be liver abscess induced by leishmania donovani. (**c**, **d**) After treatment by sodium stibogluconate, reexamination by CT scanning demonstrated absence of the lesion at the right liver lobe and shrinkage of enlarged spleen

Further Reading

Anstey NM, Jacups SP, Cain T, et al. Pulmonary manifestations of uncomplicated falciparum and vivax malaria: cough, small airways obstruction, impaired gas transfer, and increased pulmonary phagocytic activity. J Infect Dis. 2002;185(9):1326–34.

Asiedu DK, Sherman CB. Adult respiratory syndrome complicating placiparum malaria. Heart Lung. 2000;29(4):294–7.

Bae K, Jeon KN. CT finding of malaria spleen. Br J Radiol. 2006;79(946):e145–7.

Bejon P, Mwangi TW, Lowe B, et al. Helminth infection and eosinophilia and the risk of Plasmodium falciparum malaria in 1-to 6-year-old children in a malaria endemic area. PLoS Negl Trop Dis. 2008;2(1), e164.

Chai JJ, Guan LR. Kala azar and sandfly in Uygur Autonomous Region of Xinjiang. Urumqi: Xinjiang People's Publishing House; 2006.

Chen ZH, Shao XY. X-ray and enteroscopy of ulcerative coloniti. World Chin J Digest. 2000;8(3):335–6.

Cordoliani YS, Sarrazin JL, Felten D, et al. MR of cerebral malaria. AJNR Am J Neuroradiol. 1998;9(5):871–4.

Focà E, Zulli R, Buelli F, et al. P. falciparum malaria recrudescence in a cancer patient. Le Infezioni Med. 2009;17(1):33–4.

Gamanagatti S, Kandpal H. MR imaging of cerebral malaria in a child. Eur J Radiol. 2006;60(1):46–7.

Huang H, Deng YY, Lu PX. Diagnostic imaging of toxoplasmic encephalopathy in patients with AIDS. Chin J Magn Reson Imaging. 2010;5(1):353–8.

Kang X, Liu YB, Liu K. Kala azar: clinical analysis of 86 cases. Chin J Infect Chemother. 2009;9(4):241–3.

Khan W, Zakai HA, Umm-E-Asma. Clinico-pathological studies of Plasmodium falciparum and Plasmodium vivax-malaria in India and Saudi Arabia. Acta Parasitol. 2014;59(2):206–12.

Kim EM, Cho HJ, Cho CR, et al. Abdominal computed tomography findings of malaria infection with Plasmodium vivax. Am J Trop Med Hyg. 2010;83(6):1202–5.

Kumar GG, Mahadevan A, Guruprasad A, et al. Eccentric target sign in cerebral toxoplasmosis-neuropathological correlate to the imaging feature. Magn Reson Imaging. 2010;31(6):1469–72.

Li HJ, Qi S. Clinical manifestations and radiological demonstrations of AIDS complicated by nervous system infections. Chin J Magn Reson Imaging. 2010;1(5):380–8.

Lin QS. Clinical analysis serious plasmodium falciparum: a report of 45 cases. Chin Trop Med. 2005;5(2):282–5.

Lin RP, Lin BY, Yang Q. Clinical typing of dangerous plasmodium falciparum and its clinical treatment. Chin J Infect Dis. 2002;20(5): 317–8.

Liu M, Chen XG. Epidemiological study of toxoplasmosis in populations of China. Acta Parasitologica Et Medica Entomologica Sinica. 2010;17(3):184–91.

Ma YC, Ma HY, Du QX. Misdiagnosis of pulmonary toxoplasmosis as pulmonary tuberculosis: analysis of 4 cases. Chin J Tuberc Respir Dis. 2001;24(10):640.

Nickerson JP, Tong KA, Raghavan R. Imaging cerebral malaria with a susceptibility-weighted MR sequence. AJNR Am J Neuroradiol. 2009;30(6):e85–6.

Patankar TF, Karnad DR, Shetty PG, et al. Adult cerebral malaria: prognostic importance of imaging findings and correlation with postmortem findings. Radiology. 2002;224(3):811–6.

Plewes K, Royakkers AA, Hanson J, et al. Correlation of biomarkers for parasite burden and immune activation with acute kidney injury in severe falciparum malaria. Malar J. 2014;13:91.

Punsawad C, Viriyavejakul P. Nuclear factor kappa B in urine sediment: a useful indicator to detect acute kidney injury in Plasmodium falciparum malaria. Malar J. 2014;13:84.

Rafieian-Kopaei M, Nasri H, Alizadeh F, et al. Immunoglobulin a nephropathy and malaria falciparum infection; a rare association. Iran J Public Health. 2013;42(5):529–33.

Shao L, Zheng XH, Song P. Ultrasonographic diagnosis of amebic liver abscess: report of 1 case. Chin J Diagn Imaging. 2007;15(4):291–2.

Song XB, Jing XH, Bo CM. Chest X-ray signs of lungs in patients with plasmodium falciparum: a report of 86 cases. J Pract Radiol. 2004;20(7):602–4.

Sowunmi A, Ogundahunsi OA, Falade CO, et al. Gastrointestinal manifestations of acute falciparum malaria in children. Acta Trop. 2000;74(1):73–6.

Srinivas R, Agarwal R, Gupta D, et al. Severe sepsis due to severe falciparum malaria and leptospirosis co-infection treated with activated protein C. Malar J. 2007;6:42.

Tan X, Yang JB, Chen ZS, et al. Misdiagnosis of kala azar as liver cirhosis and lymphatic malignancy: report of 1 case. J Intractable Dis. 2014;13(2):204.

Vásquez AM, Tobón A. Pathogenic mechanisms in Plasmodium falciparum malaria. Biomedica. 2012;32 Suppl 1:106–20.

Viriyavejakul P, Khachonsaksumet V, Punsawad C. Liver changes in severe Plasmodium falciparum malaria: histopathology, apoptosis and nuclear factor kappa B expression. Malar J. 2014;13:106.

Wang Q, Xu L. Laboratory tests for diagnosis of ulcerative colonitis and its activities. World Chin J Digest. 2000;8(3):336–7.

Wichapoon B, Punsawad C, Chaisri U, et al. Nuclear factor kappa B in urine sediment: a useful indicator to detect acute kidney injury in Plasmodium falciparum malaria. Malar J. 2014;13:176.

Wilairatana P, Riganti M, Looareesuwan S, et al. Dyspepsia in acute falciparum malaria: a clinico-pathological correlation. Southeast Asian J Trop Med Public Health. 1992;23(4):788–94.

Wu WH, Li SW, Hu D, et al. Epidemiology and clinical features of kala azar. J Hunan Norm Univ (Med). 2013;10(3):88–93.

Xu L, Li SB, Tong Q, et al. Clinical findings of ulcerative colonitis complicated by intestinal amebiasis. J Clin Digest. 2012;24(3):167–9.

Yadav P, Sharma R, Kumar S, et al. Magnetic resonance features of cerebral malaria. Acta Radiol. 2008;49(5):566–9.

Zhan XM, Chen JP. Human parasitology. Beijing: People's Medical Publishing House; 2003.

Zhao MZ, Wu SC, Huang Y. Misdiagnosis of kala azar: report of 1 case. Chin Med Radiol Technol. 2010;26(9):1789.

8 Helminthiasis

Yulin Guo, Yi Xiao, Hong Wang, Wenya Liu, and Yonghua Tang

Helminthes are a type of multicellular invertebrate, which move in peristalsis via muscular contractions. Most of the helminthes live in nature but a small number of them parasitize in or on the surface of animals or plants. In the history of taxonomic zoology, helminthes were believed to be a special and independent group. But with the development of taxonomy, helminthes related to human body were found to be categorized into multiple phyla of invertebrata, including the phylum of Platyhelminthes, the phylum of Annelida, and the phylum of Acanthocephales. Therefore, helminthes, as scientific term, has no taxonomic significance but just a conventional use.

Medical helminthes are known as medicine related helminthes that parasitize in human body and cause helminthiasis. More than 160 species of helminthes have been identified, including 20–30 medically important species. The development of helminthes can be divided into several stages, and each developmental stage requires different environmental conditions. Based on host exchange during their development, helminthes can be categorized into two major groups.

The first is geohelminthes whose eggs or larvae develop in nature and require no intermediate host during their development. After intake of contaminated food by the eggs or larvae or access to the contaminated environment, human can be invaded and infected, such as most of the nematodes.

And the second is biohelminthes, which require intermediate host for their development and invade the final host.

Diseases caused by trematode, cestode, nematode, and acanthocephala are categorized into the field of medicine and are generally known as helminthiasis. Such helminthes parasitize in such luminal systems as digestive tract, bile duct and blood vessels. In addition, they also parasitize in parenchymal organs such as liver and brain as well as muscular tissues. Based on the parasitic site, medical helminthes do harm to the human body in varying degrees, ranging from mild symptoms like fever, diarrhea, and subjective upset to severe symptoms like unconsciousness, allergic shock and even life threatening conditions. The serious consequences of helminthiasis have gradually attracted concerns from clinicians, scholars and common people.

The rapid development of modern science plays a crucial role in diagnostic imaging of helminthiasis, whose contribution is not confined to facilitating its diagnosis. With the combined use of several radiological techniques, evidence of helminthiasis at certain sites can be directly provided for its early diagnosis. In this chapter, we introduce radiological modalities and findings about helminthiasis.

8.1 Ascariasis

Ascariasis is a chronic infectious disease caused by parasitism of ascaris lumbricoides (referred to as ascaris) in human small intestines or other organs. Ascariasis prevails widely and the patients are mostly pediatric.

8.1.1 Etiology

8.1.1.1 Morphology

Adult ascaris, like an earthworm, is cylindrical in shape that is gradually thinner from the middle towards head and tail.

Y. Guo
General Hospital, Ningxia Medical University, Ningxia, China

Y. Xiao
Shanghai Changzheng Hospital, Shanghai, China

H. Wang
The Second Affiliated Hospital, Xinjiang Medical University, Urumqi, Xinjiang Uygur Autonomous Region, China

W. Liu (✉)
Affiliated First Hospital, Xinjiang Medical University, Urumuqi, Xinjiang Uygur Autonomous Region, China
e-mail: 13999202977@163.com

Y. Tang
Ruijin Hospital, Affiliated to School of Medicine, Shanghai JiaoTong University, Shanghai, China

H. Li (ed.), *Radiology of Parasitic Diseases*, DOI 10.1007/978-94-024-0911-6_8

Its tail is blunt cone like in shape and its body is creamy white or light reddish in color, with obvious lateral white lines on both sides of its body.

8.1.1.2 Life Cycle

Ascaris does not require an intermediate host. And its life cycle includes development of eggs in natural world, migration and development of larvae in body of its host, and parasitism of adult ascaris in small intestines of its host.

8.1.2 Epidemiology

Ascariasis is the most common parasitic disease worldwide, which infects about 1/4 of the world population. Most of the patients live in the temperate and tropical regions and more serious prevalence is witnessed in under developed, warm and humid countries and regions with poor sanitation. In China, its infection rate in rural areas is higher than in urban areas, and that in children is higher than in adults.

8.1.2.1 Source of Infection

The patients with ascariasis whose feces contain fertilized eggs of ascaris are source of its infection.

8.1.2.2 Route of Transmission

Ascariasis is transmitted via intake of infective eggs of ascaris.

8.1.2.3 Susceptible Population

People are commonly susceptible to ascariasis, with a higher infection rate in rural areas and in children. And children at pre-school and school ages are especially vulnerable to the disease. Along with increase of age, its multiple infections produce immunity, which is also one of the reasons contributing to a decrease of infection rate in adults.

8.1.3 Clinical Manifestation

8.1.3.1 Larva Migrating Stage

When a small quantity of larvae migrate in lungs, patient may show no clinical symptoms. But after intake of uncooked vegetables containing a large quantity of infective eggs of ascaris or other contaminated foods within a short period of time usually causes ascaris pneumonia, asthma, and eosinophilia. Larva migrating stage has an incubation period of 7–9 days and clinically systemic and pulmonary symptoms.

8.1.3.2 Intestinal Ascariasis

Most patients show no obvious symptoms, but rare patients complain of abdominal pain and periumbilical tenderness, which are occasionally colic. The pain occurs repeatedly with varying interval. Severely infected patients show poor appetite, weight loss and anemia. Ascaris may be excreted along with feces.

8.1.3.3 Complication and Ectopic Parasitism

The complications include intestinal obstruction, biliary ascariasis, pancreatic duct ascariasis, and appendiceal ascariasis.

8.1.4 Diagnostic Examination

8.1.4.1 Laboratory Test

Routine Blood Test

In the cases with larva migration, ectopic ascariasis and complicated infection, routine blood test shows increased WBC and eosinophil counts.

Etiological Examination

Stool smear for examination is a simple way for the etiological diagnosis of ascariasis, which shows a high detection rate and is the commonly used examination for the diagnosis of intestinal ascariasis. Examination of duodenal drainage fluid provides direct evidence of biliary ascariasis when ascaris eggs are observed. In the cases with pneumonia caused by ascaris in lungs, ascaris larvae can be detected in sputum.

8.1.4.2 Radiological Examination

Ultrasound

For the patients with biliary ascariasis, abdominal B-mode ultrasound shows presence of ascaris in dilated common bile duct.

X-ray

Barium meal for X-ray examination shows the shape and quantity of ascaris. Abdominal plain X-ray is of significant diagnostic value for ascaris intestinal obstruction or peritonitis due to intestinal perforation.

Case Study 1

[Brief Case History]

A 7-year-old girl complained of paroxysmal periumbilical pain with no known causes since half a month ago. The pain could be relieved by itself, with accompanying vomiting. In the hospital, her condition was diagnosed as biliary ascariasis. After antiscolic medication, her condition was obviously improved. But after that, the patient still complained of intermittent abdominal pain, with slightly yellowish stained systemic skin mucosa and yellowish stained sclera. She was hospitalized again for further examinations. The laboratory tests showed ALT 197 U/L, AST 156 U/L, elevated total bilirubin, and elevated conjugated bilirubin.

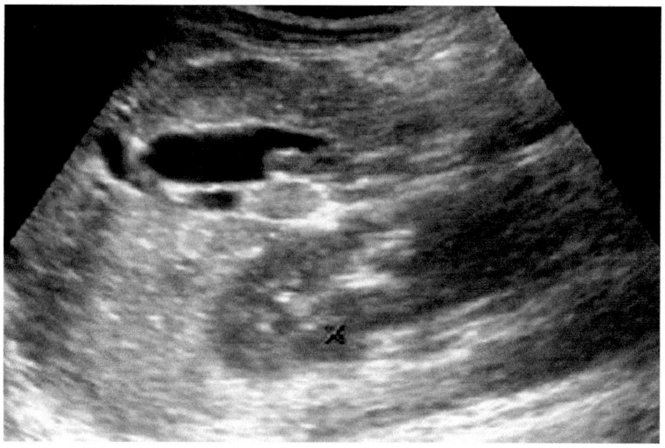

Fig. 8.1 Ultrasound revealed equally strong echo in the inferior segment of extrahepatic bile duct, and widened extrahepatic bile duct with an inner diameter of 1.2 cm

[**Radiological Demonstrations**] (See Fig. 8.1)

[**Diagnosis**] Biliary ascariasis

[**Discussion**]

The clinical manifestations of ascariasis vary based on the parasitic site and developmental stage of ascaris. When the parasitic environment in intestinal tract changes, parasitizing ascaris can leave the intestinal tract and migrate into other organs with hole to cause ectopic ascariasis. The following is several common ectopic ascariasis:

1. Biliary ascariasis.
   Biliary ascariasis is more common in children and young women. Its causes include high fever, diarrhea, pregnancy and delivery. The onset of biliary ascariasis is sudden and acute, with intense paroxysmal colic pain at the middle part of right upper quadrant, drilling sensation, upset, nausea and vomiting. Ascaris may be vomited out. Migration of ascaris into liver may cause ascaris hepatic abscess, for which surgical operation should be performed as early as possible.
2. Pancreatic duct ascariasis.
   Pancreatic duct ascariasis usually complicates biliary ascariasis, with clinical symptoms resembling to those of acute pancreatitis.
3. Appendiceal ascariasis.
   Appendiceal ascariasis is common in young children due to their comparatively wider opening at the root of appendix for access of ascaris. The diagnosis can be defined after ascaris eggs are detected by stool examination.

In patients with a past history of intestinal ascariasis, cholecystitis, calculus and biliary surgery, ascaris more commonly invades the biliary tract or even repeatedly leave and enter the biliary tract. Ultrasound revealed no obvious dilation of common bile duct at the early stage of biliary ascariasis. However, along with the progression of the condition, the common bile duct as well as the left and right hepatic ducts are subjected to slight dilation. In the dilated bile ducts, long double linear high echo strips in a width of several millimeters can be observed. The strips showed a round blunt head, well defined borderline, and pseudocoele as a fluid dark strip running through the middle part. If ascaris is alive, its wriggling is observable. But if ascaris is dead, ultrasound revealed no echo from its center and poorly defined strip, or even patchy strong echo. Multiple ascarises in bile ducts show multiple linear hyper-intense echo strips, or occasionally their curling into a mass with sound shadow. The bile sediment, tissue debris, pus embolus or dead ascarides in the lumen of bile duct may cluster into a small mass, with a higher density than bile and posterior light vague sound shadow. Such a condition is diagnosed as bile duct calculi as a part of calculi. Biliary ascariasis may be complicated by cholangitis and cholecystitis, with manifestations of enlarged gallbladder, gallbladder wall edema, thickened common bile duct wall, and extrahepatic cholestatic jaundice. The patients show slightly yellow stained skin and sclera. But in some patients, the symptoms are atypical with slight pain, which tends to be misdiagnosed, especially in the elderly patients.

Biliary ascariasis should be differentiated from the following diseases:

1. Biliary calculi
   Ultrasound shows intrahepatic strong echo, posterior sound shadow, and distal dilated intrahepatic bile duct.
2. Pneumobilia
   Ultrasound demonstrates varying shape of the strong echo, with poorly defined boundary and posterior multiple reflections.
3. Cholangitis
   Ultrasound reveals thickening and edema of the bile duct wall, and double layer echo. The dead ascaris in the bile duct shows echo from the bile duct, which is demonstrated as poorly defined bile duct. Due to the poorly defined double linear echo of ascaris, the combination with clinical manifestations facilitates its differential diagnosis from cholangitis.

Case Study 2

[**Brief Case History**]

A 42-year-old woman complained of unbearable upper abdominal blunt persistent pain for several hours with no known causes a month ago. She also reported profuse sweating, nausea, vomiting, and abdominal distension with elevated amylase. Her condition improved after being treated as acute pancreatitis. In recent days, the pain recurred with no known causes that was bearable prickling at upper abdomen. By laboratory test, EOS was 0.101, and ALT 97 U/L.

[**Radiological Demonstrations**] (See Fig. 8.2)

[**Diagnosis**] Gallbladder ascariasis and ascaris in common bile duct.

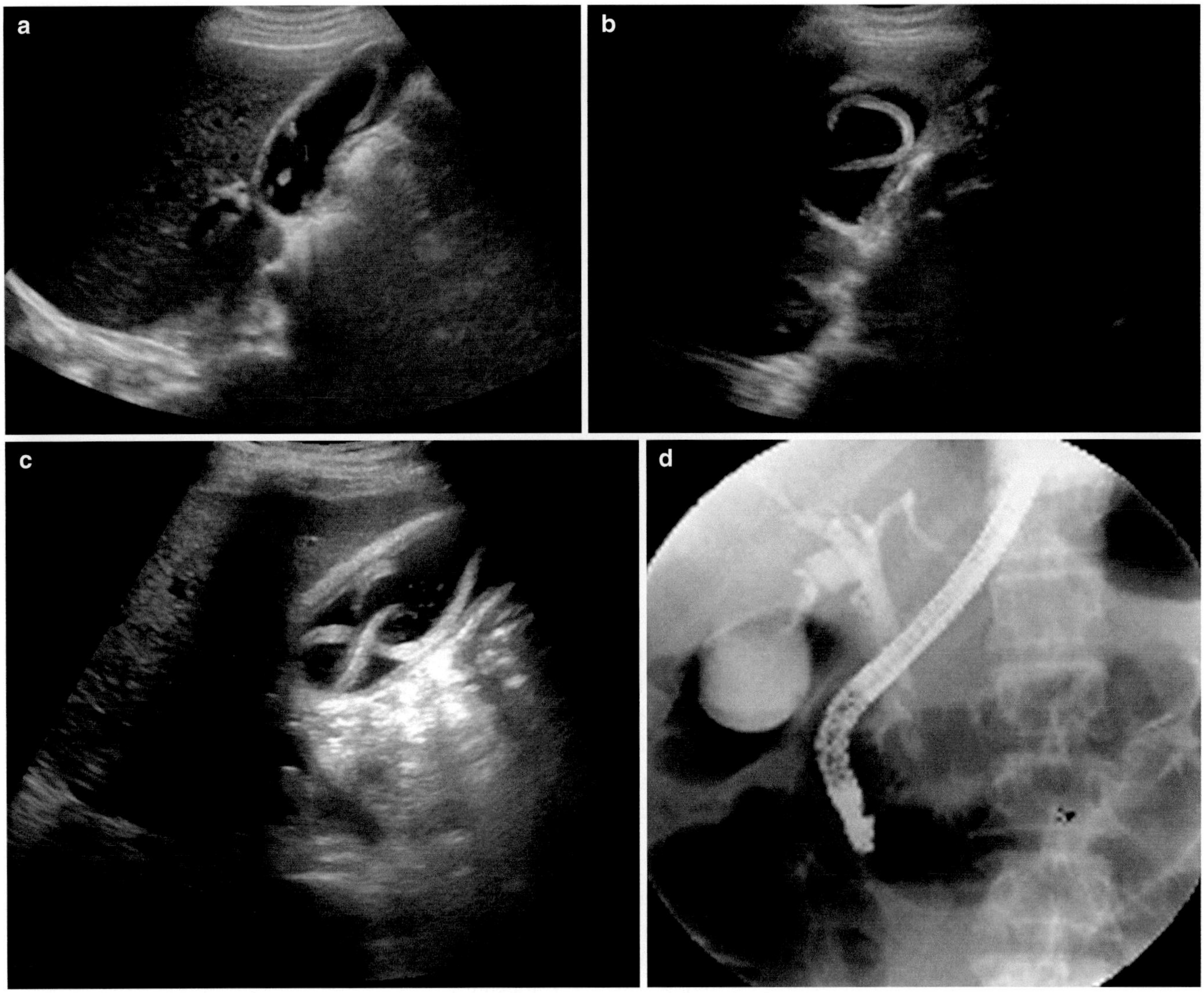

Fig. 8.2 (**a–c**) Ultrasound revealed multiple equal-sign shaped hyper-intense echo in gallbladder with observable peristalsis, thickened and coarse gallbladder wall in double side sign, and attached multiple slight hyper-intense echoes to the wall with no accompanying sound shadow that are immobile along with posture change. Multiple polypoid lesions were observed in the gallbladder, with edema of the gallbladder wall. (**d**) Endoscopy demonstrated common bile duct to be dilated. After infusion of contrast agent, the common bile duct was shown with a strip like filling defect

[**Discussion**]

By sonography, the gallbladder is revealed with ascaris in arch or curling shape with observable peristalsis. Due to non-filling of bile, the pseudocoele of ascaris is poorly defined, shown as tubular parenchymal hyper-intense echo strip by sonography.

Biliary ascariasis should be differentiated from the following conditions:

1. Gallbladder carcinoma.
   The gallbladder wall is demonstrated to be uneven and irregular with consequent irregular lumen of the gallbladder. The condition is commonly accompanied by enlarged gallbladder and its poorly defined contour.
2. Cholelithiasis
   The liver is revealed with intrahepatic hyper-intense echo, posterior sound shadow, and distally dilated intrahepatic bile duct.
3. Pneumotosis of gallbladder
   Ultrasound reveals varying shape of the hyper-intense echo, its poorly defined boundary, and multiple posterior reflections.

Case Study 3

[**Brief Case History**]

A 60-year-old man paid clinic visit to the emergency department of our hospital due to complaint of persistent upper and middle abdominal distension and pain after intake of greasy food in Aug. 2011. He also showed symptoms of chills, fever with the highest body temperature of 39.4 °C, and persistent abdominal pain. The laboratory test showed WBC 16×10^9/L, and NEUT% 81.4%. After B-mode ultrasonography, the condition was suspected to be acute cholecystitis. And anti-infection, acid inhibition and enzyme inhibition therapies were given to relieve his symptoms. Three days ago, he was hospitalized for treatment due to middle and upper abdominal upset after intake of greasy food. In addition, he showed fever, inversion to cold, yellow stained skin, and dark urine. The laboratory tests showed ALT 121 U/L, AST 87 U/L, ALP 324 U/L, GGT 1136 U/L, TBIL 161.5 μmol/L, DBIL 99.7 μmol/L, and amylase 339 U/L.

[**Radiological Demonstrations**] (See Fig. 8.3)

[**Diagnosis**] Biliary ascariasis complicated by acute pancreatitis.

[**Discussion**]

On emergency clinic visit, the patient complained of persistent middle and upper abdominal distension and pain after intake of greasy food in Aug. 2011, with accompanying chills, fever, persistent abdominal pain and jaundice. The diagnosis was finally defined after several different diagnostic examinations. Pancreatitis was etiologically attributed to biliary and alcoholic reasons as well as hyperlipidemia, inappropriate diet and other factors such as pregnancy, delivery, medication, pancreatic developmental abnormalities and trauma. Based on both international reports and reports in China, biliary disease mainly contributes to the condition. In this case, plain CT scan and ultrasound failed to reveal biliary calculi; and MRCP demonstrated no obvious cup-opening sign. The patient had a history of inappropriate diet but no history of alcoholism and hyperlipidemia. Therefore, the etiological factors such as biliary condition, alcoholism and hyperlipidemia could be excluded. Ascariasis is a common helminthic infectious disease and its etiological factor, parasitism of ascaris, commonly parasitizes in jejunum and middle ileum but rarely in the biliary system. Only when the intestinal environment for parasitism changes and/or the duodenal papillary muscle dysfunctions, the parasitic ascaris is likely to migrate upwards to duodenum and enter into biliary tract via duodenal papillary openings to cause various severe complications. The dead polypide disintegrates into The tissue fragments after death of ascaris and its eggs, as basic components, contribute to the formation of biliary calculi. Multiple radiological modalities can be chosen to diagnose biliary ascariasis, among which ultrasonography is the first choice due to its low cost and real-time imaging. The typical findings by ultrasonography include tubular echo with no accompanying sound shadow in a diameter of 3–6 mm, comparatively central hypointense echo and parallel double linear wall of hyperintense echo. However, its diagnostic value is affected by existence of intestinal gas and the experience of clinician. ERCP contributes to both diagnosis and treatment, by which the body of ascaris outside the papillary muscle can be directly observed. In addition, a singular strip or multiple strips of filling defect in the biliary tract can be shown by contrast agent. But it is an invasive examination whose application should be strictly limited. By CT scanning, only calculi formed by biliary dead ascaris can be detected, but with no specificity. Based on water imaging, MRCP is capable of favorably displaying the 3-dimensional anatomic structure of biliary system from multiple perspectives, requires no use of contrast agent, and produces no radioactive damages. By T_2WI and MRCP, the condition is revealed as tubular filling defect with hypointense signal in the biliary tract, and comparatively hyperintense signal of the liquid swallowed by the ascaris in the intestinal tract to divide the tubular ascaris into two parts, namely the double-tube sign. MRCP findings have been gradually recognized as the golden diagnostic criteria for biliary ascariasis and MRCP shows advantages in the early diagnosis and follow-up assessment of this disease.

Case Study 4

[**Brief Case History**]

A 60-year-old woman received surgery for biliary ascariasis 6 years ago. She complained of sudden unbearable drilling pain at the upper abdomen after meal 1 day ago, which was intermittent and gradually aggravating with accompanying nausea and vomiting. The laboratory test showed NEUT 13.3×10^9/L, NEUT 0.842, MONO 1.01×10^9/L. B-mode ultrasound indicated biliary ascariasis.

[**Radiological Demonstrations**] (See Fig. 8.4)

[**Diagnosis**] Biliary ascariasis.

[**Discussion**]

Biliary ascariasis is the most common complication of intestinal ascariasis, which is an acute abdominal disease caused by access of duodenal ascaris into the common bile duct via papillary openings. It occurs most commonly in children, followed by young adults. And ascaris can partly or completely migrate into the biliary system, mostly the extrahepatic bile duct and left hepatic duct. In some other cases, the ascaris may further migrate upwards into intrahepatic bile duct or interlobular intrahepatic bile ducts, but rarely into the gallbladder. Ascaris can survive in the biliary system for several days or months. If death of ascaris occurs in the biliary system, its residual body may be excreted or form the core of gallstone. CT scanning reveals the dilated bile duct caused by access of ascaris as well as various shapes of ascaris in soft tissue density or slightly higher density contrasted by low density bile in the dilated bile duct. When the sections for CT scanning are perpendicular to the bile duct, it reveals spots and nodules of soft tissue density shadows in the dilated bile duct, and occasionally ring shaped

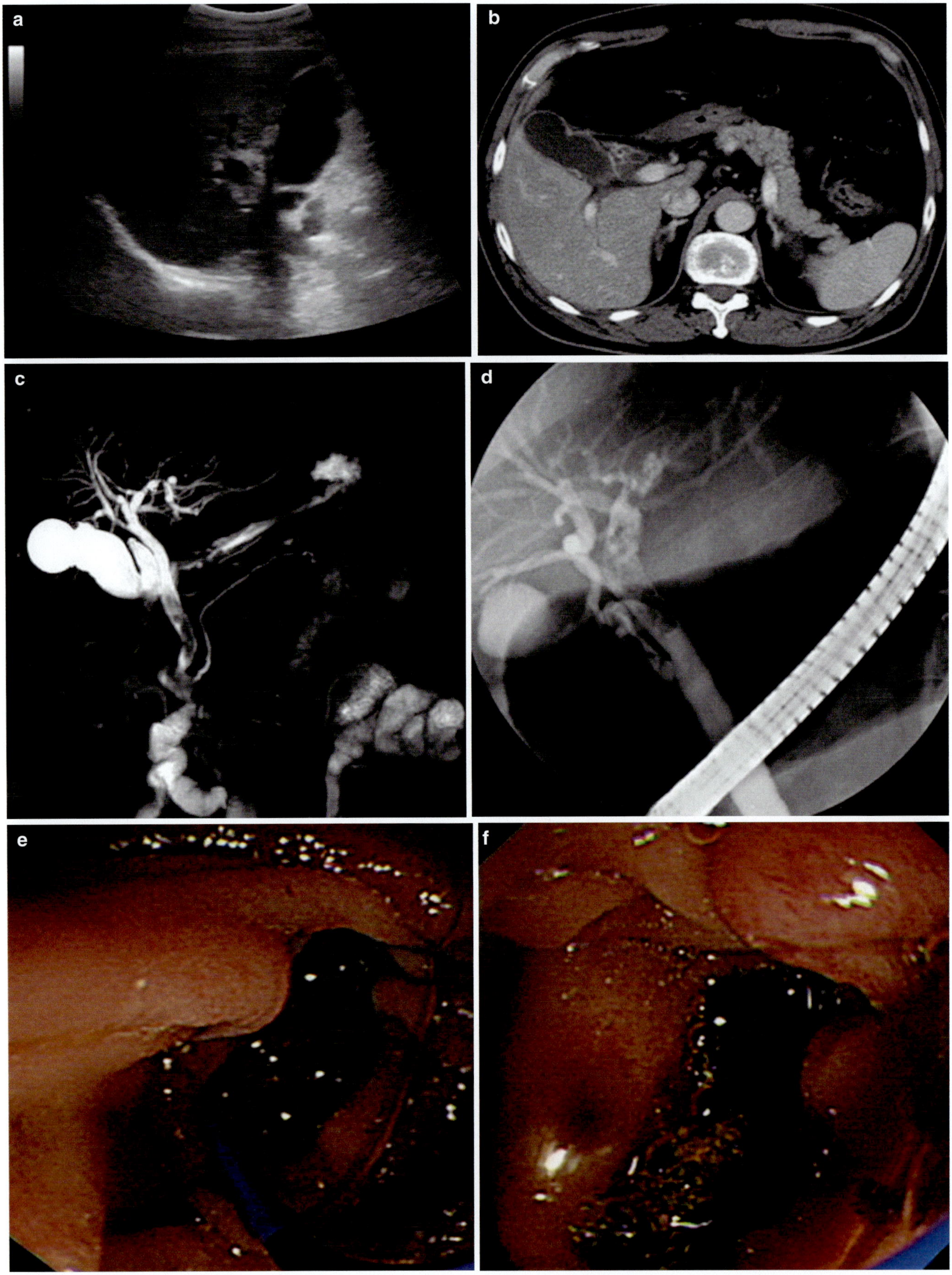
a
b
c
d
e
f

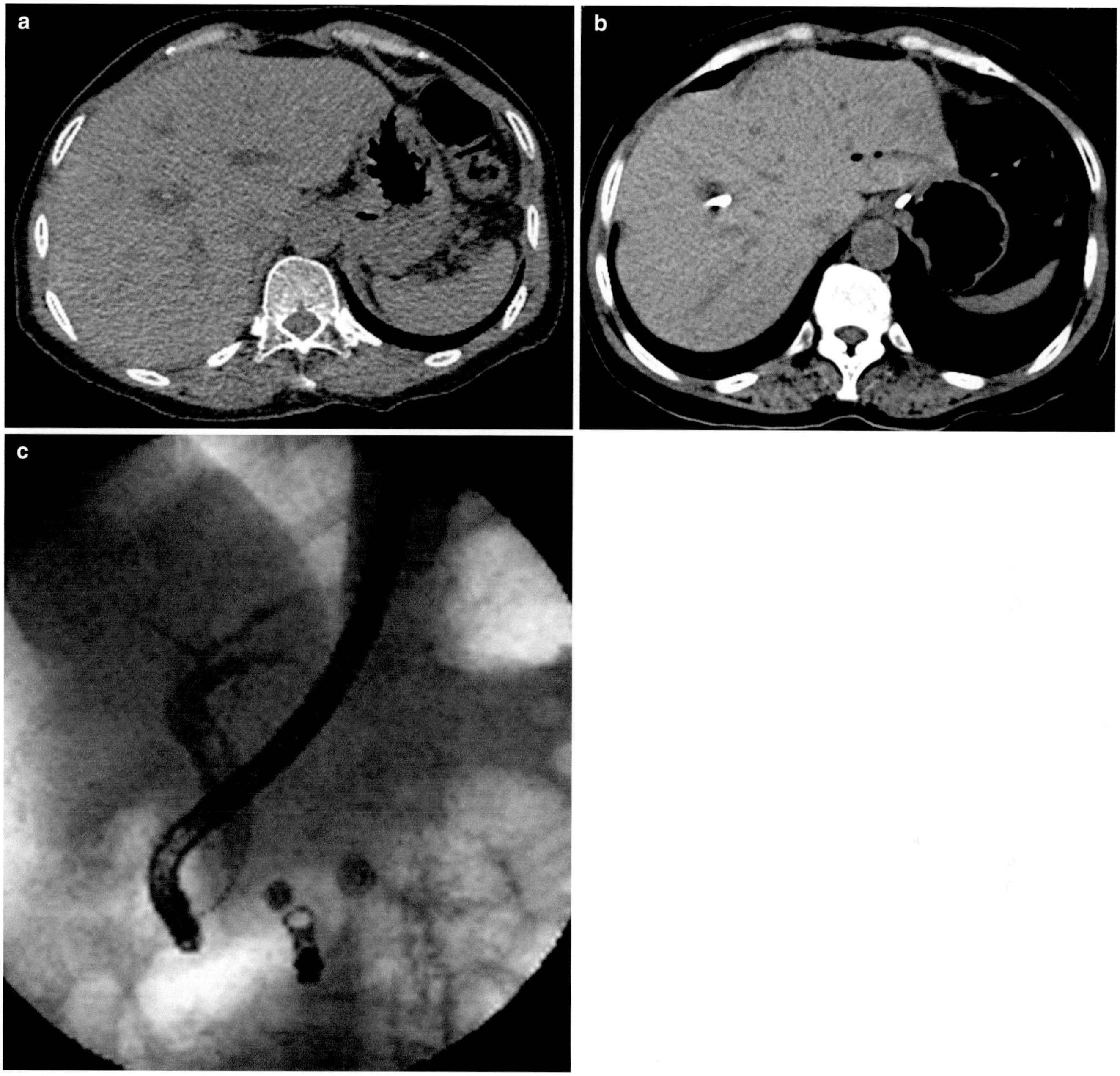

Fig. 8.4 (**a**) CT scanning demonstrated regular long strip of slightly hyper-dense shadow in the right intrahepatic bile duct with well defined boundary, and dilated bile duct. (**b**) After surgical extraction of the ascaris, the original slightly hyper-dense shadow was absent and a hyper-dense draining tubular shadow was shown. (**c**) ERCP demonstrated biliary ascaris and several filling defects were revealed in the common bile duct after infusion of contrast agent

Fig. 8.3 (**a**) Ultrasound revealed dilated extra-hepatic and intra-hepatic bile ducts, cholecystitis, enlarged gallbladder, and possible cholestasis. (**b**) Enhanced contrast CT showed ring-shaped enhancement of proximal common bile duct wall and soft tissue density opacity at the inferior duodenal ampulla. (**c**) MRCP showed plump and enlarged cholecyst, with uneven luminal signal and slightly widened cholecystic duct. The common bile duct as well as the left and right hepatic ducts were revealed to be dilated, with observable non-dilated intrahepatic bile duct branch and pancreatic duct. The common bile duct was shown with a diameter of about 9.5 mm, with no obvious local stenosis but observable strip and flake like filling defect shadows in middle and lower lumen. The duodenal ampulla inferior to the common bile duct was shown with a mass lesion in abnormal signal that was in iso-intense mixed signal. (**d–f**) ERCP showed bile duct, with no observable dilation of the extra-hepatic bile duct but long strip like filling defect shadow in a length of about 3.0 cm at the hepatic portal bile duct and the left hepatic duct. The gallbladder was visualized but the pancreatic duct was non-visualized. A columnar airbag was applied to dilate the papillary opening and long strip like blackish yellow substance was extracted. By endoscopy, the diagnosis was long strip like filling defect at the hepatic portal bile duct and the left hepatic duct. In combination to the clinical manifestations, the blackish yellow substance was most probably dead ascaris pseudocoele of ascaris. In some dilated bile ducts, circular shadow surrounded by low density bile can be demonstrated in concentric circles sign. And when the sections of CT scanning are parallel to the bile duct, it reveals strip of soft tissue shadow in the dilated bile duct. CT scan may also show various lesions complicating biliary ascariasis and biliary ascariasis complicated by liver abscess has a comparatively higher incidence, which can be diagnosed by CT scan.

Biliary ascarisis should be differentiated from the following diseases.

1. Cholangiocarcinoma.
   By plain CT scan, low density mass is manifested. And by contrast scan, no obvious enhancement is shown in the arterial phase, but marginal enhancement at the portal venous and delayed phases, which extends towards the center.
2. Cholangiolithiasis.
   The lesion is revealed in quasi-circular shape and super high density with well defined boundary.

Case Study 5

[**Brief Case History**]

A 64-year-old male received B-mode ultrasound in another hospital due to upper abdominal upset, which indicated biliary obstruction. The possibility of biliary ascariasis was not excluded. The patient experienced no abdominal pain, abdominal distension, cough and expectoration. In our hospital, abdominal contrast MRI and MRCP were ordered, both indicating biliary ascariasis. And his vital signs were stable and the patient was in generally good condition.

[**Radiological Demonstrations**] (See Fig. 8.5)

[**Diagnosis**] Biliary ascariasis.

[**Discussion**]

Biliary ascariasis is an acute abdominal disease, which is caused by mechanical stimulation due to upwards migration of ascaris into the biliary system via duodenal openings. The mechanical stimulation thus triggers contraction or spasm of Oddi sphincter to produce intense pain. Meanwhile, the ascaris induces mechanical obstruction of the bile duct that further results in increased intra-biliary pressure, obstructed biliary evacuation, and cholestasis. Thus, acute suppurative cholangitis, cholecystitis, and choleperitonitis finally occur. In addition, ascaris may die in the biliary system whose body fragments may form calculi. In this case, the patient experienced no obvious abdominal pain and the radiological findings are suspected to be fragments of dead ascaris. MRCP can more favorably demonstrate the condition of biliary ascaris.

Case Study 6

[**Brief Case History**]

A 61-year-old woman reported a medical history of gallstone for more than 10 years. One week ago, she suddenly experienced drilling pain at the part inferior to the xiphoid process with nausea and vomiting. These symptoms relieved by themselves after about 10 min. Routine blood test showed an increase of eosinophilic granulocytes and routine stool examination revealed eggs of ascaris.

[**Radiological Demonstrations**] (See Fig. 8.6)

[**Diagnosis**] Biliary ascariasis.

[**Discussion**]

Biliary ascariasis (BA) is still a common disease in rural areas and its patients are commonly teenagers and young adults with a medical history of excreting or vomitting ascaris. Ascaris commonly parasitizes in human small intestine to cause intestinal lesions. Typical adult ascaris is 15–50 cm in length and 3–6 cm in diameter, which can survive in jejunum for 1–2 years without causing any symptoms. But due to the characteristic nature of crawling and perforating, ascaris induces acute biliary conditions after its access into the biliary system. Ascaris can survive several days or months and may die in human biliary system. Its dead body can be excreted or form the core of gallstone.

B-mode ultrasound is the clinically conventional examination to diagnose biliary ascariasis. And typical ascaris is demonstrated as tubular echo with no accompanying sound shadow, comparatively low central echo, and comparatively double linear parallel high echo. Real-time ultrasonography demonstrates peristalsis of living ascaris but B-mode ultrasound fails to display ascaris in the middle-lower part of the common bile duct due to the interference by intestinal gas. Thus the finding by B-mode ultrasound may be false negative. As for X-ray radiography, T tube radiography and intravenous cholangiography are recommended to display the condition of whole biliary tract for observation of the location and quantity of ascaris. By such examinations, ascaris is demonstrated as slightly curved strip of transparent shadow with smooth boundary. CT scan reveals dilation of bile ducts induced by access of ascaris and the soft tissue density or slightly high density ascaris contrasted by low density bile in the dilated bile duct. When the sections of CT scan are perpendicular to the bile duct, it reveals nodular soft tissue shadow in the dilated bile duct, and occasionally ring shaped pseudocoele of ascaris. And when the sections of CT scan are parallel to the bile duct, it displays strip like soft tissue shadow in the dilated bile duct. With the development of MRI technology, Magnetic Resonance Cholangiopancreatography (MRCP) constitutes a favorable supplementary technology

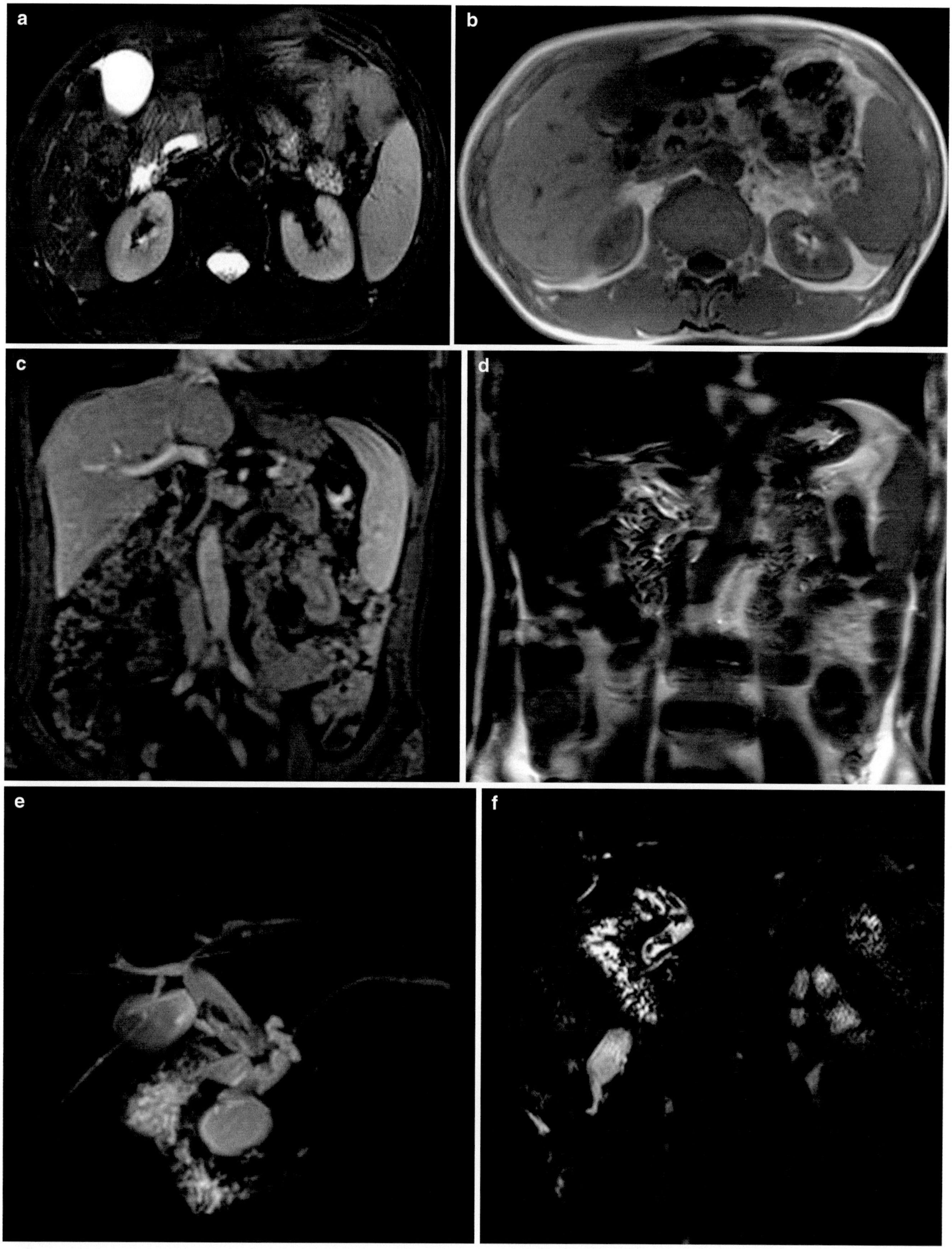

Fig. 8.5 (**a–f**) MRI and MRCP demonstrated widened inferior part of the common bile duct with a diameter of about 12 mm, in which multiple strips of high T_1WI signal and low T_2WI signal are visible. No obvious enhancement was observed by contrast imaging

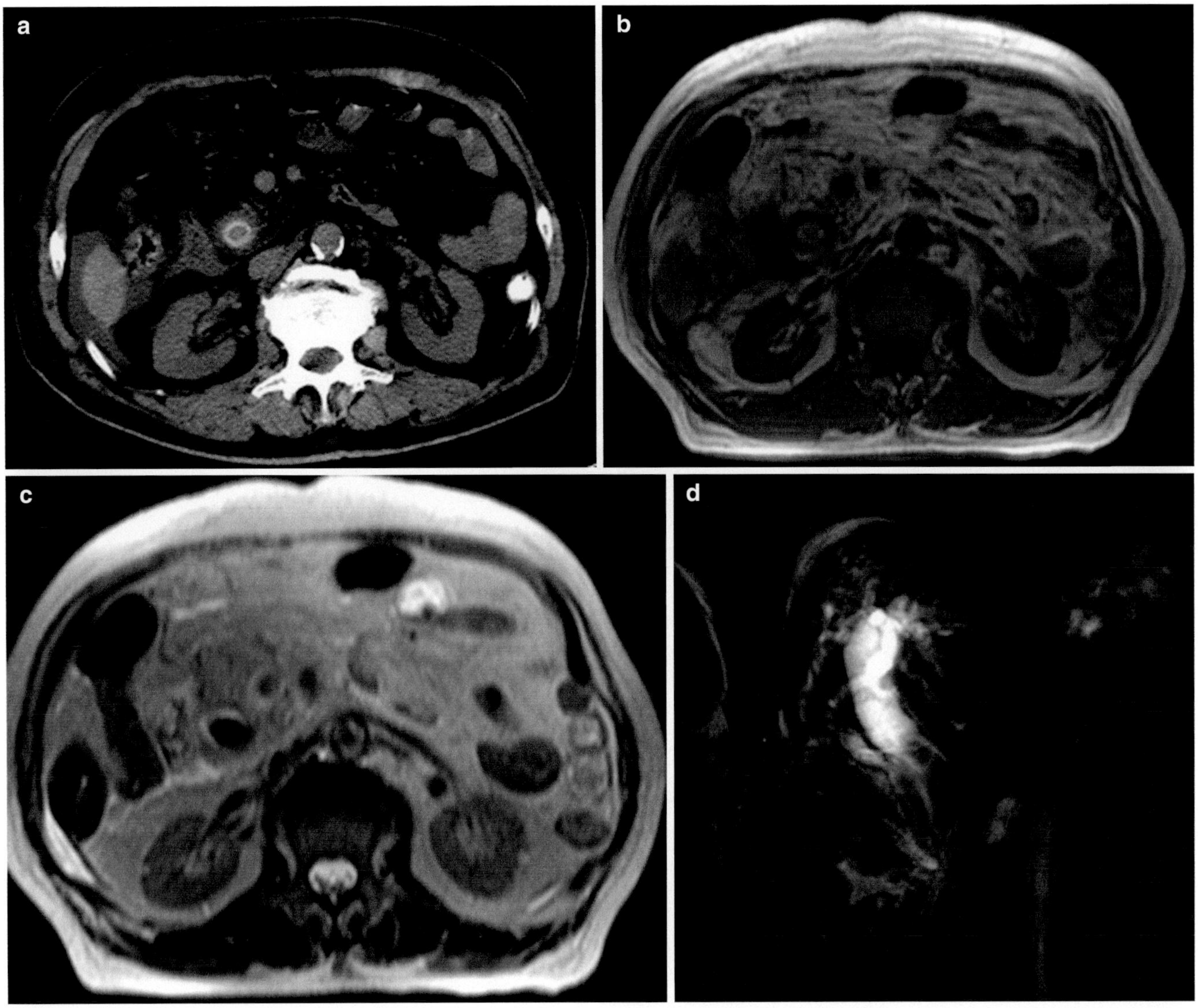

Fig. 8.6 (**a**) Plain CT scan demonstrated thickened wall of the inferior common bile duct, with poorly defined boundary and ring shaped high density shadow inside. (**b**) MR T_1WI showed widened common bile duct with nodular high signal inside. (**c**) T_2WI revealed widened common bile duct with nodular low signal inside. (**d**) MRCP displayed strip shaped filling defect at the inferior common bile duct, which was suspected to be ascaris

for the diagnosis of biliary ascariasis. On the images by thin-section MRCP, intrabiliary ascaris is clearly shown as central strip like high signal, which is the bile and digestive juice ingested into pseudocoele by ascaris; and bilateral low signals, which are the body surface of ascaris. The three parallel lines are known as the three-line sign. Therefore, the presence of three-line sign in high signal bile is the reliable evidence supporting MRCP diagnosis of biliary ascariasis.

Biliary ascariasis should be differentiated from the following diseases.

1. Cholangiolithiasis
   Most biliary calculi are round or oval in shape, demonstrated as low signal by MR T_1WI, MR T_2WI and MRCP. However, biliary ascaris is displayed as strip like slightly high signal by MR T_1WI and strip like slightly low signal by MR T_2WI. The liquid swallowed by living ascaris in human intestinal tract is revealed as central slightly high linear signal and surrounding low signal, namely the three-line sign.
2. Cholangioma
   The neoplasm grows along unilateral bile duct wall or around the wall, which is displayed as nodular shadow or thickened bile duct wall that can be enhanced by contrast scanning.
3. Other parasitic disease
   Other parasites such as clonorchis sinensis and fasciola hepatica can also induce biliary conditions. Such parasites

are smaller than ascaris and can be differentiated from ascaris based on the clinical history and laboratory tests.

4. Implanted draining tube in biliary duct
   By MRI, implanted draining tube in biliary duct may also demonstrated as the three-line sign. But the implanted tube is longer in lengthen and wider in tube diameter and tube wall, with a stiff running course. Based on the reports on clinical history, the differential diagnosis can be defined. Generally, B-mode ultrasound is the diagnostic examination of choice for the cases with clinically suspected biliary ascariasis. If negative by B-mode ultrasound, CT scan and MRI can be ordered for further diagnostic evidence.

Case Study 7

[Brief Case History]

A 9-year-old boy reported abdominal pain after satiation about 1 week ago, which was more obvious around navel and was continually relieved. Four days ago, the patient experienced vomiting with no known causes and the vomits contained no bile. Plain abdominal radiography indicated intestinal obstruction, which improved after treatment. And the patient excreted a large quantity of black dry stool. Recently, he was hospitalized again due to recurrent abdominal pain. Laboratory tests showed MONO 0.11, occult blood immunoassay (±), EB virus (+), cytomegalovirus (+).

[Radiological Demonstrations] (See Fig. 8.7)

[Diagnosis] Intestinal ascariasis.

[Discussion]

The patients with intestinal ascariasis commonly pay their clinic visit due to abdominal pain and accompanying abdominal mass. Young children as well as middle-aged and elderly women are high risk population of intestinal ascariasis. The common symptoms include periumbilical pain, poor appetite, diarrhea, and constipation; and the pediatric patients commonly report grinding teeth during sleep. Once the parasitic environment changes, such as occurrence of high fever, ascaris may curl up into a mass to obstruct the intestinal canal. Thus intense paroxysmal abdominal colic occurs, which is more obvious around navel and is accompanied by nausea, vomiting, and even vomiting out ascaris. By palpation, moving sausage like mass can be found in the abdomen. The occurrence of strangulated intestinal obstruction, intestinal twist, intestinal intussusception or penetration of ascaris through intestinal wall may further cause intestinal perforation and peritonitis, which should be treated appropriately by surgical operation.

By ultrasonography, intestinal ascariasis is displayed as dilated intestinal canal in the middle and lower abdomen, singular or multiple double linear sign in hyperintense echo on longitudinal section with no echo between the two lines, and circular sign on transverse section. In some cases, vague cords like sign in hyperintense echo may be demonstrated, which is arch, curling or twining in shape with visible peristalsis. All these ultrasonographic findings are diagnostic evidence of intestinal ascariasis. In some pediatric patients, ultrasonography demonstrates the lower abdomen with abdominal mesenteric lymphadenectasis, which may be complicated by intestinal intussusception and intestinal obstruction.

Intestinal ascariasis should be differentiated from the following:

1. Intestinal mucosa.
   The intestinal mucosa is commonly perpendicular to the intestinal wall, and its relative location to the intestinal wall is fixed.
2. Intestinal malignancy
   The intestinal wall shows uneven ring shaped thickening or parenchymal mass, with destructed mucosal surface.

8.2 Taeniasis Solium and Cysticercosis

Taeniasis solium is an intestinal taeniasis caused by parasitism of adult taenia solium in human small intestines, which is also known as taeniasis suis and taenia solium. During the whole life cycle of taenia solium, human is not only the definitive host but also the intermediate host. Parasitism of adult taenia solium in human intestines induces intestinal taeniasis solium, while parasitism of its larva in human subcutaneous tissue, muscle, brain, and other tissues and organs induces cysticercosis cellulosae.

8.2.1 Pathogen

8.2.1.1 Adult Taenia Solium

Adult taenia solium is milky white and semi-translucent, with flat back and abdomen, ribbon like appearance and a length of about 2–3 m. Its body is segmental, including scolex, neck, immature proglottid and mature proglottid. Under an electron microscope, the outer wall of adult taenia solium is a syncytial layer with cortex and parenchyma. The parenchyma is basically reticular fine fiber like mesenchyma that functions to support.

8.2.1.2 Cysticercus Cellulosae

Cysticercus cellulosae is also known as cysticercus in pigs, which is milky white, semi-translucent, and vesicular. It is 10 mm × 5 mm in size, with thin vesicular wall and vesicular fluid filling in the vesicle. a There is a white spot in size of a millet in the vesicle, which is the curling scolex. Under an electron microscope, the vesicle can be found to be composed of cortex and parenchyma.

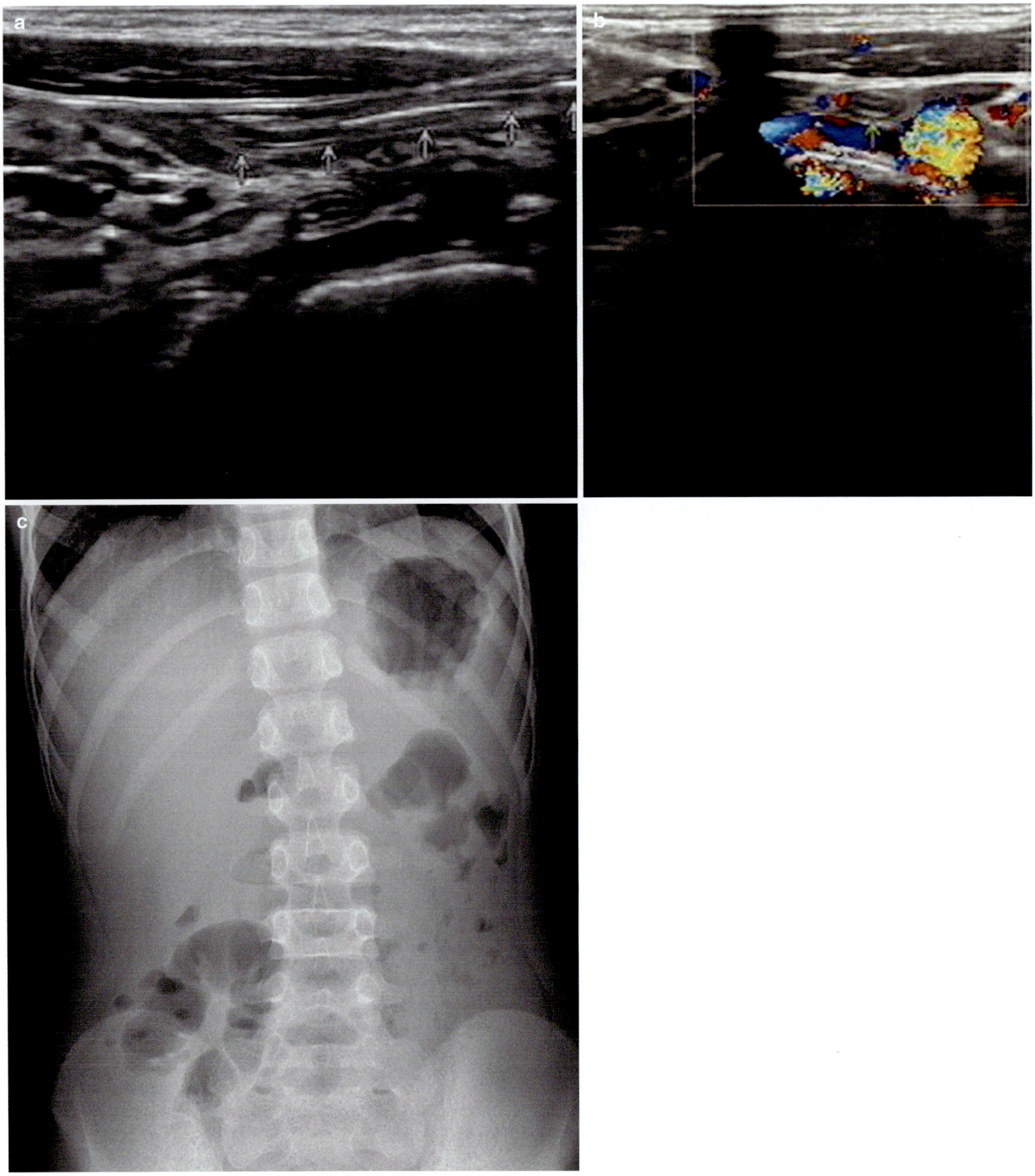

Fig. 8.7 (**a**, **b**) Ultrasonography demonstrated a fine tubular structure in the intestinal tube of the middle abdomen. The tube was displayed in a transverse diameter of 0.2–0.4 cm and a length of 10.0 cm, with thick wall, one slightly sharp end and slightly curved shape but no obvious blood flow signal. (**c**) It was demonstrated with intestinal flatulence in the abdomen and several atypical liquid levels

8.2.2 Epidemiology

8.2.2.1 Source of Infection

Human infected by adult taenia solium is the source of its infection.

8.2.2.2 Route of Transmission

Human is infected by intake of uncooked or unthoroughly cooked pork containing cysticercus cellulosae.

8.2.2.3 Susceptible Population

Populations are generally vulnerable to taenia solium. Human carrying taenia solium can acquire immunity after its infection, which protects the host against re-infection.

8.2.3 Pathogenesis

Adult taenia solium has rostellum and small hook, which can cause damages to the intestinal mucosa. It parasitizes at the duodenum 1/3 to its top, with its scolex in the intestinal villi, rostellum into intestinal wall, acetabulum in adjacent villi, and intestinal villi into the cavity of acetabulum. The host tissue within the cavity of acetabulum is injured, with cell lysis of different degrees as well as mucosal and submucosal necrocytosis. The microtrichia in its outer wall may abrade intestinal mucosa to cause damages to the intestinal mucosa. In addition, adult taenia solium occasionally penetrate the intestinal wall to cause intestinal perforation and peritonitis. In the cases of adults taenia solium intertwining into a mass, intestinal obstruction occurs.

Cysticercus cellulosae shows a relatively wider distribution in human body, possibly with destruction of local tissue, compression to its peripheral organs. Compression to lumen may cause intestinal obstruction. And its toxin can induce significant local tissue responses and systematic esosinophilia of different degrees, which further induces production of corresponding specific antibody.

8.2.4 Clinical Manifestation

8.2.4.1 Taeniasis Solium

The parasitism of adult taenia solium commonly produces no obvious symptom, but occasionally gastrointestinal symptoms such as abdominal upset, dyspepsia, abdominal distension and emaciation. The parasitism of adult taenia solium rarely causes intestinal obstruction that may be complicated by peritonitis.

8.2.4.2 Cysticercosis Cellulosae

Based on parasitic site of cysticercus cellulosae, cysticercosis cellulosae can be categorized into the following three types.

Subcutaneous and Muscular Cysticercosis

Most patients are asymptomatic. In the cases of severe infection, the patients experience muscular soreness and myasthenia, muscular swelling and numbness. Otherwise, muscular pseudohypertrophy may occur.

Cerebral Cysticercosis

Based on the recommendation on clinical typing of cysticercosis from the National Conference of Cysticercosis in China (Harbin, 2001), cerebral cysticercosis is categorized into five types, namely epileptic type, cranial hypertension type, meningeal encephalitis type, psychiatric type and cerebral ventricular type.

Ocular Cysticercosis

Ocular cysticercosis can cause uveitis, retinitis, choroiditis, purulent panophthalmitis and vitreous opacity. Otherwise, ocular cysticercosis may be complicated by cataract and glaucoma, which finally progress into blindness due to atrophy of eyeball.

8.2.5 Diagnostic Examination

8.2.5.1 Taeniasis Solium

Routine blood test sometimes shows slight eosinophilia. The positive rates are low in detecting eggs by stool test and anus swab, and the species of parasite cannot be distinguished by these examinations. The shape and quantity of uterus branch in gravid proglottid excreted along with feces can help its differentiation from taenia saginata. Enzyme-linked immunosorbent adsorption test may show the antigenic component of in the feces. And by polymerase chain reaction (PCR), species specific DNA of eggs or polypide in feces can be amplified to detect adult taenia solium in human body for diagnosis.

8.2.5.2 Cysticercosis Cellulosae

Etiological Examination

Subcutaneous nodule or cerebral tissue from the suspicious lesion should be surgically harvested for pathological examination. The finding of a white semi-translucent soy-sized oval liquid-filled cyst with a white spot in size of a millet can define the diagnosis. Cysticercus in muscle is commonly oval in shape but round in brain parenchyma. Cysticercus is comparatively large at the cranial base or cerebral ventricle, about 5–8 mm or even 4–12 cm in size, and it may appear with branches or like grapes.

Immunoassay

Immunoassays for diagnosis of cysticercosis include antibody detection, antigen detection and immune complex detection. The antibody detection can demonstrate present or past infection of cysticercosis but fails to identify symptomatic patients and the parasite load. Currently, the available antigens for antibody detection are roughly produced, such as cystic fluid antigen, scolex antigen, cystic wall antigen and whole cystic antigen. These antigens often cross react with other parasitic infections and thus show poor specificity. Early-stage immunoassays include complement fixation test, intracutaneous test, and latex agglutination test. Some of these immunoassays are simple in operation and the results can be rapidly obtained, but with poor specificity and high false positive rate. Currently, ELISA and IHA are the most widely applied immunoassays in clinical diagnosis and epidemiological studies. However, all the above mentioned immunoassays show certain false positive or false negative rates, and thus negative finding fails to definitely exclude the possibility of cysticercosis.

Radiological Examination

Ultrasound

B-mode ultrasound can define the quantity and size of subcutaneous and muscular cysticercus nodules.

CT Scan

Cranial CT scan has a positive rate of up to 80–90% in detecting cerebral cysticercosis, which shows cystic low density lesion with a diameter of less than 1 cm. And the diagnosis of cerebral cysticercosis can be mostly defined by CT scan.

MRI

Cranial MRI shows higher detection rates than CT scan in detecting the quantity, range and intrasystic scolex of cysticercus. It can also demonstrate the pathological progression of cysticercosis and provides valuable information to guide treatment.

Ophthalmoscopy and Slit Lamp

By ophthalmoscopy and slit lamp, if wriggling of cysticercus is found under the retina or in the vitreous body, the diagnosis of ocular cysticercosis can be defined.

Pathological Examination

The surgically harvested subcutaneous nodule should be routinely examined by biopsy. For the cases of cysticercosis, the pathological section is characteristic of scolex of cysticercus in the cystic cavity.

Case Study 1
[Brief Case History]

A 30-year-old man complained of headache and vomiting that aggravated for 3 days with accompanying disturbance of consciousness. Cytological examination of the cerebrospinal fluid revealed significant eosinophilia. Complement fixation test of the cyst was positive.

[Radiological Demonstrations] (See Fig. 8.8)

[Diagnosis] Cerebral cysticercosis at the stage of degeneration and edema.

[Discussion]

Cerebral cysticercosis is caused by parasitism of cysticercus in human brain after the intake of taenia solium eggs that develop into larvae in human gastric fluid. Based on its parasitic sites, cerebral cysticercosis can be categorized into brain parenchyma type, cerebral ventricular type, pia mater type and mixed type. Based on its pathological process, it can be staged into active stage, degeneration stage and inactive stage. In the active stage, cysticercus in human brain has cystic cavity, cystic wall and scolex. In the degeneration stage, cysticercus in human brain may compress its peripheral brain tissue to cause encephaledema. When death of cysticercus occurs, its cystic wall ruptures to release liquid protein, which induces inflammatory responses of brain tissue and meninx to form granuloma and small abscess. The polypide is calcified after its death.

By CT scan, cerebral cysticercosis is demonstrated as singular or multiple lesions with a diameter of lees than 2 cm in low density opacity. Enhanced scan reveals nodular or ring shaped enhancement of the lesion. And by enhanced scan, the residual scolex of cysticercus is demonstrated as spot or strip of non-enhanced opacity on the ring wall or within the ring. The lesion is shown with obvious surrounding edema and space occupying effect.

Cerebral cysticercosis should be differentiated from the following diseases.

1. **Neoplasms**
   Firstly, cerebral cysticercosis should be differentiated from astrocytoma. For astrocytoma, the shape of enhancement is various and complex. The central low density area is the necrotic cyst, and enhancement may be demonstrated as cystic with mural nodule, pseudo-cystic with mural nodule, parenchymal, ring shaped, or mixed. The circumferential wall is often irregular and heterogeneous in thickness, possibly with significantly enhanced mural nodule and obvious space occupying effect.

   Secondly, cerebral cysticercosis should be differentiated from cerebral metastasis. Cerebral metastasis is common in middle-aged and elderly populations with a

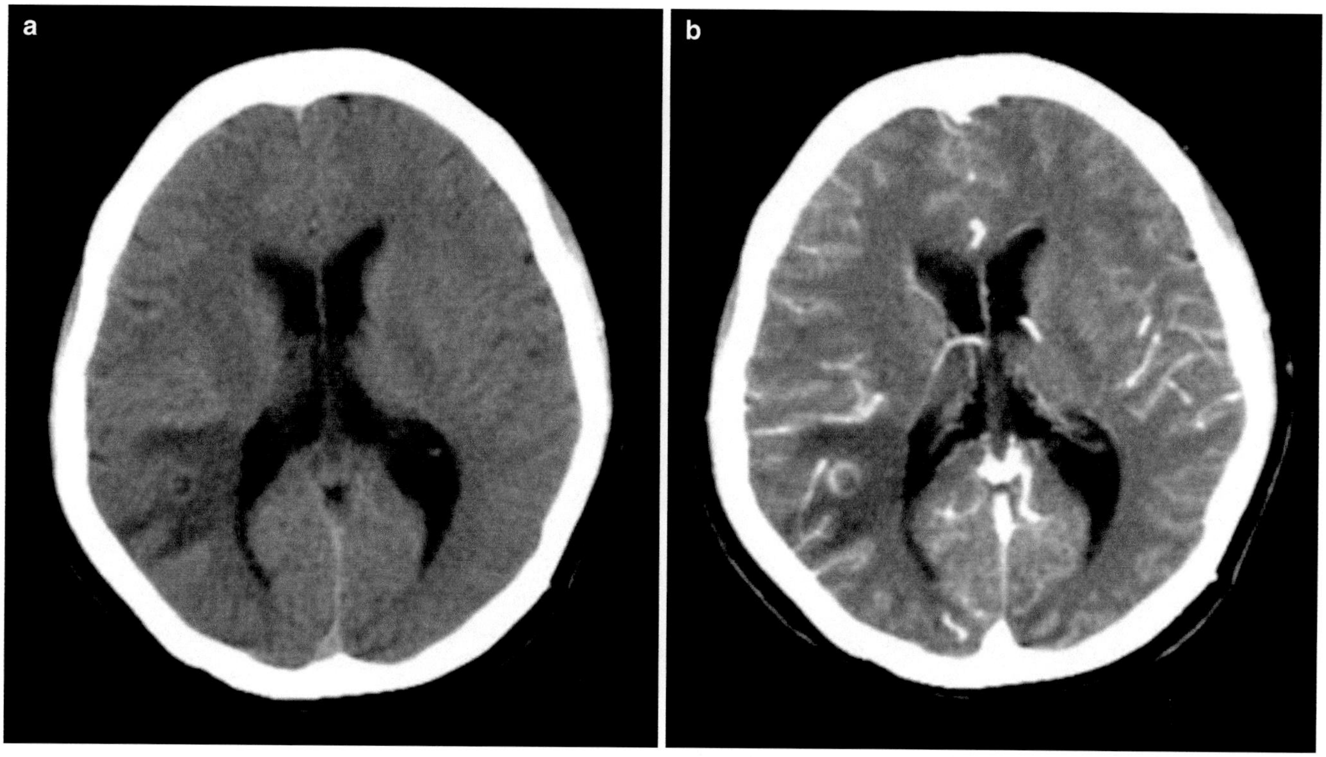

Fig. 8.8 (**a**) Plain CT scanning demonstrated a ring shaped nodule in mixed density in the right temporal lobe, with a spot of iso-dense opacity inside and surrounding flake of low density edema. (**b**) Contract CT scanning demonstrated obvious ring shaped enhancement of the nodule, no obvious enhancement of the observable spot of nodular opacity, and no enhancement of the surrounding edema

medical history of neoplasm. Enhanced scan demonstrates nodular or ring shaped enhancement. When the metastatic neoplasm is large, the lesion is commonly demonstrated as cystic and ring shaped. Finger like edema is common around the lesion, demonstrated as small lesion surrounded by large area of edema.

For some cases, cerebral cysticercosis should also be differentiated from other neoplasms, such as hemangioblastoma, ganglioglioma, anaplastic oligodendroglioma, ependymoma, and germinoma.

2. **Infective diseases**

Cerebral cysticercosis should be differentiated from brain tuberculoma, which is often caused by hematogenous dissemination of tubercle bacillus in other parts of human body. Brain tuberculoma commonly remains at the superficial brain part such as the interface of brain cortex and medulla, and more commonly occurs at the base of brain. By contrast scan, the lesion shows ring shaped, linear or small nodular enhancement which is characterized by heterogeneous thickness of the circumferential wall and no scolex of cysticercus.

Cerebral cysticercosis should also be differentiated from brain abscess, which is more common in children, young and middle aged adults. Brain abscess commonly occurs in cerebral cortex or subcortex, and the lesion is typically demonstrated as obvious ring shaped enhancement with homeogeneous thickness of the circumferential wall, central homogeneous low density area, and no scolex of cysticercus. MR diffusion-weighted imaging demonstrates abscess in obvious high signal, but ADC in low signal. The patients experience systemic toxic symptoms and signs of acute infection. Both routine blood test and cerebrospinal fluid examination reveal significant increase of WBC count, especially neutrophils.

Case Study 2

[**Brief Case History**]

A 43-year-old woman complained of headache, vomiting, epilepsy with decreased vision and disturbance of consciousness. Cytological examination of the cerebrospinal fluid revealed significant eosinophilia.

[**Radiological Demonstrations**] (See Fig. 8.9)

[**Diagnosis**] Cerebral cysticercosis (mixed type).

[**Discussion**]

Mixed type of cerebral cysticercosis has extremely complex clinical symptoms of at least two types of cerebral cysticercosis, including headache, nausea, vomiting, epilepsy, convulsion, decreased vision and disturbance of consciousness.

Mixed type of cerebral cysticercosis shows CT demonstrations of at least two types of cerebral cysticercosis,

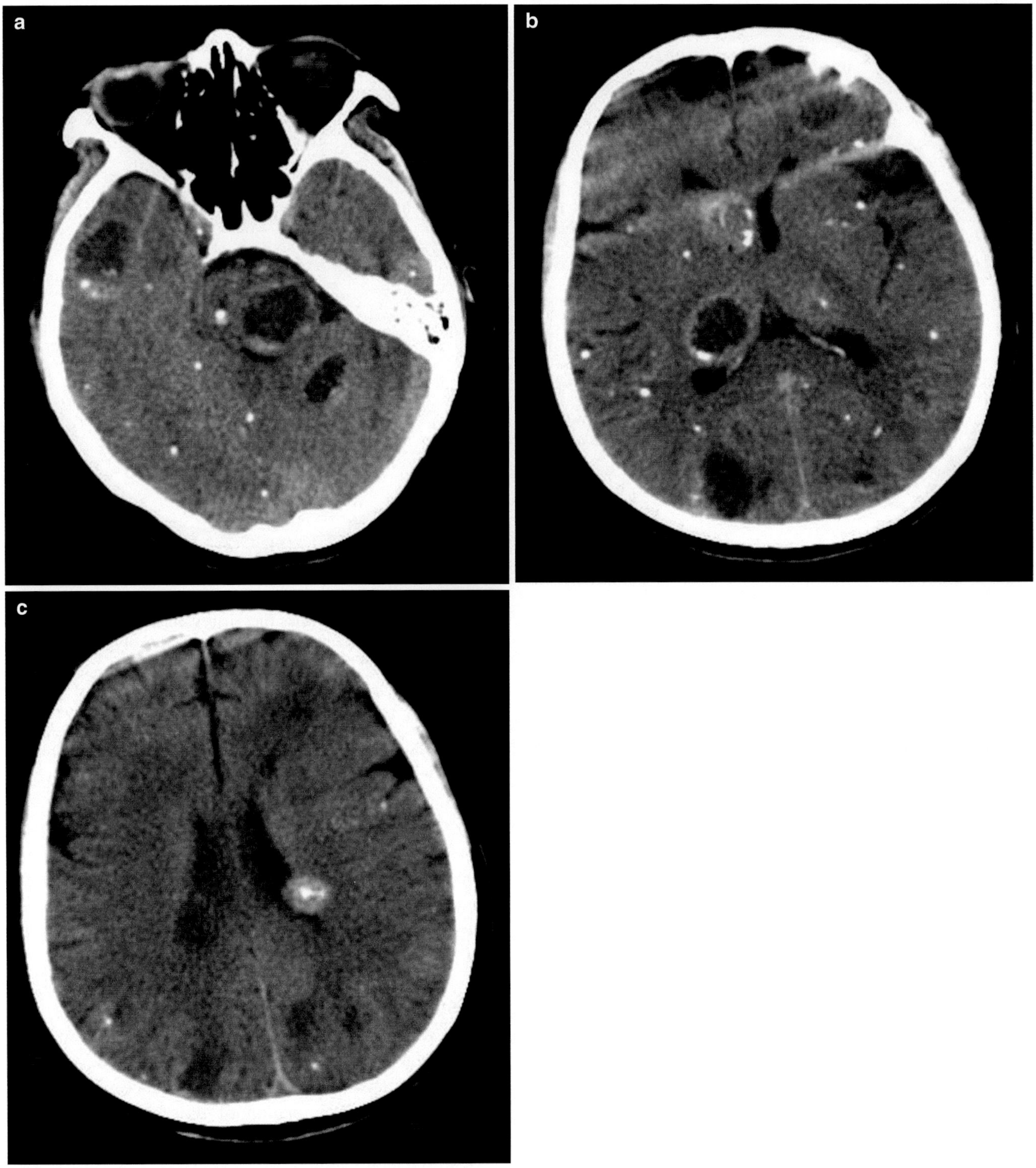

Fig. 8.9 (a–c) Plain CT scan demonstrated multiple nodular and rice grains liked high density opacities and multiple cystoid low density opacities in brain parenchyma, ventricles, and meninx, with spot of calcification inside. The lesions were shown with scattering and irregular distribution and some surrounded by flake of low density edema

including round low density lesions in different sizes with mixed existence. Some lesions may show scolex or complete calcification, and the lesions are the most commonly located in cerebral cortex or deep cortex and basal ganglia region. Generally speaking, the mixed type of cerebral cysticercosis has complex clinical manifestations and its diagnosis is challenging. CT scan is the diagnostic examination of choice that facilitates the early detection, early diagnosis and early intervention.

Cerebral cysticercosis should be differentiated from the following diseases.

1. Calcified abscess of cerebral echinococcosis
   It is a parasitic zoonosis in animal husbandry area. The calcified abscess is commonly demonstrated with arch shaped, shell like, or ring shaped wall. The two diseases can be differentiated based on the features of calcification.
2. Brain tuberculoma
   The patients with brain tuberculoma commonly have a medical history of tuberculosis at other body parts and clinical manifestations of tuberculosis. The lesion of brain tuberculoma commonly remains at the superficial brain parts such as the interface of brain cortex and medulla. Contrast scan demonstrates nodular enhancement and surrounding ring shaped enhancement, together with central spot of calcification, to form a typical target sign of brain tuberculoma. The surrounding edema has a small range. The follow-up examinations of cerebrospinal fluid and treatment is helpful for its differential diagnosis.
3. Cerebral paragonimiasis westermani
   The disease prevails in northeast China, north China, east China and Sichuan province of China, which is manifested as spot, nodular or ring shaped calcification. The calcification is surrounded by encephalomalacia and focal encephalatrophy. However, cerebral cysticercosis is often manifested as scattering rice grains like calcification.

Case Study 3
[Brief Case History]

A 23-year-old man with a history of cerebral cysticercosis for 5 years received following-up CT reexamination.

[Radiological Demonstrations] (See Fig. 8.10)

[Diagnosis] Cerebral cysticercosis in its inactive stage.

[Discussion]

During the inactive stage of cerebral cysticercosis, cysticercus is dead, with no inflammatory responses and edema but only spots of calcification in brain parenchyma. The lesions are commonly 0.2–1.0 cm in diameter and some of them may integrate into a giant plaque of calcification. The inactive stage may overlap with the degeneration stage.

CT scan demonstrates cerebral cysticercosis as multiple scattering masses and rice grains liked high density opacity with well-defined smooth boundary. The lesions scatter irregularly, with no surrounding edema. If calcification occurs at the subarachnoid space, only communicating hydrocephalus is displayed.

Cerebral cysticercosis in its inactive stage should be differentiated form the following diseases.

1. Physiologic calcification
   Physiologic calcification commonly occurs in basal ganglia region, especially in pallidum, cerebral falx, cerebellar dentate nuclei and paraplexus that distributes symmetrically. Such characteristic calcification is inconsistent with that of cerebral cysticercosis in its inactive stage.
2. Nodular sclerosis
   The calcifications are typically located at the lateral wall of paracele and around interventricular foramen, which protrude towards the paracele and are 1–15 mm in diameter.
3. Hypoparathyroidism
   The calcifications are mostly revealed at the bilateral basal nuclei or cerebellar dentate nuclei, commonly irregular in shape.
4. Brain tuberculoma
   The patients with brain tuberculoma often report a history of tuberculosis at other body parts and show clinical manifestations of tuberculosis. In the advanced stage of brain tuberculoma, the whole neoplasm may be calcified that is nodular in appearance. In some other cases, partial neoplastic wall is calcified to show continual ring shaped or shell liked calcification. Such characteristic findings help in differential diagnosis from cerebral cysticercosis in its inactive stage.

Case Study 4
[Brief Case History]

A 48-year-old man received B-mode ultrasound examination 2 months ago, which revealed a well-defined relatively low echo area in the right liver lobe in size of 38 mm × 36 mm. And no echo area was shown at its center in size of 15 mm × 14 mm. The patient experienced no symptoms of abdominal distension and pain, nausea, vomiting, dizziness, headache, cough and expectoration. He then paid clinic visit in a local hospital for further diagnosis. By examinations, the HBV was found negative (–), and no abnormalities were shown by lower abdominal CT scan and colonoscopy. He was then referred to the department of gastroenterology in our hospital on Jun. 4th, 2011 and liver function examination showed negative. On Jun. 13th, 2011, abdominal MRI revealed abnormal signal lesion at the interface between VI and V segments of the right liver, which was suspected to be parasitic infection. Subsequent examination of parasitic serum antibody revealed positive to cysticercus and the diagnosis was defined to be hepatic cysticercosis. For further treatment, the patient was hospitalized due to hepatic cysticercosis for treatment. After admission, the patient received other related examinations, including cranial CT scan (−) and ocular ultrasonography (−).

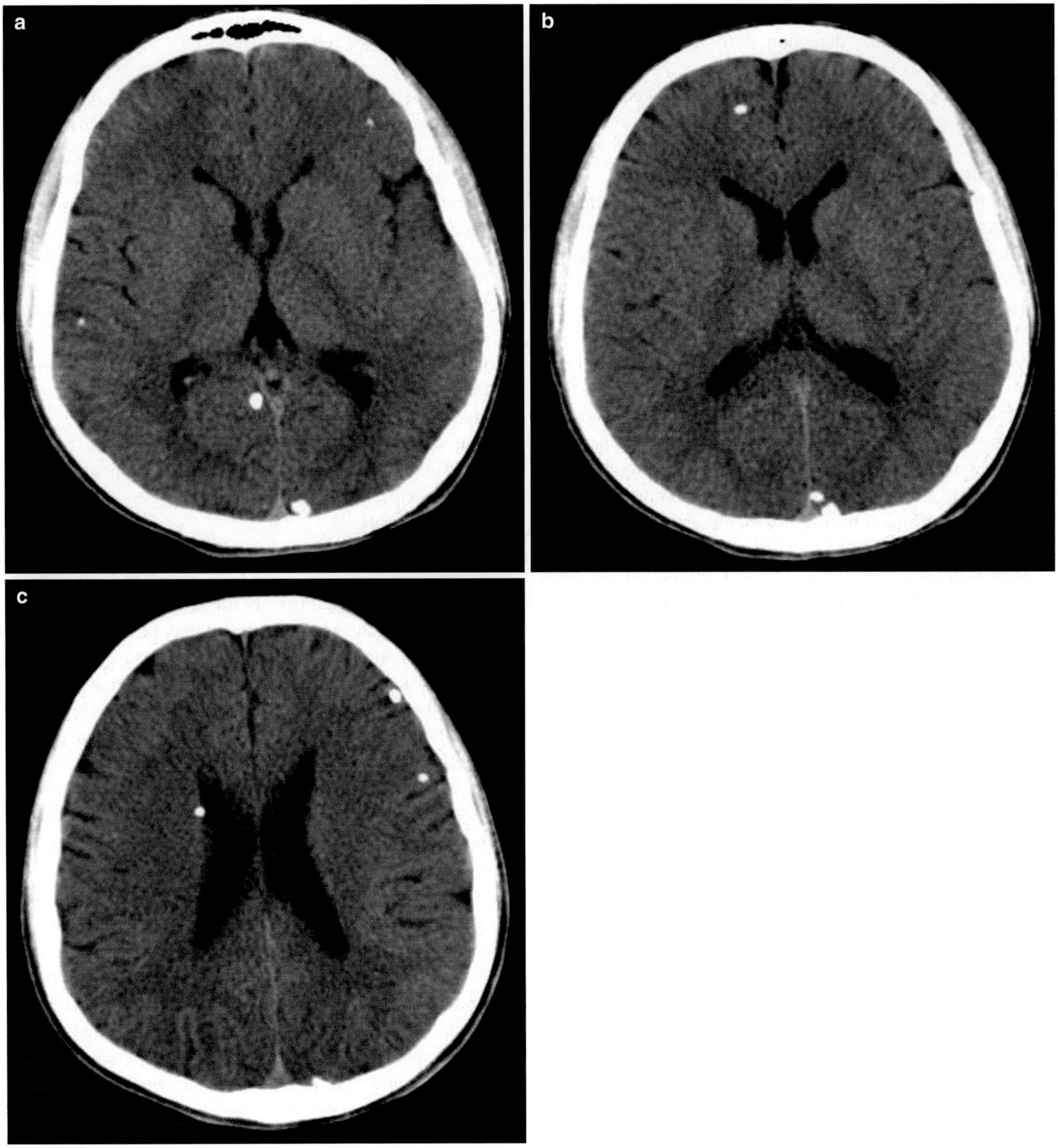

Fig. 8.10 (**a**–**c**) Plain CT scanning demonstrated multiple scattering masses and rice grains like high density opacity, with well-defined smooth boundary and no surrounding edema

Low fat diet was recommended and praziquantel was administrated as the anti-cysticercal therapy (oral intake of 3 tablets for 3 times a day and 5 days as one therapeutic course). Meanwhile, the adverser effects of praziquantel were monitored. Currently, he was discharged from the hospital due to stable vital signs and a generally good condition.

[**Radiological Demonstrations**] (See Fig. 8.11)

[**Diagnosis**] Hepatic cysticercosis.

[**Discussion**]

Cysticercosis is a disease caused by parasitism of larval cysticercus in various human tissues. Cysticercus can invade any human organ to cause corresponding symptoms,

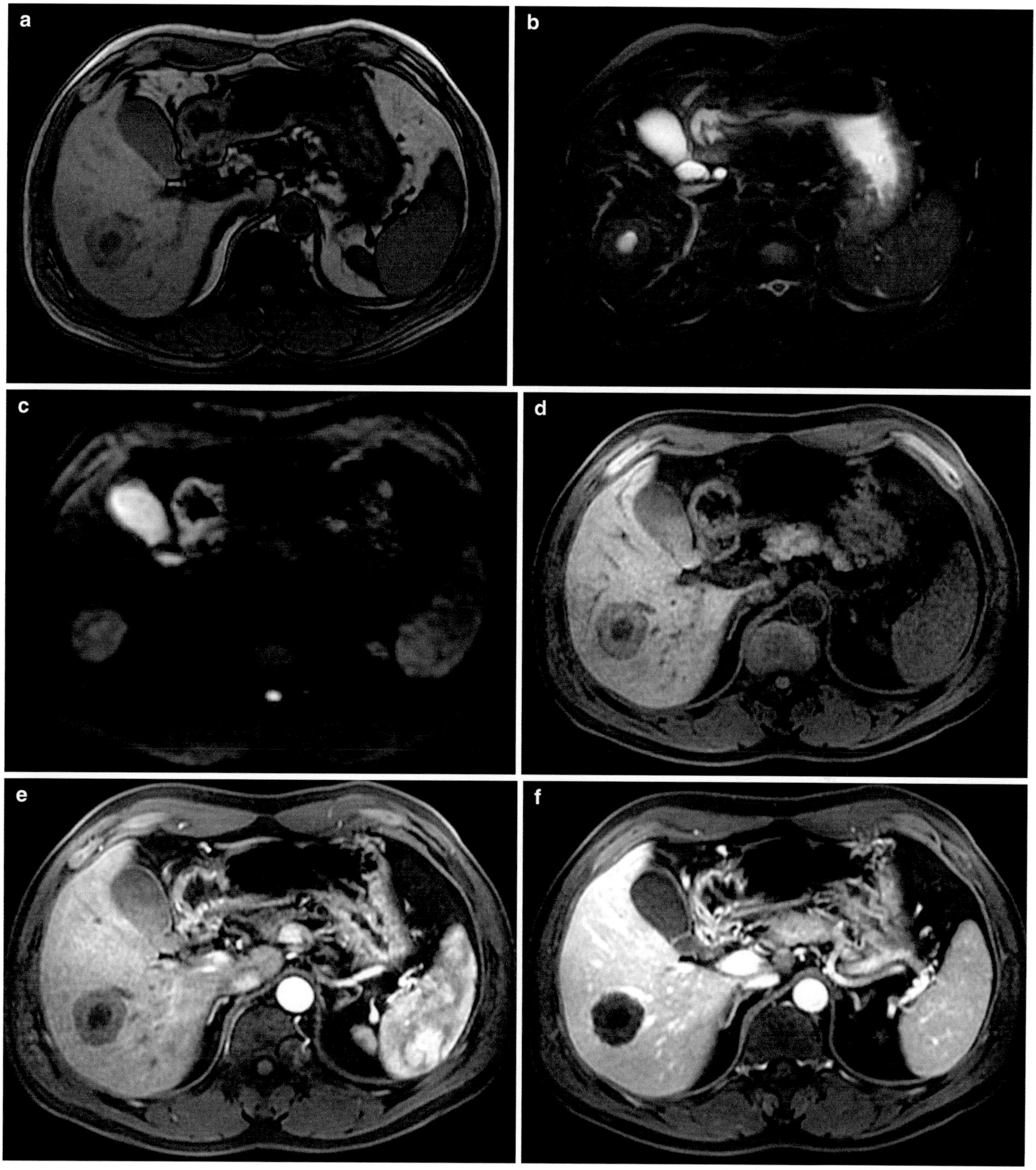

Fig. 8.11 (a–f) MRI demonstrated round lesion with abnormal signal at the interface between VI and V segments of the right liver. By T1WI, the lesion was surrounded by slightly high signal, with central homogeneous hypo-intense. T_2WI displayed the lesion in high signal, with the higher signal at the center. And DWI showed increased signal By contrast imaging, the lesion was shown with no obvious enhancement, which was persistently in relatively hypointense signal

commonly brain, subcutaneous tissue, muscle and eyes but rarely liver. After intake of taenia solium eggs or gravid proglottid, the embryonic membrane of eggs ruptures under the effect of digestive juice in small intestine to release oncosphere. The released oncosphere then penetrates into the intestinal wall and may spread into any organ or tissue along with blood or lymph flow. When oncosphere stays in the hepatic tissue to develop into cysticercus, cysticercus is enclosed by a cyst produced by host tissue. The pathological change of its peripheral tissues is basically granulomatous inflammation in local tissue with accompanying infiltration of eosinophils and formation of fibrous cyst. The MRI findings in this case are explicable by such a theory. However, in literature reports, hepatic cysticercosis commonly shows small nodule. When developed into cysticercus, scolex is calcified and the necrotic tissue around the cysticercus is further surrounded by granulation tissue. Subsequently, the fibrous tissue is subject to hyperplasia, hyaline change and infiltration of eosinophils to form nodule. At such a stage, imaging findings are atypical and the diagnosis can be preliminary made by immunological serum examination for specific antibody. But frequent following up examinations are still necessary because the diagnosis is not pathologically confirmed.

Case Study 5

[Brief Case History]

A 46-year-old woman complained of right lower back pain and right upper abdominal pain with no known causes in Jun. 2007. She reported restrained movement, vomitting, no nausea, and no fever that were not relieved after rest. In a local hospital, her condition was diagnosed as cholecystitis and was given antibiotics and painkiller. After that, her condition failed to be improved, but showed cough with a little expectoration that was occasionally bloody. Her body temperature was up to 39 °C. However, she experienced no obvious headache, nausea and vomiting. Chest CT scan in the local hospital on Jun. 27 showed left pneumothorax, space-occupying lesion at the right upper lung, and suspected nodule in the right lower lung. Lumbar CT scan revealed bone destruction and absorption. On Jul. 4, CT guided percutaneous lung puncture for biopsy demonstrated no malignancy. Cysticercus antibody in lung was positive and the patient was hospitalized twice for anti-cysticercus treatment. On Sep. 25, the patient reported fever with no known causes and no chills and was hospitalized due to a diagnosis of cysticercosis cellulosae. She also reported a history of eating measly pork.

[Radiological Demonstrations] (See Fig. 8.12)

[Diagnosis] Lung and bone cysticercosis.

[Discussion]

Cysticercosis is a disease caused by parasitism of larval cysticercus in various human tissues. Cysticercus can invade any organ within human body to cause corresponding symptoms, commonly the brain, subcutaneous tissue, muscle and eyes, but rarely the lungs and skeleton. Cysticercosis of lumbar L4, right 7th rib and sternocostal joint is even rarer. The patient of this case reported a history of eating mealsy pork and experienced cough and fever before onset of the disease. Her upper right lung lobe was radiologically revealed with a cavity and accompanying spots of cords like exudative lesion. CT-guided percutaneous lung puncture for biopsy showed no malignancy. And cysticercus antibody in lung tissue showed positive. Lumbar CT scan demonstrated worm bitten like lytic bone destruction at the L4 centrum and its appendix with slight marginal osteosclerosis. Within the bone destruction, spots of irregular high density opacity were observable. The preoperative diagnosis of cysticercosis at one site could hardly explain the findings by diagnostic examinations. And it was necessary to be differentiated from tuberculous cold abscess in lumbar vertebra and psoas major, but the lesion at the left upper lung lobe was not tuberculosis. Therefore, in combination to the clinical manifestations, the condition was suspected to be a rare infective disease. Finally, CT guided percutaneous lung puncture for biopsy defined the diagnosis of rarely occurring bone cysticercosis.

Case Study 6

[Brief Case History]

An 8-year-old boy reported a lump at the left lower abdominal wall for about more than 1 year, which was jujube like in size and hardened in texture with no local tenderness. Slight pain was felt when he changed his posture from lying to sitting. The lump was located in deep subcutaneous tissue, seemed between muscles, and slightly protruded above the skin surface in size of about 3.0 cm × 1.0 cm × 1.0 cm. It showed no adherence to the skin that was movable, with smooth surface and quite well-defined boundary. The skin surface showed no redness, no swelling, and no rupture, with normal skin temperature but no local tenderness. Urinary sediment showed mucus up to 510.18/μl as well as increased lymphocytes and neutrophils.

[Radiological Demonstrations] (See Fig. 8.13)

[Diagnosis] Cysticercosis cellulosae of abdominal wall.

[Discussion]

Based on the parasitic site, cysticercosis can be further classified into cerebral cysticercosis, dermo-muscular cysticercosis, and ocular cysticercosis. Simplex dermo-muscular cysticercosis is relatively rare, which account for 10–20 % of cysticercositic cases. On the surface of the host body, protrusion is observable whose quantity ranges from 1–2 to hundreds and thousands. The patient commonly experience no subjective symptom. CT scan demonstrates dermo-muscular cysticercosis as soft tissue density opacity with poorly-defined boundary. High-frequency ultrasonography shows characteristic signs of dermo-muscular cysticercosis, including nodules as round or oval shaped liquid dark area that is well defined with inact and smooth cystic wall; central or eccentrical hyper-intense echo mass in the cyst which is the cysticercal nodule. Other subcutaneous nodules such as

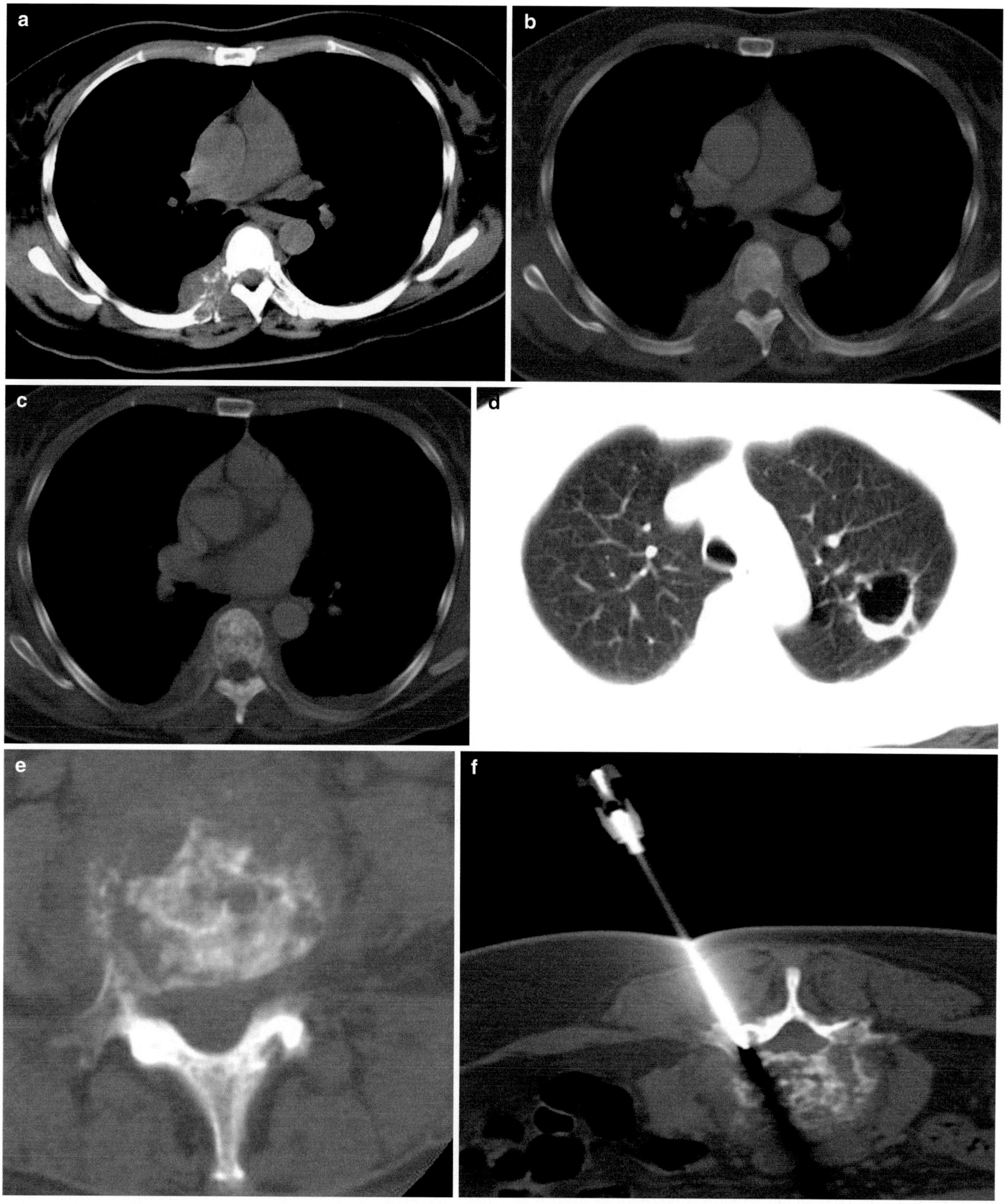

Fig. 8.12 (**a**–**d**) Chest CT revealed bone destruction of the T_7 centrum, right 7th rib and the right sternocostal joint with surrounding swelling of the soft tissue. A cavity was observable at the left upper lung lobe with liquid level inside and fibrous cords like opacity around. (**e**) Lumbar CT scan demonstrated worm-bitten like lytic bone destruction at the L_4 centrum and its appendix with slight marginal osteosclerosis. In the bone destruction, spots of irregular high density opacity were observed. The surrounding psoas major was shown with swelling and the dural sac was compressed. (**f**) By CT guided percutaneous lung puncture for biopsy demonstrated lumbar destructive lesion

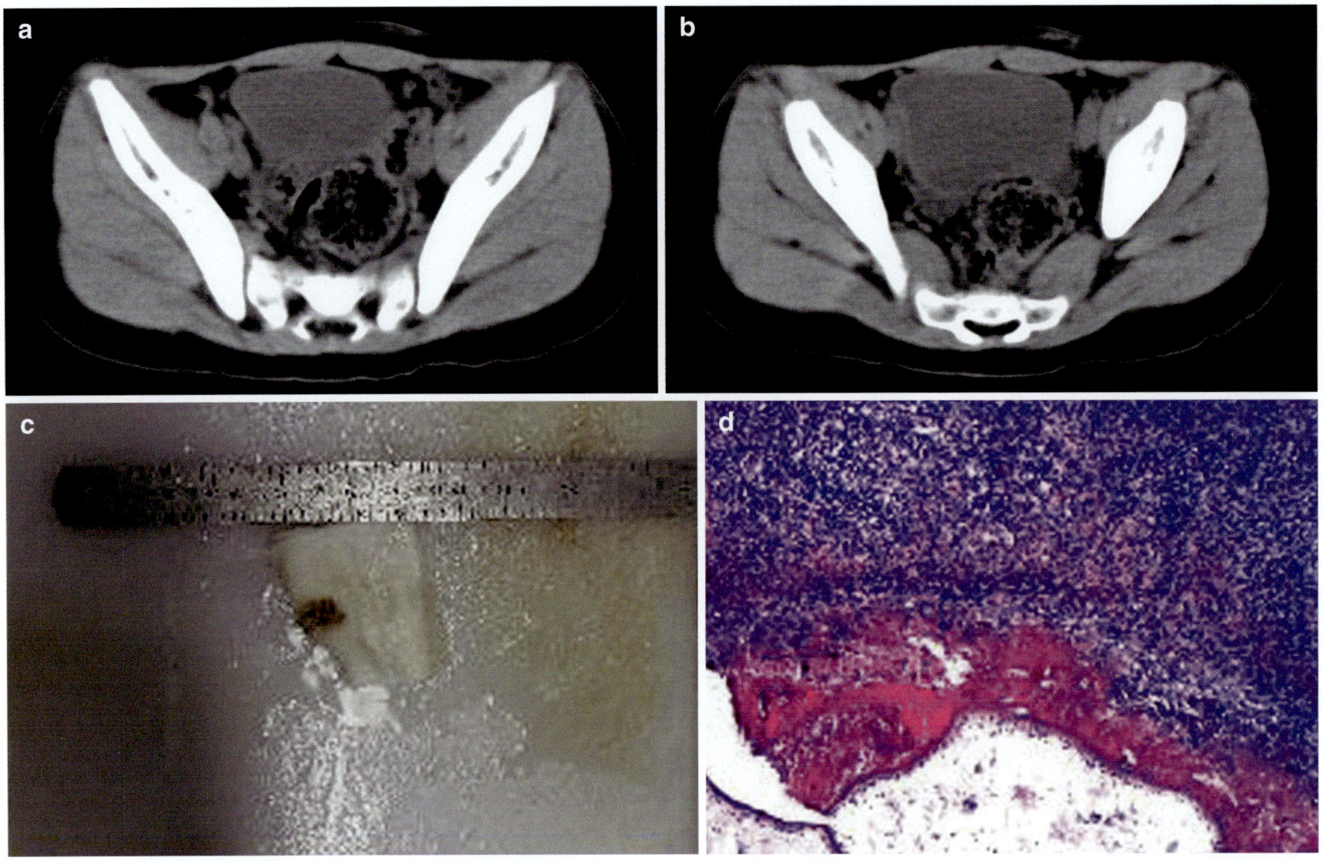

Fig. 8.13 (**a**, **b**) CT scan revealed the left lower abdominal wall with a small quantity of subcutaneous small flakes of soft tissue density opacity, with poorly defined boundary. (**c**) By naked eyes observation, grayish yellow soft tissue was observed, with the section in grayish red and grayish white. (**d**) Microscopy demonstrated cysticercosis cellulosae at the left lower abdominal wall complicated by chronic inflammation of surrounding fibrous and adipose tissue

lipomyoma and fibroma show no such ultrasonographic signs. During ultrasonographic examination of the subcutaneous nodule, the echo within the nodule should be carefully observed to avoid misdiagnosis.

Dermo-muscular cysticercosis should be differentiated from the following diseases.

1. Fibroma
   By plain CT scan, fibroma is demonstrated as well-defined nodule in regular shape with homogeneous relatively low density or almost same density as muscle. The nodule may be enveloped. And contrast scanning reveals moderate enhancement of the lesion.
2. Lipomyoma
   CT scan demonstrates lipomyoma as singular or multiple extremely low density area with intact envelope and inner septum. Contrast scan demonstrates no obvious enhancement of the lesion.
3. Hemangioma
   The limbs are the most vulnerable to hemangioma and spongy hemangioma is the most common. Plain CT scan demonstrates hemangioma as soft tissue density opacity with low density lipid, high density phlebolith and calcification. Contrast scan shows obvious enhancement of the lesion.

8.3 Sparganosis Mansoni

Sparganosis mansoni is a disease caused by parasitism of larval spirometra mansoni in human body and its clinical cases have been reported in 39 countries worldwide, mostly in Korea, Japan, Tailand and China. In China, the cases of sparganosis mansoni are mainly found in the South area, such as the provinces of Guangdong, Guangxi, Fujian and Hunan.

Sparganum mansoni can parasitize in a variety of species of hosts and its infection rate is relatively high in tadpoles, frogs, snakes and loaches. Human may be its second intermediate host, paratenic host or even definitive host. And human infection is mainly due to penetration of either sparganum or procercoid into skin or mucosa. Otherwise, human can be infected via intake of sparganum or procercoid. Specifically, its infection occurs due to: (1) placing raw frog meat on the skin lesion, diseased eyes or dental caries for medicinal reasons; (2) intake of uncooked or unthoroughly cooked frogs, snakes, and pork containing sparganum mansoni; (3) drinking unboiled water containing infected cyclops, the first intermediate host, or swallowing infected cyclops while swimming.

After invading into human body, most sparganum cannot develop into adults but are capable of migrating extensively in human body. Different types of sparganosis mansoni may occur due to different tissues and organs that are involved.

1. Subcutaneous sparganosis
   The most common symptom of sparganosis mansoni is subcutaneous lump, especially the superficial region of human trunk such as lumbodorsal part, neck, thoracic wall, abdominal wall, breasts, inguina, external genital organs (penis, scrotum, testicles, labia majora), perianus, and subcutaneous of limbs. The subcutaneous lump may also occur at eyes and oral maxillofacial region.
2. Sparganosis of the central nervous system
   Sparganosis may also occur at the brain, spinal cord and intravertebral duct. The clinical manifestations highly resemble to brain neoplasm, and thus sparganosis of the central nervous system is likely to be misdiagnosed.
3. Sparganosis of organs
   Clinical manifestations vary due to the involved organ after migration of sparganum. Sparganosis may involve abdominal organs, mesenterium, appendix and perirenal tissues. It may also invade the peritoneum due to migration along the gastrointestinal tract to cause inflammatory responses. Sparganum may even invade the thoracic cavity after its penetration through the diaphragm to involve the plura and cause pleural effusion. And its downward migration may involve the urethra and bladder.

Case Study 1

[Brief Medical History]

A 58-year-old man complained of repeated of epilepsia for 3 months that was more frequent in the past week. He reported a history of eating uncooked snakes for many years. A thorough examination of blood parasites antibody displayed sparganum mansoni antibody positive (+). Pathological report from the surgical resection for biopsy revealed granuloma of sparganosis.

[Radiological Demonstrations] (See Fig. 8.14)

[Diagnosis] Cerebral sparganosis (the left temporal occipital lobe)

[Discussion]

The diagnosis of cerebral sparganosis is based on the following key points: (1) young or middle aged adults; (2) living in a region with densely covered wateries and having a history of drinking unboiled water and eating uncooked aquatic food, frogs or snakes; (3) a prolonged illness course with varying symptoms that sometimes improved and sometimes aggravated; (4) migration of lesions or multiple lesions; (5) characteristic findings by CT or MRI; (6) a parasitic history by sparganum in other body parts, such as eyes or subcutaneous tissue.

Cerebral sparganosis should be differentiated from following diseases.

1. Bacterial brain abscess
   Cerebral sparganosis can hardly be differentiated from brain abscess when the case of sparganum mansoni shows singular cystic lesion. When multiple cystic lesions are found in the case of cerebral abscess, the radiological finding is commonly 1–3 lesions that are close with each other rather than hook together in a string-knot sign. But in the case of cerebral sparganosis, the lesions are shown as small rings that hook together.
2. Other parasitic infections
   The single-ring abscess, usually in small size, can be developed from parasitic schistosome eggs. Most of the patients report a history of living in a schistomiasis affected region and a medical history of schistosomiasis. Infection by toxoplasma gondii causes multiple single-ring small abscesses that scatter in brain tissue. They usually disappear within a short period of time after appropriate treatment. However, the lesions of cysticercosis are commonly intracerebral small vesicles rather than abscesses.
3. Neoplastic lesions
   Glioma commonly occurs in deep white matter of the brain. Contrast scanning demonstrates low grade malignancy with no obvious enhancement, but advanced grade malignancy with irregular garland shaped enhancement. In this case, the lesion is located in the superficial region with characteristic string-knot like enhancement. Lymphoma is commonly located near the midline of human body with obvious nodular enhancement by contrast scanning, which can be differentiated from cerebral sparganosis.

Case Study 2

[Brief Medical History]

A 24-year-old man complained of right limbs convulsion with no known causes that persisted for 3–4 min and relieved by itself. He then paid clinic visit in a local hospital and received fluid infusion and medication. After that, he reported improved condition. On Aug. 11th, the same symptom reccurred and persisted for about 5 min, but repeated after an interval of 4 h. The patient experienced repeated episodes of symptoms for five to six times with prolonged episodes but shorter intervals. Cranial MRI in another hospital demonstrated abnormal signal lesion in the left frontal-parietal lobe and left frontal lobe, which was suspected to be granulomous lesion. After the patient was transferred to a hospital in Shanghai, China, his blood sample was sent to National Institute for Parasitic Disease, Center for

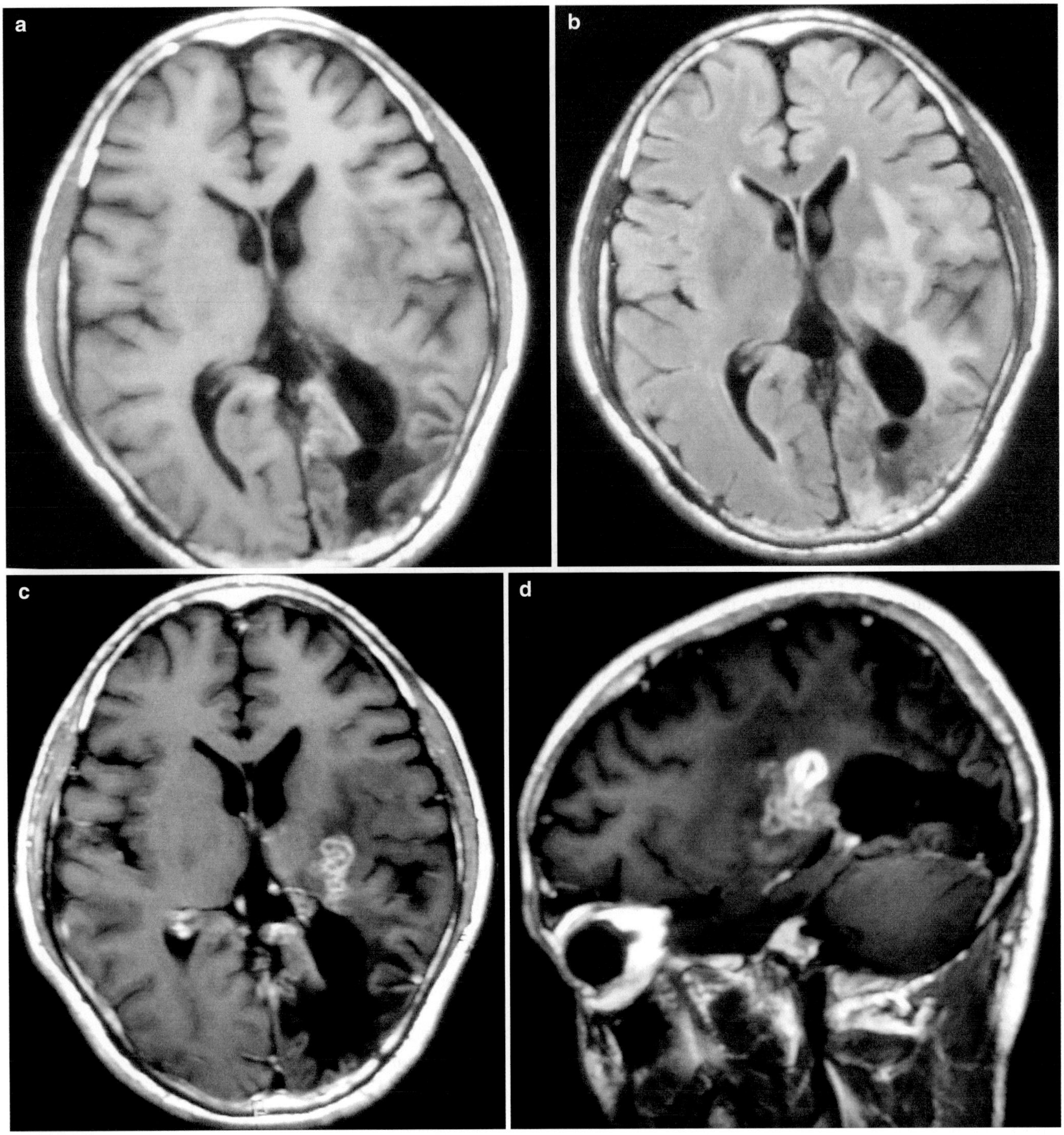

Fig. 8.14 (**a**) MR T1WI demonstrated patches of mixed low-signal area in the trigone of left lateral ventricles and the left temporal lobe as well as enlarged trigone of left lateral ventricles. (**b**) FLAIR demonstrated patches of mixed high and low signal area in the trigone of left lateral ventricles and the left temporal lobe as well as enlarged trigone of left lateral ventricles. (**c**, **d**) Contrast imaging demonstrated, in the deep part of left temporal lobe, multiple rings, hooking rings and irregular winding shaped enhancement of the lesions that appeared like a string knot. And the surrounding edema was demonstrated as low T1WI signal and malacia lesion with no enhancement. These signs indicated migration of the lesion from the left occipital lobe to the temporal lobe

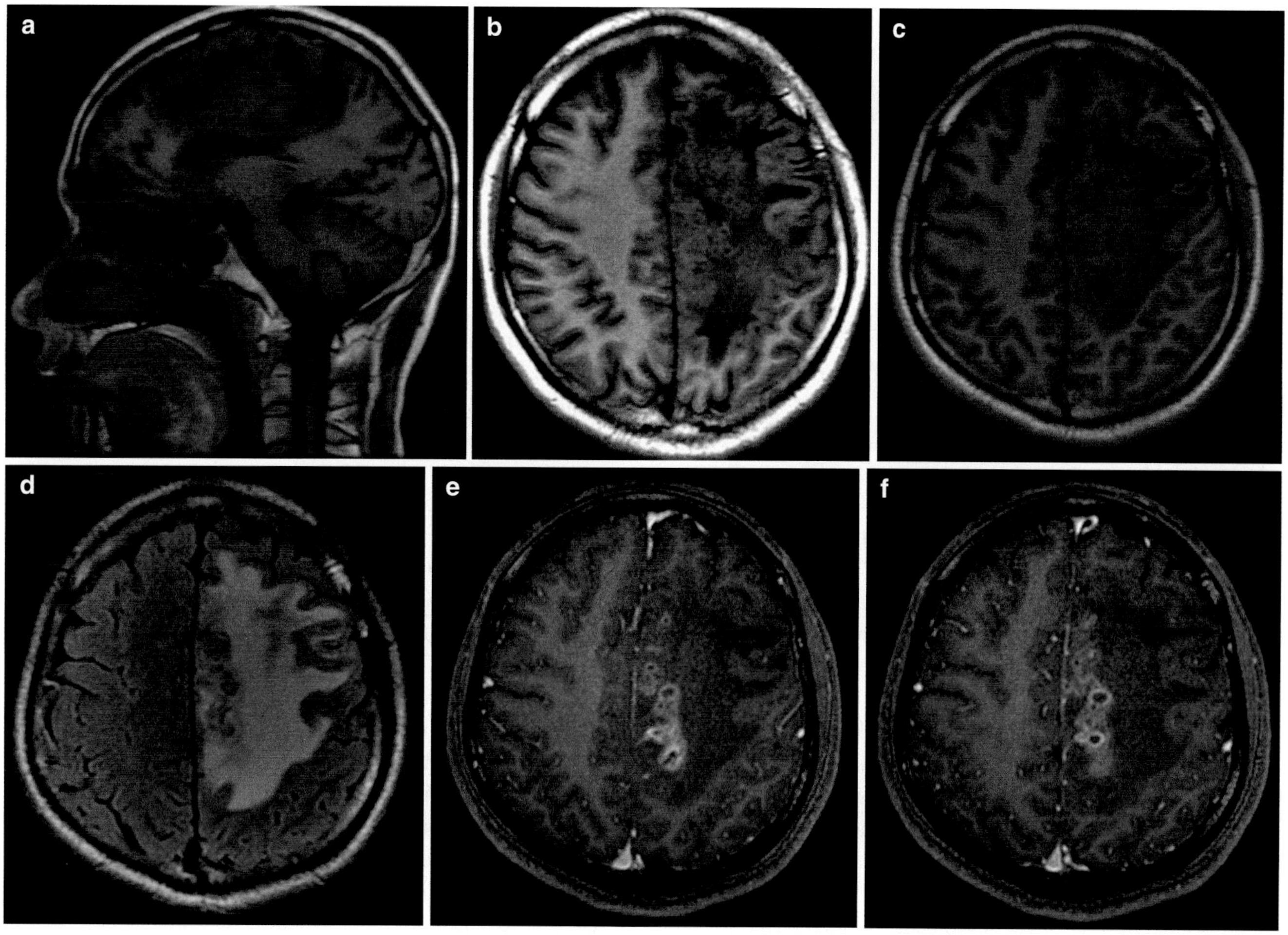

Fig. 8.15 (**a–d**) Plain MRI demonstrated flake of abnormal signal lesion in the white matter of the left frontal lobe and parietal lobe, with observable channel-like beads-shaped changes. The lesions were revealed as slightly low T1WI signal high T2WI signal, and high FLAIR signal. Spots of iso-intense signal was observed near the midline of the parietal lobe; (**e**, **f**) Contrast imaging demonstrated hooked small rings in string-knot like enhancement in the left parietal lobe, no enhancement of the surrounding edema, and no obvious migration of the midline structure

Disease Control, China and the serum antibody test showed sparganum mansoni positive. The patient was then hospitalized to the hospital with a diagnosis of sparganosis, and the patient reported a history of eating uncooked seafood years ago.

[**Radiological Demonstrations**] (See Fig. 8.15)

[**Diagnosis**] Cerebral sparganosis with secondary epilepsy.

[**Discussion**]

Cerebral sparganosis is a disease caused by parasitism of second-stage larvae of spirometra mansoni, sparganum mansoni. In recent years, due to the increasing individuals that eat uncooked or unthoroughly cooked frogs and snakes, the cases of definitively diagnosed cerebral sparganosis are increasing in China and other countries. Cerebral sparganosis is a zoonotic parasitic disease, but human is not the adaptable host to such worms since they cannot further develop into maturation after their invasion into human body. However, it mainly parasitizes in human epidermis, submucosa or superficial muscle, and rarely in the spine, spinal cord and brain. Its parasitism in human spine, spinal cord and brain causes much more serious consequences. The pathology of cerebral sparganosis is characterized by: (1) entity of sparganum with no cavity but characteristic wall; (2) scattering distribution of concentric circular or oval calcareous body and singular muscle fiber in sparganum; (3) Mixture of multiple new and old small abscesses in the brain tissue. In a previous literature report about cerebral sparganosis, the CT findings in a group of 34 cases demonstrated three diagnostically valuable signs: low density in white matter with ventricular dilation; irregular or nodular enhancement; fine needle-point like calcification. The author of this previous report believed that CT scanning is not capable of demonstrating the larvae is alive or dead, but follow-up CT reexaminations are capable of revealing the location change or progression of the nodular enhancement to suggest alive or dead larva. In a previous report of cerebral sparganosis by

MRI, the low-density by CT scan was demonstrated as slightly low T1WI signal and high T2W1 signal change in the diseased white matter, while the signal from granulomatous sparganosis is equal to the signal from the brain parenchyma. A recent report about cerebral sparganosis by MRI described fine and long channel-like lesion with beads-like changes. Contrast MRI demonstrated characteristic hooked rings in a string-knot like enhancement, which may be related to formation of slender and curving fistula along with migration of sparganum, local inflammatory responses, and formation of granuloma.

Sparganosis mansoni should be differentiated from following diseases:

1. Bacterial brain abscess
   In the cases of cerebral multi-ring abscesses, the quantity is quite small with no characteristic string-knot like enhancement.
2. Other parasitic intracerebral infections
   In the cases of infection by schistosoma eggs, the lesion is commonly single ring small abscess and the patients usually report a history related to schistosomiasis affected region. Infection by toxoplasma gondii commonly causes multiple single-ring small abscesses with a scattering distribution in the brain tissue, which may disappear within a short period of time after appropriate treatment.
3. Neoplasms
   Glioma generally occurs in the deep white matter, the low grade malignancy with no enhancement and the advanced grade malignancy with irregular garland-like enhancement. In this case, the patient reported a history of eating uncooked seafood years ago, with accompanying epilepsy. MRI demonstrated abnormal signal lesion in the left frontal-parietal lobe and the left frontal lobe, which was suspected to be granuloma. By contrast imaging, characteristic string-knot liked hooking rings enhancement was shown in the left parietal lobe. In combination to the serum antibody test by National Institute for Parasitic Disease, CDC, China, the finding of sparganum positive finally defined the diagnosis of cerebral sparganosis.

8.4 Echinococcosis

Echinococcosis, also known as hydatidosis, is a parasitic zoonosis caused by infection of larval echinococcus, which seriously threatens to human health and animal husbandry development. The disease distributes in almost all of the continents, with varying clinical manifestations due to the location and size of hydatids as well as the occurrence of complications. Since surgical intervention remains to be the main treatment for hydatidosis, radiology plays an important role in its diagnosis, differential diagnosis and preoperative assessment.

8.4.1 Etiology

With a variety of species and great variations, four types of echinococcus have been widely acknowledged: the echinococcus granulosus, the echinococcus multilocularis, the echinococcus oligarthrus and the echinococcus vogeli. The echinococcus granulosus and the echinococcus multilocularis are the major sources of human infection. The echinococcus vogeli and the echinococcus oligarthrus only exist in some areas of Central and South America and rare cases of human infection have been reported.

8.4.1.1 Echinococcus Granulosus

As the smallest tapeworm, echinococcus granulosus parasitizes in the intestine of dogs. With a length of 2–7 mm, it consists of a scolex, a neck, an immature proglottid, a mature proglottid and a gravid proglottid. The pear-shaped scolex has one rostellum and four acetabulums. And the longest and largest gravid proglottid has irregular branches and lateral branch in the uterus that is filled with eggs to be released when intraintestinal or extraintestinal uterine rupture occurs. The eggs are round with double embryonic membranes, containing radiating striations and oncospheres inside. The egg shows a strong resistance to the external environment and can survive in room-temperature water for a relatively long period of time. It cannot be inactivated by 75 % ethanol, but can be killed by 5 % sodium choloride or lime powder, which is commonly applied to disinfect the contaminated environment. The eggs are continually excreted to the external environment along with dog feces to contaminate its skin, fur, farm and animal houses. Therefore, soil, water, and vegetables are contaminated then.

The larva of echinococcus granulosus is echinococcus, which is a round or round-like cyst. The cystic wall consists of external layer of corneum and internal layer of germinal membrane, with transparent and colorless fluid filling in the cyst. The corneum is a whitish semi-transparent membrane, in appearance of sheet jelly, playing a role in protecting the germinal membrane. The germinal membrane is the body of echonococcus itself and has strong reproduction ability. It can spread to the internal bud to form brood capsules and protoscolices. The brood capsule is pedicled to attach to the germinal membrane and the detached brood capsule is the daughter cyst. On the lining of the cystic wall, the germinal membrane generates a large quantity of protoscolices, which penetrate through the cystic wall into the cystic fluid to form cystic sands. The cystic sands are small whitish particles and can be observed by naked eyes. Echinococcus can survive in human body for several decades.

8.4.1.2 Echinococcus Multilocularis

Echinococcus multilocularis and echinococcus granulosus share similarities in adult appearance and structure, but the adults of echinococcus multilocularis are smaller, with a length of 1.2–4.5 mm. Correspondingly, its scolex, rostellum and acetabulum are smaller than adults of echinococcus granulosus. The adult echinococcus multilocularis has 4–5 proglottids, and its gonopore of mature proglottid is located in the anterior part of the midline. Its uterus of gravid proglottid is a simple cyst with no lateral cyst, containing 187–404 eggs. The eggs of echinococcus multilocularis are hardly distinguishable from those of echinococcus granulosus in terms of shape and size. And the larva of echinococcus multilocularis is alveolar echinococcus, which is light-yellowish or whitish vesicle-like mass. The vesicle is round or oval in shape, with a diameter of 0.1–10 cm, containing transparent cystic fluid and protoscolices. The alveolar echinococcus develop into a large vesicle, but is sponge-liked instead that exogenously produces its daughter cysts. The external layer of corneum is thin and always incomplete, with no envelope of fibrous tissue between the whole alveolar echinococcus and the host tissue. The alveolar echinococcus generates new vesicles by exogenous budding, which grow into host tissue. The parasitized organ can almost fully occupied by generated vesicles in different sizes within 1–2 years in most cases, followed by their further extension into the surface of the parasitized organ, just like a malignant neoplasm. Therefore, the disease caused by **echinococcus multilocularis** is also known as worm cancer.

8.4.2 Epidemiology

Echinococcosis occurs worldwide, with a higher prevalence in areas with highly developed animal husbandry. In China, it occurs more commonly in Xinjiang, Qinghai, Tibet, Gansu, Ningxia, Inner Mongolia, western Xichuan and Shaanxi. With the development of tourism and the increasing number of dog keepers in urban areas, sporadic cases in urban areas of China are not rare.

8.4.2.1 Source of Infection

Dogs, wolves, foxes, jackals and other definitive hosts are major sources of its infection.

8.4.2.2 Route of Transmission

Humans can be infected via oral intake of its eggs contaminated vegetables and water or oral intake of its eggs after contacts to the contaminated animal skin and fur. In dry and windy areas, the eggs may float along with wind, whose access into human body via respiratory tract is also possible.

8.4.2.3 Susceptible Population

Human infection is mainly related to environmental contamination and inappropriate personal hygiene. Most of its patients are herdsmen, farmers and workers in skin and fur processing industry.

8.4.3 Pathogenesis and Pathological Changes

8.4.3.1 Pathogenesis

When eggs are excreted into the external environment along with dog feces, they may contaminate the farm land, vegetables, soil and water. After oral intake by human, they invade the duodenum after passing through the stomach. Under effect of digestive juice, the oncospheres are released to penetrate the intestine wall and then enter the portal venous system along with blood flow. Most of the larvae are obstructed in liver and develop into echinococcus cysts. Some may move into lung after passing through hepatic sinus and hepatic vein. Otherwise, they may further disseminate into all the organs in human body after passing through lungs. Major definitive hosts of echinococcus multilocularis are dogs and foxes, while alveolar echinococcus mainly parasitizes in human liver.

8.4.3.2 Pathological Changes

The pathological changes of echinococcosis are mainly caused by the space-occupying growth of cysts to compress the surrounding organs. The echinococcus cyst parasitizes in liver to form an internal cyst containing corneum and germinal layer. Due to the immunity of the host, immune responses occur to the foreign body in the surrounding tissue of echinococcus cyst to show infiltration of inflammatory cells and formation of fibrous connective tissue envelope by fibroblasts. With the development of larva, a fibrous cystic wall, namely the external cyst, is eventually formed. Once the echinococcus cyst is subject to external force, detachment of internal and external cysts occurs. A rapid growth of cyst and a thick cystic wall, along with other complications such as biliary fistulation or infection, impair the nutrients supply to the cyst. Subsequently, the echinococcus cyst may be subject to decreased activity, necrosis of internal cysts or daughter cysts, concentration of cystic fluid, and consolidation. After death of the echinococcus, the cyst is subject to large area of calcification.

Alveolar echinococcosis can be attributed to three major reasons, including direct invasion by alveolar echinococcus, toxic effect and mechanical compression. Due to budding and extension of alveolar echinococcus in liver parenchyma, it can directly damage and replace the liver tissue to form large-lump like lesion. At the central of such a lesion, ischemic necrosis as well as disintegration and liquefaction are always found to form a hollow cavity. During such a process, toxin is generated to further damage the liver parenchyma. The surrounding liver tissue is subject to atrophy, degeneration and even necrosis due to compression, and therefore the hepatic function is seriously impaired. If the bile duct is compressed and eroded, jaundice occurs. If alveolar echinococcosis involves branch of the hepatic portal vein, it may widely spread in the liver along with blood flow. Once it invades the hepatic veins, it may transfer to lungs and brain along with blood flow to cause related respiratory and neural symptoms such as hemoptysis, pneumothorax, epilepsy and hemiparalysis.

8.4.4 Clinical Manifestation

Echinococcosis may persist for decades in human. The clinical symptoms and signs vary according to location and size of lesions as well as occurrence of complications.

8.4.4.1 Hepatic Echinococcosis

The patients with hepatic echinococcosis may experience abdominal distension and upset as well as symptoms caused by compression. A lump is found at the right upper abdomen, which may further develop upwards to compress the thoracic cavity and cause such symptoms as reactive pleural effusion and pulmonary atelectasis. By physical examination, a lump is observable at the hepatic region, with tremor auscultated in rare cases. Alveolar echinococcus invades the human liver by proliferation, which resembles to malignancy, and the patients may experience hepatic dull pain, poor appetite and abdomen distension. In the advanced stage of the disease, if the lesion invades the biliary duct and/or blood vessels in hepatic portal area, the patients experience jaundice, portal hypertension and other signs, with hardened liver and unsmooth liver surface by physical examination.

8.4.4.2 Pulmonary Echinococcosis

The patients commonly experience such symptoms as dry cough and hemoptysis, and rarely experience coughing up sheet-jelly like substances. Choking may occasionally occur due to the overflow of cystic fluid in a large quantity. Due to the relatively loose lung tissue, echinococcus cyst grows rapidly to compress its surrounding lung tissue, inducing such symptoms as chest distress, shortness of breath, and dyspnea. Penetration of the cyst into thoracic cavity causes serious hydropneumothorax.

8.4.4.3 Cerebral Echinococcosis

Cerebral echinococcosis is commonly accompanied by hepatic and/or pulmonary echinococcosis, with clinical manifestations of epilepsy and increased intracranial pressure. In the cases with cyst under cortex, the lesion may invade the skull to cause convexobasia.

8.4.4.4 Skeletal Echinococcosis

Skeletal echinococcosis shows higher incidence at the pelvis and spine, and high incidence at long bones of extremities, skull, scapula, and ribs. The solid bone cortex and the narrow tube-shaped marrow cavity constrain the development of echinococcus cyst. Therefore, skeletal echinococcosis shows chronic progress. The patients commonly experience local soreness and painless mass, even pathological fracture and limb dysfunction in its advanced stage.

8.4.4.5 Others

Echinococcosis at the kidney, spleen, muscle, pericardium, pancreas, orbit and other parts is less likely to occur, with symptoms resembling to benign tumor.

8.4.4.6 Complications

Cystic Rupture

Penetration of echinococcus cyst into the abdominal cavity causes acute abdominal condition, showing symptoms of severe abdominal pain with shock and secondary allergic symptoms. When the cyst ruptures, its protoscolex implants into the abdominal cavity to produce secondary echinococcus cyst. When the cyst penetrates into the intrahepatic bile duct and ruptures, the cystic fragments block the bile duct to cause biliary colic and jaundice. When the cyst penetrates into the gallbladder with simultaneous penetration into the thoracic cavity with communication with bronchus, biliary bronchial fistula occurs.

Infection

Biliary infection is commonly secondary to hepatic echinococcosis, and secondary infection is the most common in the cases of pulmonary echinococcosis. Infection causes death of echinococcus, but the condition is simultaneously aggravated.

8.4.5 Diagnostic Examination

8.4.5.1 Laboratory Test

Casoni Test

The cystic fluid antigen 0.1 ml is firstly injected at the medial forearm, and skin response is then observed after 15–20 min. Those showing skin red papule at the injection site, possibly with pseudopod, are defined as positive. To terminate the use of casoni test is officially recommended in the Guidelines for Diagnosis and Treatment of Echinococcosis (2001) by World Health Organization.

Indirect Hemagglutination

Indirect hemagglutination is a relatively good serological test for the diagnosis of echinococcosis, with simple operation, low rate of false positive, and high specificity. It demonstrates a positive rate of 82% in the diagnosis of hepatic echinococcosis.

Enzyme Linked Immunosorbent Assay (ELISA)

Both the sensitivity and specificity of ELISA are higher than indirect hemagglutination, with a low rate of false positive. Diagnostic kits have been commercially available.

8.4.5.2 Radiological Examination

Ultrasound

Characterized by convenient operation, rapidity and low cost, ultrasound is especially appropriate for preliminary screening in the vast agricultural and pasturing areas. With no ionizing radiation, ultrasound is a powerful tool in dynamic follow-ups of lesions.

X-ray

X-ray has limited diagnostic value for echinococcosis, but applicable for the diagnosis of pulmonary and skeletal echinococcosis due to the obvious density contrast of lung/bone tissues to echinococcus cyst. X-ray can demonstrate the location, quantity, size, shape and complications of pulmonary and skeletal echinococcosis.

CT Scan

CT scan can be applied to examine all organs in human body and is an important diagnostic examination for echinococcosis. Plain scan can help to locate lesions and demonstrate calcifications. Contrast scan can further demonstrate the radiological signs of its complications, define the range of alveolar echinococcus cysts, and reveal small metastasis.

MR Imaging

MR imaging is a supplementary examination to CT scan for the diagnosis of echinococcosis, especially advantageous in the detection of complex lesions.

PET/CT

PET/CT is currently a powerful tool in assessing activity of alveolar echinococcus cyst, but is expensive with strong ionizing radiation. Especially in the developing countries, PET/CT has gained limited application in clinical diagnosis of echinococcosis.

Case Study 1

[**Brief Medical History**]

An Uygur woman aged 36 years complained of dull pain at the hepatic area, but no fever, nausea, and vomiting. She reported a history of living in a farm. The casoni test showed positive.

[**Radiological Demonstrations**] (See Fig. 8.16)

[**Diagnosis**] Cystic echinococcosis in the left liver lobe (Simplex cyst).

[**Discussion**]

Echinococcosis is a zoonosis, mainly prevailing in areas with highly developed animal husbandry. Cystic echinococcosis is caused by parasitism of larval echinococcus in humans or animals, commonly with the liver involved. Cystic echinococcosis in liver is common in the northwest husbandry areas of China, with typical clinical manifestations and radiological demonstrations for diagnosis. In this case, CT scan demonstrated well-defined low density lesion in the liver, with no enhancement by contrast scan, which was considered to be a benign lesion. In combination with her history of living in a husbandry area and her casoni test positive, the diagnosis was defined to be cystic echinococcosis in the liver.

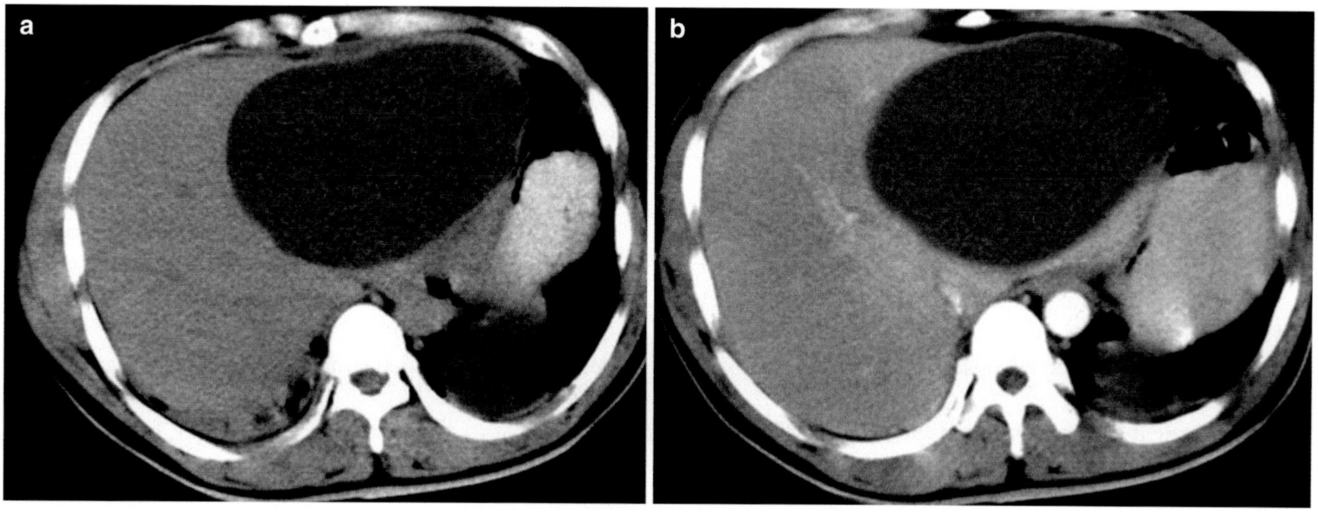

Fig. 8.16 (**a**) CT plain scan demonstrated an even density cystic lesion with a smooth and sharp boundary in the left hepatic lobe. The cystic wall was shown to be thin with even thickness and no mural nodule. The intrahepatic bile duct, blood vessels, and surrounding organs were revealed with migration due to compression. (**b**) Contrast CT scan demonstrated no enhancements of the cystic fluid and wall

Cystic echinococcosis in the liver should be differentiated from the following diseases.

1. **Hepatic cyst**
   By CT scans, the lesion of intrahepatic simplex cystic echinococcosis resembles to that of intrahepatic simplex cyst. However, the cystic wall of intrahepatic simplex cyst is extremely thin, rarely with calcification and no enhancement by contrast CT scan. The cystic wall of intrahepatic echinococcus cyst is well-defined, with higher density than surrounding liver tissues and commonly calcification. Varying degrees of enhancement of the cystic wall can be shown by contrast scan in patients with intrahepatic echinococcus cyst complicated by infection. In combination with the clinical data and immunoassays, differential diagnosis can be made.
2. **Bacterial hepatic abscess**
   The internal density of hepatic abscess is uneven in most cases, with fluid-gas level within the lesion. Calcification and small vesicles are observable in its parenchymal part. By contrast scan, no enhancements can be revealed within the lesion but obvious enhancements of the wall of hepatic abscess and the septa, with low density edema surrounding the abscess wall. In the cases of hepatic echinococcosis, the cystic wall showed no obvious enhancement by contrast scan, commonly with calcification. And the lesion is commonly multiple, with other signs detectable in the surrounding hepatic tissues and abdominal cavity. In addition, hepatic abscess causes serious systemic toxic symptoms, with obvious increase of the WBC count but casoni test negative. The casoni test negative helps its differentiation from intrahepatic cystic echinococcosis.
3. **Amoebic liver abscess**
   The wall of amoebic liver abscess may be subject to calcification, which resembles to the wall of hepatic echinococcus cyst. The calcified wall of amoebic liver abscess is commonly thick, with high density cystic fluid, while the calcified wall of echinococcus cyst is thin, with low density cystic fluid and observable polycysts and daughter cysts.

Case Study 2

[Brief Medical History]

A 27-year-old woman reported a space occupying cyst that had been detected during her physical examination 1 week ago, with no subjective upset. The Casoni test showed positive (+).

[Radiological Demonstrations] (See Fig. 8.17)

[Diagnosis] Cystic echinococcosis in the right liver lobe (Simplex cyst).

[Discussion]

Simplex cyst type of hepatic echinococcosis accounts for about 28.1 % of all the cases of hepatic echinococcosis, which is one of the most common growing modes of hepatic echinococcus cyst. Pathologically, it is caused by gradual development of oncosphere into a fluid filled cyst that gradually expands from 1 cm to above 10 cm or even larger. The quantity of fluid in the cyst is also increasing to form a ball-shaped cyst with strong tension. The cystic wall presents to be double layered due to the existence of protoscolex and germinal capsule. It is generally recognized that simplex echinococcus cyst is the early stage of parasitic echinococcus, with strong biological activity and developmental ability. The cyst may be singular or multiple, but with same pathological changes.

By CT scan, intrahepatic echinococcus cyst is demonstrated to be round or round-like with smooth cystic wall and well-defined sharp boundaries. The largest cyst is up to 20–30 cm in size, but most are larger than 3 cm between 5 and 15 cm, with the thickness of cystic wall to be 1–3 mm. The CT value of cystic fluid is −5 to 20 HU, with a potential space between the internal and external cystic wall in width of 0.5–1.5 mm. The characteristic sign of double-layer wall has important diagnostic value. Septa can be detected rarely in the cysts, and no enhancements of the cystic wall and content can be demonstrated by contrast scan.

Simplex cystic echinococcosis should be differentiated from hepatic cyst, hepatic abscess (abscess stage), and biliary cystadenoma or cystadenocarcinoma. The key points for differentiation are as the following:

1. CT scan demonstrates hepatic cyst as singular or multiple round-liked cyst in the liver, with smooth wall, well-defined boundary, no calcification of the cystic wall, and undetectable thin wall.
2. Hepatic abscess (stage of abscess) is clinically accompanied by obvious infection symptoms. By CT scan, the liver is demonstrated with round or oval shaped lesion with cystic low density and poorly-defined boundary in most cases. Contrast scan demonstrates hepatic abscess (stage of abscess) as single-ring, double-ring, triple-ring or target sign, with no enhancement of abscess liquefaction. If internal septa exist, honey comb liked change is shown.
3. CT scan demonstrates biliary cystadenoma or cystadenocarcinoma as round or oval solid cyst lesion with low density, with uneven thickness of the cystic wall or protrusion of nipple shaped soft tissue in the cyst. Contrast scan demonstrates enhancements of its solid part and fibrous septa.

In this case of simplex cystic echinococcosis, the lesion was shown with thick cystic wall, characteristic sign of double-layer wall, and spots of calcification at partial cystic wall. These findings indicated parts of echinococcus cyst remain to be the early stage of parasitic echinococcus and

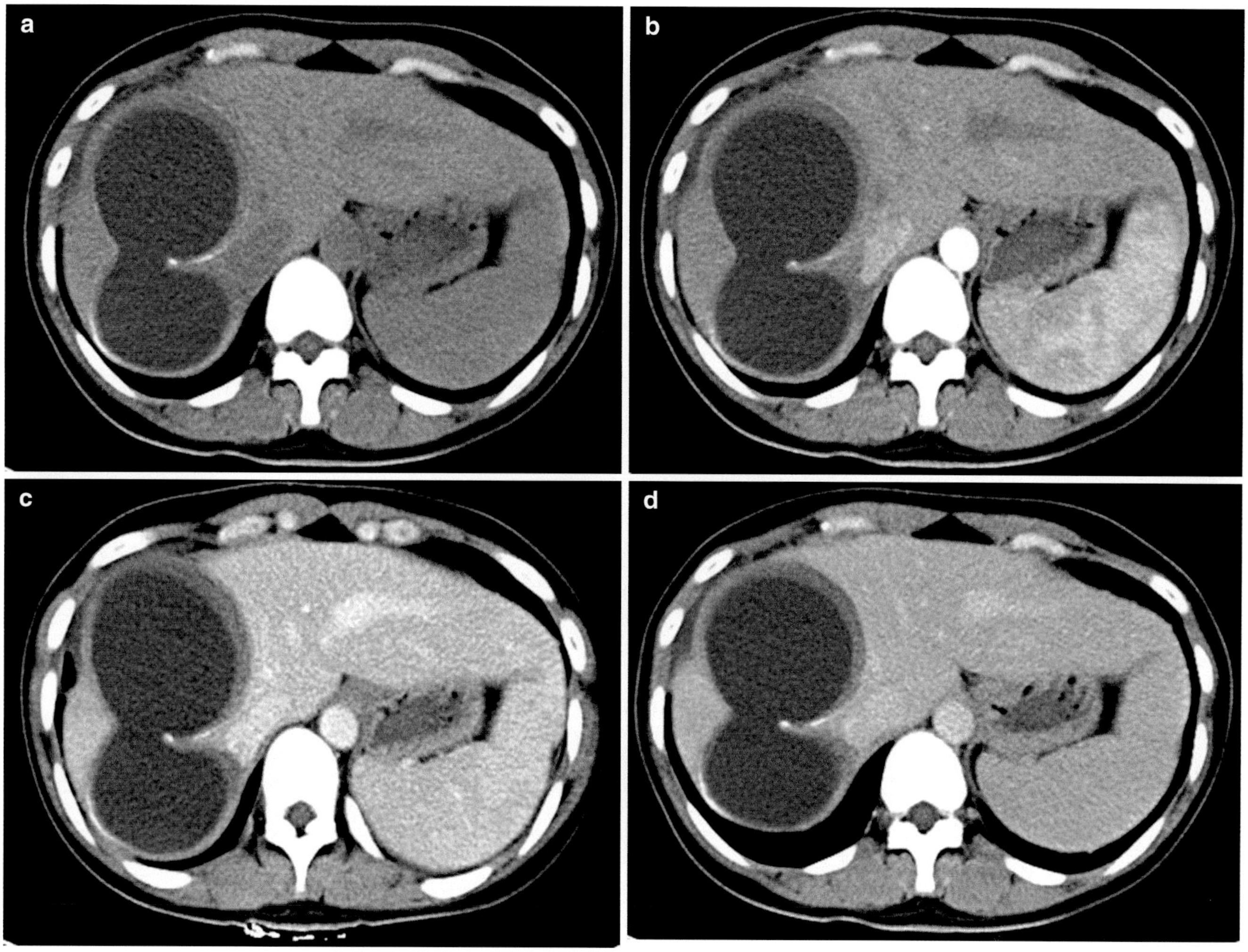

Fig. 8.17 (**a**) CT plain scan demonstrated dumbbell shaped cystic lesions with low density in the right liver lobe in a size of 5.5 cm × 7.0 cm and 5.0 cm × 4.0 cm, respectively. The two lesions are connected with partial wall to be thick and spots of calcification on parts of boundaries of the lesions. (**b**–**d**) The three phase of contrast scan revealed no enhancement of the cystic wall and content in the right liver lobe

parts are at the stage of degeneration. By contrast scan, both cystic wall and content were demonstrated with no enhancement.

Case Study 3

[**Brief Medical History**]

A 10-year-old girl was shown a cystic lesion in the liver by B-mode ultrasound more than 1 week ago. The immunoassay for echinococcus was ordered but the report was lost.

[**Radiological Demonstrations**] (See Fig. 8.18)

[**Diagnosis**] Intrahepatic cystic echinococcosis (Simplex type).

[**Discussion**]

In this case of intrahepatic cystic echinococcosis, the condition remained in its early stage. The lesions were round or round-like in shape with strong tension. Within the cyst, homogeneous water-like density was shown, with no enhancement by contrast scan. The cystic wall was thin with a thickness of 1–2 mm that failed to be demonstrated by CT scan. CT plain scan demonstrates simplex hepatic cyst as round or oval cystic lesion in the liver, with well-defined smooth sharp boundary and water like intracystic density, which resemble to the radiological findings of the case. Therefore, differential diagnosis can be hardly made based on the radiological findings. However, both the medical history and serological test are helpful for their differentiation. For simplex cyst, only surgical incision of the cyst for drainage is needed. In contrast, to harvest intracystic fluid by puncture of echinococcus cyst requires gauze with hypersonic saline to protect its peripheral area so as to prevent dissemination of the lesion via overflow of the intracystic fluid. Therefore, the surgical operation for patients with a history of living in echinococcosis affected region or serological test positive should be well prepared to prevent dissemination of the lesion.

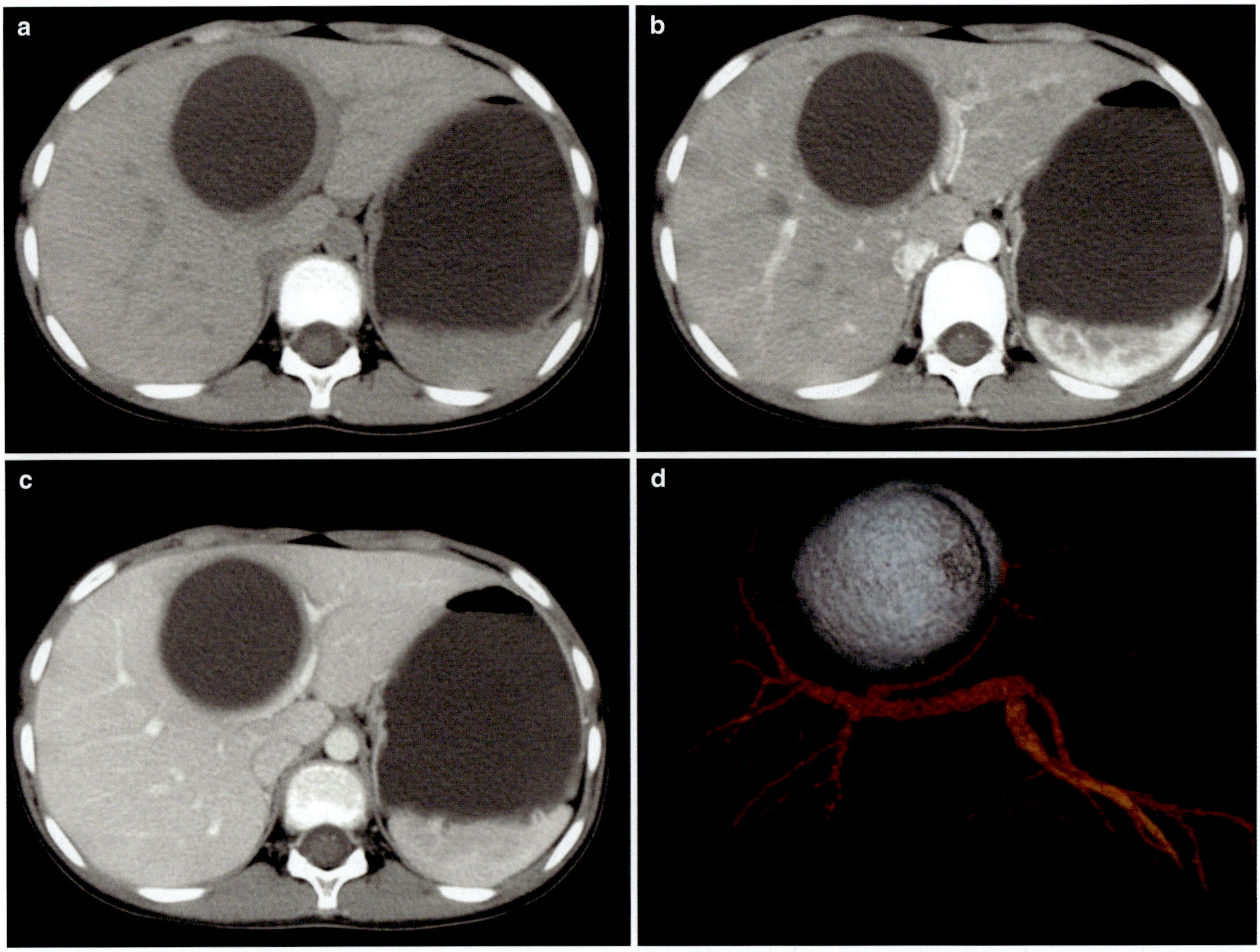

Fig. 8.18 (**a**) CT plain scan demonstrated a round liked cystic space occupying lesion with thin wall and low density from the medial segment of the left liver lobe to the anterior segment of the right liver lobe, with the largest section size of 4.5 cm×5.5 cm. The cystic wall was not shown, and the internal density of cyst was even and water like. (**b**, **c**) Contrast scan demonstrated no enhancement of the cystic wall and content as well as adherence of compressed left branch of portal vein to the lesion boundary. (**d**) VR demonstrated a ball-in-hands sign formed by the portal vein and the lesion as well as narrowed left branch of portal vein due to compression. By surgical pathology, the diagnosis of cystic echinococcosis was defined

Case Study 4

[**Brief Medical History**]

A 6-year-old girl was revealed with a cystic space occupying lesion in the liver by ultrasound for 2 weeks, and no other abnormalities by physical examination. She reported a history of living in a husbandry area for 6 years. Four-item examination for echinococcus was ordered, showing Anti-EgCF antibody (+), Anti-EgP antibody (±), Anti-EgB antibody (+) and Anti-Em2 antibody (−).

[**Radiological Demonstrations**] (See Fig. 8.19)

[**Diagnosis**] Intrahepatic cystic echinococcosis (simplex type).

[**Discussion**]

This is a typical case of singular cystic echinococcosis, with its characteristic radiological signs demonstrated by CT scan, including detectable cystic wall, shell liked calcification at the cystic wall, and double-line sign indicating separated internal and external cysts. At the early stage of singular cystic echinococcosis, the cystic wall is commonly shown to be thin, which gradually thickens with calcification along with progression of the disease. The intracystic liquid can be revealed with gradual decrease, with consequently decreased cyst tension and irregular cyst shape despite of the homogeneous water like internal density. The separated internal and external cystic walls can be radiologically shown as typical double-line sign, with no enhancement of the lesion by contrast scan. At this time, its differential diagnosis from simplex hepatic cyst presents no challenge. If the cystic wall is thick, differential diagnosis from hepatic abscess is still necessary. The wall of hepatic abscess is thick, with obvious enhancement by contrast scan. In most cases, edema with inflammatory responses can be detected around the lesion. And its differential diagnosis from intrahepatic cystic echinococcosis presents no challenge base on above radiological

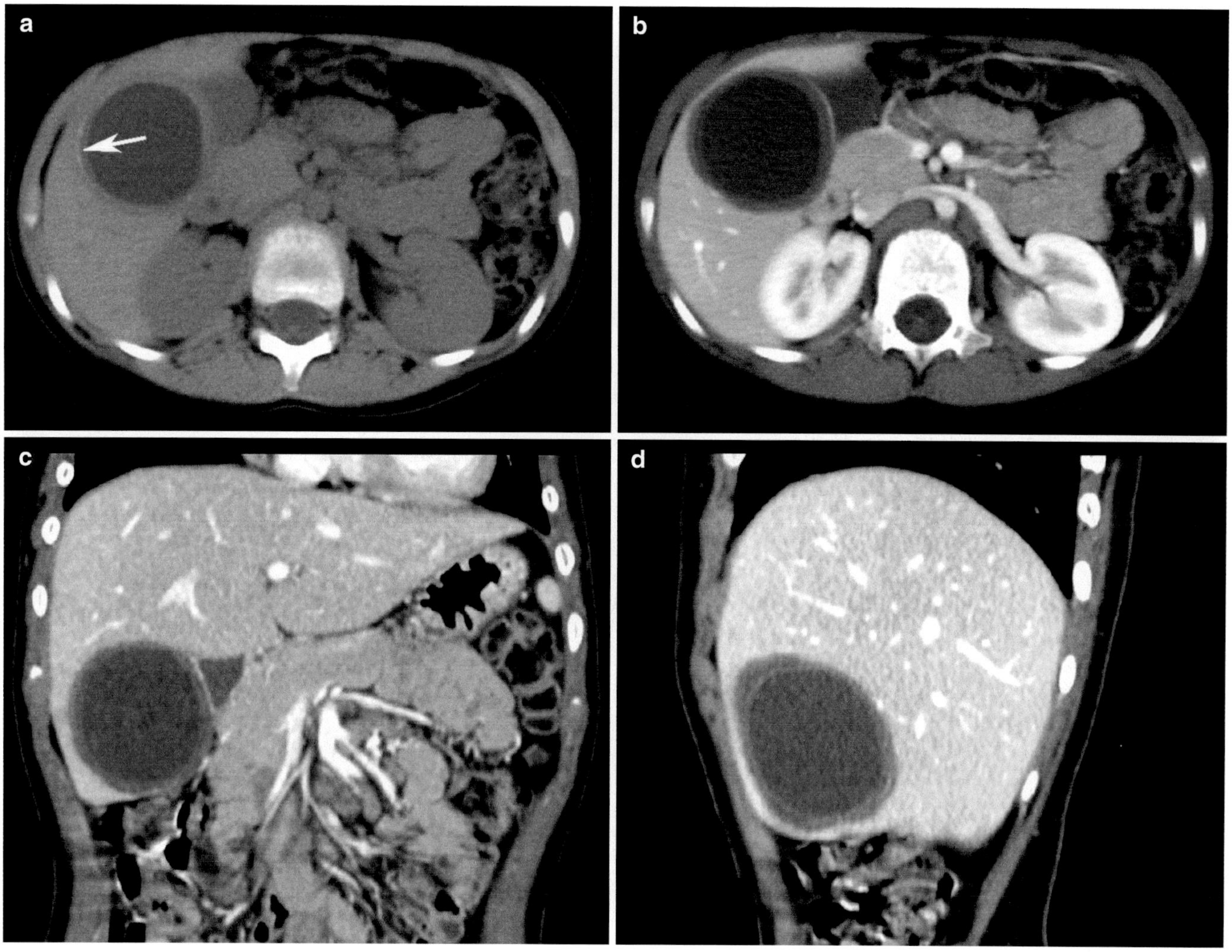

Fig. 8.19 (**a**) CT scan demonstrated an irregular cystic space occupying lesion with low density in the posterior segment of the right liver lobe, with smooth boundary. The cystic wall was shown to be thick with shell liked calcification and observable double-line sign (*white arrow*). The lesion was in a size of 4.5 cm×4.8 cm, with a intracystic CT value of around 4 HU. (**b–d**) Contrast scan demonstrated no enhancement of the lesion

signs. In some cases, intrahepatic cystic echinococcosis complicated by infection is radiologically demonstrated as hepatic abscess, with no activity of the echinococcus, necrotic tissue and pus filling in the cyst, and thickened cystic wall with enhancement. In such cases, past medical history and related laboratory tests can provide valuable reference for the diagnosis of cystic echinococcosis complicated by infection.

Case Study 5

[**Brief Medical History**]

A 53-year-old woman was revealed with a cystic space occupying lesion in the liver by physical examinations. The Casoni test showed positive (+).

[**Radiological Demonstrations**] (See Fig. 8.20)

[**Diagnosis**] Intrahepatic cystic echinococcosis (multiple daughter cysts type).

[**Discussion**]

The multi daughter cysts type accounts for about 25.3 % of hepatic echinococcosis. The multi daughter cysts type is caused by budding of numerous second generation or third generation cysts from the germinal layer of cystic wall, which are known as daughter or granddaughter cysts showing cyst-in-cyst sign. When the quantity of daughter or granddaughter cysts is small, they are arranged along the germinal layer of the mother cyst. When the quantity of daughter or granddaughter cysts is large, their arrangement is diversified. The multi daughter cysts type of echinococcosis demonstrates the strong reproducibility and biological activity of echinococcus. By CT scan, the density of daughter or granddaughter cyst is lower than that of the mother cyst, which is an important characteristic sign. Their arrangement when in a large quantity includes petals pattern, wheels pattern, honey comb pattern or grapes pattern. By contrast scan, the cystic wall and content show no enhancement.

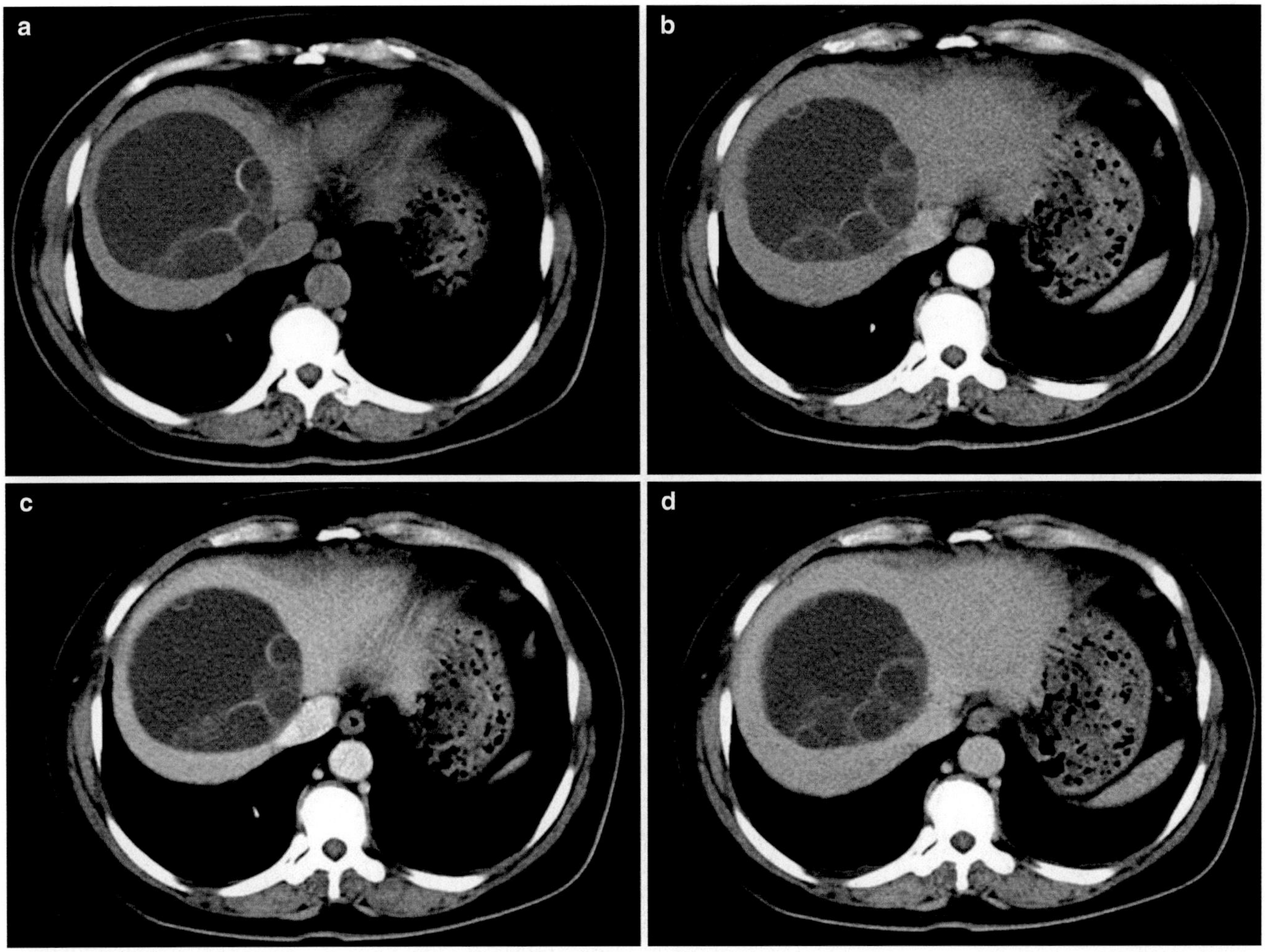

Fig. 8.20 (**a**) CT plain scan demonstrated round like cystic lesion in the right liver lobe, with well-defined boundary and even internal density. Several daughter cysts were revealed, with ring shape calcification in the wall of some daughter cysts. (**b–d**) Contrast scan demonstrated no enhancements of the cystic wall and content in the right liver lobe

As a case of multi daughter cysts type of echinococcosis, the mother cyst was shown with several small daughter cysts located along the germinal layer, namely in a small quantity. The cystic wall of some daughter cysts showed ring shaped calcification, indicating a decreased biological activity of these daughter cysts. Contrast scan demonstrated no enhancements of the cystic wall and content.

Case Study 6
[**Brief Medical History**]

A 54-year-old woman complained of chest distress and shortness of breath, with a history of living in a husbandry area. The Casoni test showed positive (+).

[**Radiological Demonstration**] (See Fig. 8.21)

[**Diagnosis**]

Hepatic echinococcosis with multiple cysts (multi daughter cysts type in the left lateral liver lobe, simplex cyst type in the right liver lobe).

[**Discussion**]

In this case, hepatic echinococcosis with multiple cysts was diagnosed, showing both multi daughter cysts type and simplex cyst type. The multiple daughter cysts in the left lateral liver lobe were revealed with arrangements of petals pattern, honeycomb pattern or grapes pattern. Simplex cyst was detected in the right liver lobe, with spots of calcification at the cystic wall that indicated degeneration of echinococcus cyst. Contrast scan showed no enhancements of the cystic wall and content.

Multi daughter cysts type of hepatic echinococcosis should be mainly differentiated from hepatic abscess (abscess stage), and biliary cystadenoma or cystadenocarcinoma. The key points for differential diagnosis include:

1. In the abscess stage, the patients with hepatic abscess often experience obvious symptoms of infection. CT scan demonstrates hepatic abscess as round or oval shaped cystic lesion with low density in the liver that is poorly defined. Contrast scan reveals single-ring, double-ring,

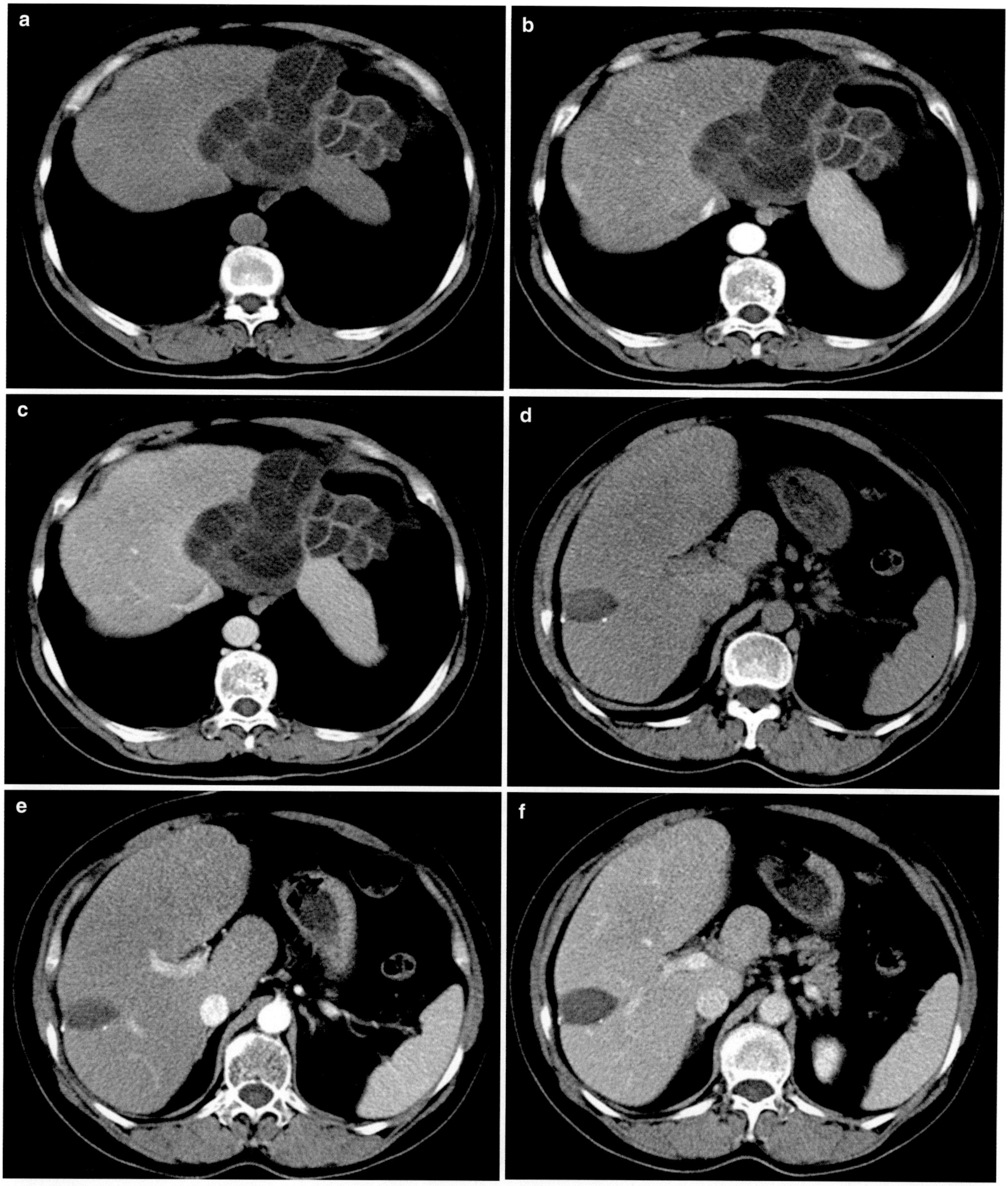

Fig. 8.21 (**a**) CT plain scan demonstrated irregular cystic lesion with low density in the left lateral liver lobe, with internal septa like several small cysts. The boundary between the lesion and its surrounding liver tissue was shown to be well defined. (**b**, **c**) Contrast scan in the arterial and portal phases revealed no enhancements of the cystic wall and content in the left lateral liver lobe. (**d**) CT plain scan showed oval like cystic lesion with low density in the right liver lobe, with well defined spots of high density calcification at its margin. (**e**, **f**) Contrast scan in the arterial and portal phases demonstrated no enhancements of the cystic wall and content in the right liver lobe

triple-ring or target sign, with no enhancements of abscess liquefaction. Honey comb pattern change may be shown if septa exist in the cyst.
2. CT scan shows biliary cystadenoma or cystadenocarcinoma as round or oval solid cystic lesion with low density in the liver. The thickness of cystic wall is revealed to be uneven or with nipple shaped soft tissue mass protruding into the cyst. Contrast scan shows enhancements of its parenchymal parts and fibrous septa.

Case Study 7

[Brief Medical History]

A 43-year-old male farm worker was revealed with a space occupying lesion in the liver by ultrasound in the physical examination. He was generally in good health, with no obvious symptoms of abdominal pain and distension, poor appetite and other upsets.

[Radiological Demonstration] (See Fig. 8.22)

[Diagnosis] Cystic echinococcosis in the right liver lobe (multi daughter cysts type).

[Discussion]

Despite no definitive history of living in husbandry area and no definitive laboratory test results, the patient showed typical radiological demonstrations of cystic echinococcosis. Although no typical calcification was detected, CT and MRI demonstrated multiple daughter cysts in the lesion, which is a typical radiological sign of cystic echinococcosis. In addition to typical sign of multiple daughter cysts, the lesion of hepatic cystic echinococcosis shows collapse of its internal

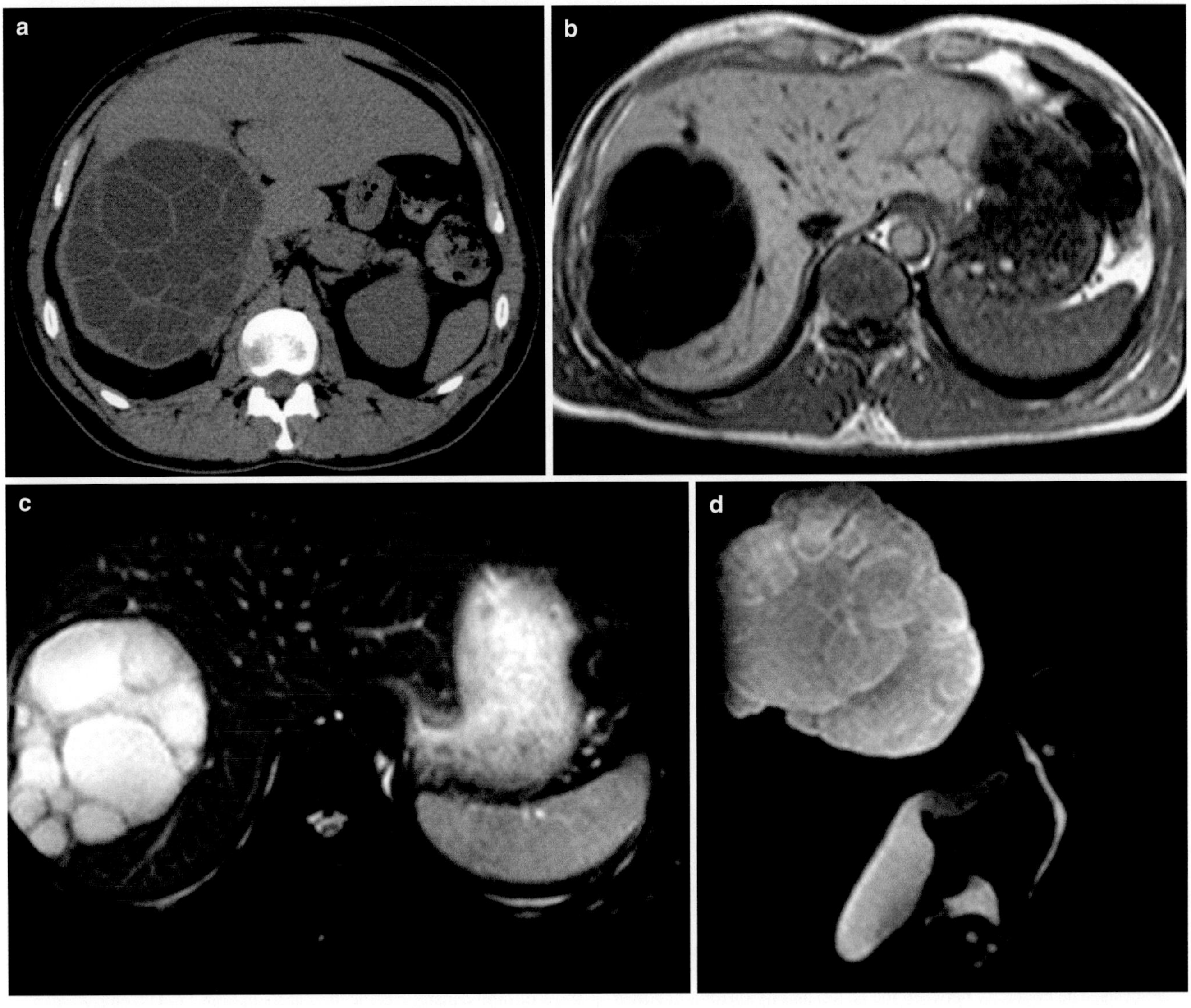

Fig. 8.22 (**a**) CT plain scan demonstrated opacity of multiple round like daughter cysts with lower density in the low density cyst in the right liver lobe. The daughter cysts were revealed in the mother cyst with an arrangement of honey comb pattern or wheel pattern. The daughter cysts were in a large quantity that filled up in the mother cyst and squeezed each other to form honey comb sign. (**b**, **c**) MR imaging demonstrated round like cystic lesion with well-defined smooth sharp boundary. The cystic wall was shown with homogeneous thickness, and fluid in low T1WI signal and high T2WI signal. (**d**) MR hydrography demonstrated rose petals pattern of the cysts, and clearly showed their relationships with bile ducts

cyst due to increased intracystic pressure, degeneration, trauma, therapy and other factors, with another typical lotus-on-lake sign.

Case Study 8

[Brief Medical History]

A 15-year-old boy complained of nausea with headache for 20 days and right lower quadrant abdominal pain for 1 week. Four-item examination for echinococcus showed Anti-EgCF antibody (+++), Anti-EgP Antibody (+++), anti EgB antibody (+++), and anti-Em2 antibody (+).

[Radiological Demonstration] (See Fig. 8.23)

[Diagnosis] Hepatic cystic echinococcosis (Multi daughter cysts type).

[Discussion]

The occurrence of daughter cysts of echinococcus, in multilocular appearance, is one of the characteristic radiological signs for the diagnosis of cystic echinococcosis. In the cases with a small quantity of daughter cysts, the lesion is radiologically demonstrated as cyst-in-cyst sign. When the quantity is large, several daughter cysts are distributed along the margin of the mother cyst, radiologically demonstrated as rose-petals sign or wheel sign. When the daughter cysts grow in size and number or even when they fully fill up the mother cyst, mulberry sign or honeycomb sign can be radiologically demonstrated. The density of daughter cystic fluid is lower than that of the mother cyst, radiologically demonstrated as higher T2WI signals of daughter cyst than mother cyst, which has important value in the diagnosis and differential diagnosis. Cystic echinococcosis with daughter cysts should be differentiated from biliary cystadenoma or cystadensarcoma, whose radiological demonstrations resemble to those of ovarian and pancreatic cystadenoma or cystadensarcoma. These include well defined neoplasm with envelope, internally multilocular, homogeneous thickness of the cystic wall and internal septa, possible calcification and wall nodules, and varying internal density due to different components of the cystic fluid. By contrast scan or imaging, the cystic wall, septa and wall nodules show enhancements. All of these radiological demonstrations help their differentiation from cystic echinococcosis with daughter cysts.

Case Study 9

[Brief Medical History]

A Kazak 28-year-old man complained of upset at the hepatic region for 1 year, with no fever, jaundice and other upsets. Laboratory test showed negative (−).

[Radiological Demonstration] (See Fig. 8.24)

[Diagnosis] Cystic echinococcosis in the right liver lobe (Multi daughter cysts type).

[Discussion] In this case, the occurrence of daughter cysts in the hepatic lesion is a characteristic sign of hepatic echinococcosis. The mother cystic fluid with higher density distributes between the daughter cysts. The daughter cysts in the cyst-in-cyst sign are small in size and quantity, and located close to the mother cyst wall, with lower density than the mother cyst. The daughter cystic wall presents particles of amorphous calcification due to deposit of calcium salt, which is another characteristic radiological sign of echinococcosis. Since cystic echinococcosis is mainly caused by mechanic compression and no obvious blood supply can be found in the cystic wall, no obvious enhancement of the lesion is the other characteristic radiological sign for the diagnosis of cystic echinococcosis.

Case Study 10

[Brief Medical History]

A 23-year-old young woman complained of abdominal distension, pain in hepatic region, poor appetite, nausea and vomiting in recent days. Laboratory tests showed increased blood eosinophile count, Casoni test positive (+) and ELISA positive (+).

[Radiological Demonstration] (See Fig. 8.25)

[Diagnosis] Hepatic cystic echinococcosis (detached internal cyst type).

[Discussion]

Based on CT scan findings, hepatic cystic echinococcosis can be classified into four types: simplex type, detached internal cyst type, multi daughter cysts type as well as consolidation and calcification type. And the specific CT findings are as the following:

1. Simplex type
   The lesion may be well-defined round or round like singular or multiple in homogeneously low density and different sizes. The intracystic CT value is close to that of water. In the cases with complicating infection, the cystic wall may be thick. Calcification of the cystic wall is common, demonstrated to be arc or shell like in shape. Contrast scan demonstrates no enhancement of the lesion.
2. Detached internal cyst type
   When the internal and external cysts are separated or when the internal cyst ruptures with the fluid spread into the space between internal and external cysts, double-ring sign is radiologically demonstrated. When the internal cyst is completely separated, collapses, or floats on the daughter cystic fluid, lily-on-water sign is radiologically demonstrated. When the internal cyst is completely detached, ribbon sign is radiologically demonstrated.
3. Multi daughter cysts type
   Multiple daughter cysts in different sizes are demonstrated in the mother cyst to show wheel pattern,

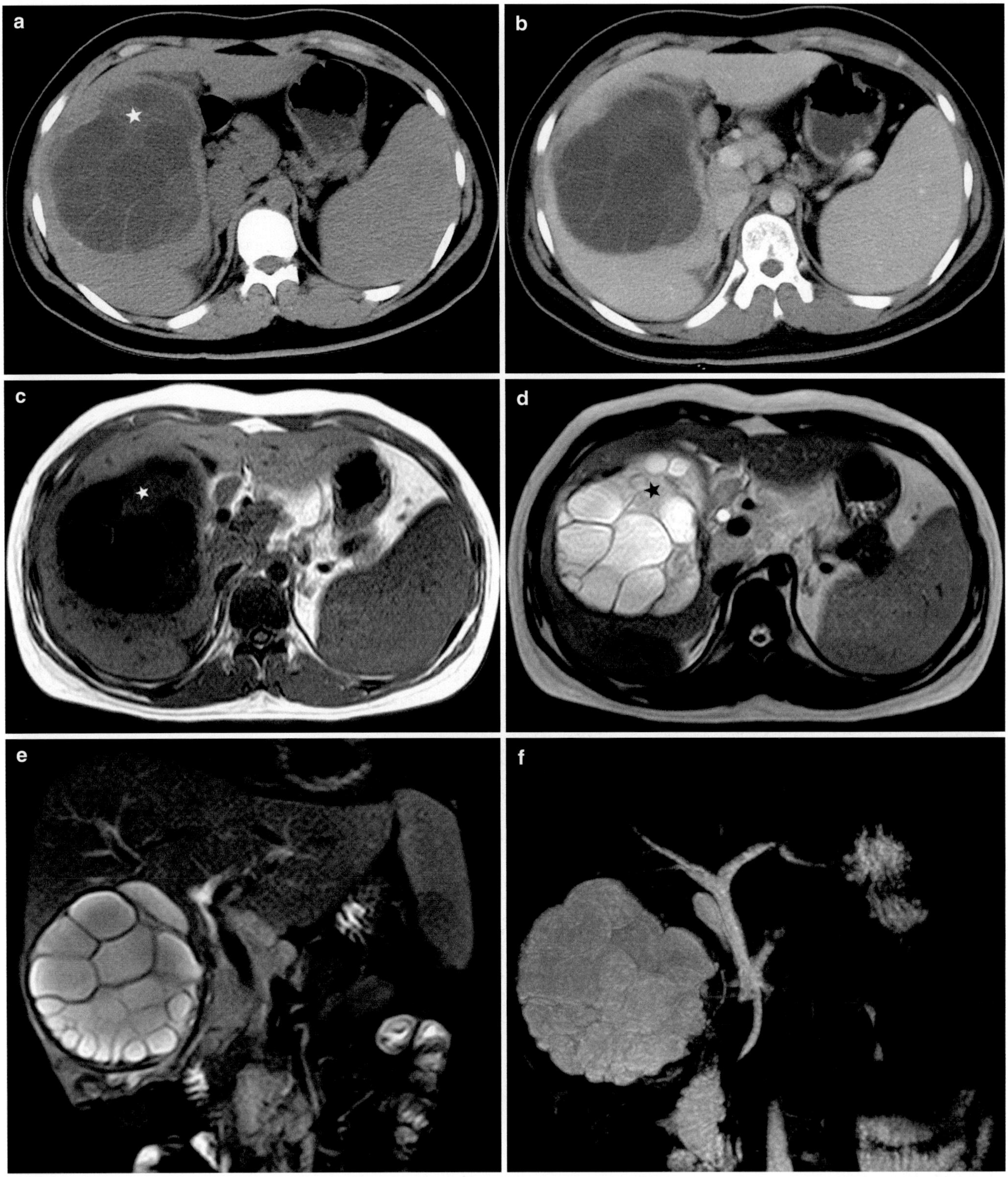

Fig. 8.23 (**a**) CT plain scan demonstrated a huge round like cystic lesion with low density in the right liver lobe, with the largest section size of 11.7 cm × 9.45 cm, heterogeneous internal density of the lesion, and detectable opacity of septa. The lesion was revealed to generate several daughter cysts, which lower intracystic fluid density than the mother cyst (*star*). (**b**) Contrast scan demonstrated no enhancement of the lesion. (**c–e**) MRI demonstrated low signal envelope in the peripheral area of the lesion, with multiple round low T1WI signals and high T2WI signals daughter cysts arranged in rose petals pattern, which was more favorably demonstrated than CT scan. A small quantity of fluid from mother cyst was observable in the daughter cysts (*star*), and the mother cyst has higher T1WI signal but slightly lower T2WI signal than the daughter cyst. (**e**) Demonstrations by Coronal T2 imaging after fat suppression. (**f**) Demonstrations by MRCP

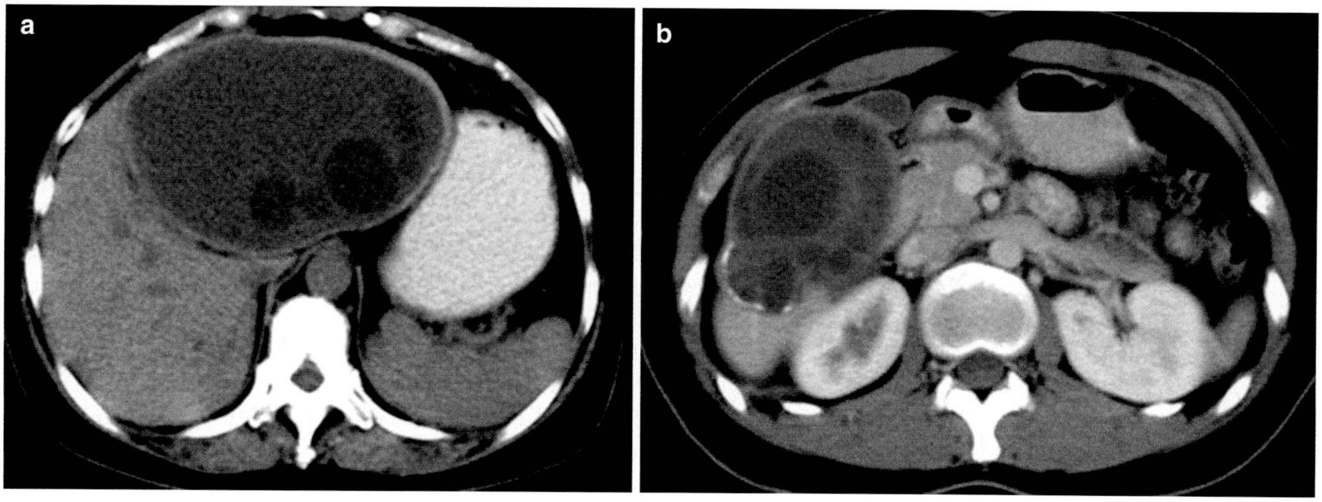

Fig. 8.24 (**a**) CT plain scan demonstrated a cyst in the right liver lobe with multiple round like daughter cysts opacity in different sizes that freely distributed in the mother cyst. The daughter cysts were revealed with lower density than the mother cyst. (**b**) Contrast scan demonstrated no enhancements of the cyst. The daughter cysts were shown to be round, rhombus like and polygon like in shape that distributed like a cluster of grapes in the mother cyst. Partial wall of the daughter cysts was revealed with calcification

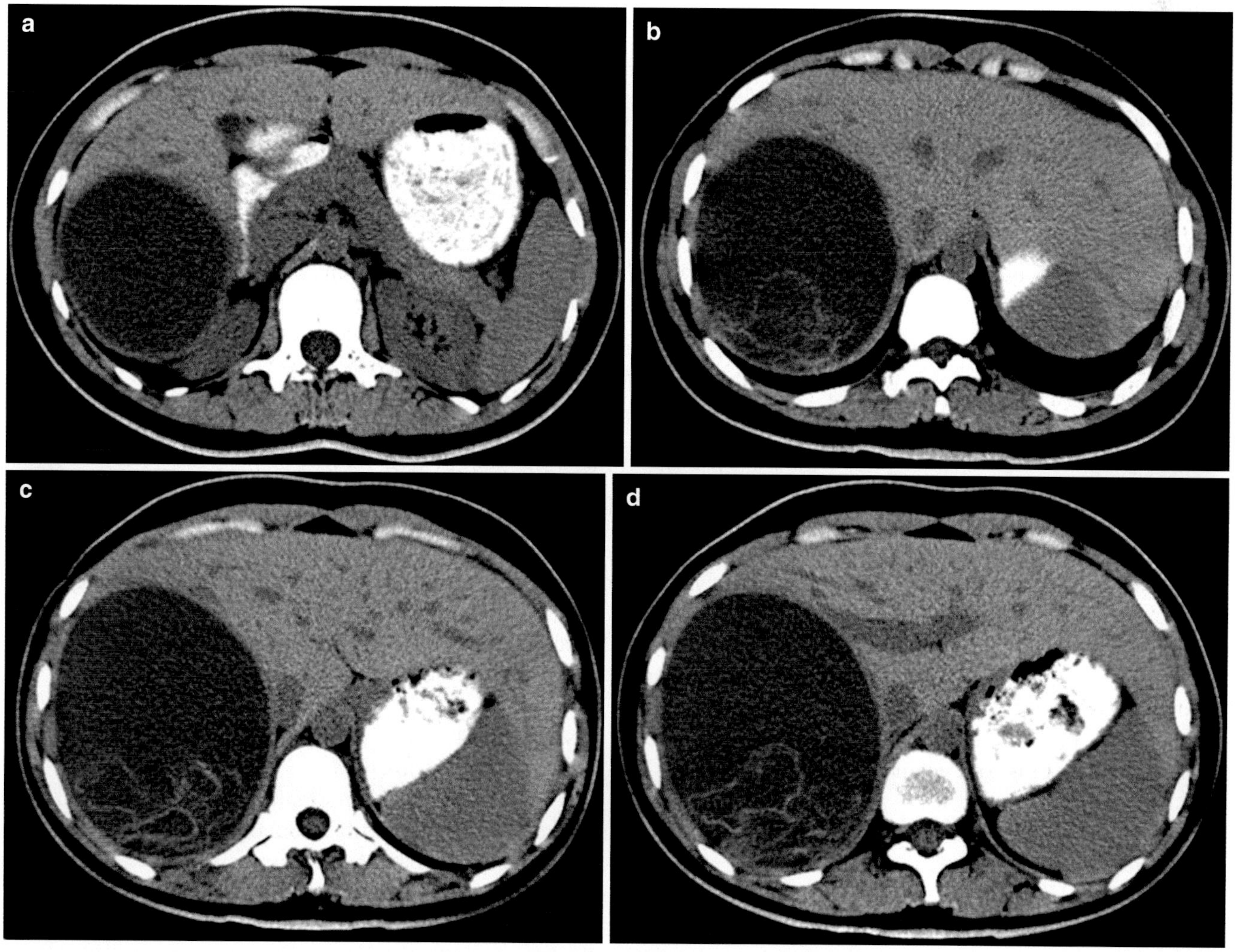

Fig. 8.25 (**a**) CT scan demonstrated singular round like cystic lesion with low density in the right liver lobe. The lesion was revealed with even density and well defined smooth intact boundary. (**b**–**d**) CT scan demonstrated rupture of the internal cysts in the large cyst in the right liver lobe with complete detachment that shown as stripes of high density opacity in a ribbon sign

honeycomb pattern and other multilocular patterns. In the cyst, mother cyst fragments, scolex and calcification of daughter cysts are always radiologically demonstrated in strips and flakes opacity.

4. Consolidation and calcification type
   Consolidation occurs due to filling of degenerated and necrotic daughter cyst fragments in the mother cyst. Calcification is demonstrated as spots, strips or shell like opacity. The lesion with predominant consolidation should be differentiated from other intrahepatic space occupying lesions.

Hepatic cystic echinococcosis should be differentiated from intrahepatic simplex cyst and hepatoangioma. Radiological signs of thick cystic wall with accompanying calcification, detachment of internal and external cysts, and daughter cysts in the mother cyst always indicate hepatic cystic echinococcosis. Hepatic cystic echinococcosis should also be differentiated from hepatic abscess. The lesion of hepatic cystic echinococcosis commonly has no surrounding edema area, which shows no enhancement by contrast scan or imaging. The past medical history, living place and contacts history are helpful for the differential diagnosis.

Case Study 11

[**Brief Medical History**]

A 25-year-old young man complained of fever for 2 days, with upset in hepatic region as well as yellowish skin and sclera. He reported a definitive history of contact to dogs.

[**Radiological Demonstration**] (See Fig. 8.26)

[**Diagnosis**] Hepatic cystic echinococcosis (Detached internal cyst type).

[**Discussion**]

According to the pathological basis and evolving process of hepatic cystic echinococcosis, the disease can be staged into 4 stages. Radiologically, the early stage of simplex cyst type is characterized by thin cystic wall. Along with the progression of the disease, the germinal layer in the lesion produces multiple daughter cysts that grow inwards. At the late stage, the daughter cysts cause collapse of internal cyst due to increased pressure and other factors to form lotus-floating-on-water pattern, and the cyst is finally filled by matrix. Different stages of the lesion are characterized by different radiological signs. According to the characteristic radiological signs, this case showed characteristic signs of detached internal cyst type of hepatic echinococcosis.

Case Study 12

[**Brief Medical History**]

A 45-year-old Kazak man complained of upper abdominal upset, poor appetite, nausea and vomiting for 2 months. He reported a history of contacts to dogs and goats.

[**Radiological Demonstration**] (See Fig. 8.27)

[**Diagnosis**] Hepatic cystic echinococcosis (Rupture of internal cyst type).

[**Discussion**]

In this case, rupture of internal cyst in liver is radiologically characterized by rupture of the internal cyst and its content within the external cyst as well as partially or completely detached internal cystic wall due to release of cystic fluid into the external cyst. CT scan demonstrated double-line sign of the cyst or wave-like folds of the internal cystic membrane in the cystic cavity, namely, the ribbon sign. In the cases with complete detachment of the internal cyst, the internal cyst collapses and crinkles to show small-lily sign. In combination to the definitive history of contacts to dogs and goats, the diagnosis was defined to be hepatic cystic echinococcosis.

Case Study 13

[**Brief Medical History**]

A 54-year-old Kazak herdswoman complained of abdominal pain and fever for 2 days, with no obvious nausea, vomiting, diarrhea, jaundice and other symptoms.

[**Radiological Demonstration**] (See Fig. 8.28)

[**Diagnosis**] Hepatic cystic echinococcosis (Ruptured internal and external cyst type).

[**Discussion**]

According to the characteristic CT findings in this case, the lesion was poorly defined from its surrounding normal hepatic tissue, with observable collapse of the internal cyst. Based on such a typical demonstration, a diagnosis of echinococcosis can be made. Based on the degree of internal cystic rupture, echinococcosis of rupture type can be further divided into two sub-types:

1. Ruptured internal cyst sub-type
   It is radiologically characterized by double-line sign or wave-like folds of internal cystic membrane in the cystic cavity, namely the ribbon sign.
2. Direct rupture sub-type
   It manifestations as rupture of both internal and external cysts, with direct discharge of its cystic fluid and other contents into abdominal cavity, into liver or under the hepatic capsule.

Case Study 14

[**Brief Medical History**]

A 35-year-old woman complained of low back pain for more than 4 months. After abdominal CT scan demonstrations indicated hepatic echinococcosis, she was treated by oral intake of albendazole until now. Four-item examination for echinococcus showed Anti-EgCF antibody (±), anti-EgP antibody (±), anti-EgB antibody (±) and anti-Em2 antibody (−).

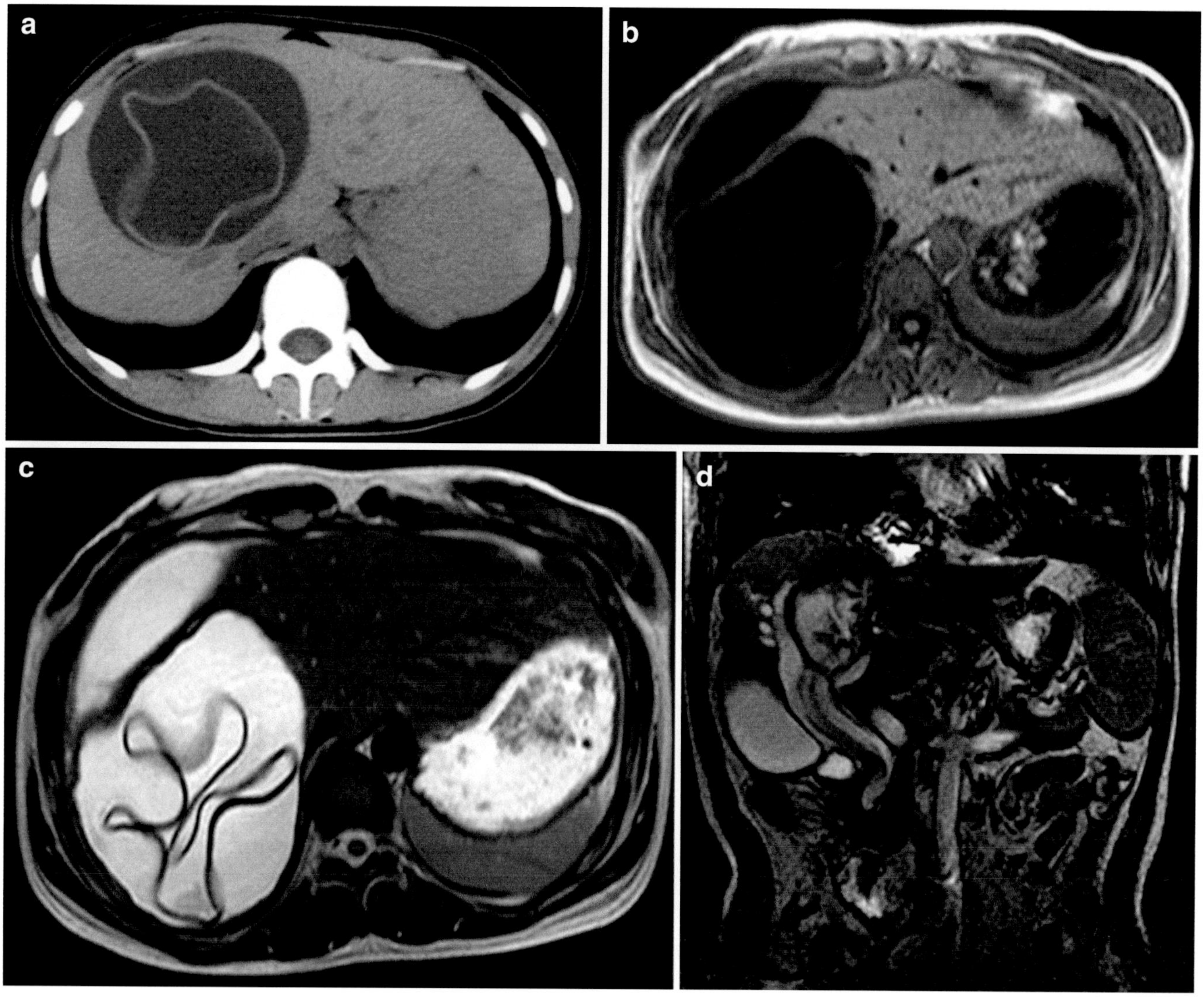

Fig. 8.26 (**a**) CT scan demonstrated a huge cystic lesion in the liver, with detached internal cyst in a ribbon sign. (**b**, **c**) MRI demonstrated uneven signal in the cyst, detached internal and external cysts, and floating of ruptured internal cyst in the cystic cavity in a ribbon sign. (**d**) Coronal MRI demonstrated rupture of echinococcus in the liver

[**Radiological Demonstration**] (See Fig. 8.29)

[**Diagnosis**] Hepatic cystic echinococcosis (Consolidation and calcification type).

[**Discussion**]

Due to rupture, degeneration or infection of the internal cyst, the fluid in echinococcus cyst is thick with increased density, with accompanying irregular calcification in the cyst. No opacity of daughter cysts is detectable, but sometimes ruptured and collapsed internal cystic wall can be observed. The activity of such echinococcus cyst is decreasing or completely lost. The external cystic wall is subject to obvious thickening in most cases due to inflammatory responses, which can be enhanced by contrast scan or imaging. Due to the absence of its original features, hepatic cystic echinococcosis (consolidation and calcification type) should be differentiated from alveolar echinococcosis and hepatic neoplasms with poor blood supply. Hepatic alveolar echinococcosis can be radiologically demonstrated as solid mass that cannot be enhanced by contrast scan in most cases. However, alveolar echinococcus shows an infiltrative growth with unsmooth boundary that is poorly defined from the surrounding normal liver tissue. The finding of characteristic vesicle indicates a diagnosis of alveolar echinococcosis. Most primary hepatic neoplasms have rich blood supply, but some lesions of hepatocellular carcinoma and cholangiocellular carcinoma show poor blood supply, which cannot be enhanced by contrast scan or imaging at their early stage. Some hepatic metastases also show poor blood supply and also should be differentiated from this type of echinococcosis. By contrast scan or imaging, hepatic neoplasms with poor blood supply always present uneven enhancement. However, the lesion of echinococcosis cannot be enhanced

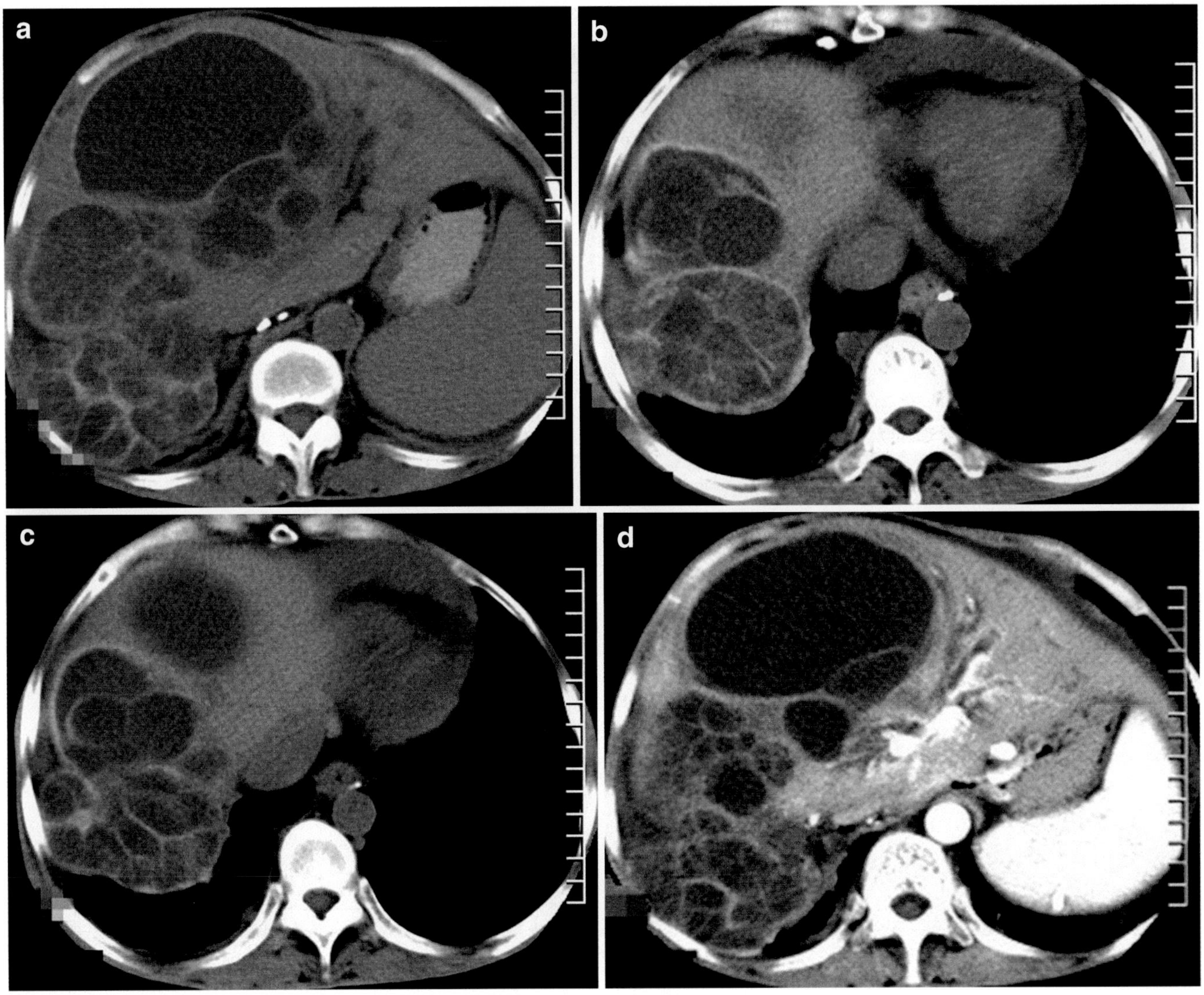

Fig. 8.27 (**a**–**c**) CT scan demonstrated uneven density of liver tissue, with multiple round like density opacities that have smooth and sharp boundaries. Some lesions were revealed with internal septa like a football. (**d**) Contrast CT scan demonstrated obvious enhancements of the normal liver tissue, and fusion of some lesions

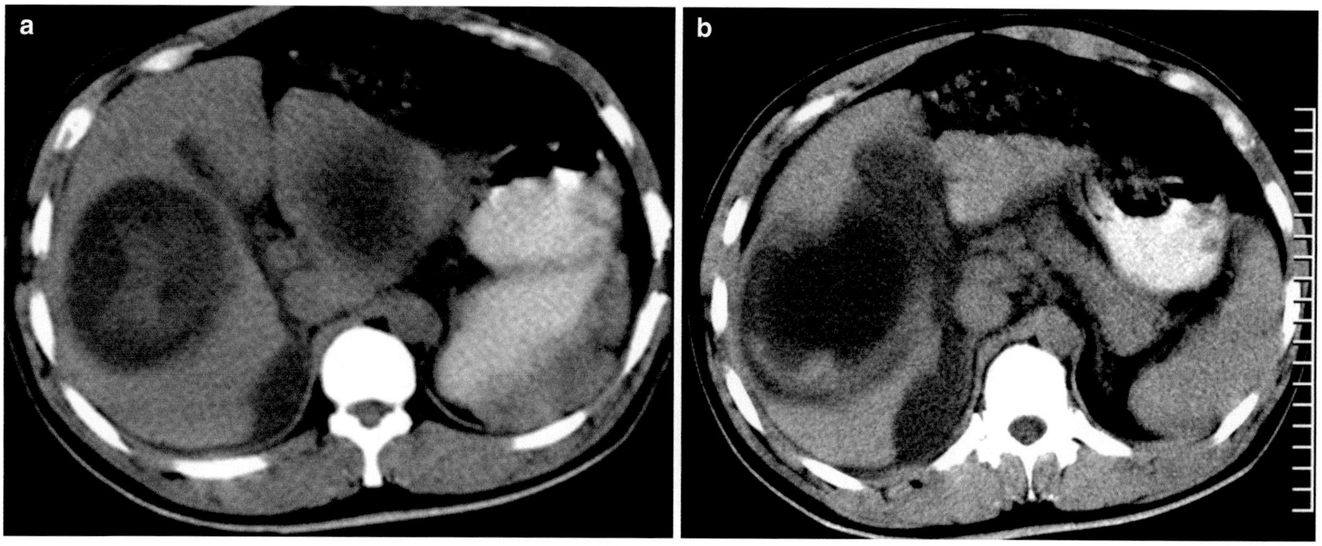

Fig. 8.28 (**a**, **b**) CT plain scan demonstrated ruptured cyst in the right liver lobe. The cyst was poorly defined, with increased density in the cystic cavity. Hydrops was detected in gallbladder fossa and around the liver

due to no blood supply to the lesion. Radiological finding of collapsed internal cyst in the lesion is helpful for the diagnosis of cystic echinococcosis.

Case Study 15

[Brief Medical History]

A 24-year-old young man complained of left upper abdominal upset for more than 6 months with no known causes. He occasionally experienced nausea but no vomiting, fever and jaundice. Laboratory test showed ALB 33.7 g/L, and A/G inverse. DBDX test of ascites showed positive (+). In addition, PALB 74 mg/L, ALT 185 U/L, AST 51 U/L as well as increased mononuclear cell count, neutrophil count and blood platelet count but decreased RBC count. He also showed HBsAg positive (+), HBeAb positive (+) and anti-HBC antibody positive (+).

[Radiological Demonstration] (See Fig. 8.30)

[Diagnosis] Hepatic cystic echinococcosis (Consolidation and calcification type).

[Discussion]

After infected by echinococcus, humans always experience allergy due to absorption of antigen in a small quantity. During puncture of cyst or surgical operation, overflow of the cystic fluid may cause skin rash, fever, tachypnea, abdominal pain, diarrhea, dizziness, delirium, coma and other allergic responses. Even death may occur due to anaphylactic shock in patients with serious condition. Complications of hepatic echinococcosis include rupture of echinococcus cyst and infection, which are main reasons for death and poor prognosis and also indications to immediate surgical excision. Ring shape, half-ring shape, irregular, flake shape or small nodule shape calcifications are characteristic radiological signs for alveolar echinococcosis.

Hepatic cystic echinococcosis should be differentiated from the following diseases:

1. Hepatic cyst
   The lesion of hepatic cyst may be singular or multiple with round or oval shape, homogeneous low density, and smooth boundary. The CT value of the lesions is close to that of water. Contrast scan demonstrates no enhancement of the lesions. in the cases with complicating hemorrhage or infection, the density may increase.
2. Hepatic neoplasm
   Hepatic cystic echinococcosis should also be differentiated from hepatic neoplasms, especially malignancies such as primary hepatic carcinoma. The lesion is poorly defined and commonly invades portal veins and hepatic veins with rapid-in and rapid-out pattern by contrast scan. In the cases of benign hepatic neoplasm, especially angiocavernoma, the lesion is well defined round or round like opacity in low density, with early-out and late-in pattern by contrast scan.
3. Hepatic abscess
   The lesion is round or round like low density opacity, with lower density cavity at the centre as well as observable gas or fluid level. The wall of abscess cavity has a higher density, with obvious ring shaped enhancement by contrast scan to show a typical double-ring sign. The central necrosis cannot be enhanced by contrast scan.

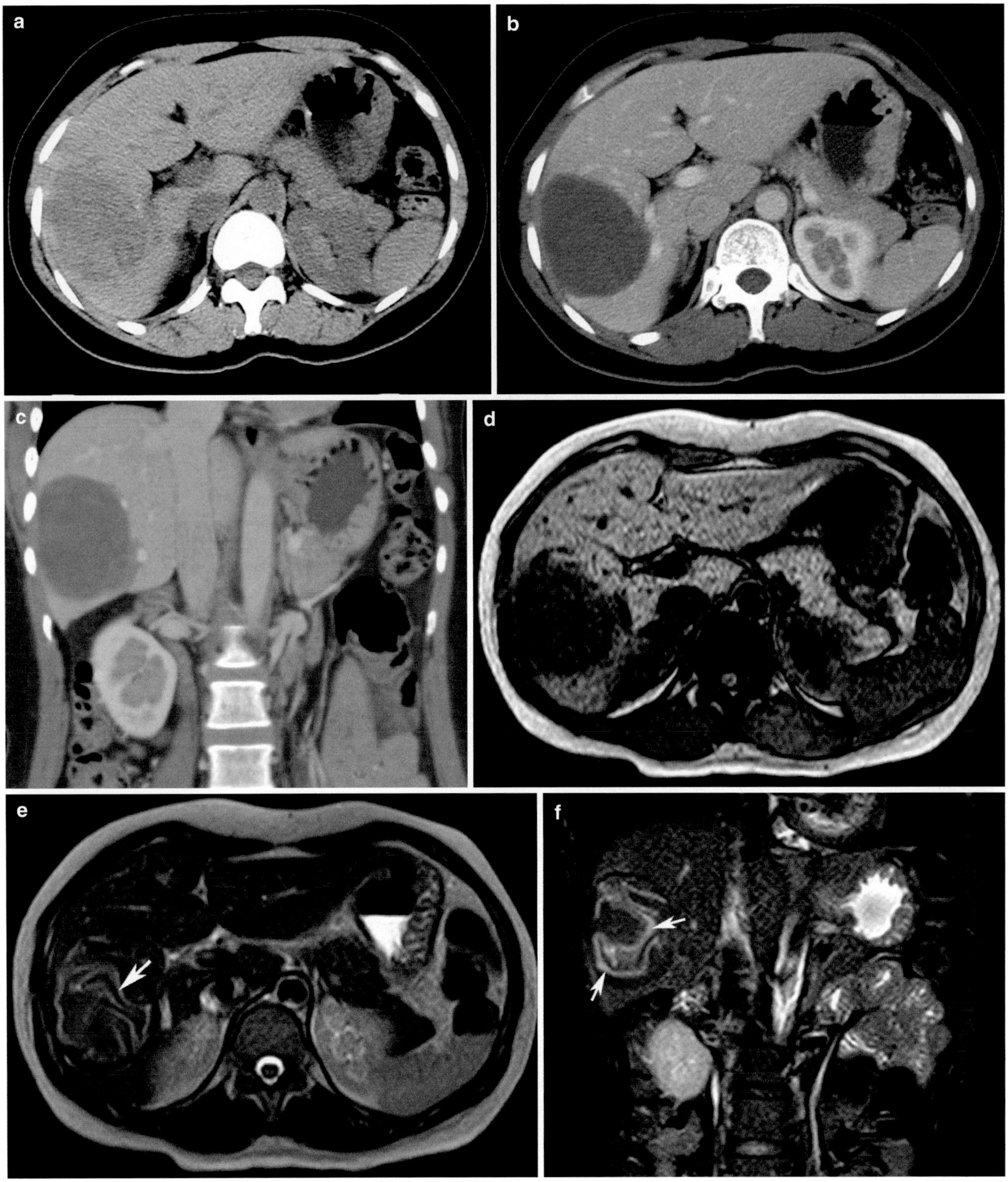

Fig. 8.29 (**a**, **b**) CT plain scan demonstrated a solid space occupying lesion with mixed density in the right liver lobe, with a CT value of 36–45 HU. (**c**) Contrast scan demonstrated no enhancement of the lesion, but better defined, with a size of about 6.63 cm × 5.57 cm. (**d**, **e**) MRI demonstrated mixed signals from the lesion, with dominant low T1WI signal and dominant slightly high T2WI signal. The cystic wall in low signal was detected around the lesion that was well defined. The lesion was shown with internal cords liked ribbon sign (*white arrow*). (**f**) Coronal T2 imaging after fat suppression demonstrated cords liked ribbon sign (*white arrow*) in the lesion

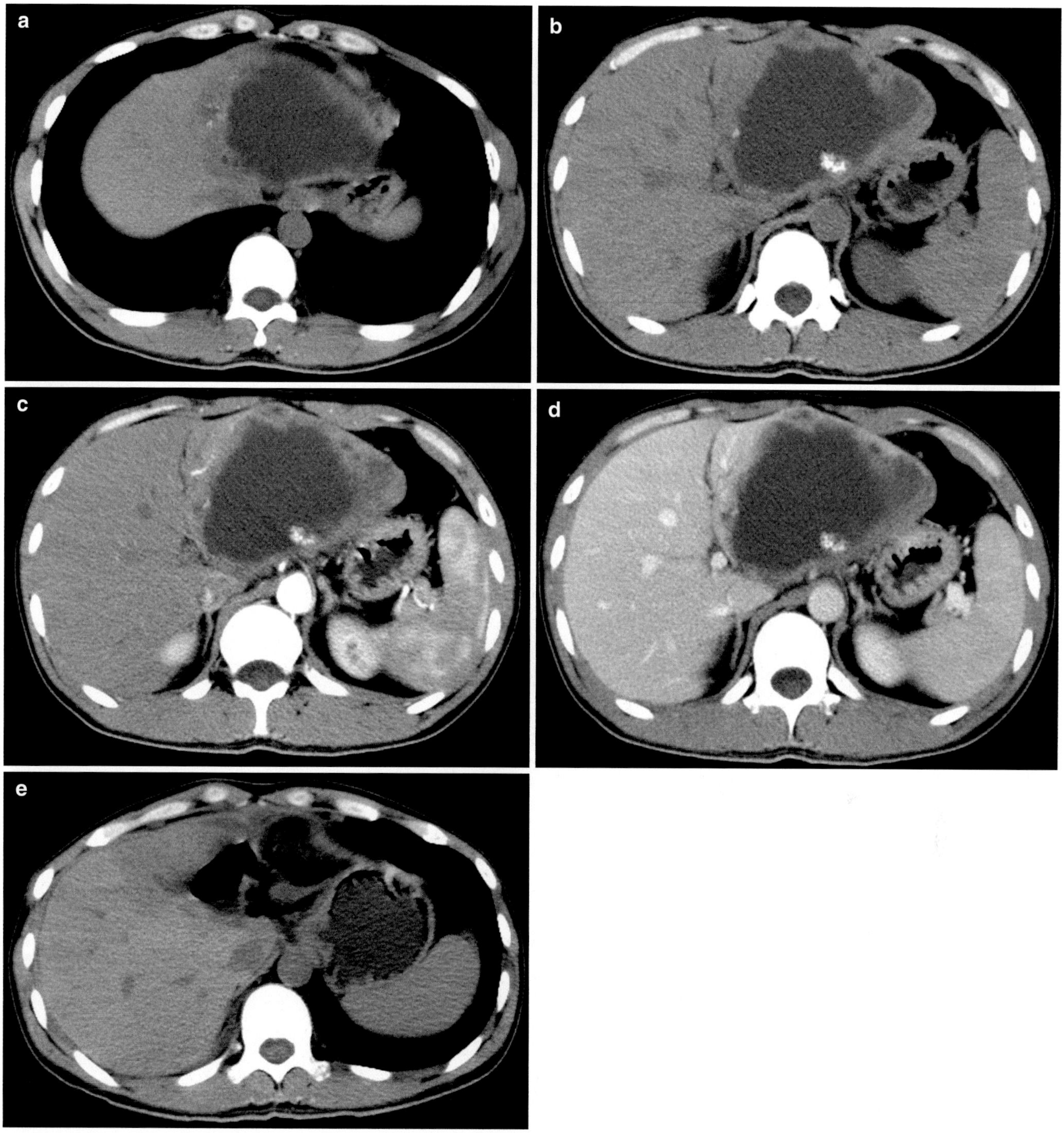

Fig. 8.30 (**a**) CT plain scan demonstrated a cystic lesion in the left live lobe that was close to the hepatic surface. (**b**) The liver was shown to be enlarged, with cystic density lesion in the left liver lobe. The cystic wall was demonstrated to be thick, with unsmooth inner margin and observable calcification. (**c**) Contrast scan in arterial phase demonstrated slight enhancement of the cystic wall in the left liver lobe. (**d**) Contrast CT scan in portal phase demonstrated cystic low density opacity in the left liver lobe. The cystic wall was demonstrated with slight but uneven enhancement, unsmooth inner margin, and multiple calcification lesions. (**e**) Three weeks after surgical operation, CT plain scan demonstrated obvious shrinkage of the cystic low density opacity in the liver, with observable gas-fluid level in the lesion

Case Study 16

[Brief Medical History]

A 42-year-old man complained of repeated upper abdominal upset for 6 months and upper abdominal dull pain for 1 month. Routine blood test showed negative. The Casoni test showed positive (+). He permanently live in a husbandry area.

[Radiological Demonstration] (See Fig. 8.31)

[Diagnosis] Hepatic cystic echinococcosis (consolidation and calcification type).

[Discussion] See Case Study 17.

Case Study 17

[Brief Medical History]

A 50-year old woman reported a hepatic cyst detected by B-mode ultrasound during the physical examination 5 years ago. She did not pay close attention to it and did not receive any reexamination. Recently, B-mode ultrasound in another physical examination showed strips and flakes of calcification within the cyst. The Casoni test showed positive (+). And she was a herdswoman permanently living in husbandry area.

[Radiological Demonstration] (See Fig. 8.32)

[Diagnosis] Hepatic cystic echinococcosis (Consolidation and calcification type).

[Discussion]

Clinically, hepatic echinococcosis is characterized by a long medical history of several years, gradual progress of the condition, and no obvious symptoms in most patients. The patients may experience elevated diaphragm caused by large cyst at the apex of liver, pulmonary symptoms such as dyspnea due to its compression to the lungs, and biliary obstructive symptoms due to compression by the cyst at the inferior liver. Secondary infection is common.

By ultrasonography, the lesion is demonstrated as well defined round or round like cyst with thick cystic wall and smooth margin. Daughter cysts are revealed within the cyst,

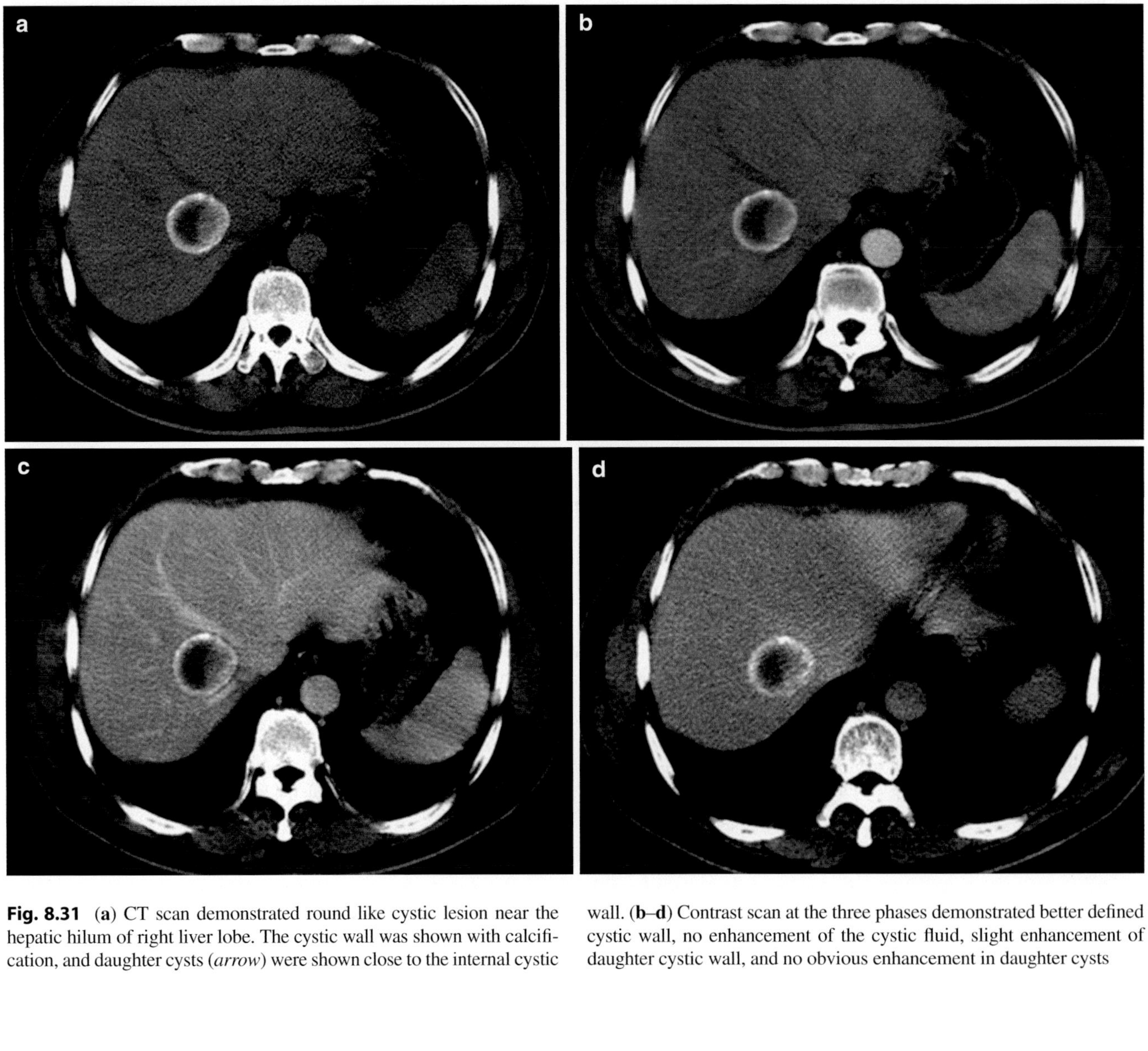

Fig. 8.31 (**a**) CT scan demonstrated round like cystic lesion near the hepatic hilum of right liver lobe. The cystic wall was shown with calcification, and daughter cysts (*arrow*) were shown close to the internal cystic wall. (**b–d**) Contrast scan at the three phases demonstrated better defined cystic wall, no enhancement of the cystic fluid, slight enhancement of daughter cystic wall, and no obvious enhancement in daughter cysts

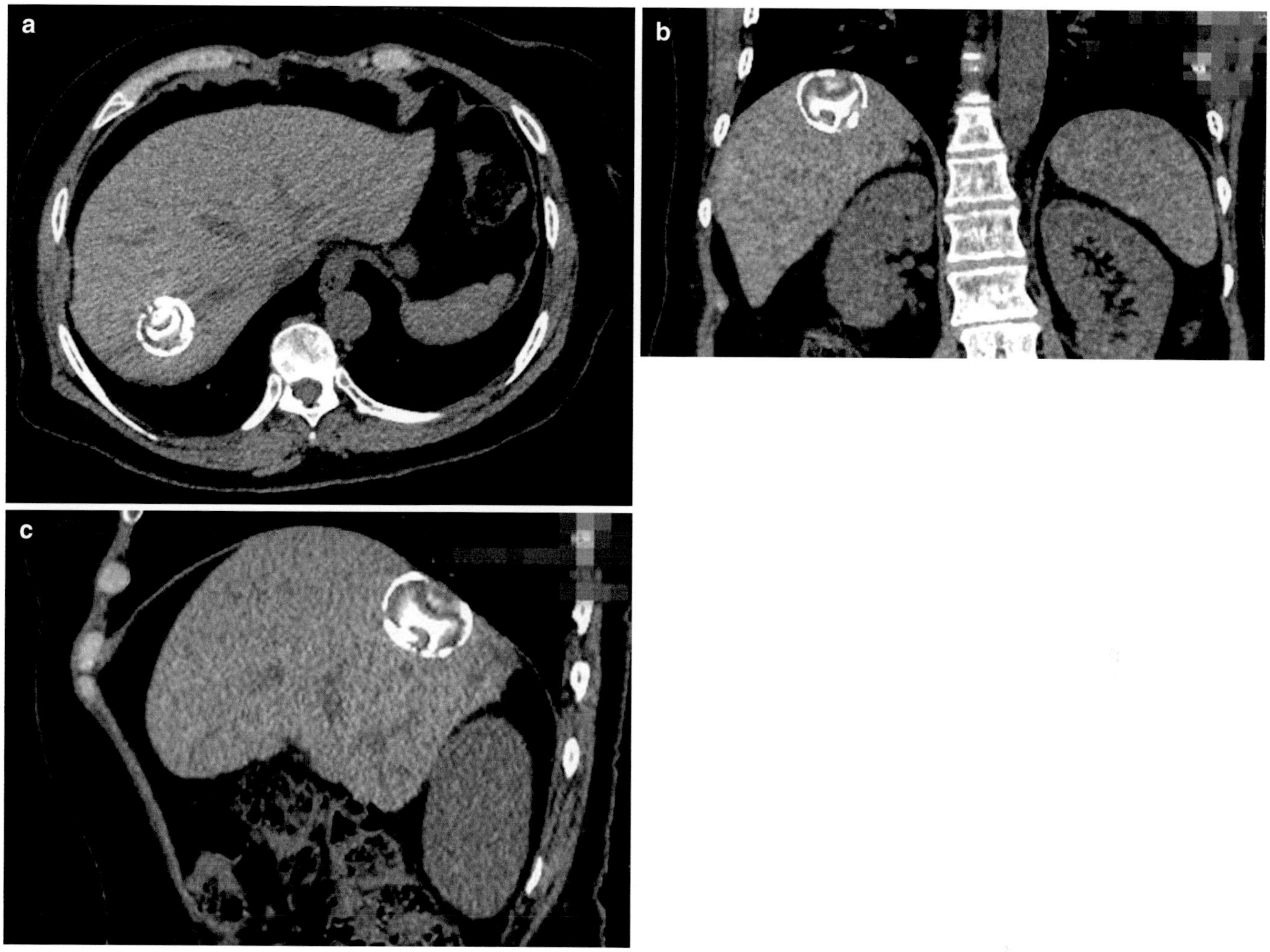

Fig. 8.32 (**a**–**c**) CT scan demonstrated round like cystic lesion in the right posterior liver lobe, with calcified cystic wall as well as strips and flakes of high density opacity in the cyst

with ring, mass or moving spot of light in them. Enhanced echo is demonstrated around the lesion.

By X-ray, plain X-ray has limited diagnostic value for echinococcosis. The occurrence of calcification can be shown at plain X-ray film, in nodular, arch or ring shape opacity. Calcification within the lesion commonly indicates death of echinococcus, but some small irregular calcifications may be malnutritional calcification of old blood clot in the lesion.

By CT scan, the lesion is more commonly detected in the right liver lobe as well defined round or round like low density area in the liver, with smooth margin. Contrast scan hardly demonstrates enhancement of the lesion, but occasionally enhancements of the cystic wall and internal septa. Within large cystic cavity, multilocular structure or daughter cysts (cyst-in-cyst sign) are detectable. The quantity and size of daughter cysts vary, most of which originate from the margin of mother cyst and sometimes arrange in wheel pattern. During the development of echinococcus cyst, the external cyst is subject to degeneration and calcium deposit to show shell like or ring like calcification with homogeneous thickness and regular appearance, which is a characteristic sign of hepatic echinococcosis. Calcification of cystic wall often indicates biological death stage of the echinococcus. Infection or damage may result in separation of the internal cyst, demonstrated as double-line sign. Completely separated, collapsed and curly internal cyst floats within the cyst, demonstrated as lotus-on-water sign. Completely detached internal cyst is radiologically demonstrated as ribbon sign.

By MR imaging, the echinococcus cyst is typically demonstrated as unilocular or multilocular round or oval like cystic lesion, with well defined boundary. The signal from its internal content is close to fluid, demonstrated as low T1WI signal and high T2WI signal with heterogeneous intracystic signal. By T1WI, the signal from daughter cyst is lower than that of the mother cyst, demonstrated as cyst-in-cyst sign. The cystic wall is demonstrated as low T1WI ring signal and low T2WI ring signal. However, MRI is less sensitive to calcification than CT scan, because MRI can hardly differentiate calcification from cystic wall in low signal. If overflow of mother cystic fluid occurs due to infection or rupture, floating-lotus sign or ribbon sign can be demonstrated, namely low T1WI and T2WI signal opacity. However, the MRI image is less favorable to CT image in terms of sensitivity and legibility.

Radiological examinations can accurately localize the lesion of hepatic echinococcosis and can help clinicians comprehensively understanding the range with lesions, detailed structure of mother cyst and daughter cysts, and its spread into abdominal cavity. By CT scan, the density and contents of cystic fluid can be observed for assessment of echinococcus activity. Both cases No. 16 and 17 showed the same history of living in echonococcosis affected region, singular lesion in the right liver lobe, and typical calcification of cystic envelope. For the case No. 16, the daughter cysts in the mother cyst (cyst-in-cyst sign) is the typical sign for the diagnosis of echinococcosis. And clinging of the daughter cysts to the margin of the mother cyst is another characteristic sign for the diagnosis of echinococcosis. For the case No. 17, the cyst envelope was intact with calcification, in addition to detected strips and flakes of calcification in the cyst, which indicated death of echinococcus.

Ultrasound and CT scan are radiological modalities of choice for clinical diagnosis of hepatic echinococcosis. CT scan is superior to ultrasound and MRI in demonstrating calcification, while MRI is more favorable in demonstrating the cystic wall and intracystic septa, but less sensitive to calcification. MRI shows advantages in multi-perspective imaging to more favorably demonstrate the relationships between lesion and its surrounding structures. In combination to epidemic history and immunoassay, hepatic echinococcosis can be diagnosed.

Hepatic echinococcosis should be differentiated from the following diseases:

1. Primary hepatic carcinoma
   CT plain scan demonstrates the lesion of primary hepatic carcinoma as low density opacity. By contrast scan, fast-in and fast-out pattern of enhancement is demonstrated. Calcified envelope is rarely detected.
2. Hepatic hamartoma
   The lesion may show cystic changes and calcification, with heterogeneous enhancement by contrast scan. However, the fluid in echinococcus cyst can hardly be enhanced by contrast scan, only the cystic wall and intracystic septa can be enhanced by contrast scan in rare cases. The calcification of hamartoma is irregularly distributed, while the calcification of hepatic echinococcosis is commonly distributed on the cystic wall.
3. Hepatic abscess
   Hepatic echinococcosis with rupture of the mother cyst and complicating infection should be differentiated from hepatic abscess. Hepatic abscess is radiologically demonstrated as irregular ring shape enhancements of the abscess wall, and the patients clinically experience fever and upper abdominal pain. However, the patients with hepatic echinococcosis commonly show no clinical toxic symptoms.
4. Non-parasitic cyst
   Both simplex hepatic cyst and multilocular hepatic cyst are rarely accompanied by calcification, which is a key point for differential diagnosis. Hepatic echinococcosis is characterized by cyst-in-cyst sign, with daughter cysts mostly clinging to the internal wall of mother cyst.

Case Study 18

[Brief Medical History]

A 38-year-old man reported a space occupying lesion in the liver detected in a physical examination by ultrasound. He had experience no obvious abdomimal upset.

[Radiological Demonstration] (See Fig. 8.33)

[Diagnosis] Hepatic cystic echinococcosis (Consolidation and calcification type).

[Discussion]

In the late stage of hepatic cystic echinococcosis, calcium salt may deposit at the cystic wall to show particles of amorphous calcification, which is a typical radiological sign of echinococcosis. Based on the features of calcification, echinococcosis can be divided into two types:

1. Cystic wall calcification
   Arch shaped calcification may be distributed continually or continuously along the cystic wall. Otherwise, calcification may also be shell like or spots like. The cyst is demonstrated with thickened cystic wall and homogeneous internal density that is close to water. The cyst is demonstrated with gas-fluid level in some cases, with no enhancement by contrast scan.
2. Intracystic calcification
   The cyst is demonstrated with patches or spots of calcification in it, otherwise with linear septa opacity in homogeneously high density. The cystic fluid is demonstrated with relatively high density, with no enhancement by contrast scan. In this case, the lesion was demonstrated with thick strips or fragments of calcification opacity, which should be categorized into intracytstic calcification type.

The lesion of amoebic liver abscess may be demonstrated with calcification of the abscess wall, which resembles to calcification of the cystic wall in the cases of hepatic echinococcosis. However, the calcified wall of amoebic liver abscess is commonly thick, and the cystic fluid shows high density. Calcification of cystic echinococcosis can be divided into cystic wall calcification and intracystic calcification.

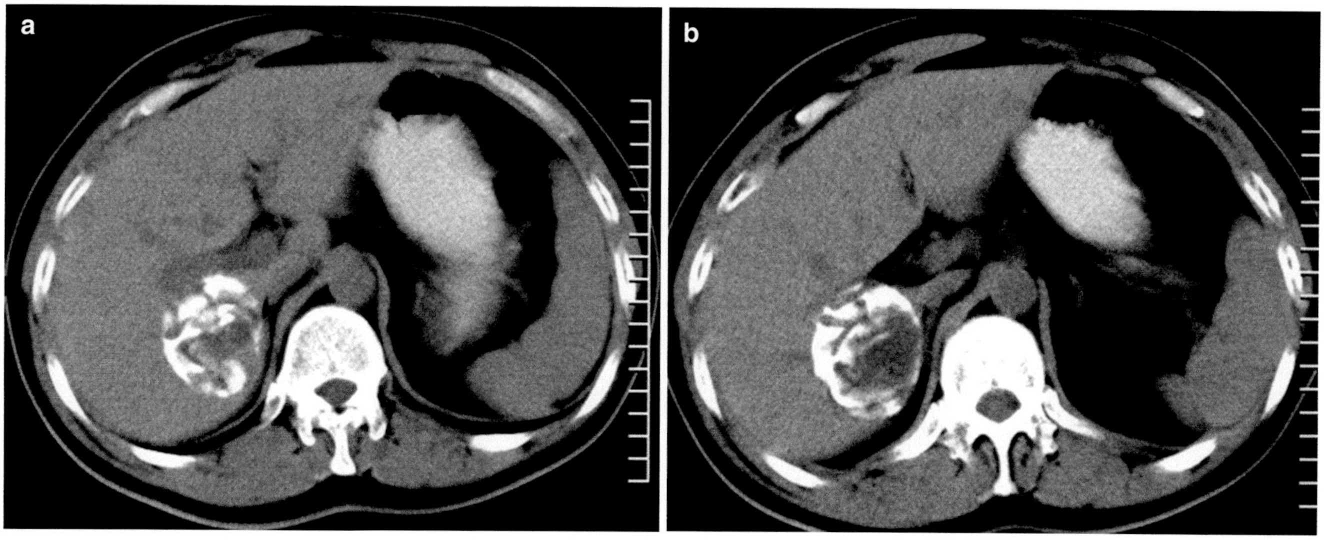

Fig. 8.33 (**a**, **b**) CT scan demonstrated an echinococcus cyst with increased heterogeneous density in the right liver lobe, with no cystic fluid in it but thick arch, strips and fragments of calcification. The lesion was well defined

Case Study 19

[Brief Medical History]

A 66-year-old woman complained of a space occupying lesion in the right liver lobe that had been detected in physical examination. The Casoni test showed positive (+).

[Radiological Demonstration] (See Fig. 8.34)

[Diagnosis] Hepatic cystic echinococcosis in the right liver lobe (Consolidation and calcification type).

[Discussion]

Echinococcosis, also known as hydatidosis, is a serious zoonosis. In China, it highly prevails in husbandry areas and agricultural-husbandry areas, including Xinjiang, Qinghai, Inner Mongolia and Ningxia. Cystic echinococcosis caused by echinococcus granulosus, and alveolar echinococcosis caused by multilocular echinococcus are the common types of echinococcosis in China. Cystic echinococcosis can be further divided into simplex cyst type, multi daughter cysts type, detached internal cyst type, consolidation and calcification type as well as mixed type. Alveolar echinococcosis can be further divided into nodule type, huge lump type, necrosis and liquefaction type, calcification type, and mixed type.

By CT scan, hepatic cystic echinococcosis (consolidation and calcification type) is characterized by spots, crescent like or ring shape calcification at the cystic wall as well as intracystic eggshell like, irregular strip or mass of calcification in plain CT scan. By contrast scan, the cystic wall and content show no enhancement. And the cases of hepatic cystic echinococcosis account for 36.3 % of hepatic echinococcosis. In the middle or late stage of hepatic echinococcosis, the echinococcus gradually declines and degenerates, with absorbed and concentrated cystic fluid. Its solid part increases with caseation, and the cystic wall thickens. In the cyst, irregular strips emerge in appearance of curve onion skin or gyrus like, which is a characteristic sign of echinococcosis. As time elapses, the quantity of calcification increases at the cystic wall and within the cyst, sometimes with complete calcification to show diversified radiological signs. Calcification indicates biological activity loss of the echinococcus.

Hepatic cystic echinococcosis (consolidation and calcification type) should be differentiated from lipiodol retention caused by interventional embolization for hepatic carcinoma, fibrolamellar hepatocellular carcinoma and hepatic cavernous hemangioma. The key points for differential diagnosis are as the following:

The patients with lipiodol retention caused by interventional embolization for hepatic carcinoma have a medical history of hepatic carcinoma and interventional embolization.

Fibrolamellar hepatocellular carcinoma is a special type of hepatic carcinoma, with spots or small round shape calcification in high density. By contrasts scan, its parenchymal part shows different degrees of enhancement, but its fibrous septa show no enhancement that indicates their low density.

Hepatic cavernous hemangioma is a benign hepatic neoplasm. By contrast scan, multiple nodular or centripetally delayed enhancement can be demonstrated in the lesion.

For this case of hepatic cystic echinococcosis (consolidation and calcification type), intracystic calcification is gyrus like in arrangement of a lump. By contrast scan, the cystic wall and contents show no enhancement.

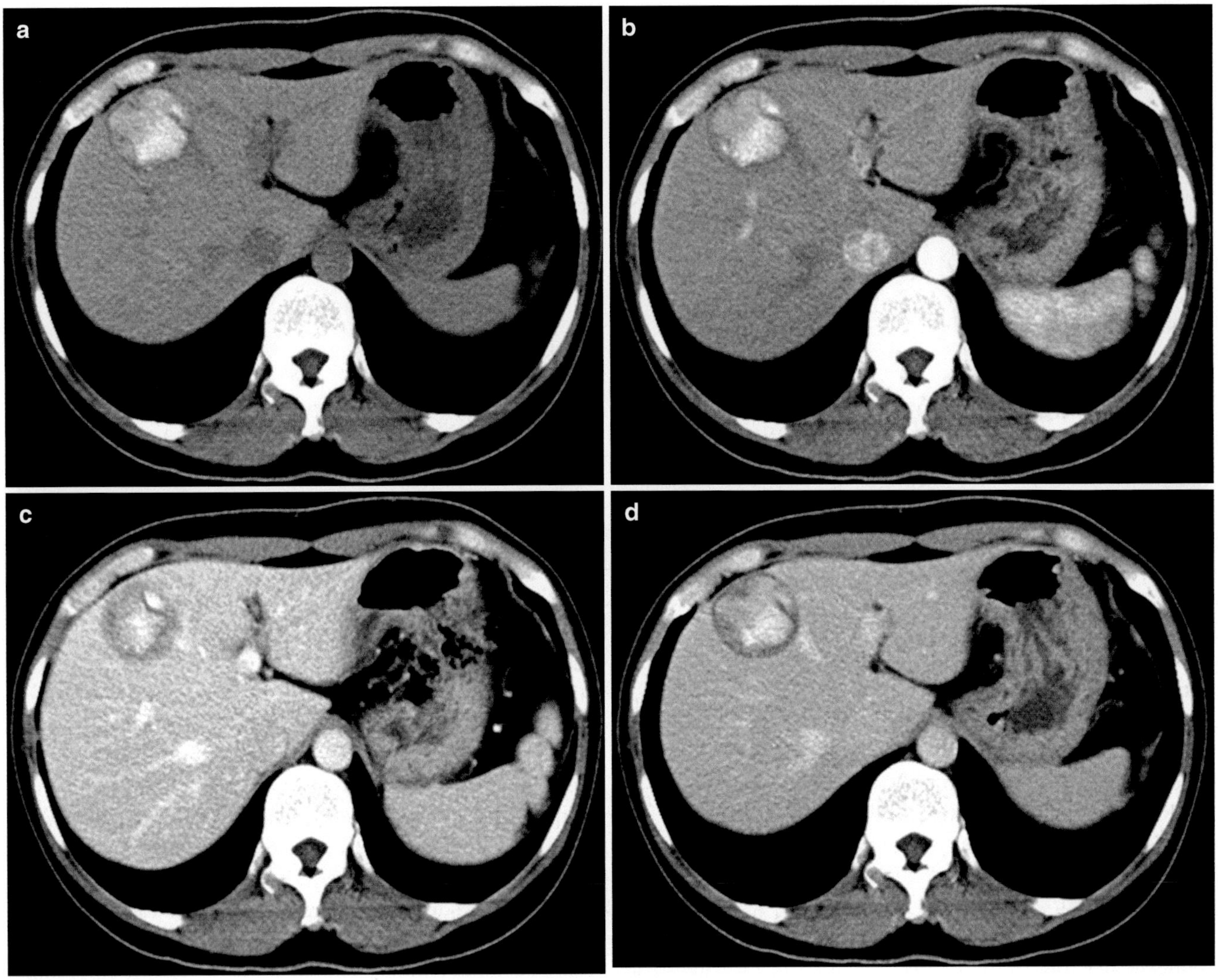

Fig. 8.34 (**a**) CT plain scan demonstrated oval like lesion with mixed high and low densities in the right liver lobe. The high density opacity was defined to be calcification that was well defined. (**b–d**) Contrast scan at all three phases demonstrated no enhancements of the cystic wall and content in the right liver lobe

Case Study 20

[Brief Medical History]

A 67-year-old man complained of upper abdominal distention with pain and upset for 6 months that aggravated for 1 month. The Casoni test showed positive (+).

[Radiological Demonstration] (See Fig. 8.35)

[Diagnosis] Alveolar echinococcosis in the right liver lobe (Calcification type).

[Discussion]

The primary lesion of alveolar echinococcosis is located in the liver in almost all the cases. After oral intake of eggs of alveolar echinococcus by human, their oncospheres are incubated in human duodenum and invade human liver via portal vein to form many small alveola with a diameter of 1–10 mm, or even larger. And then, they continually infiltrate the surrounding liver tissue by extra-alveolar budding. The alveola contain bean curd residues like echinococcal fragments, and jelly like fluid. Numerous small alveola gather to form irregularly shaped nodules, which rapidly grow into huge solid lump with its section in grayish white honeycomb like change, appearing like a neoplasm. Therefore, alveolar echinococcosis is also known as worm cancer. The lump is not enveloped by fibrous tissue, with poorly defined boundary with the surrounding liver tissue. Due to poor blood supply or degeneration of the lesion, necrosis and liquefaction often occur to form large hollow cavity with irregular shape, which may further develop into hepatic abscess due to secondary infection. And its systematic metastasis after its invasion into vascular or lymphatic vessels is one of the main characteristics of alveolar echinococcosis. During the growth and development of alveolar echinococcus, calcium tends to deposit to show particles or

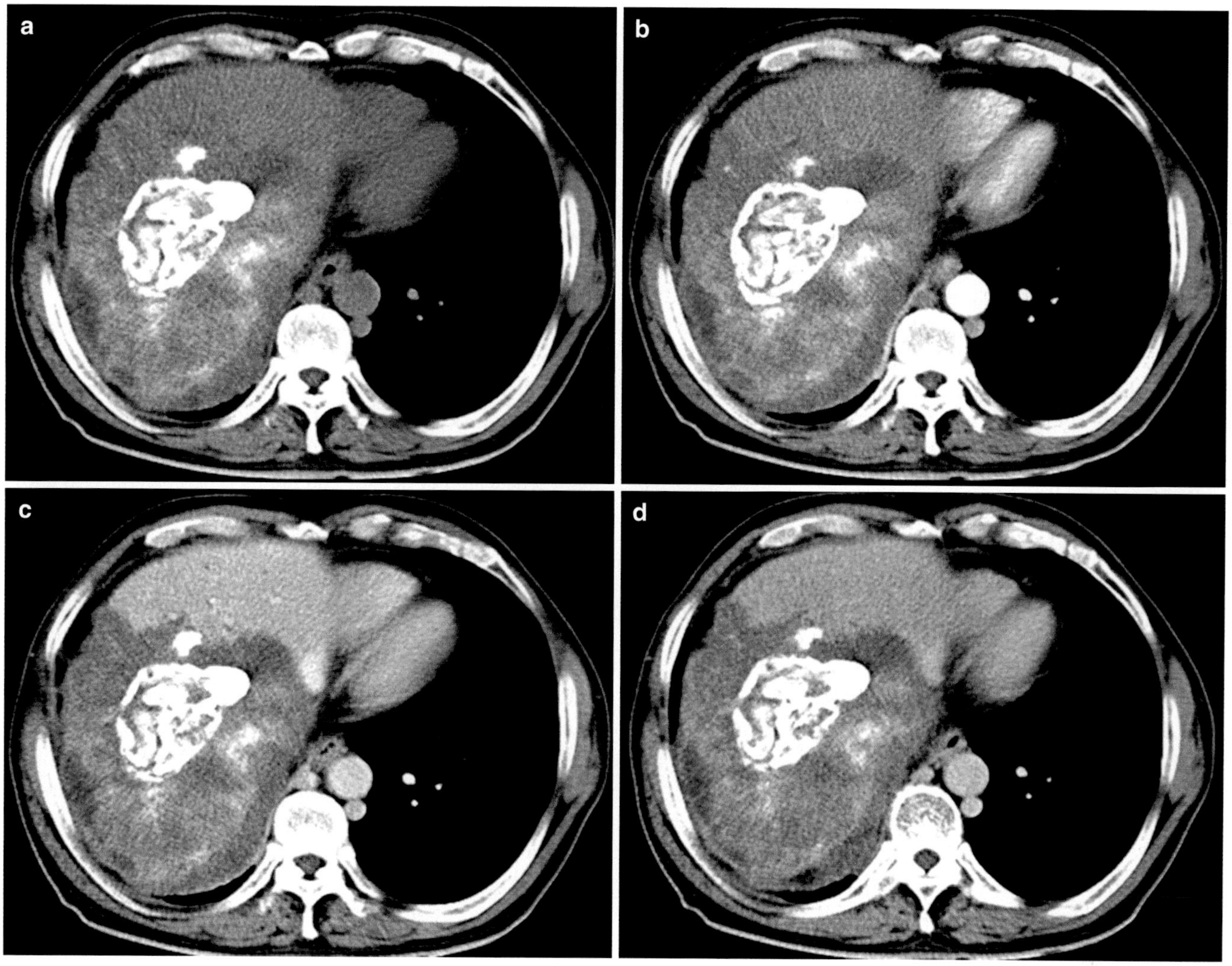

Fig. 8.35 (**a**) CT scan demonstrated a poorly defined large oval like lesion with low density in the right liver lobe. The surrounding liver parenchyma was demonstrated with decreased density and the lesion was shown with diffusely distributed patches and gyrus like calcification. (**b–d**) Contrast scan at the three phases demonstrated no enhancement of the lesion

spots of calcification in the early stage. Along with the illness course, calcifications increase and infuse into irregular flakes or lumps. Such a pathological development and evolving process of alveolar echinococcosis including infiltrative growth, necrosis, liquefaction, distant metastasis and calcification of small dense alveoli constitute the basis for radiological diagnosis. Calcification of alveolar echinococcus is one of the characteristic pathological changes of alveolar echinococcosis. In the cases of both nodular type and huge lump type during its early stage or late stage, calcification occurs. Calcification indicates degeneration and death of echinococcus in the late stage, with extensive calcium salt deposits within and at the margin of the lesion to show partial or complete calcification. No small alveoli can be shown within the lesion, which have lost its biological activity and can co-exist with human body for a long period of time.

By CT scan, the lesion of alveolar echinococcosis is demonstrated as poorly defined opacity with mixed low and high density and extensive particles or irregular shapes of calcification in a map sign. The lesion may also be demonstrated with necrosis and liquefaction, with no enhancement by contrast scan.

Alveolar echinococcosis (calcification type) should mainly be differentiated from fibrolamellar hepatocellular carcinoma and lipiodol retention caused by interventional embolization for hepatic carcinoma. The key points for differential diagnosis include enhancement of the lesion by contrast scan. Fibrolamellar hepatocellular carcinoma is a special type of hepatic carcinoma, with the lesion demonstrated as well defined low density opacity with cords like structure and necrosis by plain scan. The lesion is characterized by internal calcification, which are spots or small round in shape with high density. By contrast scan, the parenchymal part of lesion can be enhanced at the arterial phase, but the comparatively low density fibrous septa show no enhancement. At the delay phase, the central scar area shows no enhancement, which can be more favorably demonstrated with low density. The patients with lipiodol retention caused by interventional embolization for hepatic carcinoma have a medical history of interventional lipiodol embolization for hepatic carcinoma.

In this case of hepatic alveolar echinococcosis (calcification type), a large oval like low density lesion is demonstrated in the right liver lobe, with poorly defined boundary between its surrounding normal liver tissue. The lesion was demonstrated with diffusely distributed patches and gyrus like calcification. And the lesion showed no enhancement by contrast scan.

Case Study 21

[Brief Medical History]

A 59-year-old Uyghur woman complained of dull pain at the hepatic region for 1 year that aggravated for 3 days. She experienced no obvious poor appetite, fatigue or weight loss, but reported a definitive history of contacts to dogs.

[Radiological Demonstration] (See Fig. 8.36)

[Diagnosis] Hepatic alveolar echinococcosis (calcification type).

[Discussion]

Alveolar echinococcosis is caused by parasitism of echinococcus multilocularis in human or animal, with liver the most commonly involved. Its clinical manifestations and radiological demonstrations are atypical, and the condition tends to be misdiagnosed. Alveolar echinococcus grows in liver in infiltrative budding, with a large quantity of small vesicles produced in the germinal layer that resemble to a honeycomb to infiltrate its surrounding area. In this case, the patient showed a poorly defined low density lesion in the liver, with scattering calcification inside. Contrast scan showed no obvious enhancement of the lesion. In combination to the definitive history of contacts to dogs, a diagnosis of alveolar echinococcosis was suspected. Meanwhile, a differential diagnosis from intrahepatic neoplasm should be made.

Hepatic alveolar echinococcosis should be differentiated from the following diseases:

1. Hepatic carcinoma
   The lesion of both alveolar echinococcus and hepatic carcinoma grows in an infiltrative pattern to form a solid mass, which is radiologically demonstrated to be irregular, space occupying, and solid. The key points for differential diagnosis are as the following:
   The lesion of hepatic carcinoma is demonstrated as a low density lesion by CT plain scan, with poorly defined boundary and surrounding edema. However, the lesion of alveolar echinococcosis is well defined with no surrounding edema. The necrotic and liquefied area in the lesion of alveolar echinococcosis is irregular in shape with a volcanic-lava sign or map sign, while the necrotic area in the lesion of hepatic carcinoma is commonly located at the centre. By contrast scan, the lesion of alveolar hepatic echinococcosis is demonstrated with no enhancement but the lesion of hepatic carcinoma can be obviously enhanced in a dynamic rapid-in and rapid-out enhancement. Clinically, a diagnosis of hepatic carcinoma can be defined in combination with AFP test.
2. Hemangioma
   Ultrasound demonstrates the lesion of hemangioma as a round or round like lesion in sharp or well defined strong echo. Contrast CT scan demonstrates consecutive enhancement of the lesion. A differential diagnosis can be made based on Casoni test negative.

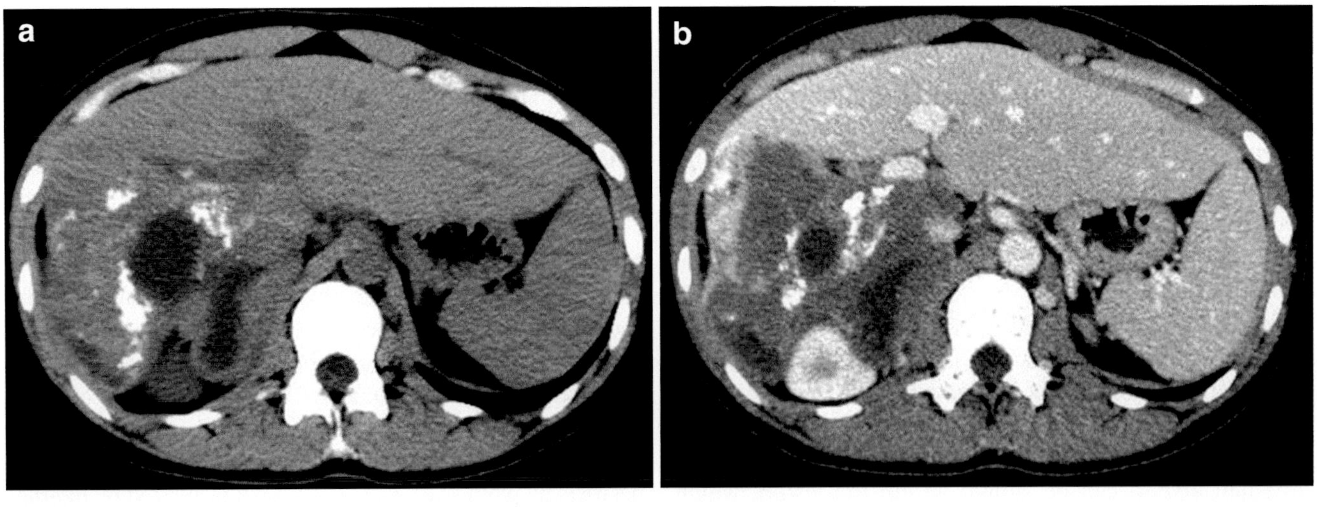

Fig. 8.36 (**a**) CT plain scan demonstrated poorly defined irregular low density lesion in the right liver lobe, with multiple calcifications and cystic changes inside. (**b**) Contrast scan demonstrated well defined lesion with no obvious enhancement

Case Study 22

[Brief Medical History]

A 56-year-old woman complained of right upper abdominal pain for 1 week. B-mode ultrasound of the abdomen showed a space occupying lesion in the liver. She reported a history of living in husbandry area for 24 years. The 4-item examination for echinococcus revealed Anti-EgCF antibody (+), Anti-EgP antibody (+), Anti-EgB antibody (+) and Anti-Em2 antibody (+).

[Radiological Demonstration] (See Fig. 8.37)

[Diagnosis] Hepatic alveolar echinococcosis.

[Discussion]

Resembling to neoplasm, the lesion of hepatic alveolar echinococcosis is characterized by infiltrative growth after budding from the germinal layer to spread to its surrounding area. By contrast scan, no enhancement of the internal lesion can be demonstrated, but the lesion is well defined due to the enhanced hepatic parenchyma around it. MRCP can more favorably demonstrate signs of small vesicles. Based on presence of liquefaction and necrosis in the lesion, the lesion can be further divided into consolidation type, mixed type and pseudocyst type. Due to inflammatory responses and necrosis, secondary calcium sedimentation occurs, showing spots, particles, tree rings and characteristic small rings calcification. Pseudocyst type of alveolar echinococcosis should be differentiated from simplex cystic echinococcosis. Generally, the differential diagnosis can be made. But some cases of alveolar echinococcosis with more liquefaction and necrosis in the lesion and less solid component at the margin tend to be misdiagnosed. Due to the characteristic infiltrative growth of alveolar echinococcus with unsmooth margin, the presence of small vesicles can exclude the diagnosis of cystic echinococcus no matter how large the liquefaction is. The mixed type and consolidation type of alveolar echinococcosis should be differentiated from primary neoplasm, and the findings by contrast scan are of great importance. By contrast scan, most hepatic malignancies are demonstrated with heterogeneous internal enhancement and sometimes abnormal blood supply to the neoplasm. However, by contrast scan, the lesion of alveolar echinococcosis shows no internal enhancement but slight marginal enhancement..

Case Study 23

[Brief Medical History]

A 40-year-old man reported a space occupying lesion in the liver for 1 week.

[Radiological Demonstration] (See Fig. 8.38)

[Diagnosis] Hepatic alveolar echinococcosis with intrahepatic metastasis.

[Discussion]

An infiltration area exist between the lesion of hepatic alveolar echinococcosis and normal liver tissue, which is dominantly proliferated fibrous connective tissue. This area is related to the proliferative and infiltrative activity of the lesion. Some lesions are revealed with obvious marginal enhancement by contrast scan, which tend to be misdiagnosed as hepatic carcinoma with poor blood supply. In some cases, the intrahepatic metastatic lesion of alveolar echinococcosis can be also demonstrated as marginal ring shape enhancement, which presents difficulty for its differential diagnosis from hepatic metastatic carcinoma. By MRCP, the small vesicles sign as well as the indirect signs of shrunken and sunken lesion near hepatic margin, and compensatory enlargement of the healthy liver lobe all facilitate the diagnosis of alveolar echinococcosis.

Case Study 24

[Brief Medical History]

A 59-year-old woman complained of dull pain at the hepatic region for 1 year that aggravated for 3 days. She was administered therapy for hepatic carcinoma, with no obvious therapeutic efficacy.

[Radiological Demonstration] (See Fig. 8.39)

[Diagnosis] Hepatic alveolar echinococcosis (liquefaction and necrosis type).

[Discussion]

Resembling to hepatic carcinoma, the lesion of hepatic alveolar echinococcosis is characterized by infiltrative growth, with occurrence of hollow cavity due to ischemia, degeneration, necrosis and liquefaction. The lesion may also show no obvious envelope but commonly calcification. Small spots with multiple layers concentric circle like calcification are characteristic radiological sign. MR imaging is less sensitive to calcification than CT scan. In this case, no obvious enhancement of the lesion is demonstrated by contrast scan and the patient showed no therapeutic response to anti-neoplastic therapy. Based on these facts, hepatic alveolar echinococcosis is suspected.

Case Study 25

[Brief Medical History]

A 45-year-old man complained of right upper abdominal distension for 1 year, and B-mode ultrasound showed demonstrating space occupying lesion in the liver. He reported a history of living in husbandry area.

[Radiological Demonstration] (See Fig. 8.40)

[Diagnosis] Hepatic alveolar echinococcosis (huge lump type).

[Discussion]

In this case, MR imaging demonstrated poorly defined irregular lump like lesion in the hepatic parenchyma with uneven internal signal and no envelope. The lesion was shown with low T1WI signal and mixed lower T2WI signal.

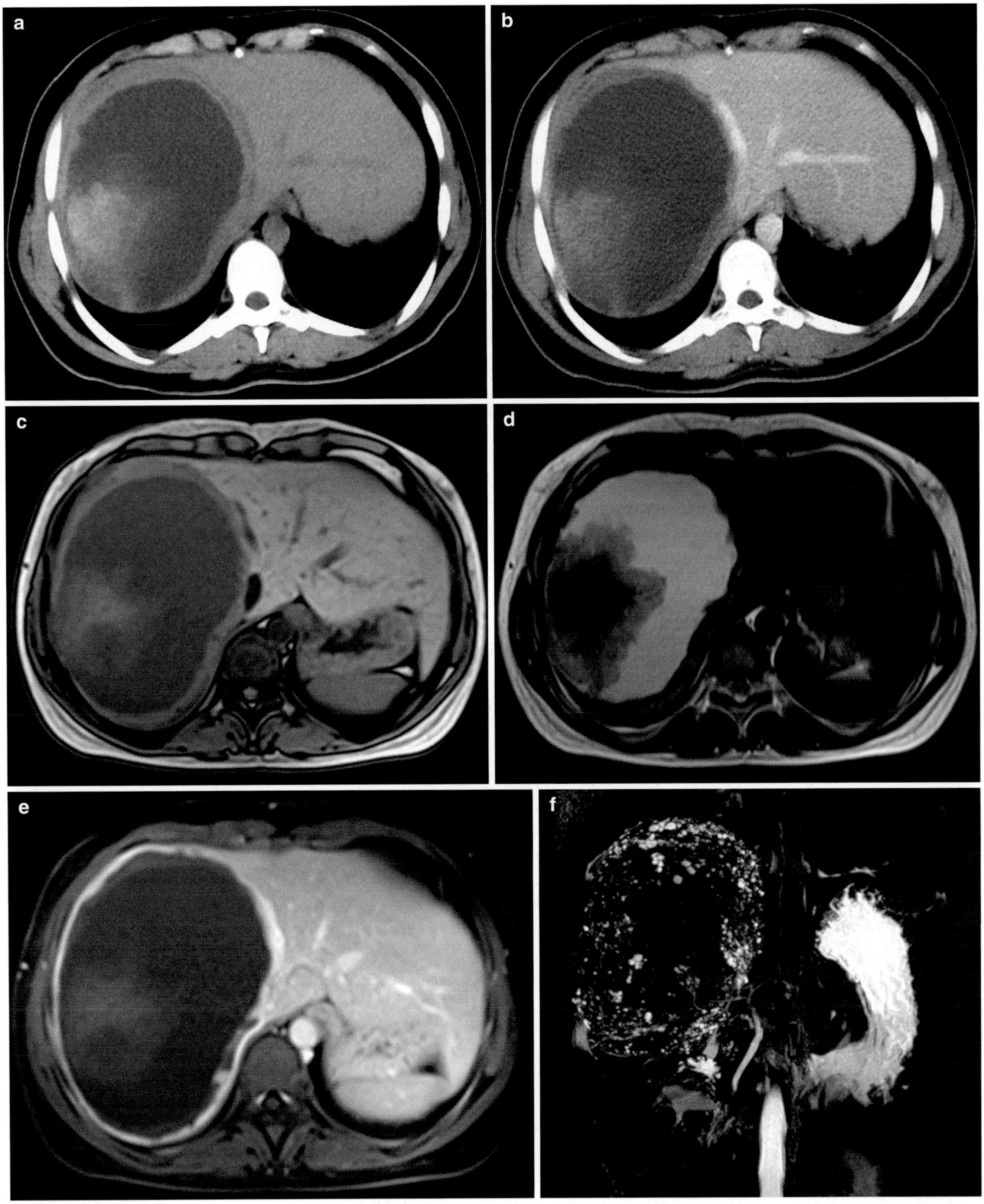

Fig. 8.37 (**a**) CT plain scan demonstrated a huge solid-cystic space occupying lesion with mixed density in the right liver lobe, with amorphous calcification in its solid part. (**b**) Contrast scan demonstrated no obvious enhancement of the lesion, with more clearly defined boundary. (**c**, **d**) MR imaging demonstrated mixed T1WI signal and T2WI signal in the lesion. (**e**) Contrast imaging demonstrated marginal enhancement of the lesion. (**f**) MRCP demonstrated multiple vesicles around the lesion

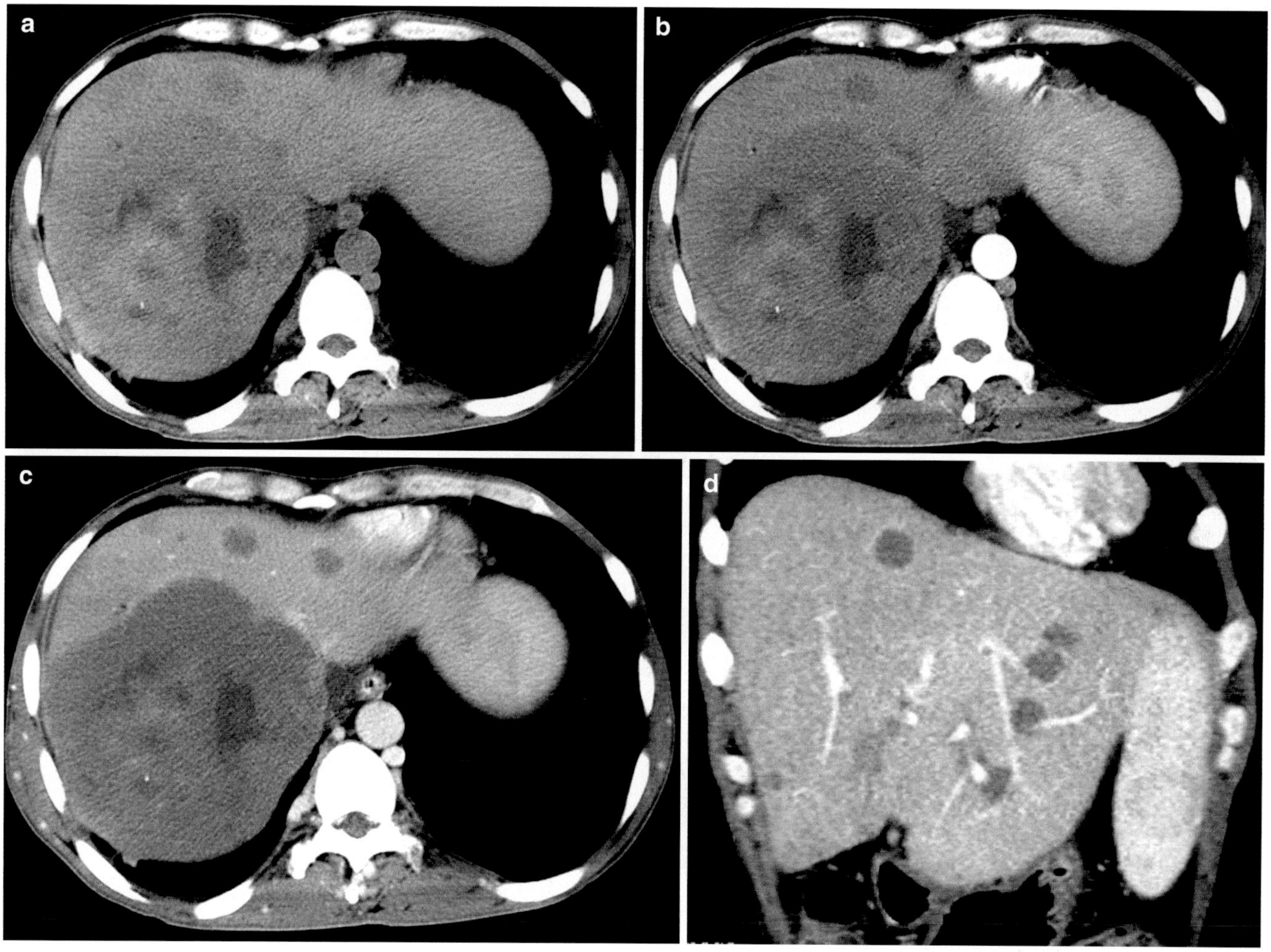

Fig. 8.38 (**a–d**) CT scan demonstrated a huge space occupying poorly defined lesion with mixed density in the right liver lobe. Contrast scan demonstrated marginal enhancement of the lesion, which was especially obvious at the portal venous phase. The other hepatic parenchyma was demonstrated with multiple round like low density nodules, which showed ring shape enhancement by contrast scan

Contrast imaging demonstrated no enhancement of the lump. And biliary obstruction and migration of adjacent biliary duct due to compression were revealed. In combination to his history of living in husbandry area, hepatic alveolar echinococcosis should be firstly considered. Meanwhile, it also should be differentiated from other special types of hepatic neoplasms.

Case Study 26

[**Brief Medical History**]

A 44-year-old man complained of difficult bowel movements with frequent urination for 3 months. B-mode ultrasound revealed a space occupying lesion in the pelvic cavity. The Casoni test was positive (+).

[**Radiological Demonstration**] (See Fig. 8.41)

[**Diagnosis**] Hepatic cystic echinococcosis in the right liver lobe (detached internal cyst type) and pelvic cystic echinococcosis (multi daughter cysts type).

[**Discussion**]

Detached internal cyst type of cystic echinococcosis accounts for 5.4 % of all the cases of hepatic echinococcosis. Due to external physical force or internal environmental change (e.g. changes in external and internal pressure), the internal cyst of echinococcus may be subject to rupture and detachment. At this time, the internal cyst is partially or completely separated from the cystic wall, demonstrated as curved strips floating on the cystic fluid. Detached internal cyst is one of the typical signs indicating decreased biological activity and degeneration of echinococcus. CT scan shows varying signs due to different degrees of internal cyst separation from the cystic wall. For the cases with slight or partial separation, the space between the internal and external cysts gradually widens, with part of the internal cyst still attached to the cystic wall. For the cases of complete separation, the internal cyst detaches from the cystic wall, showing lily-on-water sign. Otherwise, the internal cyst is free from the cystic wall and floats on the cystic fluid to show a ribbon

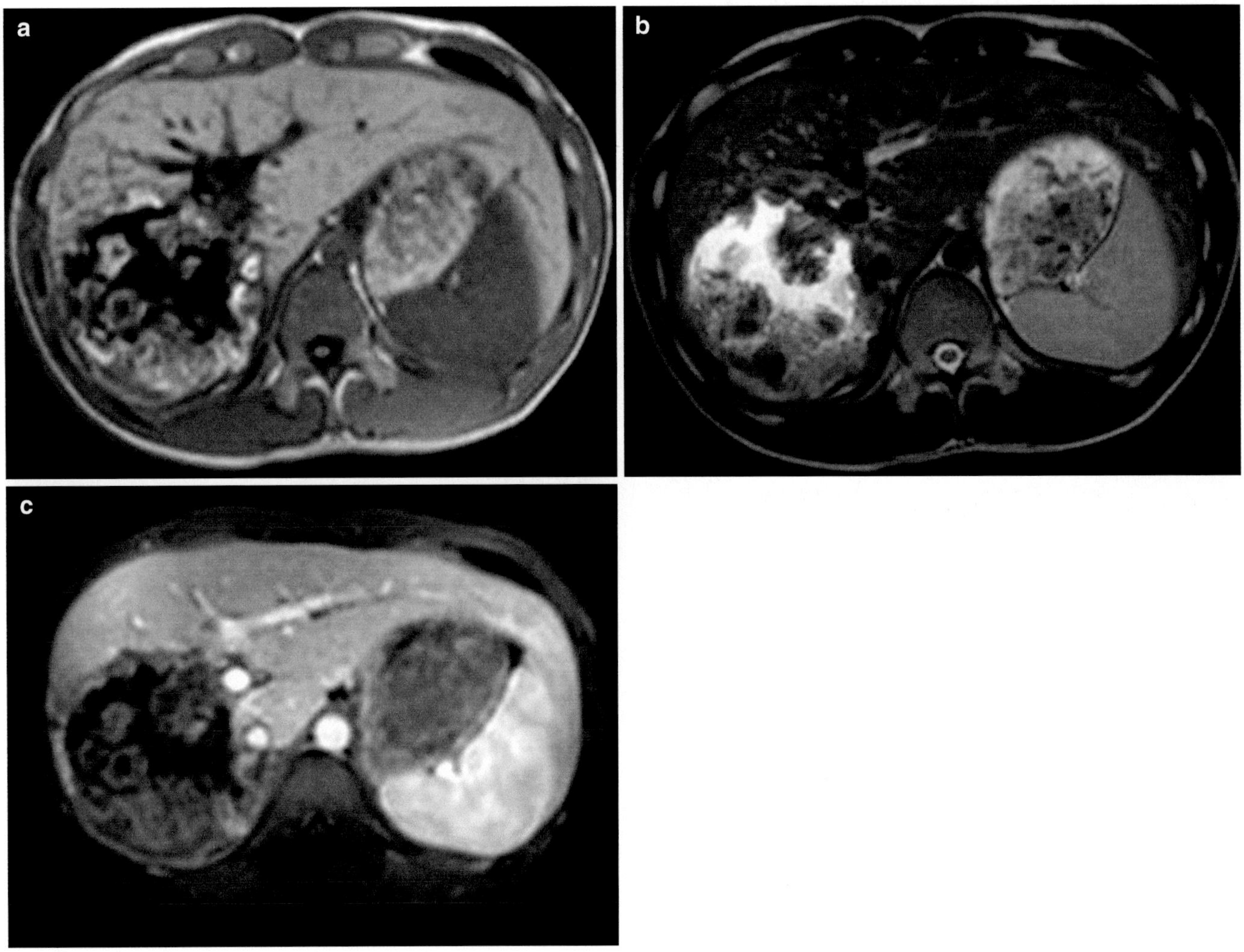

Fig. 8.39 (**a**, **b**) MR imaging demonstrated a lesion of alveolar echinococcosis in the right liver lobe, with internal liquefaction lesion in volcanic-lava sign. The lesion was demonstrated with low T1WI signal and high T2WI signal, with patches of calcification inside. (**c**) Contrast scan demonstrated well defined lesion with no enhancement

sign. By contrast scan, the cystic wall and content show no enhancement.

Detached internal cyst type of cystic echinococcosis should be mainly differentiated from hepatic abscess (abscess period), biliary cystadenoma or cystadenocarcinoma.

In this case of cystic echinococcosis, the lesion can be categorized into partially detached internal cyst type. The right liver lobe was shown with low density lesion with uneven internal density and lower density close to its margin. Cotton like high density opacity was detected at the centre, indicating ruptured internal cystic wall in the lesion with multi daught cysts floats in the cystic fluid to show atypical ribbon sign. The pelvic cavity was revealed with oval like low density lesion, with septa inside, indicating multi daughter cysts type of cystic echinococcosis. Contrast scan showed no enhancements of the cystic wall and content.

Case Study 27

[**Brief Medical History**]

A 37-year-old woman complained of enlarged abdomen for 7 years and abdominal distension for 1 week. Based on B-mode ultrasound and CT scan, the diagnosis is considered to be abdominal myxadenoma. The 4-item examination for echinococcus showed Anti-EgCF antibody (++), Anti-EgP antibody (++), Anti-EgB antibody (++) and Anti-Em2 antibody (+).

[**Radiological Demonstration**] (See Fig. 8.42)

[**Diagnosis**] Cystic echinococcosis (multi daughter cysts type) at the liver, abdominal cavity and pelvic cavity.

[**Discussion**]

In this case, the condition is caused by cystic rupture of multi daughter cysts type in the posterior segment of right liver lobe, which further leads to dissemination of the lesion into the abdominopelvic cavity. The abdominopelvic cavity

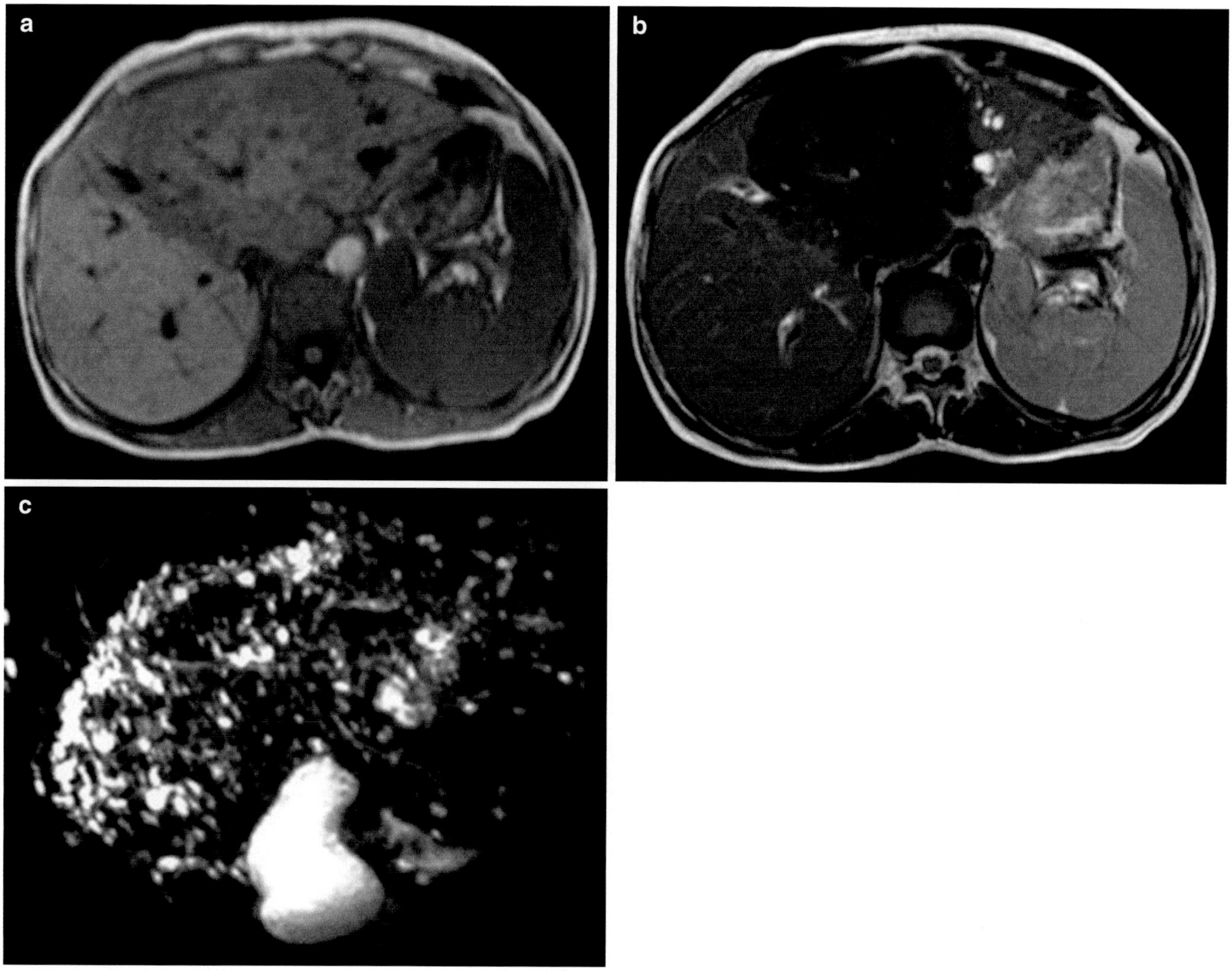

Fig. 8.40 (a–c) MR imaging demonstrated poorly defined irregular lump like lesion in the liver with uneven internal signal. The lesions was revealed with low T1WI signal and lower T2WI signal. The adjacent biliary duct was shown with migration due to compression and dilation

was demonstrated to be occupied by thin-wall cystic low density lesions and the part of lesion in the pelvic cavity was shown with soft tissue density. These findings confused the clinician in its diagnosis. In combination to reconstructed coronal and sagittal CT scan images, the cystic wall of at the right marginal liver was revealed to be continual, with partial internal cystic wall protruding into the abdominal cavity. The marginal calcification on the lesion of right liver margin is also helpful for the diagnosis. This type of condition should be differentiated from ovarian and appendiceal cystadenoma or cystadenocarcinoma. Cystadenocarcinoma may show extensive spread and metastasis in the abdominopelvic cavity, with cystic or soft tissue density masses of different sizes occurring on the surface of peritoneum, omentum and organ surface. Contrast scan demonstrates enhancements of solid components and the cystic wall, which facilitates its differential diagnosis from cystic echinococcosis.

Case Study 28

[**Brief Medical History**]

A 31-year-old Uyghur woman reported her finding of an abdominal lump 2 months ago, with protruding abdominal wall. She reported a history of contact to dogs.

[**Radiological Demonstration**] (See Fig. 8.43)

[**Diagnosis**] Cystic echinococcosis in abdominal and pelvic cavity (multi daughter cysts type).

[**Discussion**]

The lesion of cystic echinococcosis in abdominal and pelvic cavity are often distributed in the space between abdominal and pelvic organs or in the spaces close to the pelvic wall to compress its neighbouring organs. Cystic echinococcosis is commonly complicated by cystic echinococcosis of abdominal organs. According to the morphology of abdominal and pelvic lesions, the lesion can be divided into five types: simplex cyst type, multi daughter cysts type, cystic

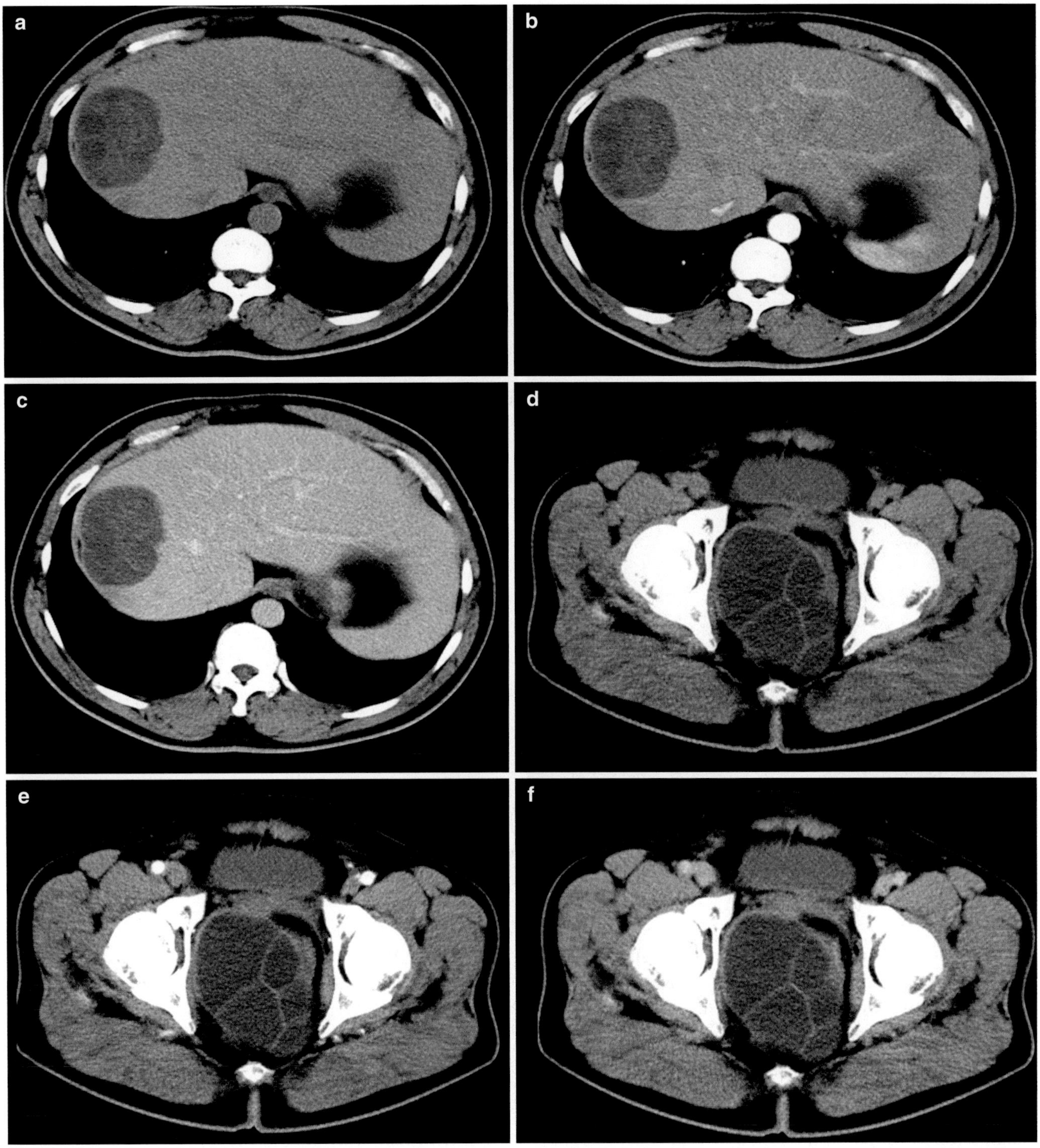

Fig. 8.41 (**a**) CT plain scan demonstrated well defined oval like cystic low density lesion in the right liver lobe, with uneven internal density and lower density close to its margin. The lesion was shown with cotton like slightly high density opacity at the center and septa, demonstrated as cyst-in-cyst sign. The upper part was revealed with curved strips floating on the cystic fluid. (**b**, **c**) Contrast scan demonstrated at the arterial and portal venous phases no enhancements of the cystic wall and content in the right liver lobe. (**d**) CT plain scan demonstrated well defined oval like cystic low density lesion in the pelvic cavity, with septa inside. (**e**, **f**) Contrast scan at the arterial and portal venous phases demonstrated no enhancements of the cystic wall and content in the pelvic cavity

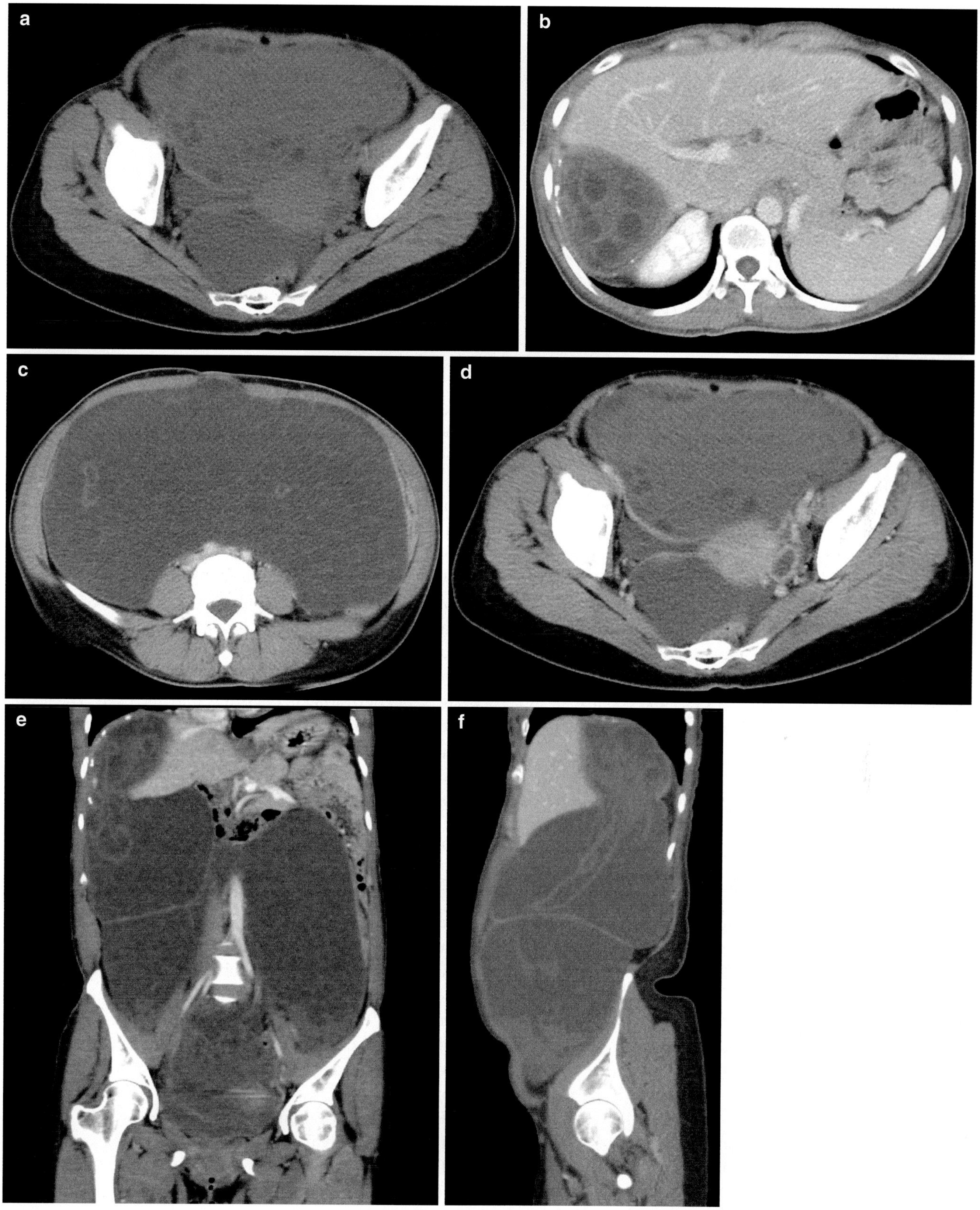

Fig. 8.42 (**a**–**f**) CT scan demonstrated a huge cystic-solid space occupying lesion with dominantly fluid density in the abdominopelvic cavity, with multiple septa opacity inside. The lesion was revealed to penetrate into the perihepatic space, and the part in the perihepatic space showed irregular marginal calcification. The upper and middle parts of the lesion was demonstrated in cystic low density, while the part in the pelvic cavity in soft tissue density. Contrast scan demonstrated no obvious enhancement of the lesion

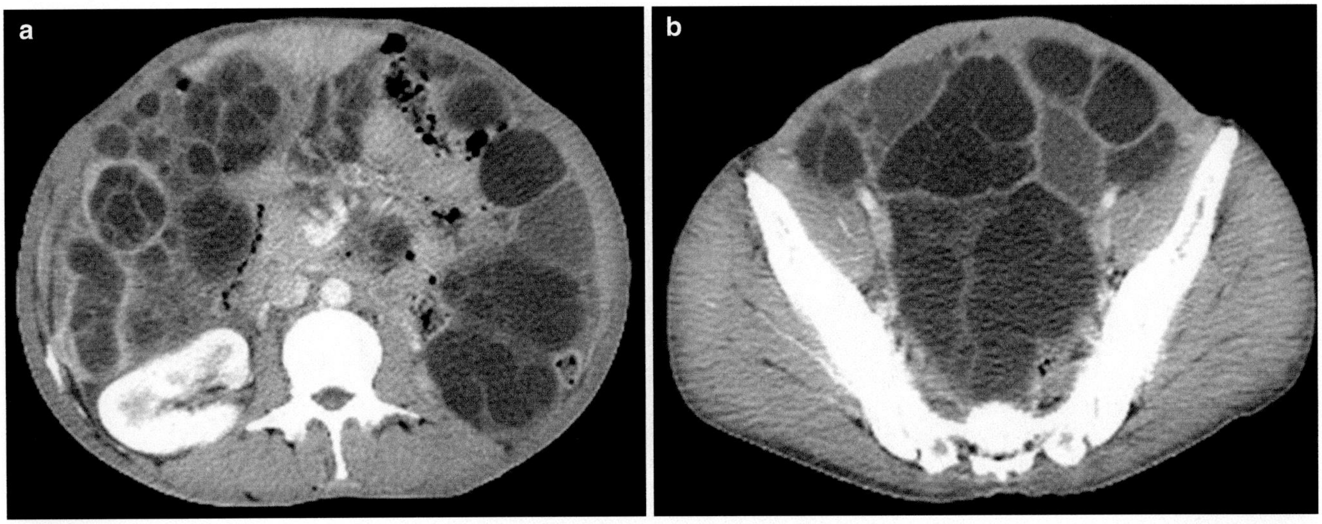

Fig. 8.43 (**a**, **b**) CT scan demonstrated multiple round or round like low density lesions in between intestinal loops, colonic groove, and posterior peritoneum. The cystic wall was shown to be smooth with even thickness, but no enhancement by contrast scan

wall calcification type, rupture or infection type, and posterior peritoneal space type.

Abdominal and pelvic cystic echinococcosis should be differentiated from the following diseases:

1. Cystic lesion in abdominal solid organ
   Such lesions include hepatic abscess and renal abscess, commonly to be well defined but rarely with calcification. The lesion shows no daughter cysts and collapse of internal cyst, which are characteristic signs of cystic echinococcosis.
2. Cystic lesion in abdominal cavity
   Such lesions include congenital mesenteric abscess, mesonephric duct cyst, cystic teratoma and lymphatic cyst. The lesion is commonly well defined with thin wall. No obvious calcification at the cystic wall, no obvious daughter cyst and no obvious ribbon sign can be detected in the cyst. Contrast scan demonstrates no obvious enhancement.

Case Study 29
[**Brief Medical History**]

A 51-year-old woman reported a history of surgical operation for hepatic cystic echinococcosis 7 years ago. Currently, she found a cystic space occupying lesion in the pelvic cavity 1 week ago.

[**Radiological Demonstration**] (See Fig. 8.44)

[**Diagnosis**] Pelvic cystic echinococcosis (multi daughter cysts type).

[**Discussion**]

Pelvic cystic echinococcosis is commonly secondary to dissemination of echinococcus after rupture of hepatic lesion. The lesion is commonly demonstrated to be large, resembling to the lesion of hepatic cystic echinococcosis with shell like calcification of the cystic wall, lower daughter cysts density than mother cyst, and rupture of internal cyst to show ribbon sign or water-snake sign. Such lesion can be accurately diagnosed. Singular cystic echinococcosis in female pelvic cavity should be differentiated from ovarian cystadenoma. Ovarian neoplasm is unilocular or multilocular, with loculus in different sizes. The cystic wall and intracystic septa of ovarian neoplasm is thin and even. The lesion of serous cystadenoma can be shown with nodules on the cystic wall. Contrast scan demonstrates enhancements of the cystic wall, internal septa and wall nodules. However, the lesion of echinococcosis shows no enhancement of the cystic wall and no nodules on the intracystic wall, which is helpful for their differential diagnosis.

Case Study 30
[**Brief Medical History**]

A 49-year-old man complained of repeated lower abdominal distension for 1 year, and urination urgency and frequency that aggravated during nights for 2 months. He permanently lived in husbandry area. Routine blood test was normal. The Casoni test showed positive (+). And colonoscopic examination showed negative.

[**Radiological Demonstration**] (See Fig. 8.45)

[**Diagnosis**] Pelvic cystic echinococcosis (Mixed type).

[**Discussion**]

The patients with echinococcosis mainly in abdominal and pelvic cavity may experience abdominal distension and pain. In the cases with large lesion, its adjacent organ is subject to migration due to compression, showing corresponding compressive symptoms.

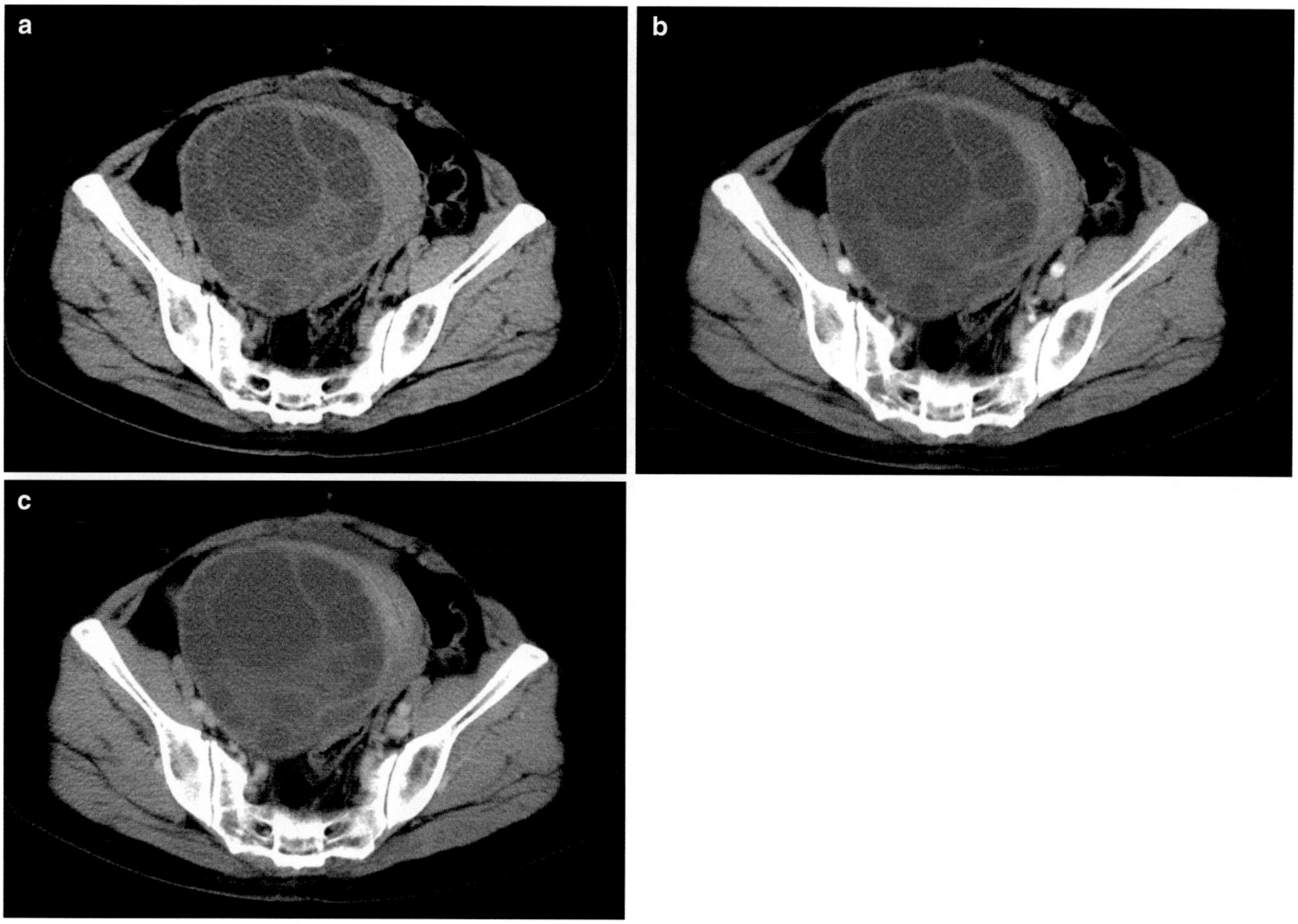

Fig. 8.44 (**a–c**) CT plain and contrast scans demonstrated a cystic space occupying lesion in the right pelvic cavity, with the largest section in size of 10 cm × 12 cm. The lesion was shown with multiple daughter cysts and multiple septa inside, adhering to the right uterine wall. The uterus was revealed with leftward migration due to its compression. Contrast scan showed no enhancement of the lesion

Pelvic and abdominal echinococcosis commonly occurs in the space of abdominal and pelvic cavity as well as space in the mesenterium. By CT scan, the lesion is demonstrated as round or round like opacity with cystic density, with well defined and smooth boundary. Contrast scan basically shows no enhancement of the lesion, but possible enhancement of the cystic wall and intracystic septa. For this case, multiple cystic masses are demonstrated in the pelvic cavity, and the large masses show obvious calcification of the cystic wall. Multiple daughter cysts can be detected in large lesions, which can be more clearly demonstrated after enhancement of the cystic wall by contrast scan.

Pelvic echinococcosis should be differentiated from other cystic and solid cystic lesions in abdominal and pelvic cavity.

1. Intermesenteric cutaneous cyst
   The lesion is unilocular or multilocular, commonly in females at the child bearing age. The intracystic content is complex, commonly with no accompanying calcification of the cystic wall.
2. Enterogenic cyst
   The lesion is unilocular or multilocular, with thin, smooth, sharp but rarely calcified cystic wall.
3. Ovarian serous cystadenoma
   The lesion is unilocular or multilocular, and all the lesions are isolated with no cyst-in-cyst sign.
4. Ovarian serous cystadenocarcinoma
   The lesion is solid cystic mass, with irregularly enhanced solid components by contrast scan. In the cases of echinococcosis, only the cystic wall can be enhanced by contrast scan with no enhancement of the intracystic content and no calcification on the cystic wall. Generally speaking, its differential diagnosis from other pelvic cystic lesions, the cyst-in-cyst sign in the cases of echinococcosis is a typical radiological demonstration.

Case Study 31

[**Brief Medical History**]

A 38-year-old woman reported a space occupying lesion in the spleen that was detected by ultrasound 5 days ago. She

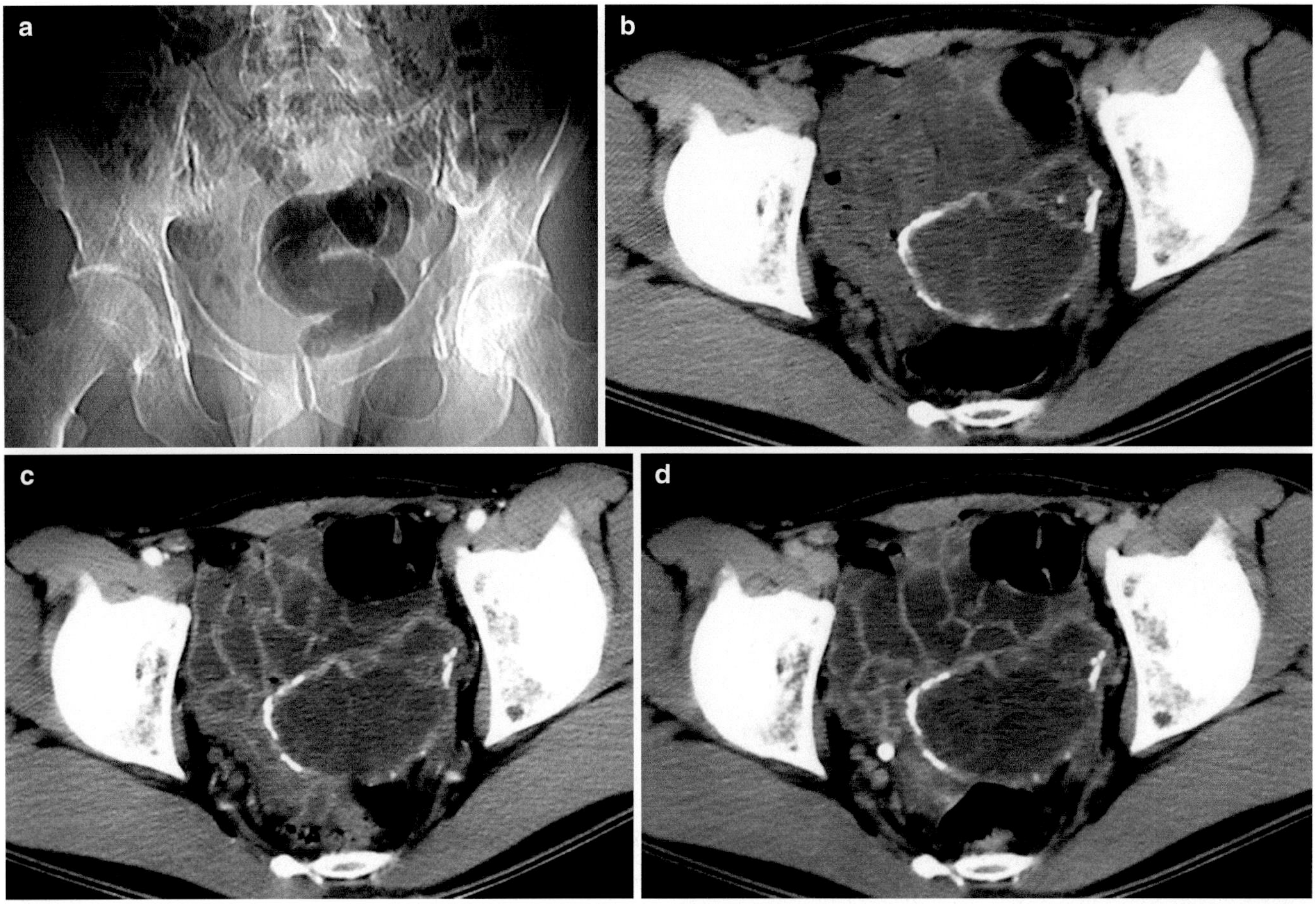

Fig. 8.45 (**a–d**) CT scan demonstrated multiple cystic masses in the pelvic cavity, with the larger ones showing obvious calcification on the cystic wall. Multiple daughter cysts were revealed in the large cyst. Contrast CT scan demonstrated obvious enhancement of the cystic wall (**c**: the arterial phase; **d**: the delayed phase)

experienced no obvious fever, abdominal pain, nausea, vomiting and other upsets.

[**Radiological Demonstration**] (See Fig. 8.46)

[**Diagnosis**] Splenic cystic echinococcosis (Multi daughter cysts type).

[**Discussion**]

Resembling to the lesion of hepatic echinococcosis, the lesion of splenic echinococcosis is also radiologically demonstrated to be space occupying in the parenchymal organ. However splenic echinococcosis has a much lower incidence rate than hepatic echinococcosis, because echinococcus is infected via the gastrointestinal tract and gains its access into blood flow at the duodenum. In such a way, it firstly invades the liver along with blood reflux in the portal vein. Therefore, the incidence rate of hepatic echinococcosis shows the highest among all the body parts.

Splenic echinococcosis should be differentiated from the following diseases.

1. Splenic cyst
   The lesion of splenic cyst is commonly demonstrated as well defined low density lesions with obvious calcification.

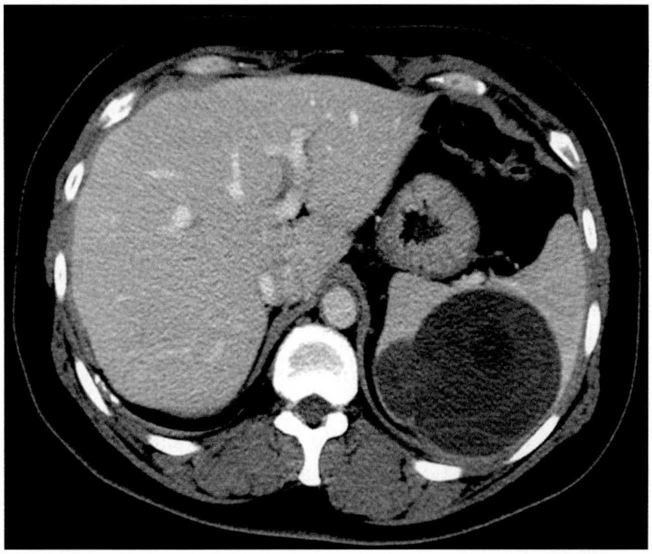

Fig. 8.46 CT scan demonstrated round like cystic lesion in the spleen, with thin cystic wall and well defined boundary. Contrast scan demonstrated no enhancement but daughter cysts in lower density than the mother cyst

No daughter cysts or collapsed internal cyst can be detected within the lesion.

2. Splenic hemangioma

   The key points for differential diagnosis include gradual enhancement of the lesion by contrast scan, and almost the same lesion density to splenic parenchyma by at the delayed phase of contrast scan.

3. Splenic metastasis

   The lesion is commonly enhanced by contrast scan, and most of the patients have a medical history of primary malignancy.

Case Study 32

[**Brief Medical History**]

A 46-year-old woman living in a husbandry area complained of upper abdominal distension, pain and upset, poor appetite, nausea and vomiting. Laboratory tests showed increased eosinophile count, Casoni test positive (+), and ELISA positive (+).

[**Radiological Demonstration**] (See Fig. 8.47)

[**Diagnosis**] Hepatic and splenic echinococcosis.

[**Discussion**]

Echinococcosis, also known as hydatidosis, mainly prevails in husbandry areas. Hepatic and pulmonary echinococcosis are more common, and splenic echinococcosis rarely occurs. Splenic echinococcosis commonly coexists with hepatic or pulmonary echinococcosis, showing no obvious clinical manifestations and more commonly occurring in middle-aged and young populations. Despite no symptoms in its early stage, a lump can be palpable in the left upper abdomen along with the enlargement of the lesion, with accompanying abdominal distension and pain. If the lesion

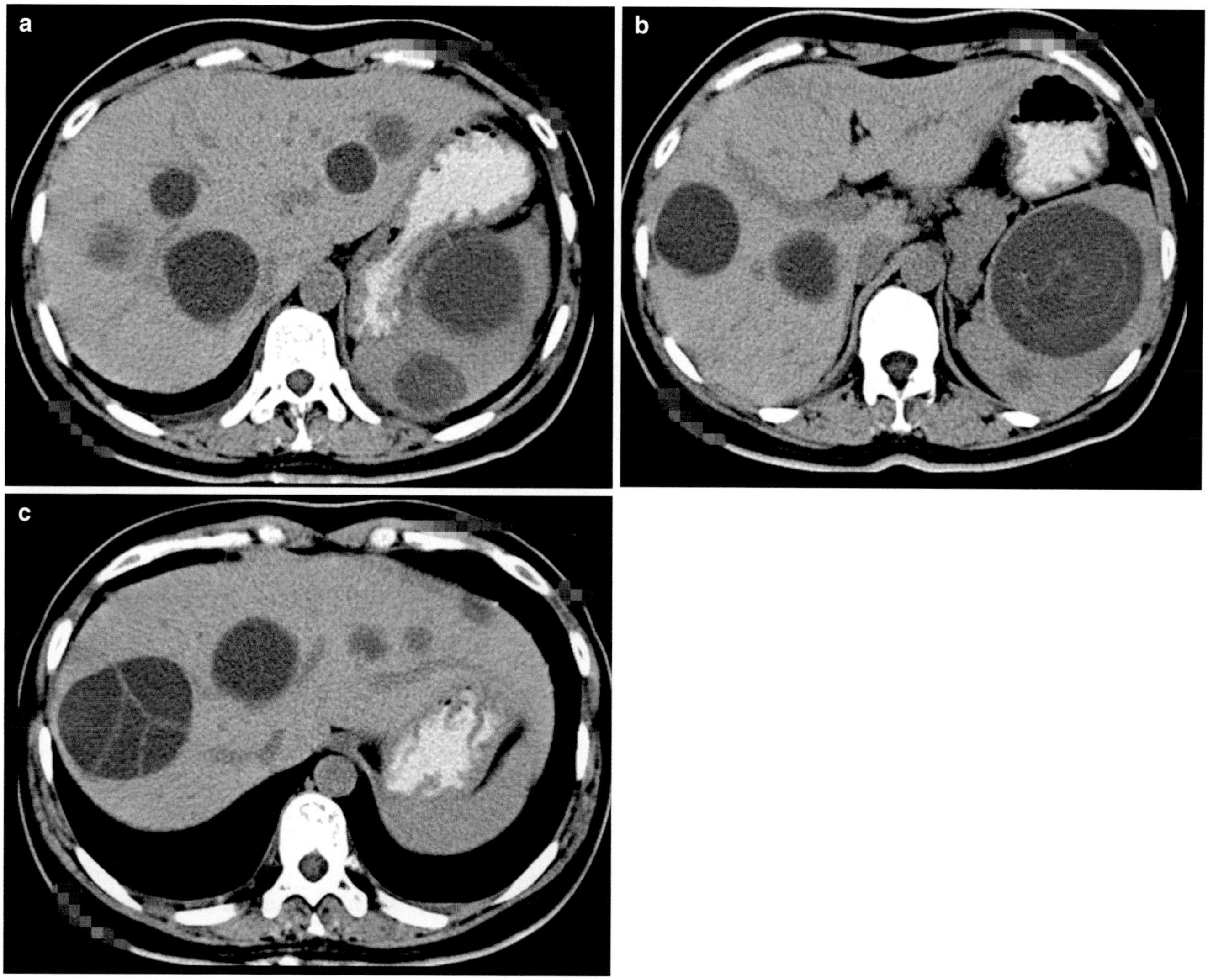

Fig. 8.47 (**a**) CT scan demonstrated multiple round like cystic lesions of different sizes in the liver and the spleen. The lesions were shown with even density, smooth and clearly defined boundary and slightly thickened cystic wall. (**b**) The lesions in the spleen were revealed with complete detachment of the internal cyst, which was demonstrated as strips of high density opacity in ribbon sign. (**c**) The liver was visualized with multiple round like low density areas, with the large one showing wheel liked cyst-in-cyst sign

located at the upper splenic margin, the diaphragm is elevated, with subsequent respiratory symptoms.

CT scan demonstrates singular or multiple well defined round like cystic lesion(s), with or with no arch or shell like calcification of the cystic wall. Contrast scan demonstrates no enhancement of the intracystic content. In the cases with multiple daughter cysts of different sizes in the mother cyst, the lesions are multilocular, appearing like a wheel or honeycomb with characteristic cyst-in-cyst sign. In the cases with separated internal and external cysts or access of cystic fluid into the space between internal and external cysts due to rupture of the internal cyst, the lesions are demonstrated as lotus-on-water sign or ribbon sign.

Splenic cystic echinococcosis should be differentiated from the following diseases.

1. Splenic abscess
   The lesion of splenic abscess can be detected with no surrounding edema, with no obvious enhancement by contrast scan. In typical cases of splenic abscess, the lesion can be revealed with gas-fluid level, in combination to clinical data of high fever, chills, abdominal pain, tenderness at the left upper abdomen and increased WBC count, the differential diagnosis can be made.
2. Pancreatic pseudocyst
   The invasion of pancreatic pseudocyst into spleen shows similar lesion to splenic echinococcosis, with emergence of septa. However, the patients with pancreatic pseudocyst commonly have a medical history of acute or chronic pancreatitis.
3. Splenic cystic lymphangioma
   Splenic cystic lymphangioma is commonly asymptomatic, with singular or multiple lesion(s) and possible thick septa inside that do not involve the liver in most cases. The lesion of splenic echinococcosis commonly co-exists with hepatic echinococcosis, with separation of the internal and external cysts and multilocular appearance due to occurrence of daughter cysts in the mother cyst.
4. Splenic arterial aneurysm
   Ring shape calcification in the cases of splenic arterial aneurysm resembles to the calcification of cystic echinococcosis, but with enhancement of the lesion in the cases of splenic arterial aneurysm by contrast scan.
5. Old splenic hematoma
   Old splenic hematoma can be distinguished from echinococcosis based on the medical history of trauma and the date that bleeding occurs. In combination to the resident region and history of contact, the differential diagnosis can be made with the help of different CT scan demonstrations, clinical manifestations and laboratory tests findings.

Case Study 33

[Brief Medical History]

A 49-year-old Kazak man was admitted to the hospital due to complaints of persistent right upper abdominal pain for 1 month, with jaundice for 2 months. The patient lived in a husbandry area, with a history of contact to goats and dogs. He had received two surgical operations for hepatic echinococcosis in another hospital. The 4-item examination for echinococcus showed Anti-EgCF antibody (+++), Anti-EgP antibody (+++), Anti-EgB antibody (+++) and Anti-Em2 antibody (+).

[Radiological Demonstration] (See Fig. 8.48)

[Diagnosis] Splenic echinococcosis.

[Discussion]

Splenic cyst is demonstrated as a cystic lesion in the splenic tissue, which can be classified into parasitic cyst (e.g. cystic echinococcosis) and non-parasitic cyst. Splenic cystic echinococcosis is commonly concurrent with hepatic or pulmonary cystic echinococcosis, which prevails in husbandry area. In this case, the patient had received two surgical operations for hepatic cystic echinococcosis, and had a history of living in a husbandry area. CT scan and MR imaging demonstrated round cystic lesion, with septa inside. Both clinically and radiologically, the findings indicated typical splenic cystic echinococcosis. Splenic echinococcosis should be differentiated from splenic abscess, pancreatic pseudocyst and splenic cystic lymphangioma.

Case Study 34

[Brief Medical History]

A 31-year-old man was admitted to the hospital due to complaints of persistent right upper abdominal pain for 1 month, and jaundice for 2 months. He permanently lived in a husbandry area, with a history of contact to goats and dogs. The Casoni test showed positive. And he had received two surgical operations for hepatic cystic echinococcosis in another hospital.

[Radiological Demonstration] (See Fig. 8.49)

[Diagnosis] Pancreatic echinococcosis.

[Discussion]

Echinococcosis highly prevails in husbandry area, commonly occurring in the liver and lungs. The incidence rate of pancreatic echinococcosis is extremely low, being about 0.2–2 % with about 50 % of pancreatic echinococcosis occurring at the pancreatic head. In this case, the patient experienced obvious cystic enlargement of the pancreatic head, which subsequently causes biliary obstruction due to its compression to the biliary duct and systemic yellowish skin. In addition, the patient had received two surgical operations for hepatic echinococcosis, and the condition should be preliminary diagnosed as pancreatic echinococcosis. If the echinococcus cyst is small with no obvious compression to the

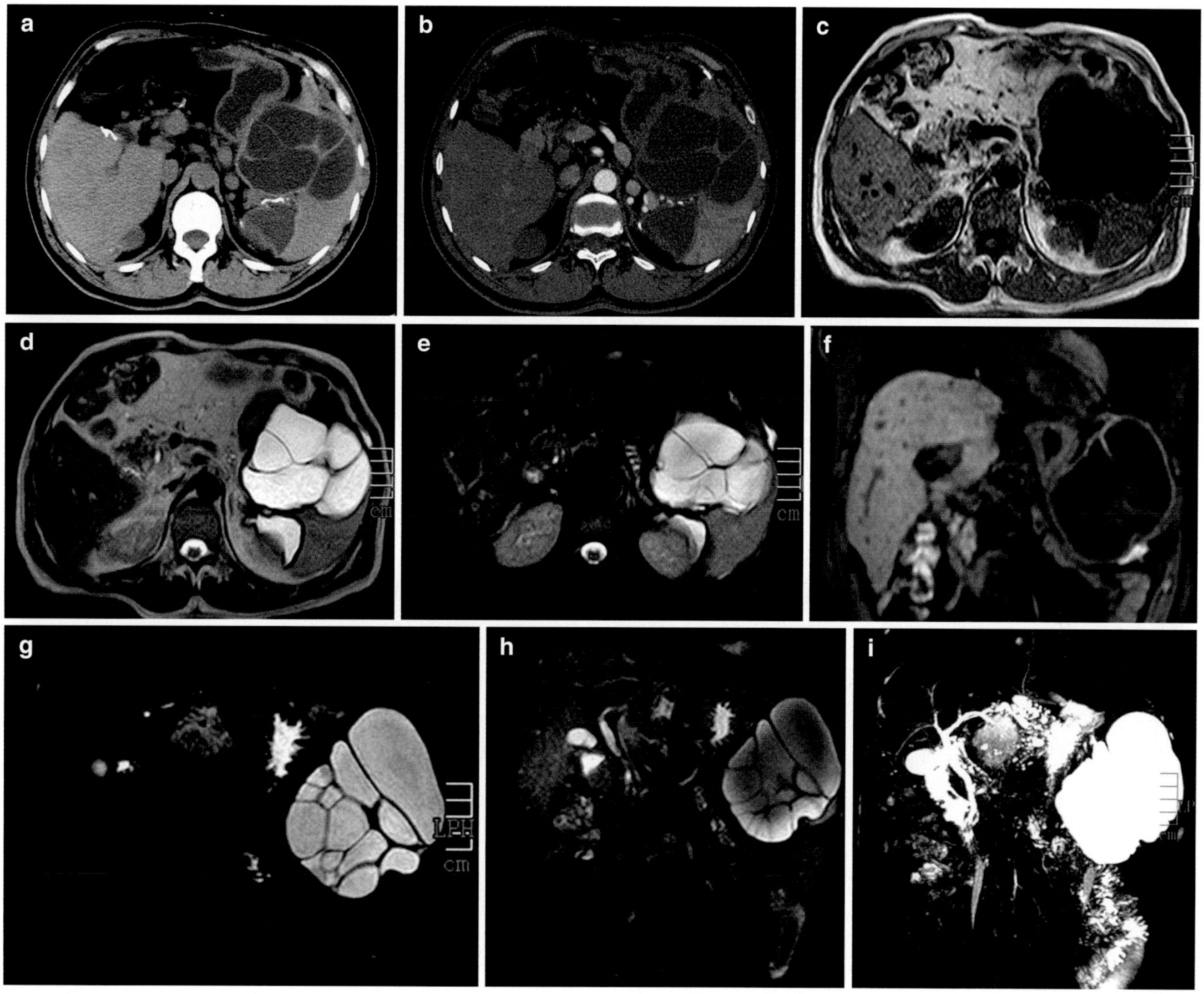

Fig. 8.48 (**a**, **b**) CT scan demonstrated a round like cystic lesion with a size of 7.0 cm×5.9 cm in the spleen, with internal septa, regular shape, and well defined boundary. Contrast scan revealed no enhancement of the lesion, but more favorably revealed lesion. (**c–i**) MR imaging demonstrated multiple cystic lesions in the spleen, with the large one in a size of 7.5 cm×6.2 cm with long T1 long T2 signals and multiple septa inside. By fat suppression sequence of MR imaging, the lesion was shown with high signal, with smooth intact and well-defined boundary

common bile duct, no positive physical sign can be detected, which presents difficulty for its early diagnosis. Pancreatic echinococcosis should be differentiated from the following diseases: (1) Cystic dilation of the common bile duct; (2) Pancreatic cystadenoma; and (3) Pancreatic tuberculosis. Although CT scan is helpful for its diagnosis, the diagnosis should be defined in combination to the medical history and laboratory tests.

Case Study 35

[**Brief Medical History**]

A 60-year-old woman complained of lower back and abdominal distension and pain for 1 year. She had received surgical operation for hepatic echinococcosis.

[**Radiological Demonstration**] (See Fig. 8.50)

[**Diagnosis**] Renal cystic echinococcosis.

[**Discussion**]

CT scan and MRI imaging demonstrations of renal cystic echinococcosis resemble to those of hepatic cystic echinococcosis. The lesion of renal cystic echinococcosis can be divided into three types: simplex cyst type, multi daughter cysts type as well as rupture and infection type. Typical lesion of cystic echinococcosis is demonstrated with low T1WI signal and high T2WI signal. The lesion of cystic echinococcosis with complicating infection shows corresponding signal change. The cystic wall of echinococcus is homogeneous in thickness, with characteristic sign of low T2WI signal. In the cases with complicating rupture, the external cystic wall thickens with uneven internal signal, and

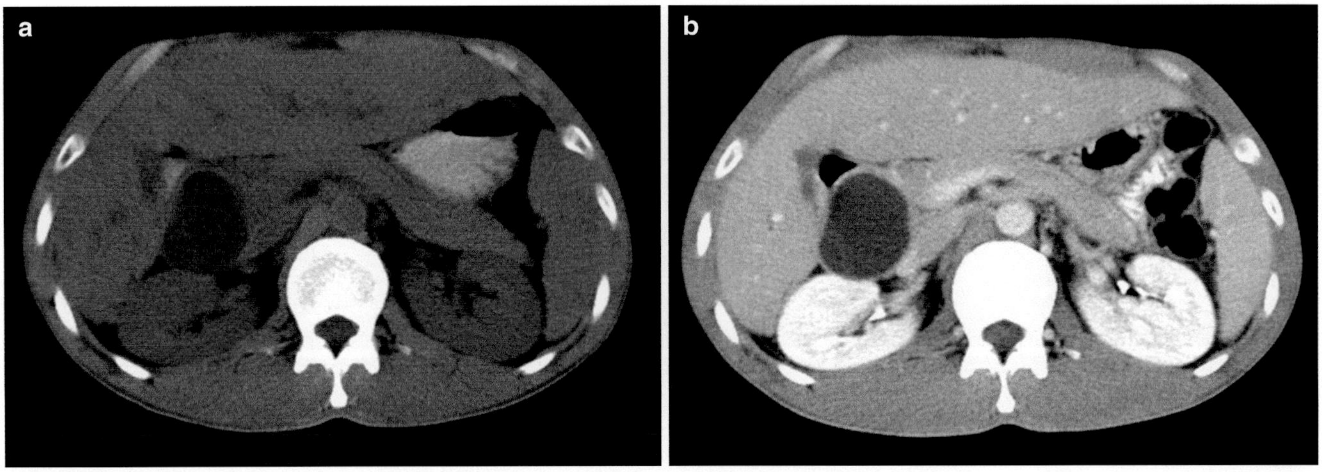

Fig. 8.49 (**a**) CT plain scan demonstrated a round like cystic lesion with a size of 5.0 cm × 3.9 cm in the pancreatic head, with regular shape and well defined boundary. The duodenum was shown to be compressed with deformation. (**b**) The lesion was revealed with no enhancement by contrast scan, but was more favorably demonstrated

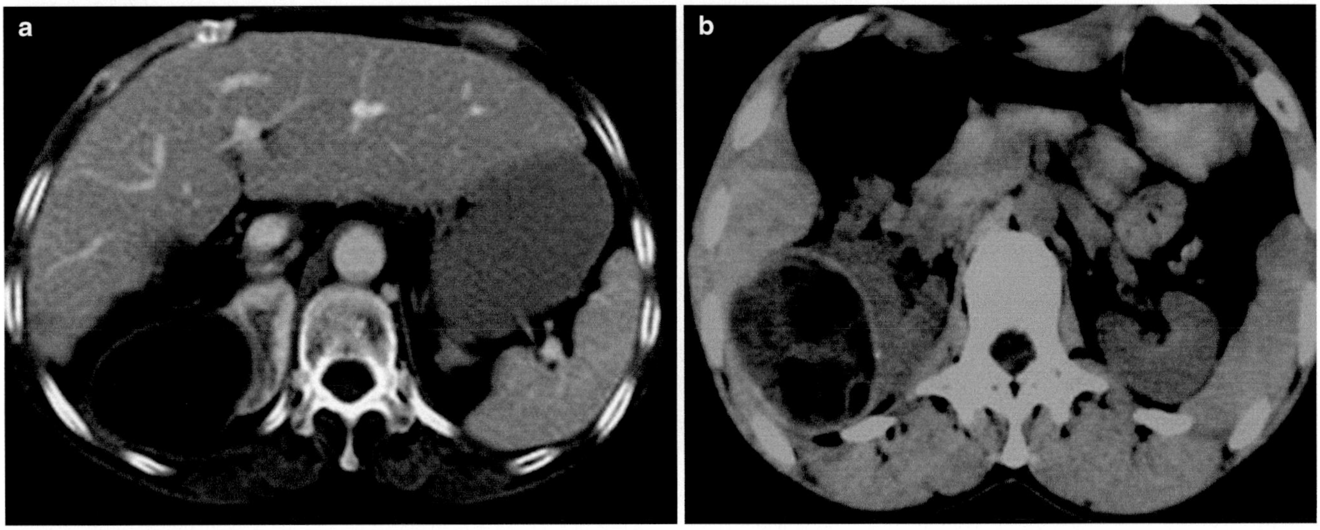

Fig. 8.50 (**a**) CT scan demonstrated a round like cystic lesion with smooth boundary in the right renal parenchyma, with no enhancement by contrast scan. (**b**) In the multilocular echinococcus cyst in the right kidney, multiple daughter cysts were shown to scatter in the mother cyst, which squeezed to each other in mulberry like pattern

the ruptured internal cyst floats in the cystic fluid, demonstrated as ribbon sign. In the cases with daughter cysts in the mother cyst, the characteristic rose-petals sign is radiologically demonstrated. By contrast scan, the cystic wall shows no enhancement in the cases of renal cystic echinococcosis. However, the cystic wall and its surrounding tissue are enhanced in the cases of renal cystic echinococcosis complicated by infection. MRU can clearly demonstrate the presence of echinococcus cyst, its rupture and its penetration into the ureter.

Renal cystic echinococcosis should be differentiated from the following diseases:

1. Renal cyst
   The lesion of renal cyst is commonly demonstrated as well defined low density opacity, with no obvious calcification and no obvious daughter cysts and collapsed internal cyst.
2. Renal complex cyst
   The lesion is demonstrated with uneven internal density and no daughter cysts inside.
3. Renal cystic carcinoma
   The key points for differential diagnosis include enhancement of the lesion by contrast scan and no obvious daughter cysts.

Case Study 36

[Brief Medical History]

A 30-year-old woman reported a cystic space occupying lesion in the brain that had been detected 5 days ago during her physical examination. She experienced occasional headache but no obvious nausea, vomiting, disturbance of consciousness, limb sensory disturbance and dyskinesia.

For case detail and figures, please refer to fig 26.27 of the book (Hongjun Li, Radiology of Infectious diseases, Volume 2, 2015)

[Diagnosis] Cerebral cystic echinococcosis (simplex cyst type).

[Discussion]

The brain tissue is soft with abundant blood supply, which facilitates the growth of echinococcus. The large lesions are demonstrated with space occupying effect and surrounding edema. The internal cyst wall is the echinococcus itself, while the external cyst is a pseudocapsule formed by proliferation of glial cells in brain tissue that is extremely thin and can hardly demonstrated by MR imaging. The cystic lesion is singular or multiple, and singular lesion is more common, with low T1WI signal and high T2WI signal. And the cystic wall is commonly demonstrated as continuous homogeneous low signal opacity, with no enhancement by contrast imaging. In the lesions with multiple daughter cysts, the daughter cysts are commonly arranged along the mother cystic wall that are demonstrated as rose-petals sign or wheel sign with low T1WI signal and high T2WI signal. By T1WI, the daughter cysts show lower signal than the mother cyst that is close to signal intensity of water, and no demonstration of the daughter cystic wall. Due to the discrepancy in signal intensity, the arrangement pattern of daughter cysts in the mother cyst can be demonstrated and described. By contrast imaging, the lesion shows slight marginal enhancement. If the cyst is subject to rupture and infection, the cystic wall irregularly thickens and the cystic fluid shows increased signal, with obvious ring shape enhancement of the lesion by contrast imaging.

Cerebral cystic echinococcosis should be differentiated from the following diseases:

1. Epidermoid cyst
   The lesion of epidermoid cyst is commonly demonstrated as low density opacity, but with a CT value close to adipose tissue. If with bleeding or calcification deposits in the lesion, high density can be demonstrated. In some cases, the cystic wall shows arch shape calcification. The location of lesions is also help for the differential diagnosis, and the lesion of epidermoid cyst is commonly located in the cerebellopontine angle, saddle area and middle cranial fossa. The cases with the lesion of epidermoid cyst in hemicerebrum account for less than 10 % of cerebral cystic echinococcosis.
2. Cerebral metastasis
   The lesion of cerebral metastasis is commonly located in the area supplied by the middle cerebral artery or under the cortex, with high T2WI signal. By contrast scan, homogeneous enhancement of the lesion is demonstrated, but no enhancement of necrotic and liquefied area of the lesion and no multiple small vesicles. In combination to the clinical history, the differential diagnosis can be made.
3. Arachnoid cyst
   Two thirds of the patients with arachnoid cyst show the lesion in the lateral fissure, and the lesion is rarely located on the cerebral convex surface in a shape of semicircle or biconvex. Its density is equal to that of cerebrospinal fluid. And the cystic wall is demonstrated to be thin with smooth boundary and no enhancement. All the lesions are accompanied by restrained thinner or protruding endosteal lamella.

Case Study 37

[Brief Medical History]

A 5-year-old Kazak boy was hospitalized due to chief complaints of paroxymal headache, dizziness and vomiting for 1 year that aggravated for 1 month. He lived in a husbandry area, with a history of close contact to dogs and goats.

[Radiological Demonstration] (See Fig. 8.51)

[Diagnosis] Cystic echinococcosis in the left frontotemporal lobe, complicated by hydrocephalus and interstitial cerebral edema.

[Discussion]

The lesion of cerebral cystic echinococcosis (simplex cyst type) is commonly singular, but occasionally multiple. The soft brain tissue and rich blood supply facilitate the growth of echinococcus. CT plain scan and MR plain imaging demonstrate the lesion as round or egg shape cystic mass. By CT plain scan, the density within the lesion is homogeneous that is close to the density of cerebrospinal fluid. And the cystic wall is demonstrated with eggshell like calcification. MR plain imaging demonstrates the lesion to be round like with smooth and well defined boundary in low T1WI signal and high T2WI signal. The cystic wall is demonstrated with homogeneous thickness and persistently equal or relatively lower T2WI signal by T2WI. The large lesion may be accompanied by space occupying effect, which is demonstrated as deformed, narrowed or occluded local ventricle due to compression. Contrast scan demonstrates no abnormal enhancement of most cystic margin, but slightly abnormal enhancement of the cystic margin when the lesion is complicated by infection. When singular or multiple cystic lesion(s) is/are detected with calcification of the cystic wall in the brain, cystic echinococcosis should be firstly considered in combination to the histories of living in husbandry area and contact to dogs and goats.

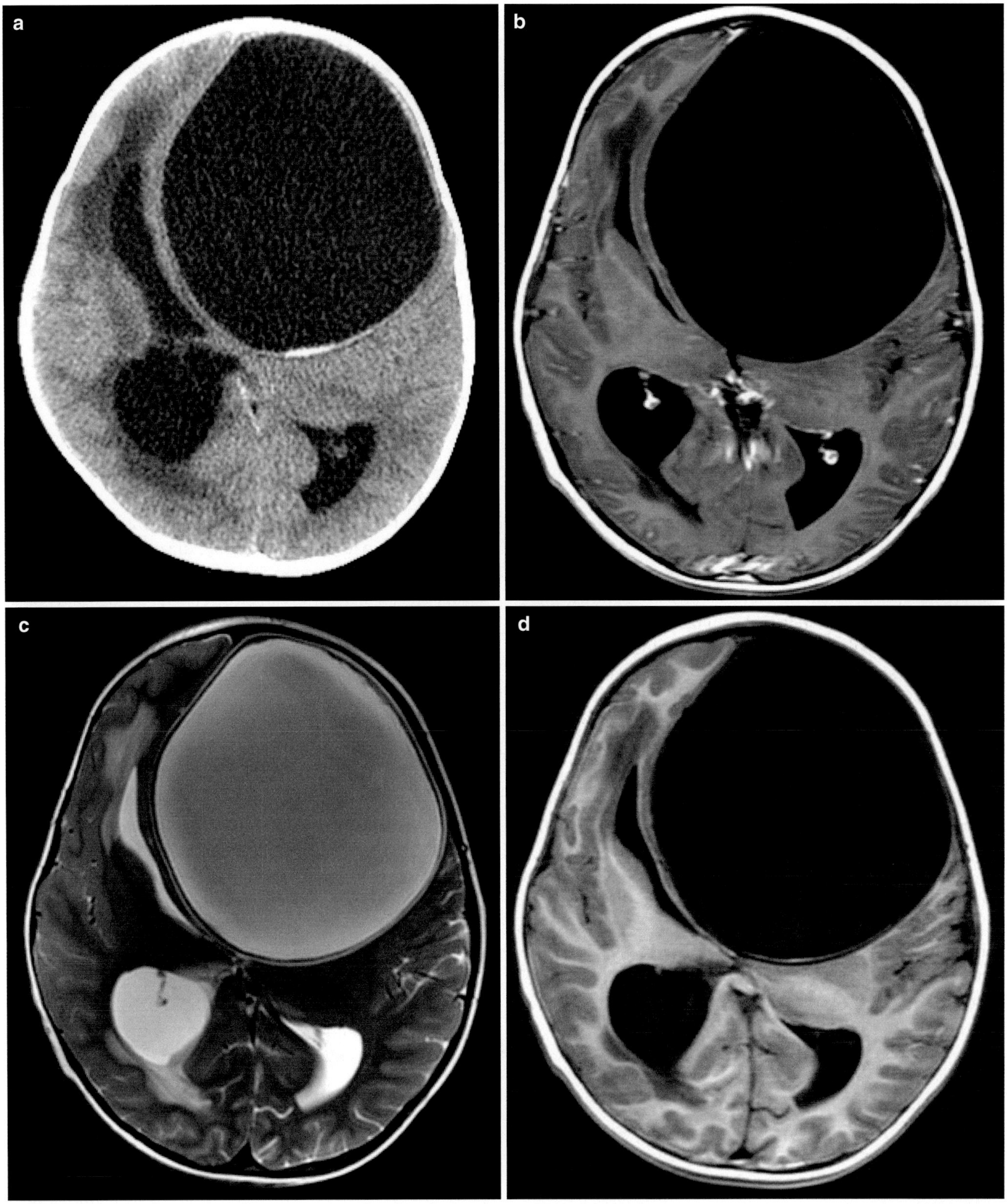

Fig. 8.51 (**a**–**d**) The left frontotemporal lobe was demonstrated with a huge cystic space occupying lesion, with obvious space occupying effect. CT plain scan demonstrated water like density in the lesion, equal density of the cystic wall, and eggshell like high density calcification opacity. MR plain scanning demonstrated low T1WI signal and high T2WI signal in the lesion, and equal signal of the cystic wall in homogeneous thickness. Contrast MR imaging demonstrated no obvious enhancement of the cystic cavity and wall

Cerebral cystic echinococcosis (Simplex cyst type) should be differentiated from singular cerebral cyst, cerebral abscess, capillary astrocytoma and other cerebral cystic neoplasm. Cerebral cyst is a common benign lesion in the brain, with clinical manifestations to be affected by its location. Most patients are asymptomatic, often with the lesion detected in physical examination. CT plain scan demonstrates the lesion to be cystic round like, with its density close to the cerebrospinal fluid. Its boundary is smooth, showing no calcification in most cases and no obvious abnormal enhancement of the lesion by contrast scan. MR plain imaging demonstrates the lesion in low T1WI signal and high T2WI signal, with no abnormal enhancement of the cystic wall by contrast imaging. In the stages of suppuration and envelope formation in the cases of cerebral abscess, CT plain scan demonstrates the abscess wall in ring shaped equal density, and the abscess cavity in slightly lower density or water like low density. Gas-fluid level may be detected in some cases. By contrast scan, the smooth, intact and homogeneously thick abscess wall shows obvious enhancement. MR plain imaging demonstrates low T1WI signals in the abscess cavity and surrounding edema, while equal signal of the ring shape septa, the abscess wall, between the cavity and edema. MR plain imaging also demonstrates high T2WI signal in the abscess cavity and surrounding edema, while equal or low signal from the abscess wall. Contrast imaging demonstrates obvious enhancement of the abscess wall, but no abnormal enhancement of the abscess content. The lining of the abscess wall is smooth, with no nodule in most cases. By DWI imaging of a high b value obvious high signal of the lesion is demonstrated since the diffusion of water molecules is constrained by thick pus within abscess cavity, which is a characteristic radiological sign of cerebral abscess. The lesion of capillary astrocytoma is a solid cystic astrocytoma, which commonly occurs at the infratentorial area of children and young adults with a slow development. The lesion is quite well defined, often with cystic changes. The patients may experience focal neurological disorders or non-specific physical signs such as microcephalus, headache, endocrine disorder and increased intracranial pressure. CT plain scan demonstrates the lesion with dominantly water like low density, with obvious abnormal enhancement of it solid part by contrast scan. MR plain imaging demonstrates the lesion with dominantly cystic components, with mixed solid content. The cystic components are demonstrated with low T1WI signal and high T2WI signal, while the solid part with almost equal signal. Contrast imaging demonstrates slight or moderate abnormal enhancement of its solid part.

Case Study 38

[**Brief Medical History**]

A 38-year-old man was hospitalized due to the chief complaints of post-operative for hepatic echinococcosis for 13 years and blurry vision for 7 days. The patient had received surgical removal of hepatic echinococcus 13 years ago and experienced blurry vision of both eyes for 7 days with more serious condition of the right eye. The symptom aggravated when the eyes were exposed to intense light, with accompanying double vision. Physical examination showed no special positive signs. The 4-item examination for echinococcus showed Anti-EgCF antibody (++), Anti-EgP antibody (+), Anti-EgB antibody (++) and Anti-Em2 antibody (–).

[**Radiological Demonstration**] (See Fig. 8.52)

[**Diagnosis**] Cerebral cystic echinococcosis in the right frontotemporal area (multi daughter cysts type).

[**Discussion**]

When multiple daughter cysts occur in the mother cyst of echinococcus, a cyst-in-cyst sign or internal septa sign can be detected, with the daughter cysts commonly arranged along the mother cyst wall in rose-petals or wheel pattern. CT plain scan demonstrates lower density of the daughter cysts than the mother cyst. MR plain imaging demonstrates low T1WI signal and high T2WI signal, with lower T1WI signal of the daughter cysts than the mother cyst that is close to water density. The daughter cystic wall is extremely thin that is poorly demonstrated. But due to the signal intensity discrepancy between the daughter cysts and the mother cyst, the arrangement pattern of daughter cysts in the mother cyst can be clearly demonstrated. By contrast imaging, no abnormal enhancement is demonstrated on the margin of cystic wall in most cases. In the cases with concurrent rupture and infection, the cystic wall is subject to irregular thickening, with increased signal in cyst fluid. By contrast imaging, obvious ring shape abnormal enhancement is demonstrated. The multi daughter cysts type of cystic echinococcosis shows characteristic radiological signs, although the multi daughter cysts type of cerebral cystic echinococcosis rarely occurs, the diagnosis can be made in combination to the radiological signs and medical history.

The multi daughter cysts type of cerebral cystic echinococcosis should be differentiated from cerebral abscess and capillary astrocytoma. In the stages of suppuration and envelope formation in the cases of cerebral abscess, CT plain scan demonstrates the abscess wall with ring shape equal density, and the abscess cavity with sightly lower density or water like density. In some cases, the abscess cavity is shown with gas-fluid level. Contrast scan demonstrates obvious enhancement of the abscess wall, which is smooth and intace with homogeneous thickness. MR plain imaging demonstrates low T1WI signals in the abscess cavity and the surrounding edema. And the abscess wall between the cavity and edema is demonstrated as ring shape septa with equal signal. MR plain imaging demonstrates high T2WI signal in the abscess cavity and the surrounding edema, while the abscess wall in equal or low signal. Contrast imaging demonstrates obvious enhancement of the abscess wall, but no abnormal enhancement of the abscess content. The lining of

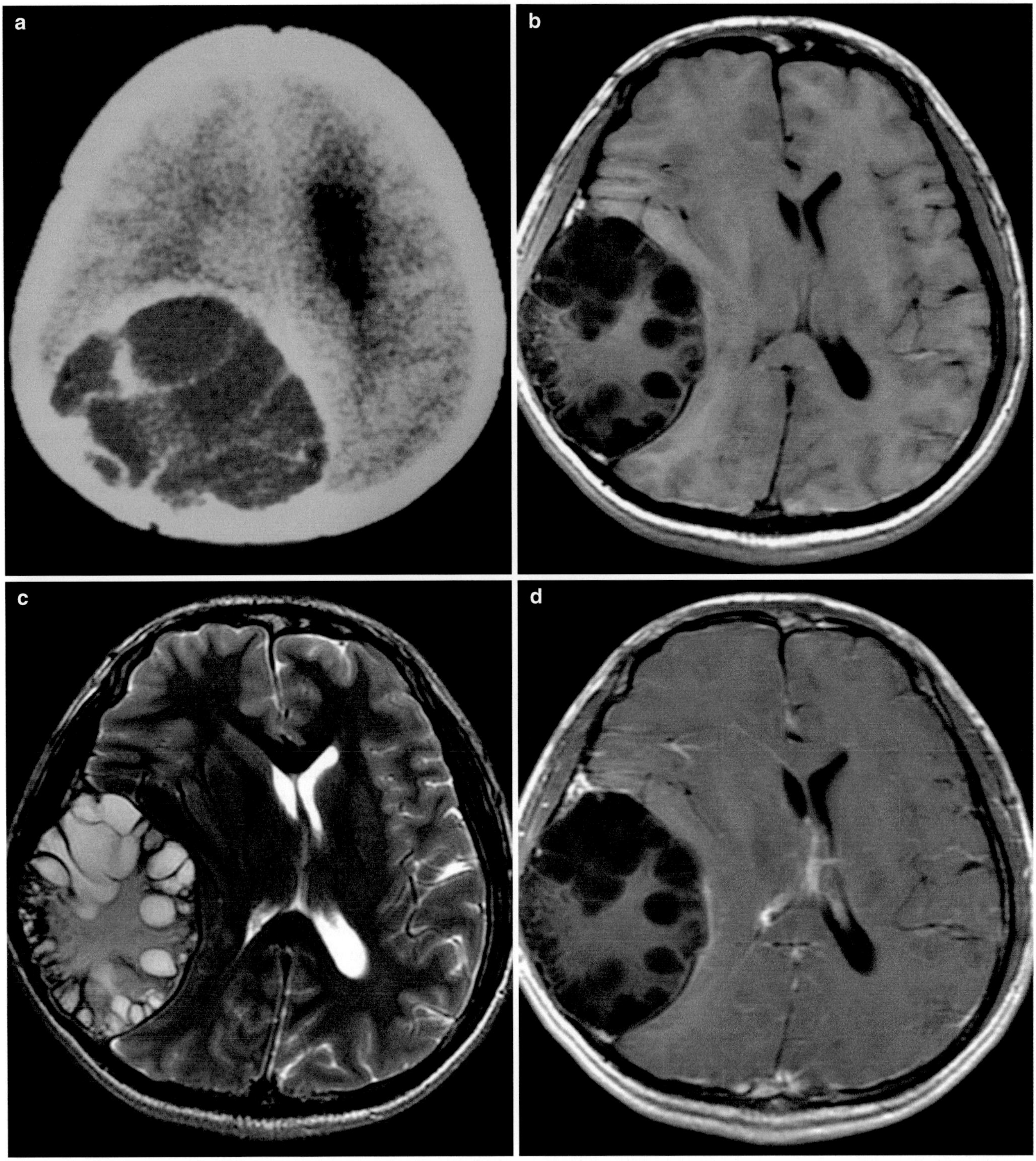

Fig. 8.52 (**a–d**) The right temporoparietal area was demonstrated with a cystic lesion in irregular shape, with obvious space occupying effect. CT plain scan demonstrated the lesion with dominantly low density and internal multiple septa opacity in equal density. MR plain imaging demonstrated T1WI signal of multiple daughter cysts, with the mother cyst in low signal and the daughter cysts in lower signal. The lesion was demonstrated with dominantly high T2WI signal, with the mother cyst in high signal and the daughter cysts in higher signal. Contrast MR imaging demonstrated no obvious abnormal enhancement of the lesion

abscess wall is smooth, with no nodule in most cases. The multilocular septa sometimes show abnormal enhancement by contrast imaging within the abscess cavity. By DWI of high b value, the lesion of cerebral abscess is demonstrated as obvious high signal since diffusion of water molecules is constrained by thick pus within the abscess cavity, which is a characteristic sign of cerebral abscess. The lesion of capillary astrocytoma is solid cystic, which is commonly located in the infratentorial area of children and young adults with a slow development. The lesion is quite well defined, often with cystic changes. The patients may experience focal neurological disorders or non-specific physical signs such as microcephalus, headache, endocrine disorder and increased intracranial pressure. CT plain scan demonstrates the lesion with dominantly water like low density, with obvious abnormal enhancement of its solid part by contrast scan. MR plain imaging demonstrates the lesion with dominantly cystic components in low T1WI signal and high T2WI signal, with mixed solid content in almost equal signal. Contrast imaging demonstrates the solid part of lesion with slight or moderate abnormal enhancement.

Case Study 39

[Brief Medical History]

A 35-year-old man was hospitalized due to chief complaints of intermittent episodes of epileptic seizure for 6 months. The patient suddenly experienced numbness of the left hand about 1 year ago, which then spread to the left forearm but he did not pay close attention to it. About 6 months ago, he suddenly experienced coma due to the onset of epileptic seizure, and paid his clinic visit in a local hospital. He then received therapy for cervical hyperplasia, but showed no therapeutic response, during which epileptic seizure occurred for several times. Physical examination showed no special positive physical signs.

[Radiological Demonstration] (See Fig. 8.53)

[Diagnosis] Alveolar echinococcosis in the right fronto-parietal lobe.

[Discussion]

Cerebral alveolar echinococcosis commonly occurs in adults, and the lesion is commonly located at the supratentorial area and the areas supplied by the middle meningeal artery. In almost all the cases of cerebral alveolar echinococcosis, the lesion is secondary to hematogenous dissemination of hepatic alveolar echinococcus, which may be singular or multiple. Pathologically, the lesion is characterized by bilateral budding with dominant exogenous budding to produce numerous small vesicles and its section in honeycomb like change. The lesion is commonly located in or under the cortex, with a rapid growth resembling to malignancy due to the soft brain tissue and rich blood supply. CT plain scan demonstrates the lesion in soft tissue density, often with accompanying high density calcification inside, obvious space occupying effect and surrounding edema. MR plain imaging demonstrates the lesion with equal or slightly higher T1WI signal and mixed T2WI signal with dominant low signal. By T2WI, most of the lesions are demonstrated with multiple small vesicles in high signal and different sizes, which is a characteristic sign demonstrated by MR hydrography. In some cases, multiple sand shape low signal calcifications can be detected, which is the main reason for its low T2WI signal. Unlike alveolar echinococcosis in other body parts, the lesion of cerebral alveolar echinococcosis shows irregular peripheral enhancement by contrast imaging, which is a characteristic sign of cerebral alveolar echinococcosis. The underlying mechanism of such an enhancement may be related to the inflammatory responses around the lesion and destruction of blood-cerebrospinal fluid barrier. In recent years, with the development of magnetic resonance imaging, new technologies have been applied in the diagnosis and differential diagnosis of cerebral alveolar echinococcosis. By 1 H-Magnetic Resonance Spectroscopy (^{1}H-MRS), the lesion of cerebral alveolar echinococcosis is characterized by varying decreases of NAA peak, Cho peak and Cr peak as well as obvious increase of the Lipd peak, with or with no accompanying Lac peak, which are related to the destruction and proliferation of the lesion as well as its surrounding inflammatory responses. ^{1}H-MRS, to a certain degree, can help to reveal the pathological changes of alveolar echinococcus, which further increases our knowledge about the pathophysiological changes of the disease. Cerebral alveolar echinococcosis shows typical radiological signs, including irregular ring enhancement of the lesion with marginal small vesicles by contrast CT scan, high T2WI signal of multiple small vesicles in varying sizes on the background of low T2WI signal by MR imaging, and characteristic signs by ^{1}H-MRS. In combination to the history of contact to source of infection and laboratory tests, the diagnosis of cerebral alveolar echinococcosis can be made.

Cerebral alveolar echinococcosis should be differentiated from intracranial tuberculoma, metastases and grade III and IV astrocytoma. Intracranial tuberculoma is caused by hematogenous dissemination of pulmonary and other parts of tuberculosis and commonly occurs in children and young adults. Tuberculoma is often located in cortex with rich blood supply, demonstrated in nodular or lobulated pattern. The lesion is surrounded by fibrous envelope with internal caseous necrosis, which may occasionally be complicated by calcification. Singular lesion is found in 80 % of the cases with intracranial tuberculoma, and multiple lesions in 20 % of the cases. CT plain scan demonstrates the lesion with equal density, high density or mixed density, accompanied by calcification. MR imaging demonstrates the lesion with low T1WI signal inside, and equal signal in the envelope. MR imaging demonstrates heterogeneous low T2WI signal

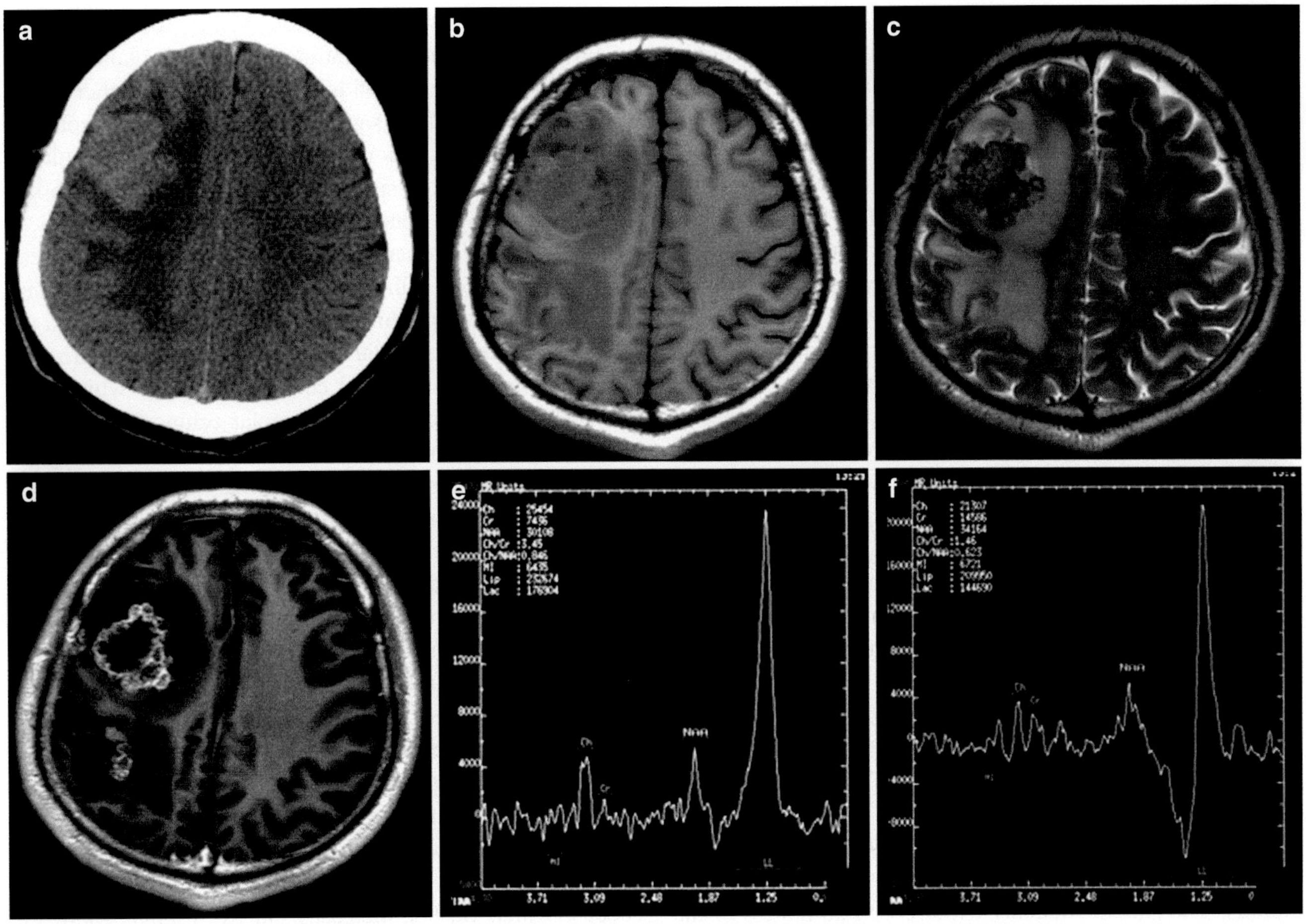

Fig. 8.53 (a–f) The right frontoparietal lobe was demonstrated with two solid space occupying lesions in irregular shape, with surrounding halo flakes of edema opacity. CT plain scan demonstrated slightly high density of the lesion in the right frontal lobe, with dominant equal signal in mixed T1WI signal and dominant low signal in mixed T2WI signal by MR plain imaging. The lesion was revealed with multiple small vesicles in different sizes and high signal inside. Contrast imaging demonstrated irregular ring shape enhancement of the lesion, with incomplete ring shape enhancement of the small vesicles in most cases. MRS demonstrated varying decreases of NAA peak, Cho peak and Cr peak as well as obvious increase of Lipd peak, with or with no accompanying Lac peak

in most areas, with low signal in the envelope. The accompanying calcification is demonstrated as low T1WI and T2WI signal. The lesion with caseous necrosis is demonstrated with ring or small nodule abnormal enhancement by contrast imaging. Cerebral metastasis often occurs in middle aged and elderly individuals aged 40–60 years, and 70–80 % of patients with cerebral metastasis show multiple lesions. The lesion is more commonly located in supratentorial area at the interface of cortex and medulla. The malignancy is often subject to central necrosis, cystic change and bleeding, occasionally with calcification. Obvious edema occurs around the metastasis whose severity is related to the type of malignancy. CT plain scan demonstrates the lesion with uneven density and nodule like or irregular ring shape. Most of the lesions are accompanied by surrounding edema, and small lesions are commonly accompanied by extensive surrounding edema, which is a characteristic sign of cerebral metastasis. MR plain imaging demonstrates most of the lesions with low T1WI signal and high T2WI signal. Since malignancy is originated from different pathological tissues and structures, the signals can be diversified. Extensive edema is commonly detected around the lesion in most cases, with obvious space occupying effect. Contrast imaging demonstrates obvious abnormal enhancement of the lesion, in diversified patterns of enhancement, such as nodular pattern, ring shape pattern and wreath pattern. The internal wall may even be accompanied by irregular mural nodules. Astrocytoma is the most common intracranial primary neoplasm, and grade III and IV astrocytomas show obvious space occupying effect. The neoplasm may spread contralaterally along the callosum, with accompanying surrounding edema and complicating bleeding, calcification and cystic change in most lesions. Most of grade III and IV astrocytomas are irregular in shape or wreath like, with mixed T1WI signal that dominantly low signal by MR plain imaging. The lesion with concurrent bleeding shows interval high T1WI signal. MR T2WI demonstrates the lesion in uneven high signal, the lesion with concurrent liquefaction, necrosis and cystic change in water like high signal, and the lesion with concurrent bleeding in varying signal according to different stages of hematoma. Contrast imaging demonstrates the lesion with obvious abnormal enhancement that may be patch like, linear, wreath like or nodular. The necrotic and hemorrhagic areas show no enhancement. ^{1}H-MRS may demonstrate grade III and IV astrocytomas as well as metastases, due to the concurrent necrosis, elevated Lip peak. Due to the highly heterogeneous malignancy cells, high ratio of cellular nucleus to cytoplasm, and more active transportation of the cytomembrane, Cho peak shows an obvious increase, which is different from the lesion of cerebral alveolar echinococcosis.

Case Study 40

[Brief Medical History]

A 31-year-old man complained of headache, nausea and vomiting for 2 months that aggravated for 7 days. He reported a definitive history of contact to dogs.

[Radiological Demonstration] (See Fig. 8.54)

[Diagnosis] Cerebral alveolar echinococcosis.

[Discussion]

The lesions of cerebral alveolar echinococcosis show infiltrative growth, with poorly defined boundary, space occupying effect and surrounding edema in most cases. MR imaging demonstrates the lesion in equal T1WI signals and dominantly low T2WI signal. The small vesicles or vesicular nest is poorly defined by MR T2WI with slightly higher signal, but well defined by MR hydrography. Multiple sand like low signal calcifications can be detected in some cases. Unlike alveolar echinococcosis at the other body parts, the lesion of cerebral alveolar echinococcosis can be irregularly enhanced by contrast imaging, which is a characteristic sign of the disease.

Cerebral alveolar echinococcosis should be differentiated from the following diseases:

1. Cerebral tuberculoma
   The patients commonly experience low grade fever, night sweat, fatigue, emaciation and other toxic symptoms induced by tuberculosis. Radiologically, the condition is characterized by involved basilar cistern, with nodular and ring shape abnormal enhancements in the brain. The part of lesion near the meninges shows obvious abnormal enhancement and thickening.
2. Cerebral metastases
   The lesion os cerebral metastases are commonly distributed in the subcortical areas supplied by the middle cerebral artery, with high T2WI signal. Contrast imaging demonstrates homogeneous enhancement of the lesion in most cases, but the liquefied and necrotic part show no enhancement and no multiple small vesicles sign is detectable. In combination to the clinical history, the diagnosis can be made.
3. Other neoplasms with concurrent bleeding
   The patients report no history of contact to dogs, and clinical laboratory test shows negative to echinococcus.

Case Study 41

[Brief Medical History]

A 32-year-old Kazak man was hospitalized due to the chief complaints of hepatic echinococcosis for 4 years and sudden seizures of the left body for 4 days that aggravated for 1 day.

[Radiological Demonstration] (See Fig. 8.55)

[Diagnosis] Alveolar echinococcosis in the right fronto-parietal lobe.

[Discussion]

Cerebral alveolar echinococcosis is more common in adults, with lesion in the supratentorial area and the areas supplied by the middle meningeal artery. All the lesions of cerebral alveolar echinococcosis are secondary to hematogenous dissemination of hepatic alveolar echinococcosis, which may be singular or multiple. Pathologically, the lesion is characterized by growth of bilateral budding with dominantly exogeneou budding, numerous mall vesicles and its section with honeycomb like change. It commonly occurs in or under the cortex, with rapid growth due to the soft brain tissue and rich blood supply that resembles to malignancy. CT plain scan demonstrates the lesion in soft tissue density, commonly with high density calcification opacity inside, obvious space occupying effect and surrounding edema. MR plain imaging demonstrates the lesion with equal or slightly high T1WI signal and mixed T2WI signal with dominantly low signal. Most of the lesion are demonstrated with multiple high signal vesicles with different sizes inside by T2WI. And multiple small vesicle by MR hydrography is a characteristic sign of cerebral alveolar echinococcosis. In some cases, multiple sand like low signal calcifications are detectable, which is believed to be the main reason for low T2WI signal. Unlike alveolar echinococcosis at other body parts, cerebral alveolar echinococcosis shows irregular peripheral enhancement of the lesion by contrast imaging, which is also a characteristic sign of cerebral alveolar echinococcosis. MR hydrography (MRH) was originally applied to detect lesions in pancreatic biliary ducts, which has been introduced to examine other human body organs. MR hydrography has important diagnostic value for cerebral alveolar echinococcosis, which can favorably demonstrate the detailed structures of the lesion as well as its relationship with the neighbouring involved organs such as ventricle that cannot be demonstrated by MRI. MR hydrography can also favorably demonstrate small vesicles in and around the lesion, which is the key point for the differential diagnosis of

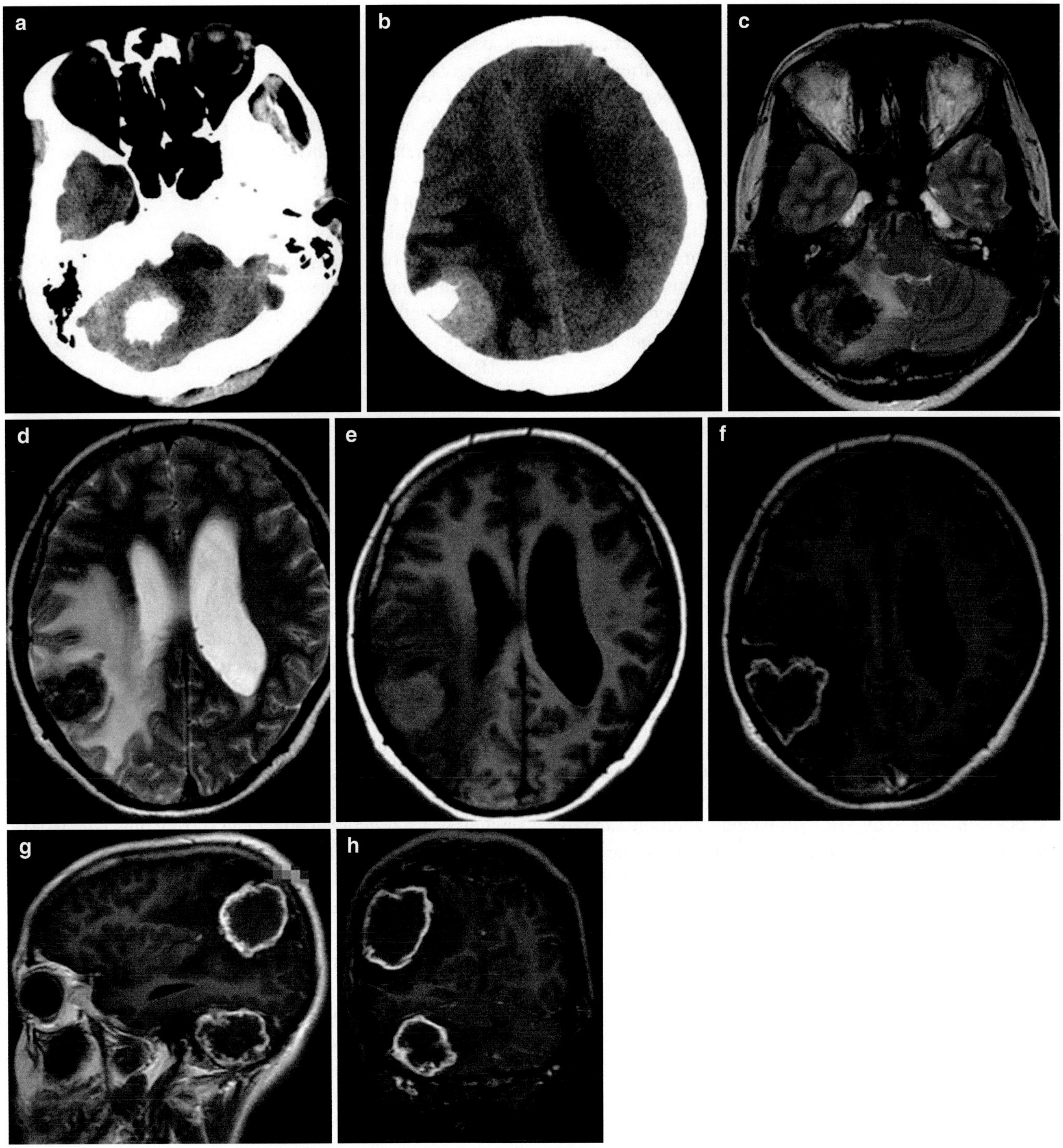

Fig. 8.54 (**a**, **b**) CT plain scan demonstrated multiple high density lesions in the right cerebellar hemisphere and parietal lobe, with surrounding flakes of edema and observable space occupying effect. The lateral ventricle was revealed with enlargement. (**c**–**e**) MR imaging demonstrated equal T1 and long T2 signals of the above lesions. (**f**–**h**) Contrast MR imaging demonstrated ring shape enhancements of the above lesions

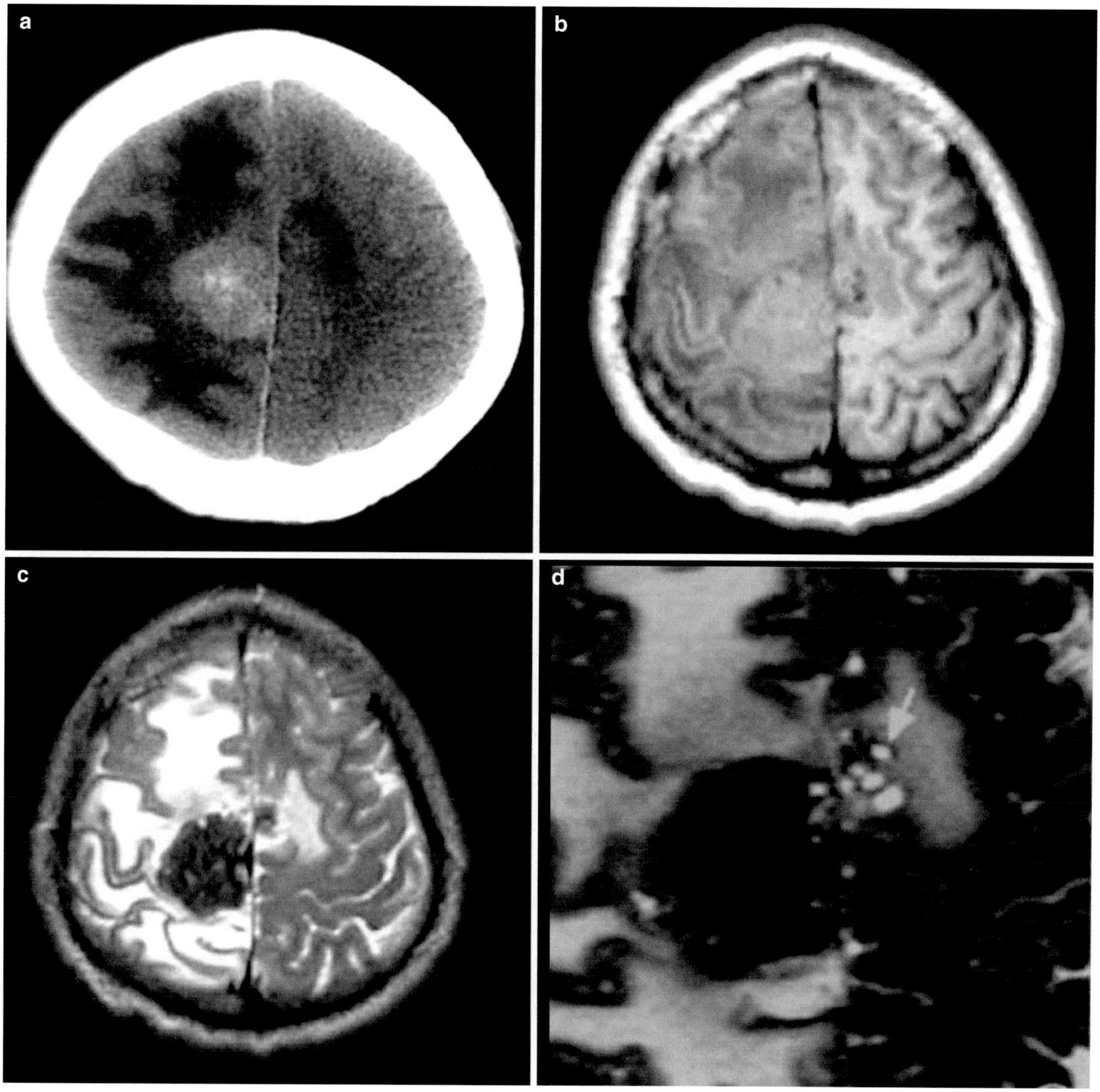

Fig. 8.55 (**a–d**) The right frontoparietal lobe was demonstrated with ball like lesion that partially grew contralaterally with obvious surrounding edema. CT plain scan demonstrated predominant high density, with high density calcification inside. MR plain imaging demonstrated equal T1WI signal and dominantly low T2WI signal of the lesion. MR hydrography demonstrated multiple small vesicles (*white arrow*) with high density and different sizes within and around the lesion

cerebral alveolar echinococcosis from other diseases because the lesion of cerebral tuberculoma and metastases shows no small vesicles. MR hydrography is superior to other radiological modalities in increasing the detection rate of cerebral alveolar echinococcosis and in decreasing the rate of misdiagnosis. If applied in combination to routine MR imaging, MR hydrography can provide more information about the characteristic pathological changes of cerebral alveolar echinococcosis so as to facilitate its diagnosis.

Cerebral alveolar echinococcosis should be differentiated from cerebral tuberculoma, metastases as well as grade III and IV astrocytomas.

Cerebral tuberculoma is caused by hematogenous dissemination of pulmonary and other parts of tuberculosis, which is commonly located in the cortex with rich blood supply and commonly affects children and young adults. The lesion is demonstrated to be nodular or lobulated, with surrounding envelope, internal caseous necrosis, and occasionally accompanying calcification. In 80 % of patients with cerebral tuberculoma, the lesion is singular, while the other 20 %, multiple lesions. CT plain scan demonstrates the lesion with equal density, high density or mixed density, with accompanying calcification. MRI plain imaging demonstrates low T1WI signal in the lesion, and equal T1WI signal in its envelope. MR plain imaging demonstrates mostly uneven low T2WI signal, and low T2WI signal in the envelope. The accompanying calcification is demonstrated as low T1WI and T2WI signals in most areas. If the lesion shows concurrent caseous necrosis, the lesion is demonstrated with ring shape or small nodule abnormal enhancement by contrast imaging.

Cerebral metastases is more common in middle-aged and elderly populations aged 40–60 years, and 70–80 % of the patients show multiple lesions. The lesion is more commonly located in the supratentorial area at the interface between cortex and medulla. The neoplasm is subject to central necrosis, cystic change and bleeding, occasionally with calcification. Its surrounding edema is obvious, whose severity is related to the type of neoplasm. CT plain scan demonstrates the neoplasm in uneven density with a nodular or irregular ring shape, which is commonly accompanied by surrounding edema. In the cases of small neoplasm, extensive edema is detectable, which is a characteristic sign of cerebral metastases. MR plain imaging demonstrates cerebral metastases in low T1WI signal and high T2WI signal. Since the neoplasms are originated from different pathological tissues and structures, their signals on MRI are diversified. Edema around the lesion in most cases is extensive, showing obvious space occupying effect. By contrast imaging, obvious abnormal enhancement is demonstrated in diversifying enhancement patterns, such as nodule pattern, ring pattern and wreath pattern. The internal cystic wall may also show accompanying irregular shape mural nodules.

Astrocytoma is the most common intracranial primary neoplasm, and the grade type III and IV astrocytomas show obvious space occupying effect, which grow contralaterally along the callosum and are commonly accompanied by surrounding edema. Concurrent bleeding, calcification and cystic change are also common. Most of grade III and IV astrocytomas are irregular in shape or wreath like. By MR plain imaging, the lesion is demonstrated as mixed T1WI signals with dominantly low signal, and the lesions with bleeding show high T1WI signal. MR plain imaging demonstrates uneven high T2WI signal, and the lesions with liquefaction, necrosis and cystic change show water like high T2WI signal. The lesion with concurrent bleeding show signal with corresponding changes along with duration of hematoma. Contrast imaging demonstrates obvious abnormal enhancement of the lesion in patch pattern, linear pattern, wreath pattern or nodule pattern, but no abnormal enhancement of its necrotic and hemorrhagic area.

Case Study 42

[Brief Medical History]

A 54-year-old man, Hui nationality, was hospitalized due to the chief complaints of progressive enlargement of neck for 2 years. The patient reported the finding of a cervical lump with no known causes 10 years ago, the diagnosis of hyperplasia of thyroid gland was made in a local hospital. He then did not pay close attention to the lump and received no special examination and treatment. In the recent 2 years, the cervical lump gradually grew larger with no known causes. And he experienced a bad cold and cough with whitish flakes of substances. After treatment for 1 week, his condition was not improved.

[Radiological Demonstration] (See Fig. 8.56)

[Diagnosis] Cystic echinococcosis (multi daughter cysts type) in bilateral thyroid lobes, neck base and left clavicular fossa.

[Discussion]

Cystic echinococcosis rarely occurs in the thyroid gland, which is commonly secondary to hepatic or pulmonary echinococcosis and commonly shows the multi daughter cysts

Fig. 8.56 (**a**–**f**) The thyroid gland was demonstrated to be abnormal in shape and structure, with obvious enlargement. Multiple cystic lesions in different sizes were detected in the bilateral thyroid lobes, neck base and left clavicular fossa. CT plain scan demonstrated as low density with multiple septa inside. MR plain imaging demonstrated multiple cystic lesions of different sizes, with some daughter cysts arranged in necklace pattern at the coronal section. The lesions were demonstrated in low T1WI signal and high T2WI signal. And the cystic wall was demonstrated with equal T1WI signal but slightly lower T2WI signal. Contrast imaging demonstrated obvious abnormal enhancement of partial cystic wall, but no enhancement of the cystic fluid

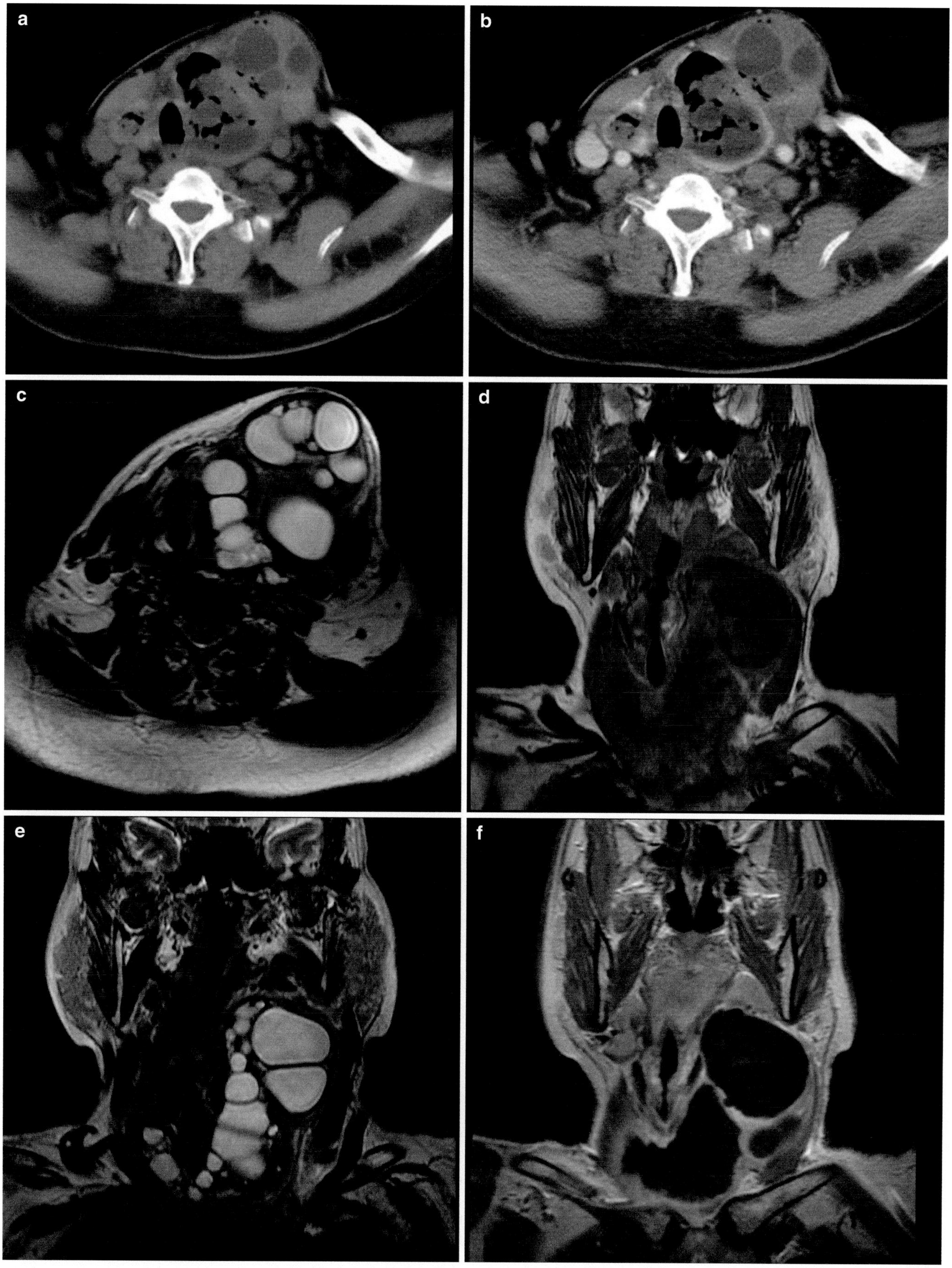
a
b
c
d
e
f

type. By CT scan and MR imaging, multiple daughter cysts in the mother cyst can be detected, with lower density mother cyst than daughter cysts by CT scan. By MR imaging, the T1WI signal of the mother cyst is lower than that of the daughter cysts, and the T2WI signal of the mother cyst is higher than that of the daughter cysts, with the daughter cysts well defined. Septa opacity is demonstrated between the mother and daughter cysts and between the daughter cysts. When mother cyst contains daughter cysts, rose-petals sign is demonstrated, which is consistent with the sign demonstrated in the cases of hepatic echinococcosis (multi daughter cysts type). The lesion is commonly not enhanced by contrast scan. If with concurrent infection, the lesion shows irregular ring shape enhancement of the cystic wall by contrast scan. The patients showing cystic lesion in the thyroid gland with multi daughter cysts should firstly report the medical history. If a history of contact to the source of infection of echinococcus can e defined, the diagnosis can be made in combination to related laboratory tests and typical radiological signs.

In such a case, the condition should be differentiated from thyroid cyst, hemangioma in cervical soft tissue and cystic lymphangioma in cervical soft tissue.

Despite no obvious symptoms in most cases of thyroid cyst, MR imaging commonly demonstrates the lesion of thyroid cyst to be round like, with low T1WI signal and high T2WI signal. The lesion shows smooth and intact margin, with its envelope in equal signal and no septa opacity. Contrast imaging demonstrates no enhancement of the lesion.

Hemangioma is a common benign neoplasm occurring in cervical soft tissue, round like or lobulated, and is often located in subcutaneous soft tissue. MR plain imaging demonstrates the lesion with equal or slightly high T1WI signal and mixed high T2WI signal, with mixed strips or spots of low signal in the lesion. Contrast imaging demonstrates the lesion with moderate uneven enhancement. Along with gradual filling of the contrast reagent, the lesion shows progressive enhancement.

Cystic lymphangioma in cervical soft tissue commonly occurs in children, and the lesion grows slowly as painless soft semi-solid cervical lump. Typically, MR imaging demonstrates the lesion to be round like or irregular in shaped, with low T1WI signal and high T2WI signal. The lesion is well defined with homogeneous signal, which grows in a crawling manner along the space between soft tissues. If with concurrent bleeding, the lesion is demonstrated with fluid-fluid level and bleeding signals of different stages, with no abnormal enhancement by contrast imaging.

Case Study 43

[Brief Medical History]

A 29-year-old woman was detected with a space occupying lesion in the chest by chest X-ray during her physical examination 1 day ago. She experienced occasional cough, and reported a history of living in husbandry area.

[Radiological Demonstration] (See Fig. 8.57)

[Diagnosis] Pulmonary cystic echinococcosis.

[Discussion]

Echinococcosis occurs at any part of human body, which is caused by access of the pathogenic larva of echinococcus to blood flow and its subsequent invasion of organ. Hepatic echinococcosis is the most common, but other extrahepatic organs such as lungs, spleen and brain can also be involved. Its typical radiological signs include singular or multiple round or round like fluid low density lesion(s) that is(are) occasionally lobulated, calcification of partial cystic wall, a few daughter cysts in the lesion, and low density daughter cysts than mother cyst. The ruptured cyst may be demonstrated with crescent sign, lotus-on-water sign, and ribbon sign.

Pulmonary cystic echinococcosis should be differentiated from the following diseases:

1. Pulmonary abscess
   The clinical symptoms of pulmonary abscess are commonly serious. CT scan demonstrates uneven thickness of the abscess wall, and no lace like change in the lesion.

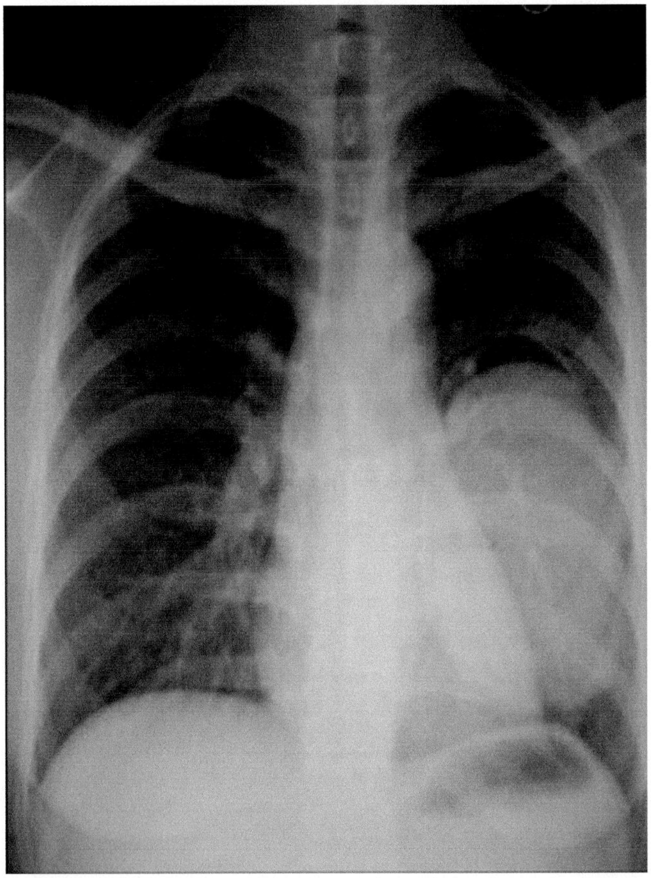

Fig. 8.57 Chest X-ray demonstrated a round lump in the left lower lung, with smooth and clearly-defined boundary and even density. Its upper margin was demonstrated with crescent shape gas opacity

2. Cavity tuberculosis
   The lesion of tuberculoma is commonly located in the posterior segment of upper lung lobe and the dorsal segment of lower lung lobe, with surrounding satellite lesions. The lesions show uneven density, with spots of calcification inside. The lesion of cystic echinococcosis may be located at any part of lungs, commonly at the lateral field of the lower lung with homogeneous density. Along with the progression of the condition, the diagnosis can be defined.
3. Bronchial cyst
   Simplex cyst lesion of pulmonary cystic echinococcosis should be differentiated from bronchial cyst, which is demonstrated as cystic lesion containing gas. If with secondary infection, the lesion of bronchial cyst is demonstrated with increased intracystic density, whose communication with bronchus induces formation of fluid level. However, the simplex cyst lesion of pulmonary echinococcosis is demonstrated as a cystic lump in water like density and smooth margin. The lotus-on-water sign can be demonstrated within the fluid level.

Case Study 44

[Brief Medical History]

A 58-year-old herdswoman complained of chest distress and pain as well as cough with a small quantity of whitish sputum. She permanently lived in husbandary area. The erythrocyte sedimentation rate (ESC) was tested to be 25 mm/h.

[Radiological Demonstration] (See Fig. 8.58)

[Diagnosis] Pulmonary cystic echinococcosis.

[Discussion] See Case Study 48.

Case Study 45

[Brief Medical History]

A 16-year-old boy was detected with nodules in the lung by physical examination. He experienced no obvious upset, The Casoni test showed positive (+). And he had a definitive history of contact to dogs.

[Radiological Demonstration] (See Fig. 8.59)

[Diagnosis] Pulmonary cystic echinococcosis.

[Discussion] See Case Study 48.

[Case Study 46]

[Brief Medical History]

A 10-year-old girl complained of chest distress and pain for 2 months, and dry cough and fever with the highest body temperature of up to 39.5 °C. The Casoni test showed positive (+). She permanently lived in husbandry area.

[Radiological Demonstration] (See Fig. 8.60)

[Diagnosis] Pulmonary cystic echinococcosis.

[Discussion] See Case Study 48.

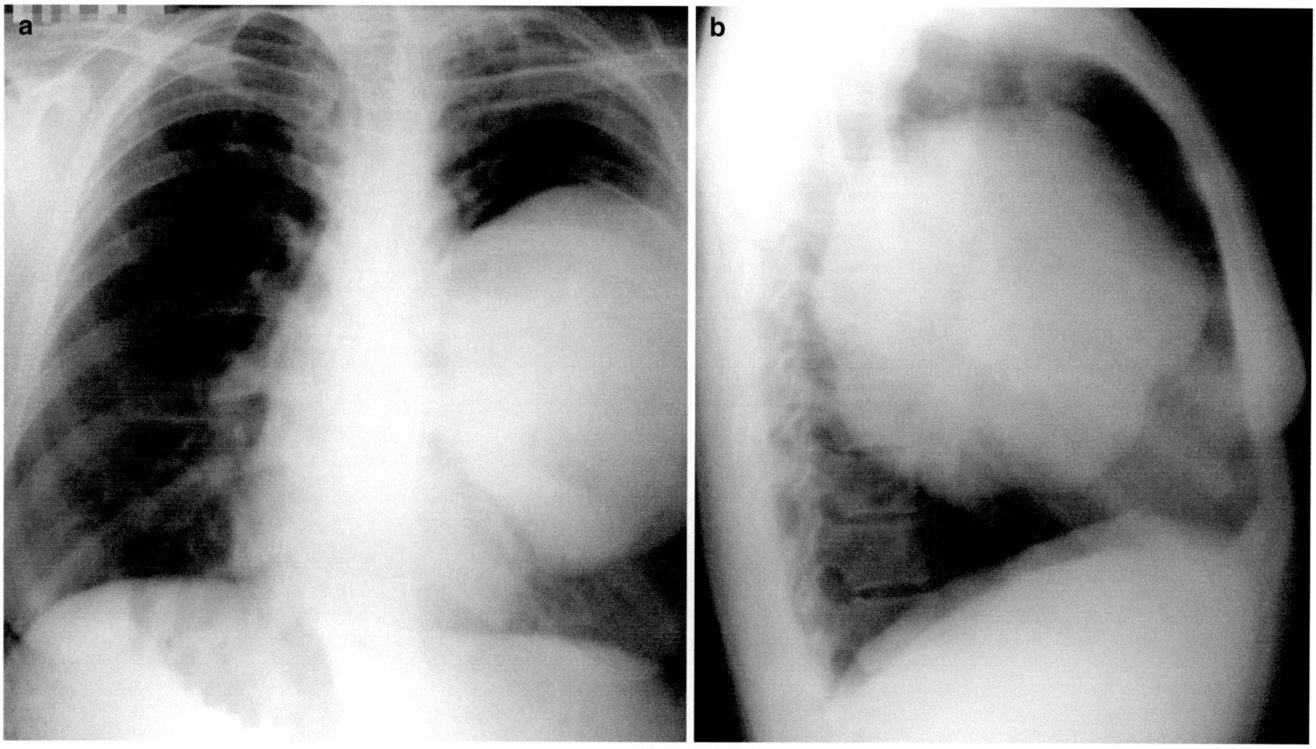

Fig. 8.58 (**a**, **b**) Chest X-ray demonstrated a huge round like soft tissue lump in the left upper lung lobe, with smooth and clearly-defined boundary. The lesion was shown to be lobulate. And atelectasis was observable in the lingual segment of the left upper lung lobe

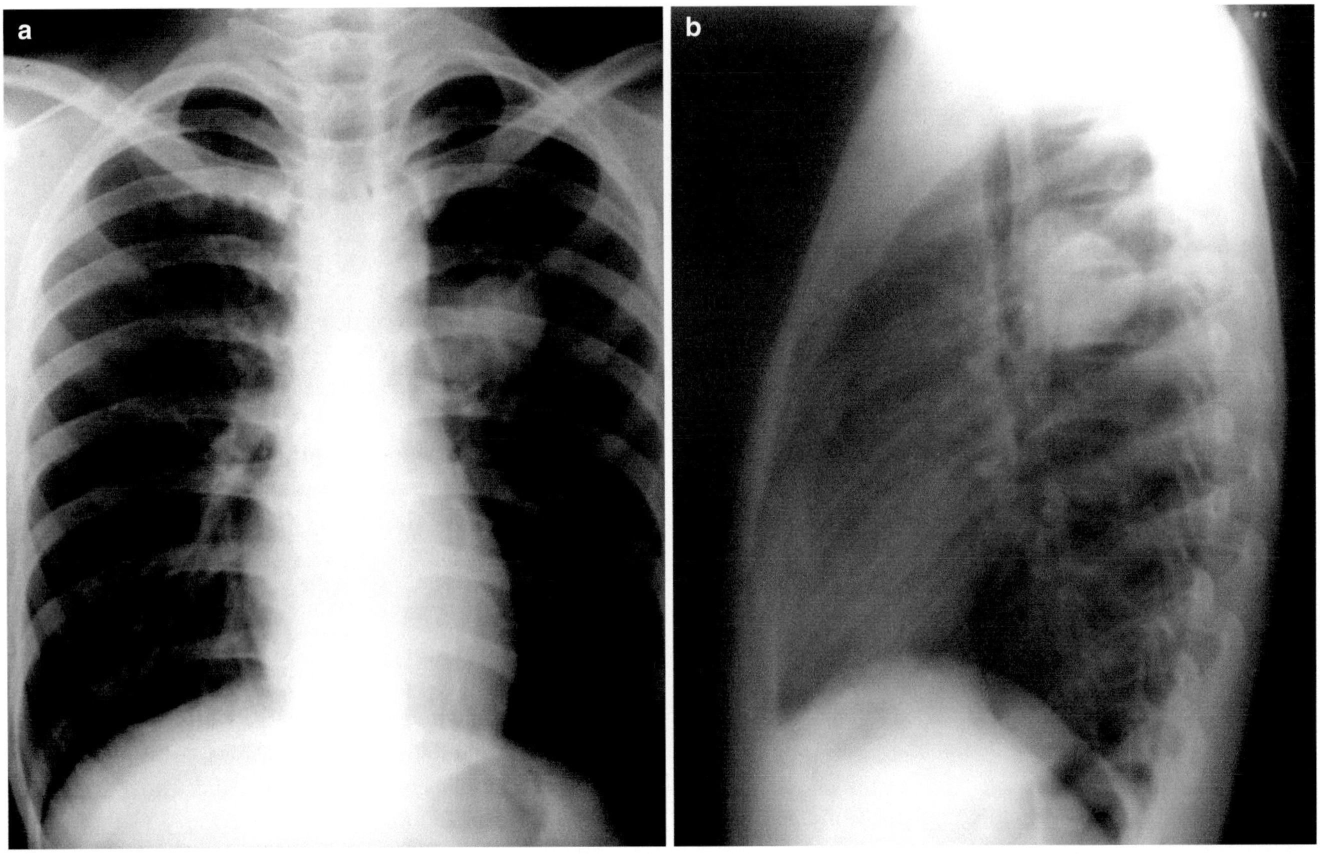

Fig. 8.59 (**a**, **b**) Anterior-posterior chest X-ray demonstrated round like soft tissue lump in the apical posterior segment of left upper lung lobe, with smooth and well defined boundary

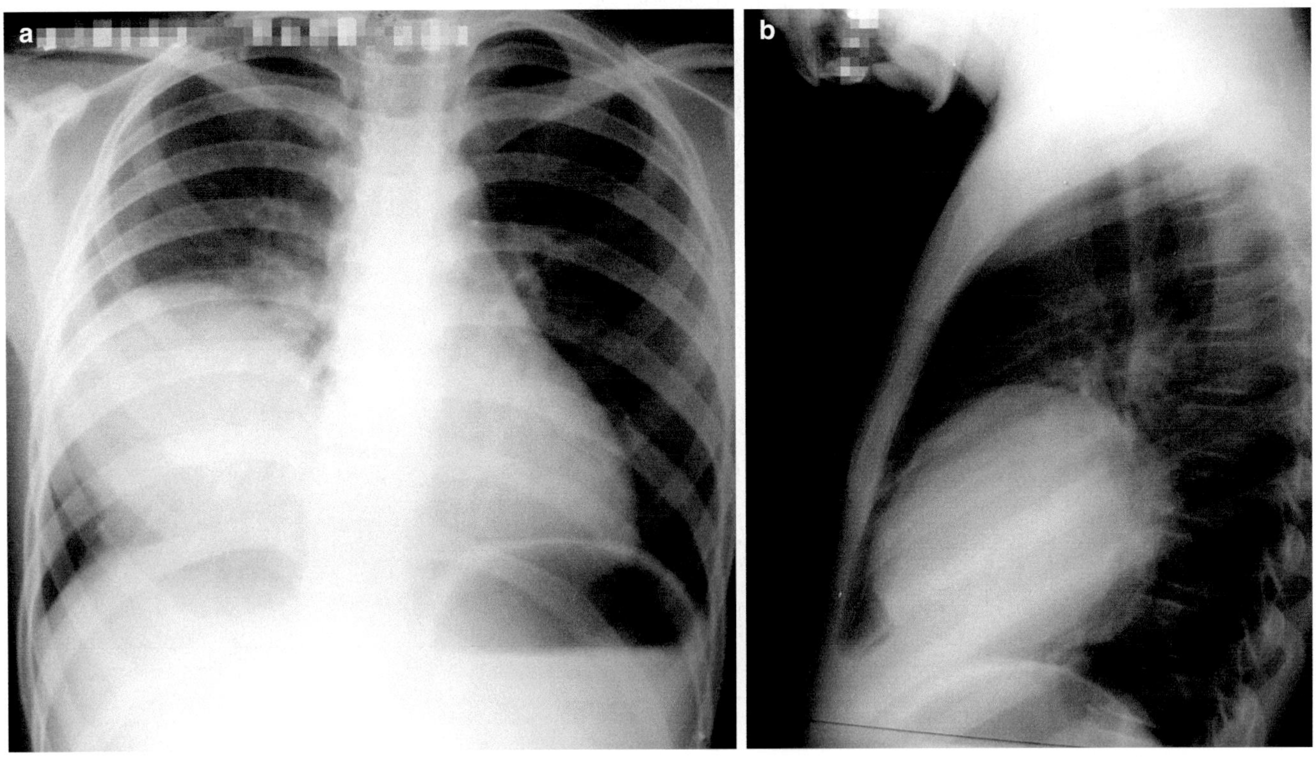

Fig. 8.60 (**a**, **b**) Anterior-posterior chest X-ray demonstrated a huge round like soft tissue lump in the right middle lung lobe, with smooth and clearly defined boundary. The lesion was shown to be lobulated

Case Study 47

[Brief Medical History]

A 51-year-old woman was detected with nodules in the lung in physical examination. She experienced no obvious upset. Laboratory tests showed negative. And she permanently lived in husbandry area.

[Radiological Demonstration] (See Fig. 8.61)

[Diagnosis] Pulmonary cystic echinococcosis.

[Discussion] See Case Study 48.

Case Study 48

[Brief Medical History]

A 30-year-old man complained of chest pain, cough, hemoptysis and fever. Laboratory tests showed WBC count 25×10^9/L, eosinophilic count 2.1×10^9/L, and ESR 45 mm/h. He had a history of traveling in husbandry area 1 year ago.

[Radiological Demonstration] (See Fig. 8.62)

[Diagnosis] Pulmonary cystic echinococcosis.

[Discussion]

The pathomechanism of hepatic echinococcosis is the gradual development of oncosphere into a enveloped cyst in the liver, which grows slowly and enlarges gradually to compress its surrounding tissue. The compressed surrounding tissue is subject to atrophy to form a fibrous layer. The cystic inner wall produces the germinal layer that grows into the cystic cavity, and scolex grows from the inner wall of germinal layer. Migration and invasion of such scolex into other body parts induce occurrence of secondary cyst.

Clinically, the patients usually have a long medial history, with gradual progression of the lesion. Most of the patients are asymptomatic. The large lesion in the lung may compress the lung, mediastinum and diaphragm to produce corresponding symptoms such as dyspnea. The small solitary nodules in the lung generally induce no obvious symptoms. Secondary infection is a common symptom. In Cases No. 42–46, echinococcosis was defined by postoperative biopsy. In combination to epidemiological and clinical medical histories, the diagnosis of this group of cases can be made.

By chest X-ray, the lesion of pulmonary echinococcosis is commonly singular or multiple round or round like. In the early stage, the lesion is small with low density and poorly defined boundary. But the lesion with a diameter of above 2 cm can be well defined. The huge echinococcus cyst in the lung is demonstrated by chest X-ray fluoroscopy to be vertically deformed along with deep breathing, which is known as the breathing sign of echinococcosis. The enlarged cyst pushes its surrounding lung tissue, which is demonstrated as ball-in-hands sign. If the cystic wall communicates with bronchus, crescent sign or double-bow sign is shown. When

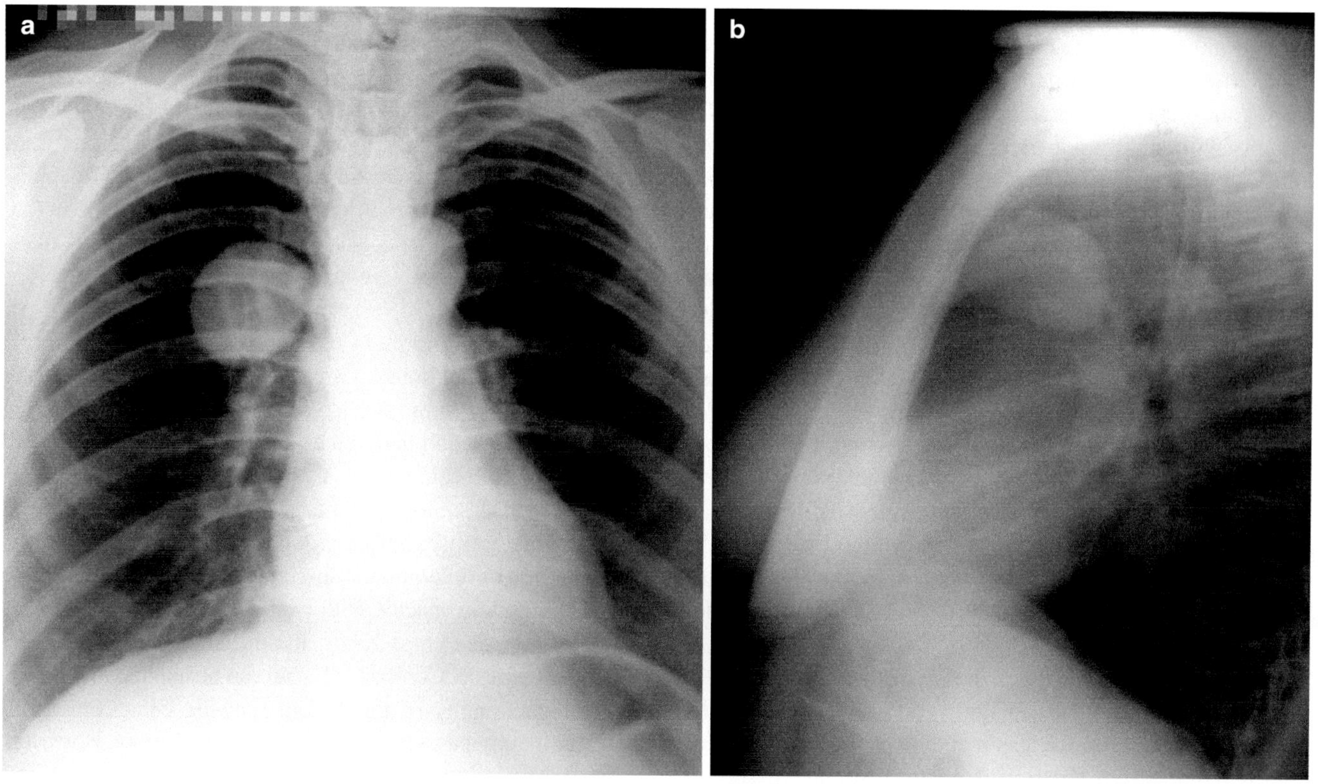

Fig. 8.61 (**a**, **b**) Anterior-posterior chest X-ray demonstrated round like soft tissue density lump in the anterior segment of right upper lung lobe, with smooth and clearly-defined boundary

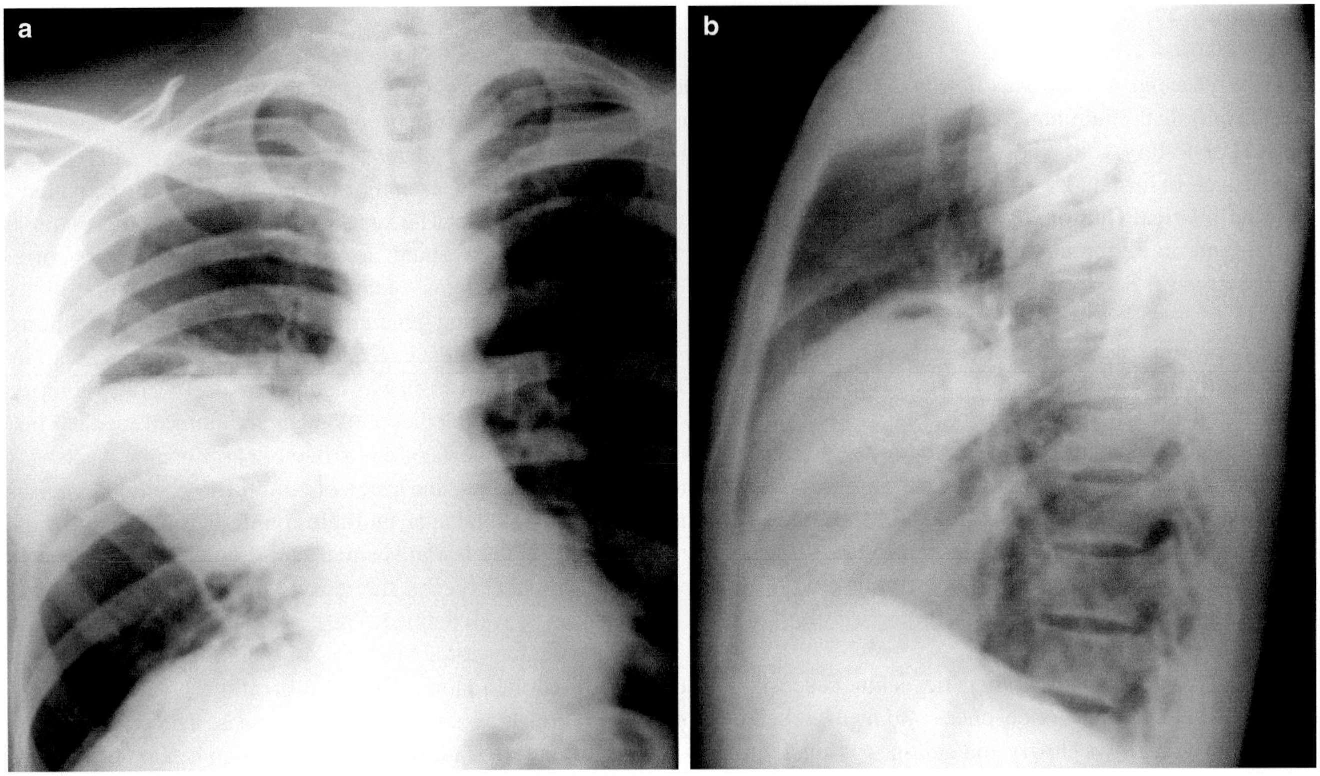

Fig. 8.62 (**a**, **b**) Anterior-posterior chest X-ray demonstrated irregular soft tissue lump in the right middle lung lobe, with smooth and clearly defined boundary. The lesion was shown to be lobulated at the margin, with fluid level inside. The distal part of the lesion was demonstrated with patches of high density opacity

the internal cyst ruptures and collapses to float in the cystic fluid, lotus-on-water sign is demonstrated. In this group of cases, all the lesions were detected by chest X-ray. The patients with small nodules experienced no obvious symptoms, and the lesions were detected during routine physical examination. The patients with large lesions experienced compressive pulmonary symptoms (e.g. Case Study 42 and 44). In Case Study 42, the patient experienced compressive pulmonary atelectasis. In Case Study 46, the lesion is subject to rupture of internal cyst with gas-fluid level. The adjacent lung tissue is subject to inflammatory changes with subsequent clinical symptoms.

By CT scan, the lung is demonstrated with cystic lump with a CT value close to water density. The lesion is shown with smooth margin and homogeneous density. If with cavity, the lesion can be demonstrated with crescent like or sickle like gas density opacity. Contrast scan may demonstrate ring shape enhancement of the cystic wall or enhancement of the cystic wall.

MR imaging demonstrates the cystic fluid with long T1 and long T2 signals. MR imaging can help to identify the relationships between the lesion and the neighbouring mediastinum, diaphragm, heart and major vascular vessels. In addition, MRI is superior to chest X ray and CT scan in demonstrating the internal and external cysts of the lesion.

Chest X-ray and CT scan are of the choice for diagnosis of pulmonary cystic echinococcosis. CT plain and contrast scans are superior to chest X-ray in demonstrating the internal lesion structure, while MR imaging can more favorably demonstrate the cystic wall and intracystic septa. In combination to epidemiologic history and immunoassay, the diagnosis can be made.

Pulmonary echinococcosis should be differentiated from the following diseases.

1. Pulmonary carcinoma
   The lesion of pulmonary carcinoma is demonstrated as solitary nodule or lump with coarse margin and the neoplasm signs of lobulation and spikes. However, the lesion of pulmonary cystic echinococcosis is demonstrated with benign signs, such as smooth and intace margin of the lesion.
2. Inflammatory pseudotumor
   The lesion os inflammatory pseudotumor is demonstrated as solitary nodule opacity, with well defined and smooth margin. By routine chest X-ray, the differential diagnosis can hardly be made. Contrast CT scan can be applied to observe the blood supply of the lesion, showing enhancement of lesion in most cases of inflammatory pseudotumor. However, as a cystic lesion, the lesion of pulmonary cystic echinococcosis shows no enhancement of its internal structure but slight enhancement of the cystic wall by contrast CT scan.

3. Pulmonary sarcoma
   The lesion of pulmonary sarcoma resembles to the lesion of pulmonary carcinoma. The differential diagnosis can be made based on findings by contrast CT scan.
4. Tuberculoma
   The lesion of tuberculoma may show satellite lesions, with caseous substances inside by CT scan. However, the lesion of pulmonary cystic echinococcosis is demonstrated with predominantly cystic fluid inside.
5. Pulmonary metastases
   The lesion of pulmonary metastases is more commonly multiple, but rarely singular, which can be enhanced by contrast CT scan. In differential diagnosis of pulmonary cystic echinococcosis from other pulmonary diseases with solitary round like lesion, CT plain and contrast scans can more favorably demonstrate the internal structure of the lesion so as to increase the accuracy of diagnosis.

Case Study 49

[**Brief Medical History**]

A 47-year-old woman was detected with hepatic and pulmonary echinococcosis by CT scan during her physical examination 1 month ago.

[**Radiological Demonstration**] (See Fig. 8.63)

[**Diagnosis**] Pulmonary cystic echinococcosis in the right upper lung lobe.

[**Discussion**]

In this case, the diagnosis was simplex cyst of pulmonary echinococcosis. The lesion was shown with homogeneous internal density as well as smooth and sharp margin. During routine physical examination, the lesion of hepatic cystic echinococcosis was also detected. If with no medical history, the condition can also be suspected to be pulmonary cyst with fluid. The patients with congenital pulmonary cyst experience repeated respiratory infection, with the lesion compressing the bronchi to show pulmonary atelectasis. If

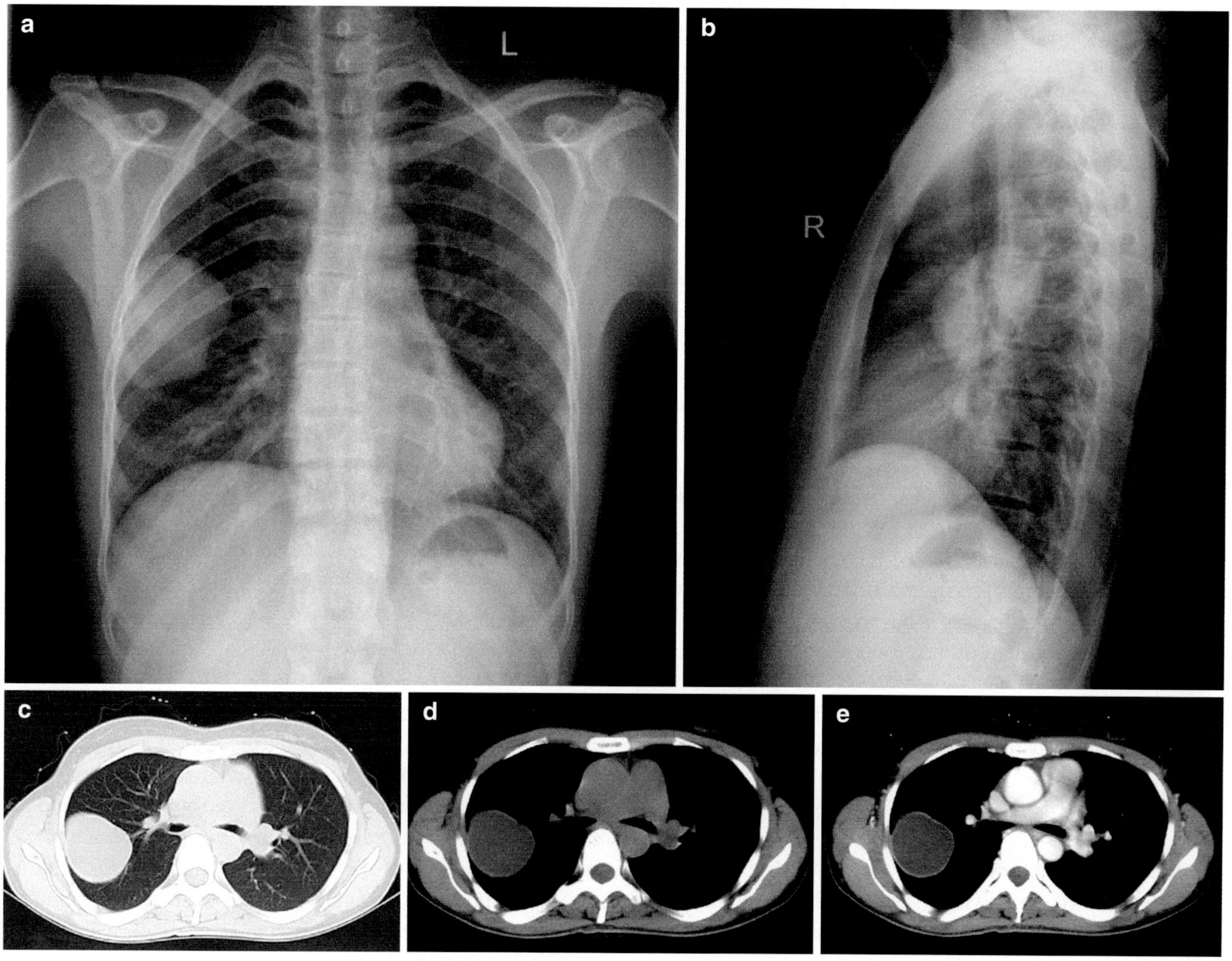

Fig. 8.63 (**a**, **b**) Anterior-posterior chest X-ray demonstrated a ball shape high density lesion in the middle-outer area of the right middle lung field, with clearly defined and smooth margin as well as even internal density. (**c**, **d**) CT plain scan demonstrated round like cystic lesion with low density in the posterior segment of righter upper lung lobe, with the largest section in size of 5 cm×6 cm. The lesion was shown with homogeneous fluid density in the lesion. (**e**) Contrast scan demonstrated no obvious enhancement of the lesion

pulmonary cyst penetrates into the bronchus, gas-fluid level emerges, but showing no lotus-on-water sign. Pulmonary cystic echinococcosis should be differentiated from peripheral pulmonary carcinoma. The lesion of peripheral pulmonary carcinoma is demonstrated as dense lump in most cases, with uneven density and irregular shape. Its margin is blunter than that of echinococcus cyst, with signs of lobulation, spikes and collapsed pleura. Enlarged lymph nodes are commonly detected at the lung hilum. The lesions of pulmonary hematogenous metastases are commonly multiple in cotton lump like appearance that is more commonly located in the lateral area of lung fields. During short-term follows-up, the lesions can be found with rapid growth.

Case Study 50

[Brief Medical History]

A 18-year-old Uyghur young woman complained of upper abdominal upset for 4 months. The 4-item examination for echinococcus showed Anti-EgCF antibody negative (−), Anti-EgP antibody negative (−), Anti-EgB antibody negative (−) and Anti-Em2 antibody negative (−).

[Radiological Demonstration] (See Fig. 8.64)

[Diagnosis] Pulmonary cystic echinococcosis in the left upper lung lobe.

[Discussion]

Echinococcosis is an edemic and zoonotic parasitosis, with dogs as its main source of infection. After oral intake of the eggs of echinococcus by human, the eggs are incubated into oncospheres in human duodenum, with subsequent invasion into the portal system via mesenteric veins. They firstly settle in liver to develop into adults echinococcus to cause hepatic echinococcosis, and hepatic echinococcosis shows the highest incidence rate in all types of echinococcosis. The oncospheres can also pass through the liver to invade the lung along with the blood flow in the pulmonary artery, and pulmonary echinococcosis shows the second highest incidence rate in all types of echinococcosis. The total incidence rates of hepatic and pulmonary echinococcosis account for 80–90% of all echinococcosis. In this case, the lesion was singular cyst with homogeneous internal density as well as smooth and sharp margin of the cystic wall. The patient had a medical history of pulmonary echinococcosis. Based on these findings, the diagnosis can be accurately made. If no medical history, the condition should be differentiated from bronchial cyst.

Case Study 51

[Brief Medical History]

A 26-year-old woman reported a history of hepatic space occupying lesion for 16 years. She experienced intermittent fever for 1 month.

[Radiological Demonstration] (See Fig. 8.65)

[Diagnosis] Multiple cystic echinococcosis in the right lung (multi daughter cysts type).

[Discussion]

This is a case of pulmonary echinococcosis, with multiple daughter cysts in the lesion in petals sign. The density of daughter cysts is always lower than that of the mother cyst, which is a characteristic sign of cystic echinococcosis of multi daughter cyst type.

References

1. Jiang DD, Taxipulati, Fang KH. CT diagnosis of pulmonary echinococcosis. J Imaging Diagn Interv Radiol. 1998;7(1): 10–12.
2. Wen H, Xu MQ. Practical echinococcosis. Beijing: Science Press; 2007.

Case Study 52

[Brief Medical History]

A 52-year-old woman, Hui nationality, complained of right upper abdominal pain for more than 4 months. The 4-item examination for echinococcus showed Anti-EgCF antibody negative (−), Anti-EgP antibody negative (−), Anti-EgB antibody negative (−) and Anti-Em2 antibody negative (−).

[Radiological Demonstration] (See Fig. 8.66)

[Diagnosis] Pulmonary cystic echinococcosis.

[Discussion]

In this case, a huge cystic lesion was detected, with daughter cysts inside in cyst-in-cyst sign. The Cystic wall was partially thick, with smooth and intact margin. By contrast scan, the cystic fluid showed no obvious enhancement, but the cystic wall was enhanced. In this case, typical signs were detected, including cyst-in-cyst sign and thick cystic wall, based on which the diagnosis can be accurately made.

Case Study 53

[Brief Medical History]

A 9-year-old boy was hospitalized due to complaints of fever and cough for more than 1 month, with the highest body temperature of 39.6 °C. He experienced nausea, vomiting, cough with a small quantity of whitish sputum. Ig antibody of pulmonary echinococcus showed positive. Pathologic examination showed pinkish staining of the structureless lamella cystic wall of the lesion in the right upper lung as well as infiltrates of lymphocytes and neutrophilic granulocytes in surrounding areas of partial cystic wall, which indicated inflammatory responses.

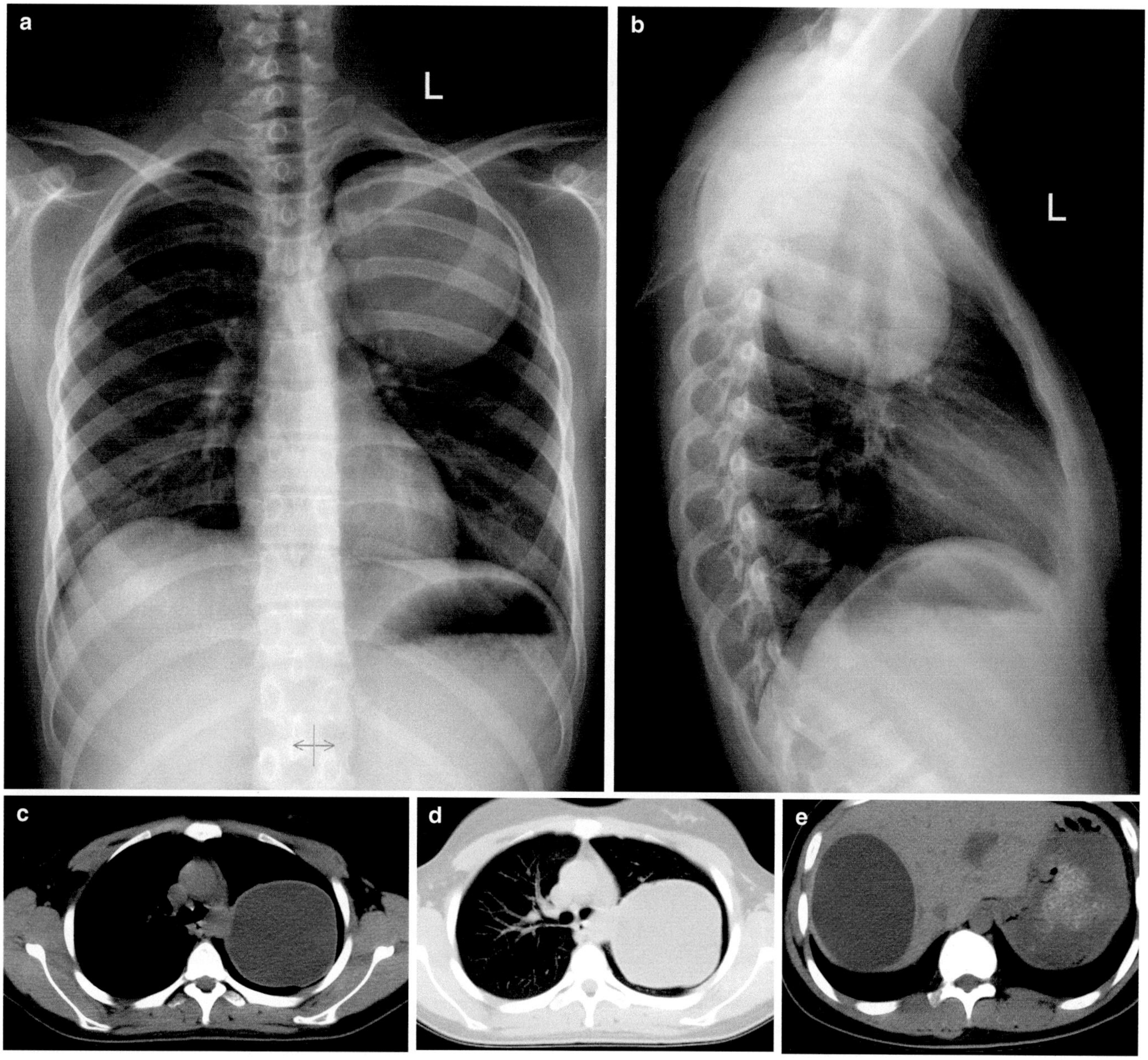

Fig. 8.64 (**a**, **b**) Anterior-posterior chest X-ray demonstrated a huge ball shape high density lesion in the left upper lung lobe, with a size of about 10.8 cm×9.1 cm, homogeneous internal density, as well as well defined and sharp margin. (**c**, **d**) CT plain scan demonstrated a huge cystic lesion in the left upper lung lobe, with its large section in size of 8.43 cm×7.91 cm, homogeneous internal density, a CT value of 12 Hu and smooth cystic wall. The cystic wall was shown to be slightly thickened, with smooth and intact margin. (**e**) A huge cystic space occupying lesion was shown in the parenchyma of right liver lobe, with a size of about 9.67 cm×7.38 cm, homogeneous internal density, and well defined boundary

[**Radiological Demonstration**] (See Fig. 8.67)

[**Diagnosis**] Pulmonary cystic echinococcosis in the right upper lung lobe (simplex cyst type). Concurrently ruptured internal and external cysts were demonstrated with communication to the bronchus, with air flowing into the internal and external cysts.

[**Discussion**]

When echinococcus granulosus gains its access into human body, its larva firstly settle in the right heart along with blood flow in the inferior vena cava, and then spreads to lungs along with the pulmonary artery. It gradually develops into echinococcus cyst in lungs to cause pulmonary cystic echinococcosis. The lesion is more commonly located in the right lung lobe than in the left lung lobe, and is more commonly located in the lower lung lobe than in the upper lung lobe. Its development and pathological changes resemble to those of hepatic cystic echinococcosis, with the simplex cyst type to be more common, which accounts for 80–90% of all the types of echinococcosis.

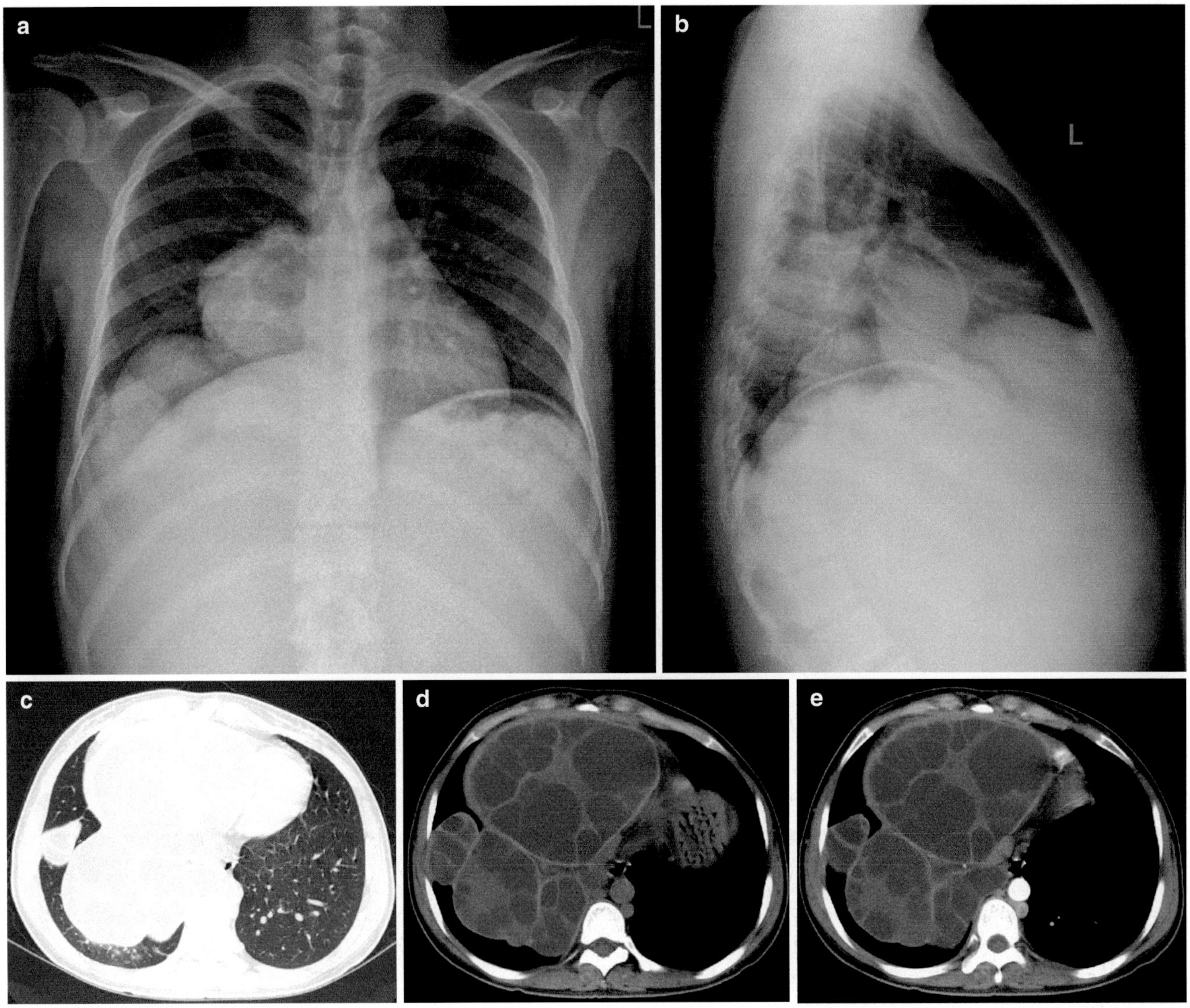

Fig. 8.65 (**a**, **b**) Anterior-posterior and lateral chest X-rays demonstrated multiple semi-round high density opacity in the right middle lower lung field, with smooth and sharp margin. (**c**, **d**) CT plain scan demonstrated multiple round like space occupying lesions of different sizes in the right lower thoracic cavity, with the largest lesion in size of 8 cm × 10.6 cm. The lesions were shown with round like lower density daughter cysts opacity of different sizes. (**e**) Contrast scan demonstrated no definitive enhancement of the cystic wall and content

The multi daughter cysts type is the second common type of echinococcosis, but the consolidation and calcification type is rare. As the lung tissue is loose with little resistance and rich blood supply, the echinococcus cyst in lungs grows rapidly. Despite no obvious clinical manifestations in the early stage, the cyst is commonly large, with a length of up to 10–20 cm and commonly 5–10 cm. When the surrounding lung tissue is compressed by the cyst, the patients show symptoms and physical signs, such as chest pain, cough, hemoptysis and respiratory dysfunction. In rare cases, echinococcus can penetrate into the bronchus to cause pulmonary consolidation or atelectasis. The patients may cough up sheet jelly like internal cyst of the echinococcus, which is a typical sign of pulmonary cystic echinococcosis and has diagnostic value. Pulmonary echinococcosis is commonly complicated by infection or the patients show symptoms of pleural effusion and empyema due to rupture of the echinococcus cyst. In some serious cases, allergic shock occurs. In some cases, the echinococcus in liver apex directly penetrate the diaphragm into thoracic cavity or lungs to form continuous lesion from infradiaphragm to supradiaphragm.

The cases of pulmonary echinococcosis account for 20 % of all echinococcosis cases, with its incidence rate being only second to hepatic echinococcosis. It is mainly caused by echinococcus granulosus, accounting for 95 % of the cystic echinococcosis cases. Pulmonary echinococcosis can be further divided into cystic and alveolar echinococcosis. By

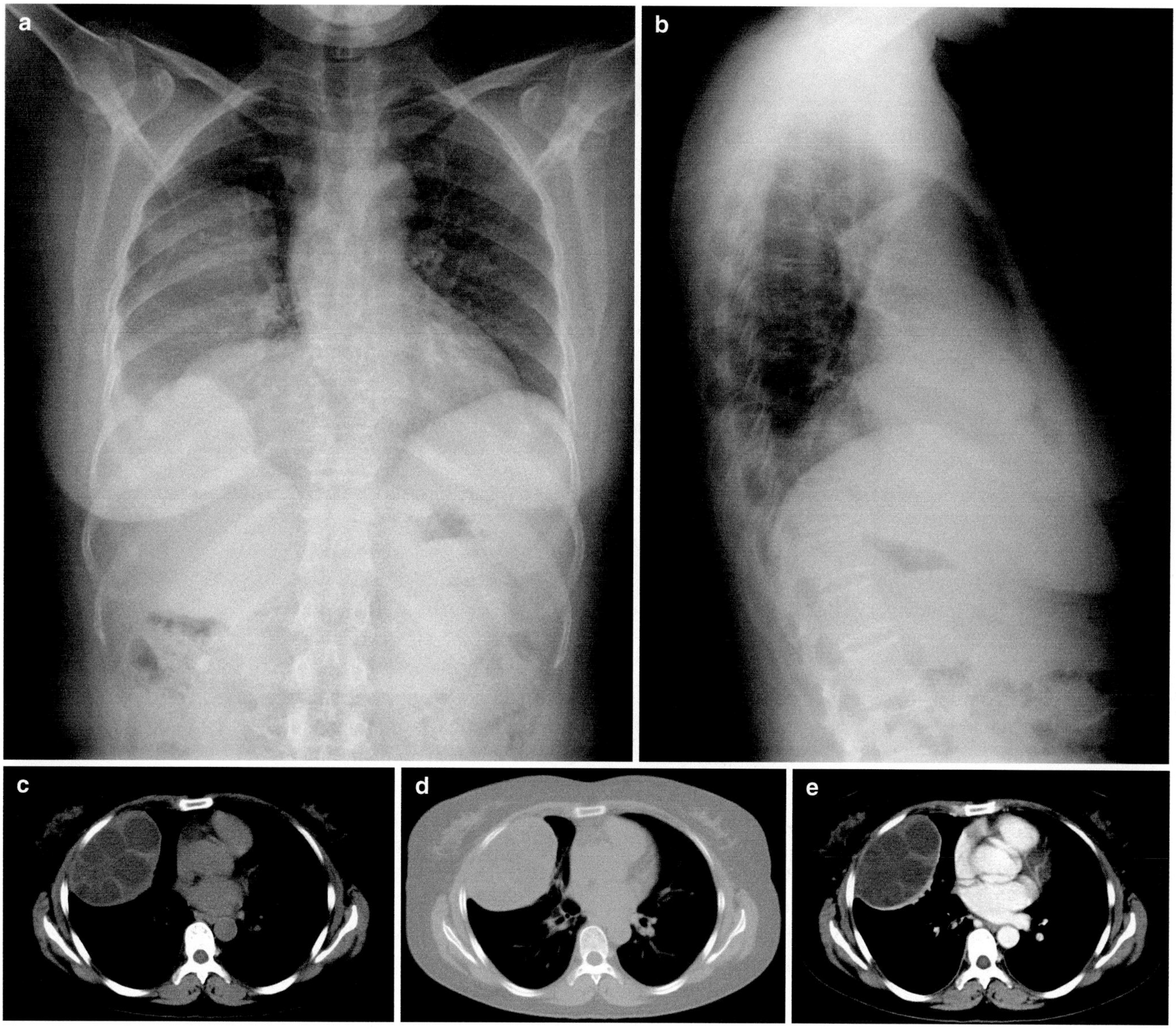

Fig. 8.66 (**a**, **b**) Anterior-posterior and lateral chest X-rays demonstrated a huge oval lesion in the left upper middle lung field, with a size of about 11.2 cm×10.1 cm. The lesion was shown with homogeneous internal density, well defined and smooth margin. (**c**, **d**) CT scan demonstrated multiple huge round like cystic lesions with low density in the anterior segment of right upper lung lobe and in the right middle lung lobe, with the largest section in size of about 10.3 cm×7.4 cm. The lesions were revealed to be multilocular with a intralocular CT value of 2–10 Hu, and slightly thick cystic wall with well defined margin. (**e**) Contrast scan demonstrated slight enhancements of the cystic wall and internal septa but no obvious enhancement of the cystic content

CT scan, pulmonary cystic echinococcosis is demonstrated with following signs:

1. Simplex cyst type
   The lesion is demonstrated as round or oval in shape, with smooth, sharp and well defined boundary. The cystic wall is demonstrated with a thickness of 1–3 mm, with detectable calcification in the late stage and an internal density of 5–20 Hu.
2. Multi daughter cysts type
   The mother cyst is demonstrated with multiple uniform shape daughter cysts of different sizes inside with an arrangement of wheel pattern, radiation pattern or honeycomb pattern.
3. Rupture of echinococcus type
   When the cyst ruptures, the characteristic lotus-on-water sign, crescent sign or gas-fluid level is demonstrated.

Pulmonary alveolar echinococcus is secondary to hepatic lesion in most cases, and primary pulmonary alveolar echinococcosis is rare.

Pulmonary cystic echinococcosis with concurrent rupture of internal and external cysts should mainly be differentiated

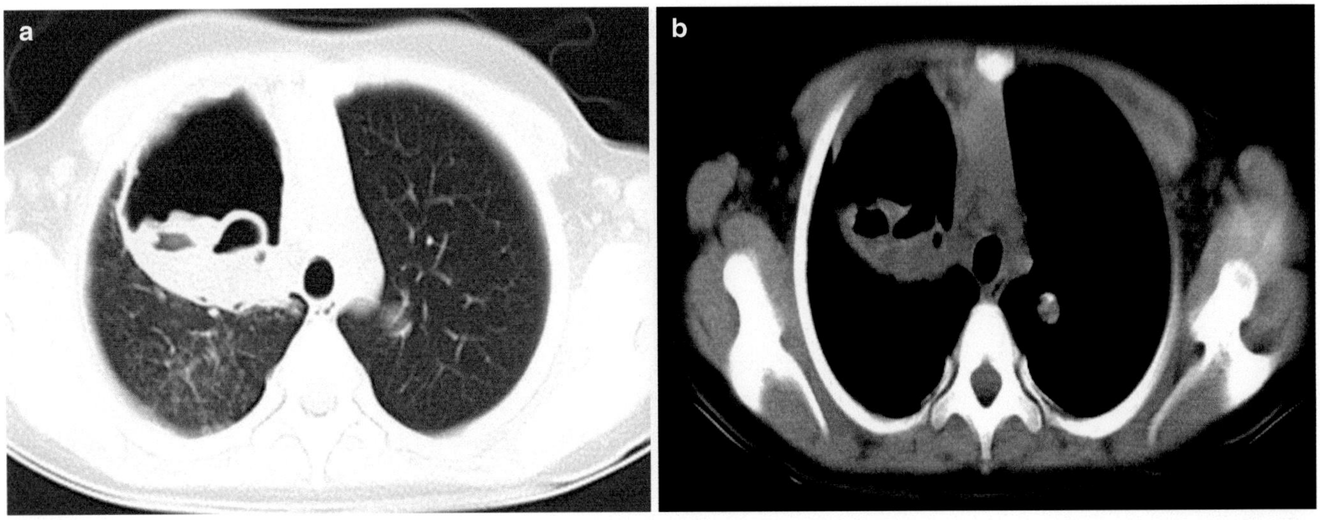

Fig. 8.67 (**a**, **b**) CT scan demonstrated round like cavity in the right upper lung lobe, with smooth and well defined margin as well as heterogeneous thickness of the cystic wall. The lesion was shown with, typical lotus-on-water sign and gas-fluid level, with patches of effusive opacity scattering around the lesion

from fibrocavernous pulmonary tuberculosis, pulmonary abscess, peripheral lung cancer (cavity type) and bronchial cyst.

CT scan demonstrates the lesion of chronic fibrocavernous pulmonary tuberculosis as cavity in lung segment or lobe, with no fluid level inside. The cavity is surrounded by many cords like dense opacity, commonly with calcification, thickened and twisted lung markings, and dilated bronchus. The isolateral and/or contralateral lung field of the lesion shows old and new lesions caused by bronchial dissemination, with greatly varying internal densities and calcification. The mediastinum migrates towards the affected side; the pleura is thickened; and the corresponding thoracic wall collapses.

CT scan demonstrates the lesion of acute pulmonary abscess as solid opacity in the lung segment or lobe, with necrosis and liquefaction inside. The lesion is accompanied by pleural effusion in a small quantity, local thickening of the pleura, and even empyema and pyopneumothorax. CT scan demonstrates the lesion of chronic pulmonary abscess as thick wall cavity with well defined internal and external abscess wall. The lesion is accompanied by chronic inflammation of the neighbouring lung issue, dilated bronchus, newly formed lesion caused by dissemination and fibrosis of old lesions.

CT scan demonstrates the lesion of peripheral lung carcinoma (carcinomatous cavity) as heterogeneous thickness of the cavity wall, unsmooth or nodular inner cavity wall, and well defined wave like or lobulated external cavity wall that is mostly central.

CT scan demonstrates the lesion of bronchial cyst with secondary infection as cystic lesion with gas or gas-fluid level, commonly around the lung hilum and in both lower lungs. The cystic wall is subject to regular or irregular thickening, and the size and shape of the cyst may change along with the development of infection. The lesion is commonly accompanied by surrounding restrained emphysema or other congenital malformations, such as sequestration and congenital diaphragmatic hernia.

In this case, the condition was diagnosed to be pulmonary cystic echinococcosis in the right upper lung lobe (simplex cyst type) with concurrent rupture of internal and external cysts. A cavity was shown in the right upper lung lobe, with lotus-on-water sign and gas-fluid level.

Case Study 54
[Brief Medical History]

A 79-year-old man was hospitalized due to the chief complaints of cough and expectoration for 1 week after catching a bad cold. He experienced intermittent fever, with the highest body temperature of 39.5 °C. Laboratory tests showed neutrophilic count 0.755, lymphocyte count 0.129, and monocyte count 0.101. Ig antibody of pulmonary echinococcus was demonstrated positive.

[Radiological Demonstration] (See Fig. 8.68)

[Diagnosis] Pulmonary cystic echinococcosis in the right upper lung (simplex cyst type) complicated by infection.

[Discussion]

When echinococcus granulosus invades human body, its larva firstly settles in the right heart along with blood flow in the inferior vena cava, and then disseminated to both lungs along with blood flow in the pulmonary artery. In the lungs, the larva gradually grows into adult to cause echinococcosis. Its lesion is more commonly located in the right lung lobe than in the left lung lobe, and is more commonly

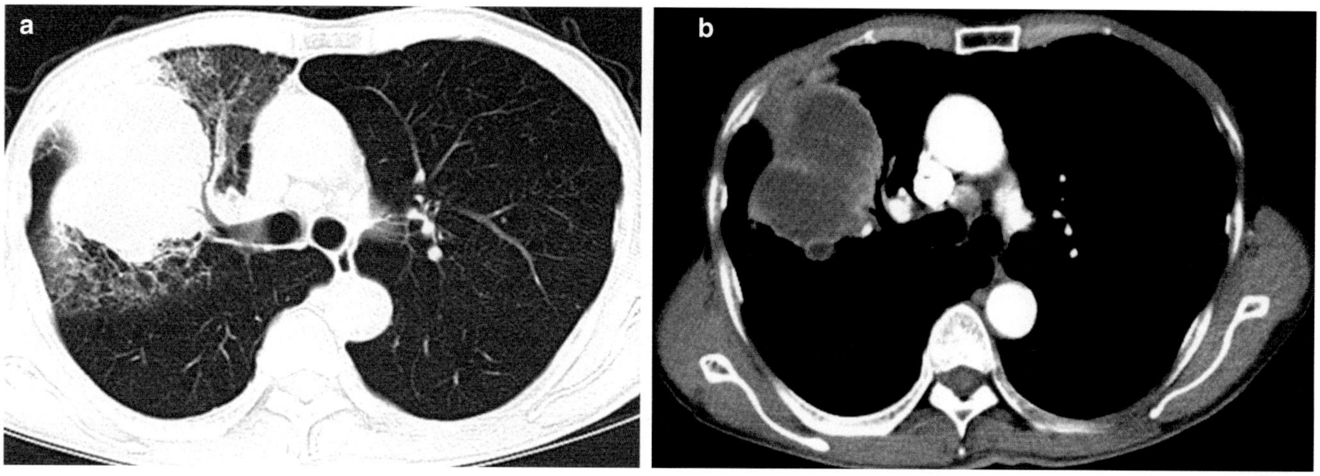

Fig. 8.68 (**a**, **b**) Contrasts CT scan demonstrated round like cystic lesion in the right upper lung lobe, with smooth and clearly defined boundary. The cyst wall was thin with a homogeneous thickness. The lesion was shown with surrounding diffuse patches of effusion and the local pleura was shown to be thickened

located in the lower lung lobe than in the upper lung lobe. And its development and pathological changes resemble to that of hepatic echinococcosis. Simplex cyst type is the most common type of pulmonary cystic echinococcosis, accounting for 80–90 %; followed by the multi daughter cysts type, but the consolidation and calcification type is rare. As pulmonary tissue is loose with little resistance and rich blood supply, the echinococcus cyst grows rapidly. Despite no obvious clinical manifestation in the early stage, the cyst is quite large, with a length of up to 10–20 cm, commonly 5–10 cm. When surrounding pulmonary tissue is compressed by the enlarging cyst, the patients may experience symptoms and physical signs such as chest pain, cough, hemoptysis and respiratory dysfunction. In rare cases, the echinococcus cyst may penetrate into the bronchus to cause pulmonary consolidation or atelectasis. The patients may cough up sheet jelly like internal cyst echinococcus, which is a characteristic sign and has diagnostic value. Pulmonary cystic echinococcosis may be complicated by infection or cystic rupture to cause pleural effusion and empyema. In some serious cases, even allergic shock occurs. Sometimes, echinococcus in liver apex may penetrate the diaphragm into the thoracic cavity or lungs to cause continuous lesion from infradiaphragm to supradiaphragm.

Pulmonary cystic echinococcosis complicated by infection should mainly be differentiated from acute pulmonary abscess, bronchial cyst, congenital adenomatoid malformation and pulmonary sequestration.

CT scan demonstrates the lesion of acute pulmonary abscess as necrosis and liquefaction in consolidation opacity of pulmonary segment or lobe, with accompanying pleural effusion in a small quantity, local pleural thickening, and even empyema and pyopneumothorax.

CT scan demonstrates the lesion of bronchial cyst to be cystic, which is commonly located around the pulmonary hilum and in bilateral lower lungs, with smooth cystic wall and homogeneous internal density. The lesion is commonly accompanied by surrounding restrained emphysema or other congenital malformations, such as sequestration and congenital diaphragmatic hernia.

CT scan demonstrates the lesion of congenital adenomatoid malformation as multiple cystic or solid cystic lesions that are supplied by the pulmonary artery. The lesion is always complicated by developmental abnormality of other body parts, such as aplasia of eyes and kidney.

CT scan demonstrates the lesion of pulmonary sequestration to be cystic with fixed location. It is characterized by smooth cystic wall, homogenous density and abnormal blood supply by aorta.

Case Study 55

[**Brief Medical History**]

A 81-year-old man complained of cough and expectoration with fever for 1 week. He had a past medical history of hepatic echinococcosis. CT scan showed space occupying lesion in the right middle lung lobe.

[**Radiological Demonstration**] (See Fig. 8.69)

[**Diagnosis**] Pulmonary cystic echinococcosis in the right middle lung lobe and anterior segment of the right lower lung lobe complicated by infection.

[**Discussion**]

The differential diagnosis of pulmonary cystic echinococcosis complicated by infection from pulmonary abscess is difficult. With the cyst filled by pus, enhancements of the cystic wall and content can be demonstrated. Its surrounding lung tissue shows inflammatory responses with pulmonary consolidation and atelectasis, which blurs the original boundary

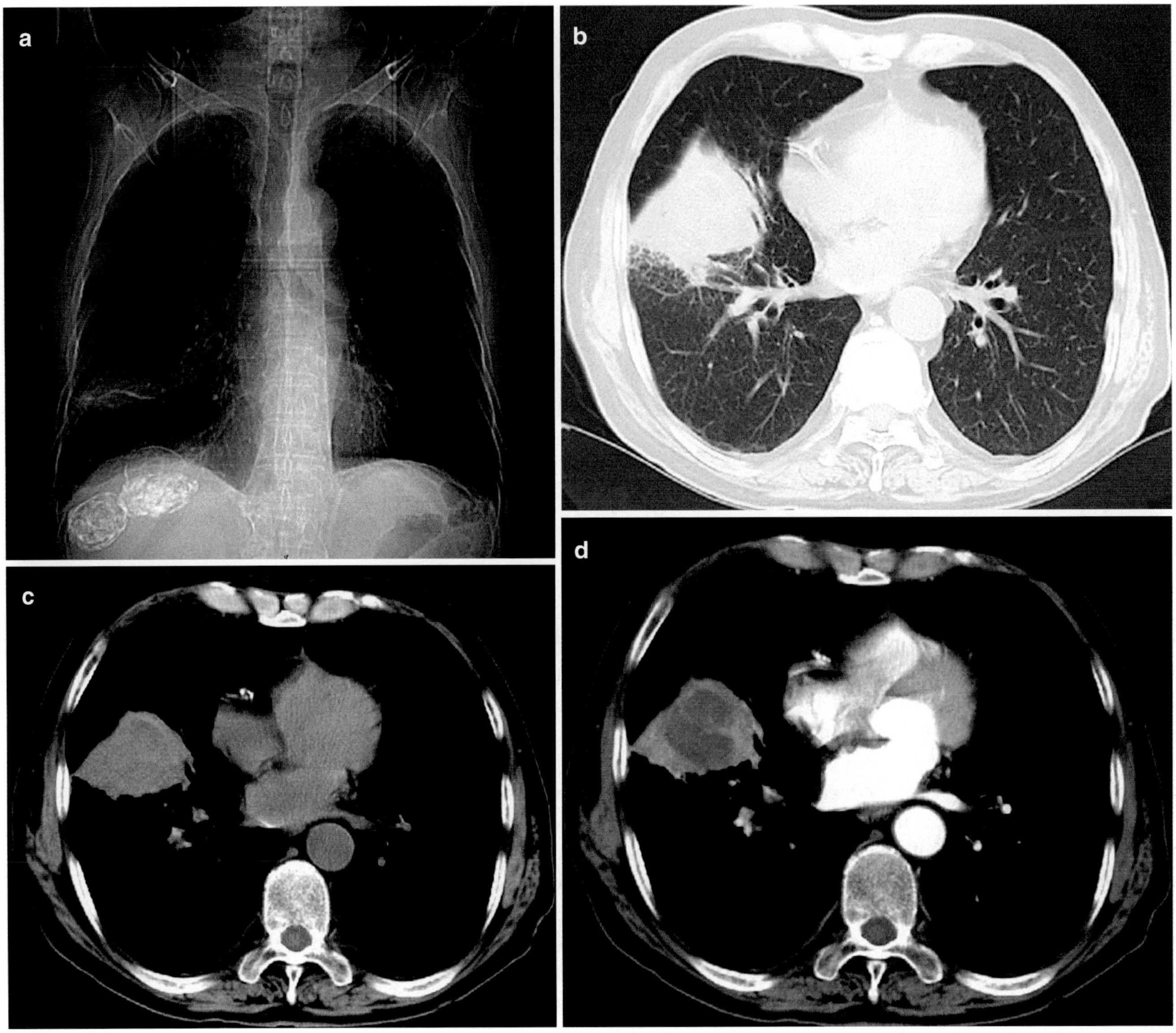

Fig. 8.69 (**a–d**) CT scan demonstrated a dense opacity from the lateral segment of the right middle lung lobe to the anterior segment of the right lower lung lobe, with a poorly defined round like low density opacity in the lesion. The lesion was shown with shell like calcification at its margin, a size of 4.64 cm × 4.44 cm, surrounding patches of dense opacity, and air bronchogram inside. By contrast scan, the lesion showed septa like enhancement of its margin and inner content, but its surrounding lung tissue with inflammatory responses showed no enhancement

between the lesion and its surrounding lung tissue. In combination to the past medical history or detectable echinococcus cyst in other body parts, the diagnosis can be defined.

Case Study 56

[Brief Medical History]

A 40-year-old man complained of upper abdominal pain and upset for 2 years that aggravated for 1 month.

[Radiological Demonstration] (See Fig. 8.70)

[Diagnosis] Pulmonary alveolar echinococcosis.

[Discussion]

In this case, the lesion was located in the lateral basal segment of right lower lung lobe, with irregular calcification inside. The lesion was shown with a density close to the observable lesion in liver, with no definitive enhancement by contrast scan. The condition should be differentiated from pulmonary tuberculoma, the lesion of which is commonly detected in the apical posterior segment of upper lung lobe and dorsal segment of lower lung lobe. The lesion of pulmonary tuberculoma shows more regular margin than the lesion of pulmonary alveolar echinococcosis, with nodule like calcification inside and surrounding cords and satellite lesions.

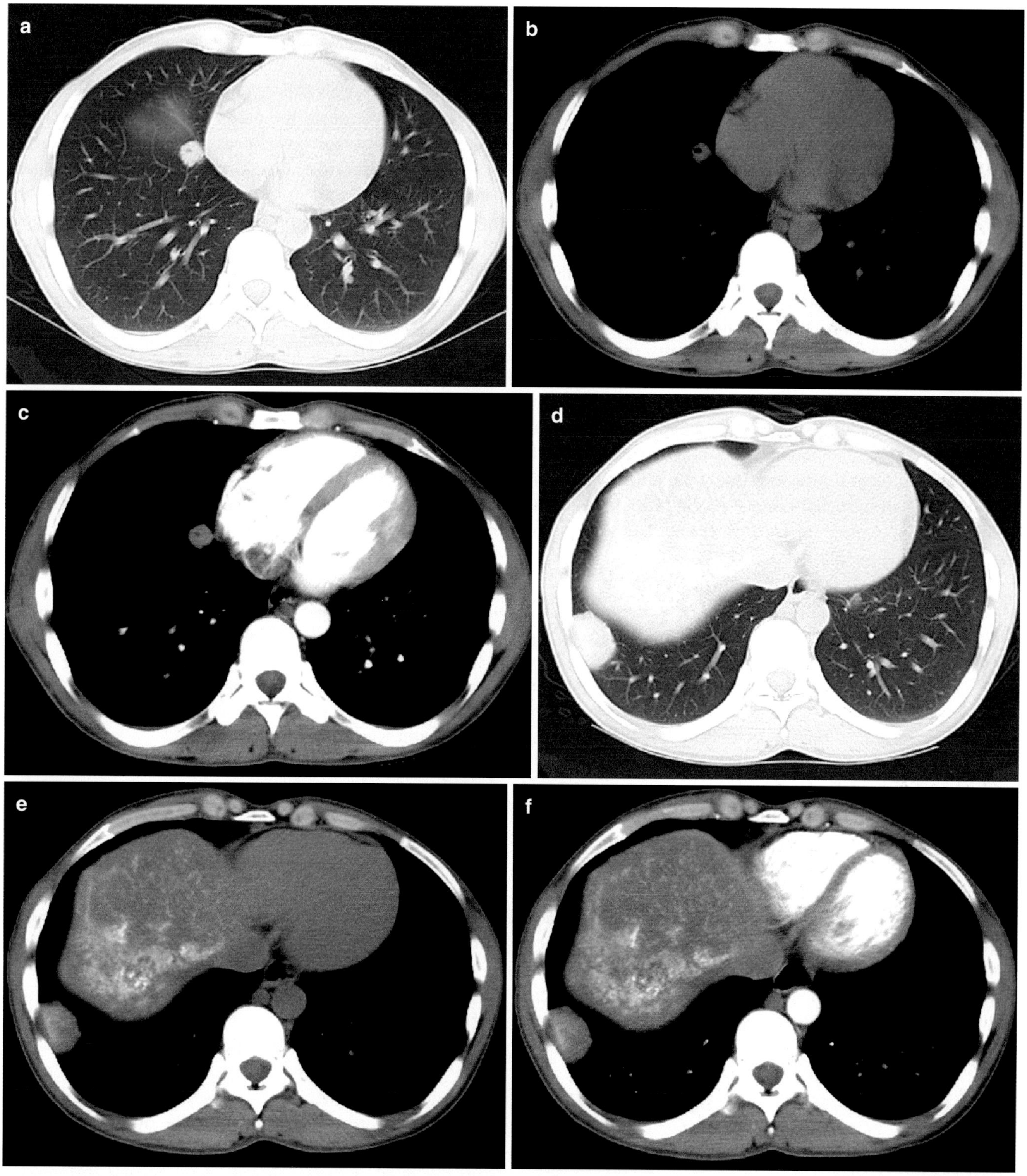

Fig. 8.70 (**a**–**f**) CT scan demonstrated subpleural round like space occupying lesion in the lateral basal segment of right lower lung lobe, with the large lesion in a size of 3.48 cm×3.20 cm. The lesion was shown with heterogeneous internal density, with irregular patches of slightly high density opacity. Contrast scan demonstrated no definitive enhancement

Case Study 57

[Brief Medical History]

A 39-year-old man complained of cough, chest distress and chest pain for 2 years. He had a history of living in husbandry area.

For case detail and figures, please refer to fig 26.22 of the book (Hongjun Li, Radiology of Infectious diseases, Volume 2, 2015)

[Diagnosis] Pulmonary alveolar echinococcosis.

[Discussion]

The lesion of pulmonary alveolar echinococcosis is demonstrated to be multiple in the lung, which is more in the lateral area of lung field. The lesion shows small nodules or small patches of soft tissue density opacity, with slightly blurry boundary. Within the lesion, calcification, liquefaction and cavity can often be detected. The lesion may penetrate upwards from the liver apex into the diaphragm to show pneumonia like blurry opacity, with accompanying pleural effusion. In combination to the history of living in husbandry area, the diagnosis of echinococcosis can be made after differential diagnosis from space occupying neoplasm.

Case Study 58

[Brief Medical History]

A 44-year-old man reported a medical history of surgical operation for hepatic alveolar echinococcosis 4 years ago. The 4-item examination for echinococcus showed Anti EgCF antibody negative (−), anti EgP antibody (±), anti EgB antibody (±) and anti Em2 antibody (±).

[Radiological Demonstration] (See Fig. 8.71)

[Diagnosis] Postoperative recurrence of hepatic alveolar echinococcosis with penetration into the thoracic cavity.

[Discussion]

In this case, the lesion was located in the diaphragmatic angle that right anterior to the heart, with poorly defined boundary from its neighboring diaphragm, pericardium and left liver lobe. The lesion also involved the anterior thoracic wall. Coronal and sagittal reconstructed images can more favorably demonstrate the range and origin of the lesion, which facilitates the diagnosis of echinococcosis.

Case Study 59

[Brief Medical History]

A 45-year-old man, Hui nationality, was hospitalized due to the chief complaints of palpitation, shortness of breath and chest pain after working hard for 1 year that aggravated for 1 month. The 4-item examination for echinococcus showed Anti-EgCF antibody negative (−), Anti-EgP antibody negative (−), Anti-EgB antibody negative (−) and Anti-Em2 antibody negative (−).

[Radiological Demonstration] (See Fig. 8.72)

[Diagnosis] Cardiac cystic echinococcosis.

[Discussion]

Cardiac and pericardial echinococcosis is rare, accounting for 0.5–2.0% of all the cases of echinococcosis. After oral intake of eggs of echinococcus granulosus (in dog feces) by human, the eggs, after digestion in gastric juice, are incubated into oncospheres in the duodenum, which then penetrate the intestinal wall into the portal vein. After double filtrations by the hepatic and pulmonary capillaries, a small quantity of oncospheres remain to flow into the systemic circulation, with 0.5–5.0% of them into the coronary artery. The incidence rates of primary myocardial and pericardial echinococcosis are low. The lesion of pericardial echinococcosis is commonly located in the left heart, which is originated from blood flow. The deposited oncospheres in the myocardium may grow toward the cardiac or pericardial cavity. Due to the dense myocardial fibre, the oncospheres grow slowly, showing no obvious clinical symptoms. Cardiac and pericardial echinococcosis show no characteristic clinical manifestations. Chest X-ray may demonstrate enlarged heart or partially protruding cardiac margin. Heart ultrasound and CT scan can be applied for their diagnosis, but qualitative diagnosis can hardly made. Surgical operation for removal of cardiac echinococcus should be immediately performed after diagnosis. And overflow of the cystic fluid after the cyst ruptures may cause serious consequences, such as allergic responses, extensive spread and dissemination, and secondary infection.

Case Study 60

[Brief Medical History]

A 24-year-old young man complained of palpitation and dizziness for 2 years that aggravated for 3 days. Echocardiography demonstrated cystic space occupying lesion in the left heart ventricle.

[Radiological Demonstration] (See Fig. 8.73)

[Diagnosis] Cardiac cystic echinococcosis.

[Discussion]

Cardiac echinococcosis rarely occurs. Chest X-ray demonstrates partially protruding cardiac margin. Fluoroscopy demonstrates that the lesion does not move along with heartbeat, but is movable during heartbeat. Multiaxial observation demonstrates its inseparability from the heart, which helps its location and differential diagnosis from pulmonary echinococcosis. CT scan can clearly demonstrate the type and radiological sign of the lesion. Contrast scan demonstrates obvious enhancement of the heart but no enhancement of the lesion, which helps to define location and nature of the lesion. Its characteristic signs include septa in the cystic lesion or daughter cysts in the lesion.

Cardiac cystic echinococcosis should be differentiated from pericardial cyst or pericardial effusion.

The lesion of cardiac alveolar echinococcosis is demonstrated as solid lump with internal calcification. In combination to detected alveolar echinococcus in the liver or other body parts, the diagnosis can be defined. In addition to accurate localization, MR imaging can clearly demonstrate the echinococcus itself and its relationship with neighboring structures. By electrocardiography and respiratory gating technology, MR imaging can accurately demonstrate the conditions of cardiac valves, interventricular septum and myocardium.

Case Study 61

[**Brief Medical History**]

A 32-year-old Kazak man complained of chest upset for 2 years that aggravated for 1 month. The 4-item examination for echinococcus showed Anti-EgCF antibody positive (+++), Anti-EgP antibody positive (+++), Anti-EgB antibody positive (+++) and Anti-Em2 antibody positive (+).

[**Radiological Demonstration**] (See Fig. 8.74)

[**Diagnosis**] Costal echinococcosis in the left 6th, 7th and 8th ribs.

[**Discussion**]

The occurrence of costal echinococcosis is extremely rare with slow development, which is commonly diagnosed based on radiological findings. In its early stage, the lesion is demonstrated as solitary swelling osteolytic lesion in the ribs, which then develops into disruptive lesion in ribs and spines, with local soft tissue lump. Extraskeletal conditions occasionally occur, such as compressive atrophy of ribs caused by pleural cystic echinococcosis. It should be differentiated from osteolytic lesions, such as osteoclastoma, round cytoma, osteolytic metastasis and neurofibroma.

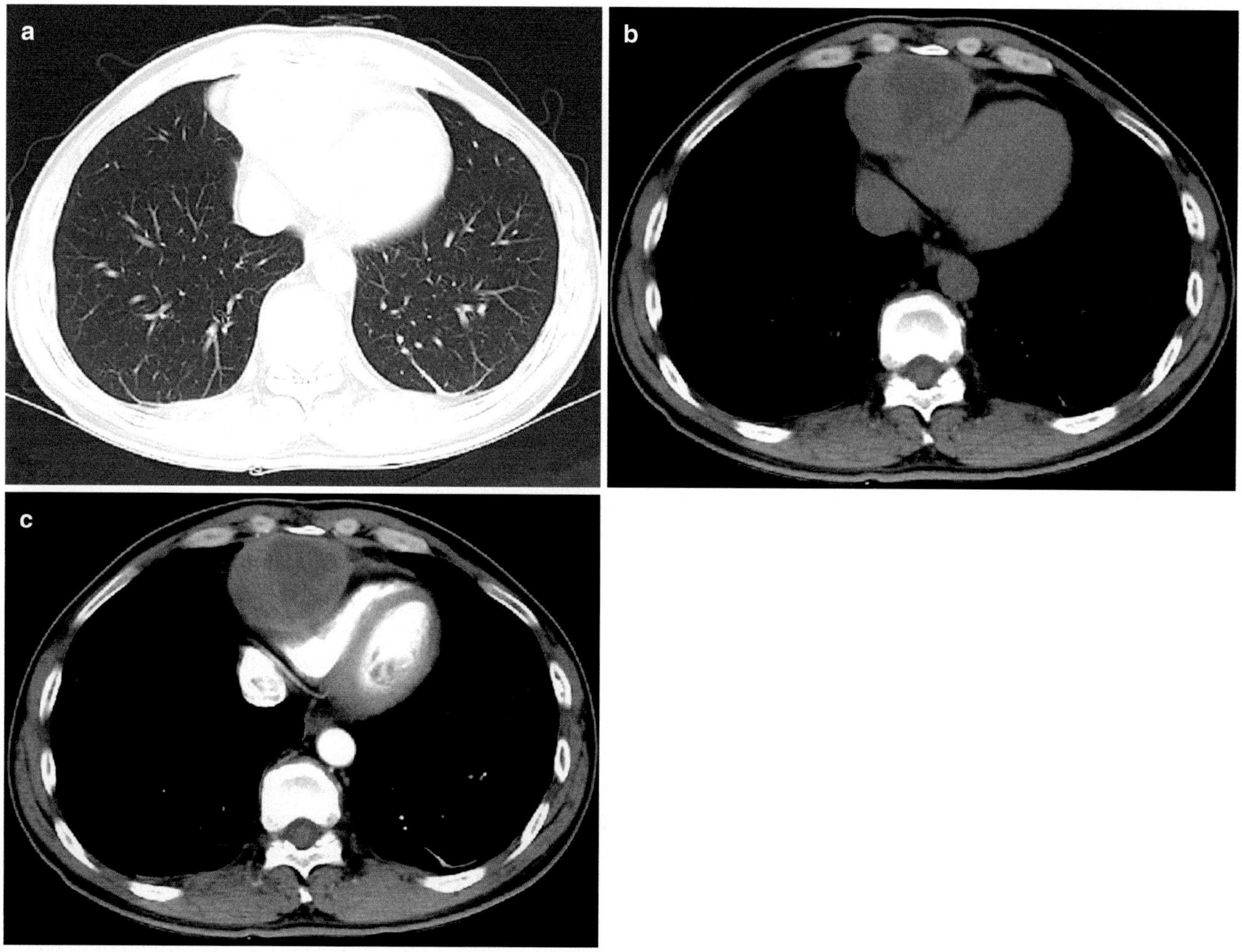

Fig. 8.71 (**a–e**) CT scan demonstrated a space occupying lesion with mixed density in the space right anterior to the heart, with a size of 7.09 cm × 5.09 cm. The lesion was shown with heterogeneous internal density and flakes of low density opacity. The lesion was poorly defined from the neighboring diaphragm, pericardium and left liver lobe, with the thoracic wall involved. Contrast scan demonstrated no enhancement of the lesion

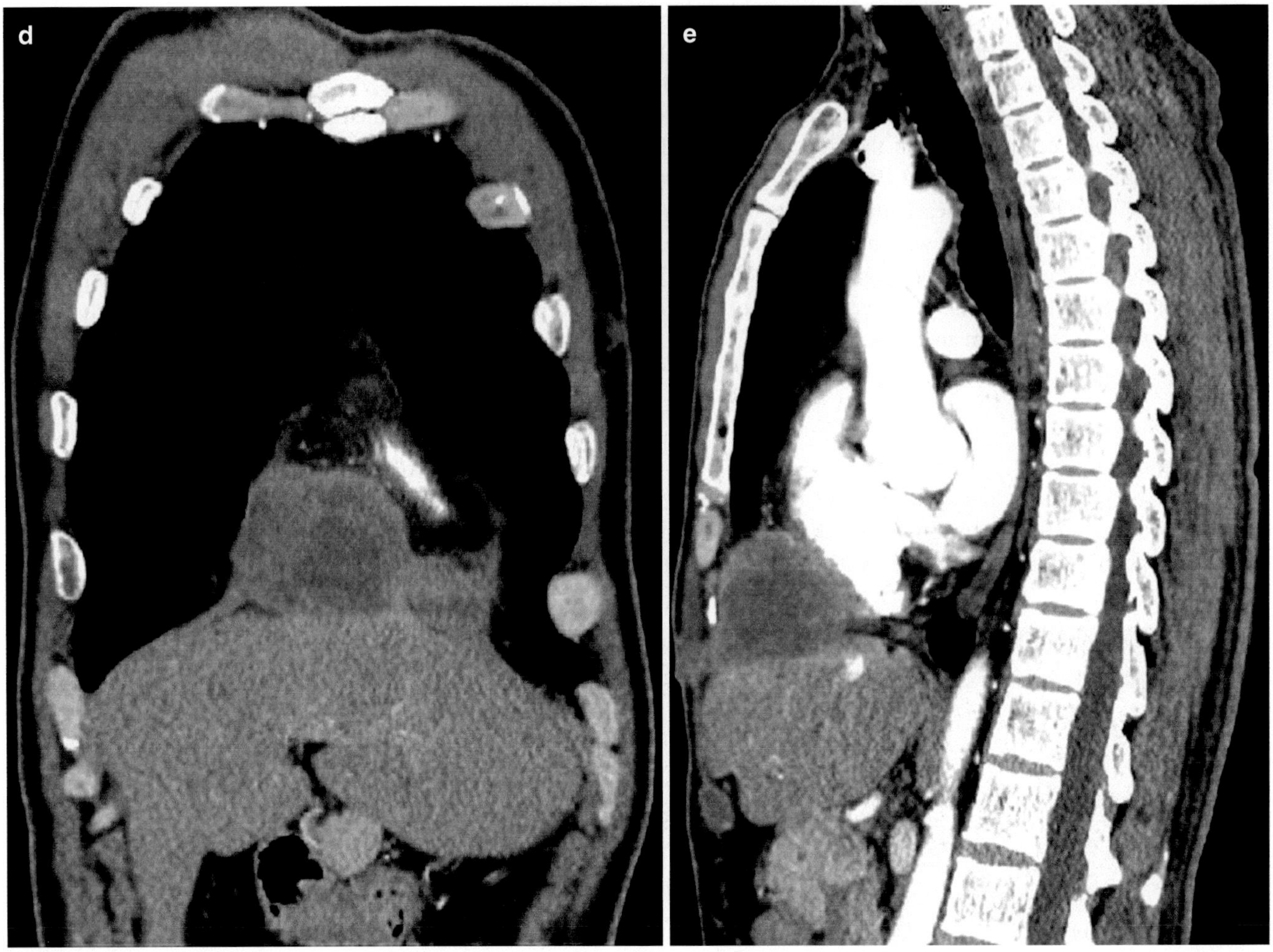

Fig. 8.71 (continued)

Case Study 62

[**Brief Medical History**]

A 25-year-old woman, Han nationality, was hospitalized due to the chief complaints of right lower limb weakness for 15 days. She experienced right lower limb weakness with no known causes 15 days ago and had difficulty climbing up and down stairs that gradually aggravated. She also experienced pain and weakness of the right lower limb for 5 days, with no obvious reasons for its aggravation and relief. Physical examination showed no special positive sign.

[**Radiological Demonstration**] (See Fig. 8.75)

[**Diagnosis**] Cystic echinococcosis in the appendiceal area of 4th lumbar vertebra and in the bilateral paravertebral soft tissue.

[**Discussion**]

The cases of spinal cystic echinococcosis account for more than 60 % of all the cases of skeletal echinococcosis, with thoracic and lumbar vertebrae more commonly involved but the cervical vertebrae rarely involved. The lesion of spinal echinococcosis may be singular, with multiple neighbouring vertebrae involved but with the intervertebral disc not involved. Singular or adjacent multiple vertebra(e) is/are subject to cystic bone destruction. CT plain scan demonstrates the lesion as multiple round like low density opacity in different sizes. And MR imaging demonstrates the lesion in low T1WI signal and high T2WI signal. The lesion is shown to be multilocular or multiple daughter cysts in different sizes filling in the mother cyst. The lesion of spinal echinococcosis commonly grows into the spinal canal, with invasions into the spinal canal, appendiceal area of vertebra, and the paravertebral soft tissue. Due to the spatial restriction, the spinal echinococcus tends to grow into its surrounding soft tissue, showing atypical rose-petals sign. Spinal MR hydrography can more clearly demonstrate the lesion of cystic echinococcosis in the spine, especially the internal mother cyst and surrounding daughter cysts. Spinal cystic echinococcosis is a common skeletal echinococcosis, with its radiological demonstrations resembling to those of echinococcosis in other common body parts, and its diagnosis can be made.

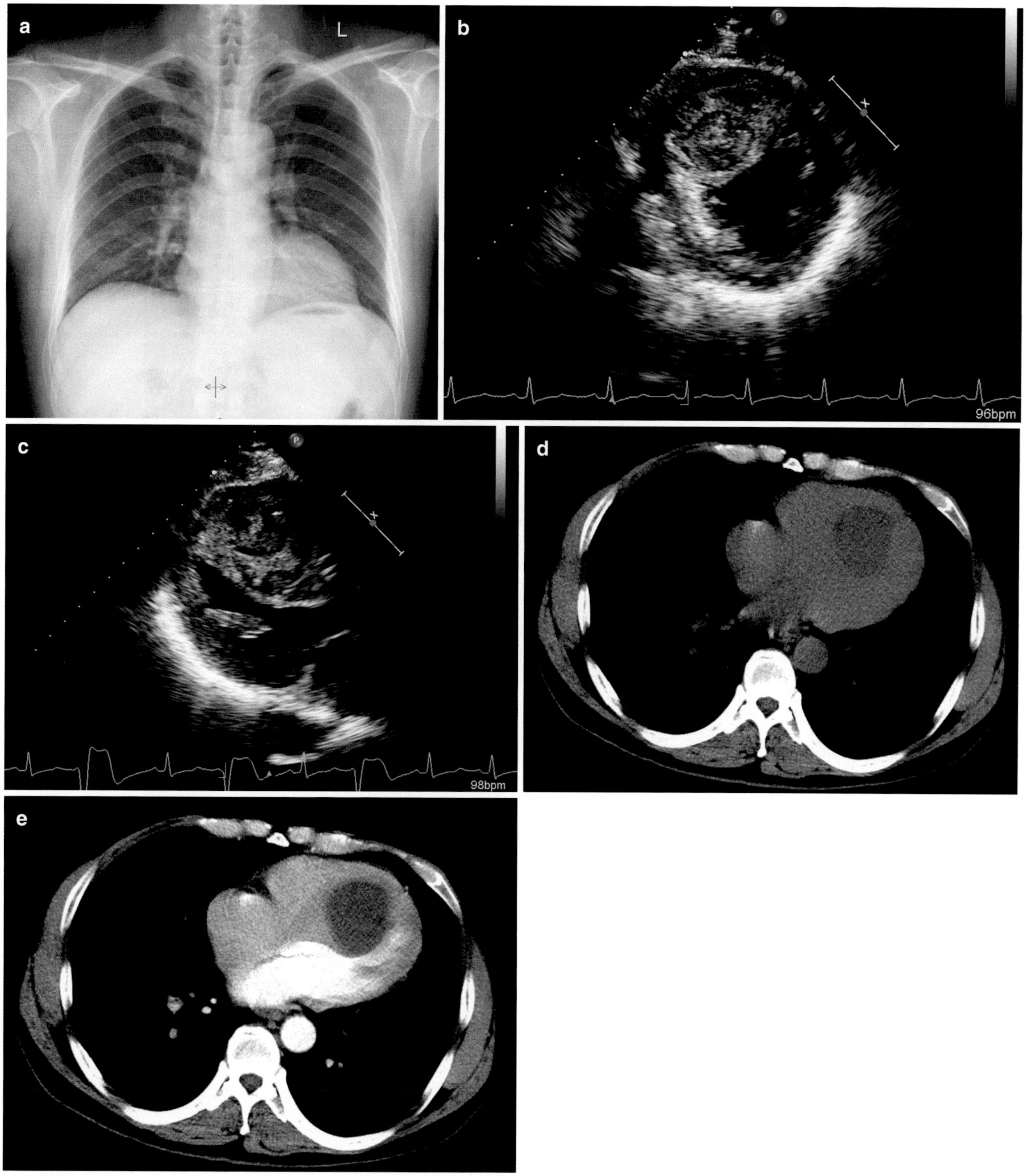

Fig. 8.72 (**a**) Chest X-ray demonstrated enlarged heart opacity with blunt cardiac apex. (**b**, **c**) B-mode ultrasound demonstrated a cystic lesion in the enlarged left ventricular cavity in a size of 5.3 cm × 4.7 cm × 4.3 cm that was attached to the lateral wall of left cardiac ventricle, and poor valve closure. (**d**, **e**) CT scan demonstrated enlarged heart opacity and a round like lesion with cystic density in the left ventricle in a size of 5.4 cm × 4.8 cm. The lesion showed homogeneous density and well defined boundary, with a CT value of 25–31 Hu. Contrast scan demonstrated no obvious enhancement of the lesion. The left and right ventricles were compressed to show arch shape imprint. Contrast scan demonstrated no definitive enhancement of the lesion

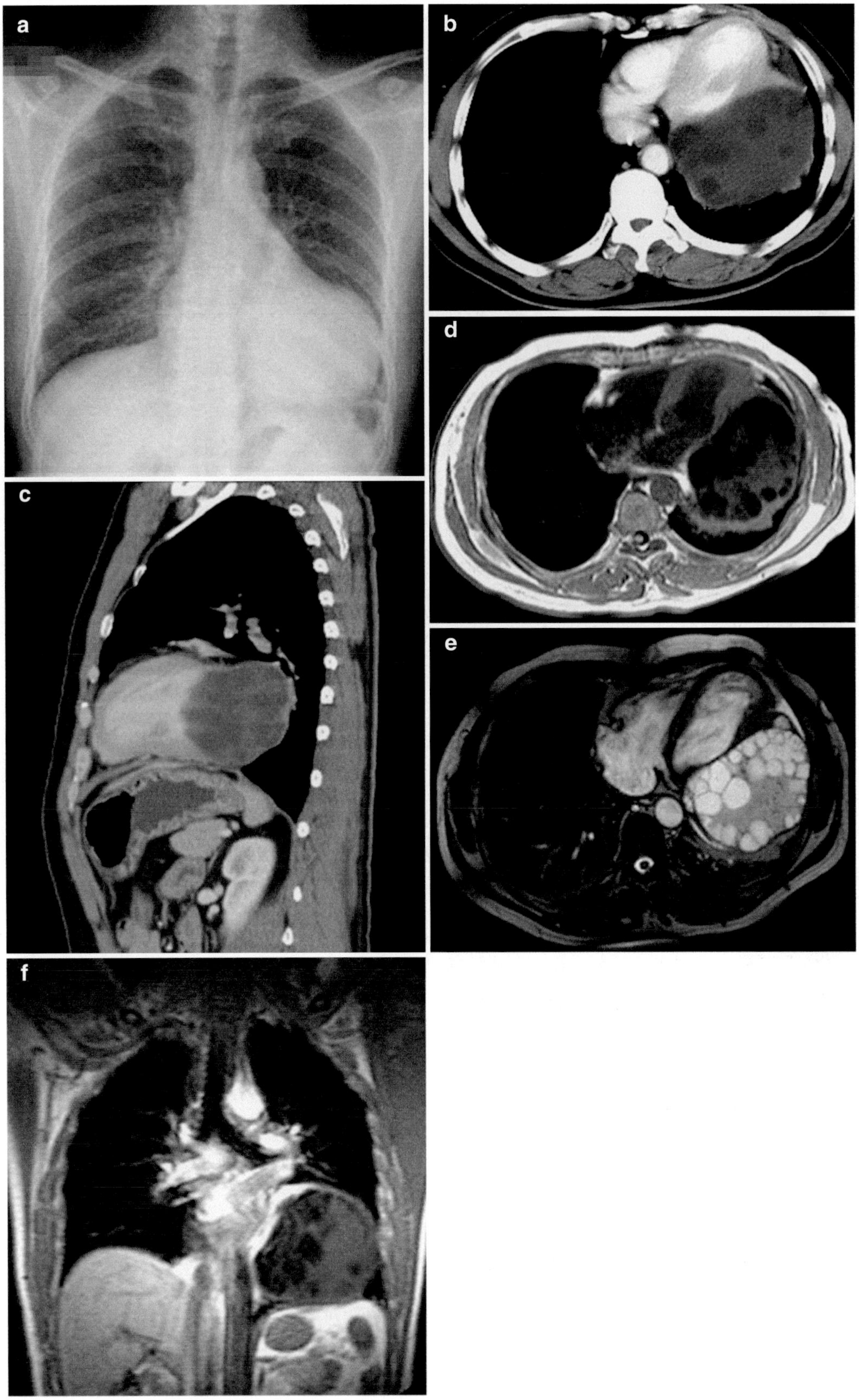

Fig. 8.73 (**a**) Chest X-ray demonstrated enlarged heart opacity with predominantly leftward enlargement of the left ventricle and increased density. (**b**, **c**) CT scan demonstrated cystic lump on the left ventricular wall, with scattering of daughter cysts in different sizes in the mother cystic fluid and higher cystic fluid density in the mother cyst than the daughter cysts. (**d**–**f**) MR imaging demonstrated multiple well defined cystic lesions on the left ventricular wall, with water like signal in the daughter cysts and equal signal in the mother cyst (Reprint with permission from Hongjun Li, Radiology of Infectious diseases, Volume 2, 2015)

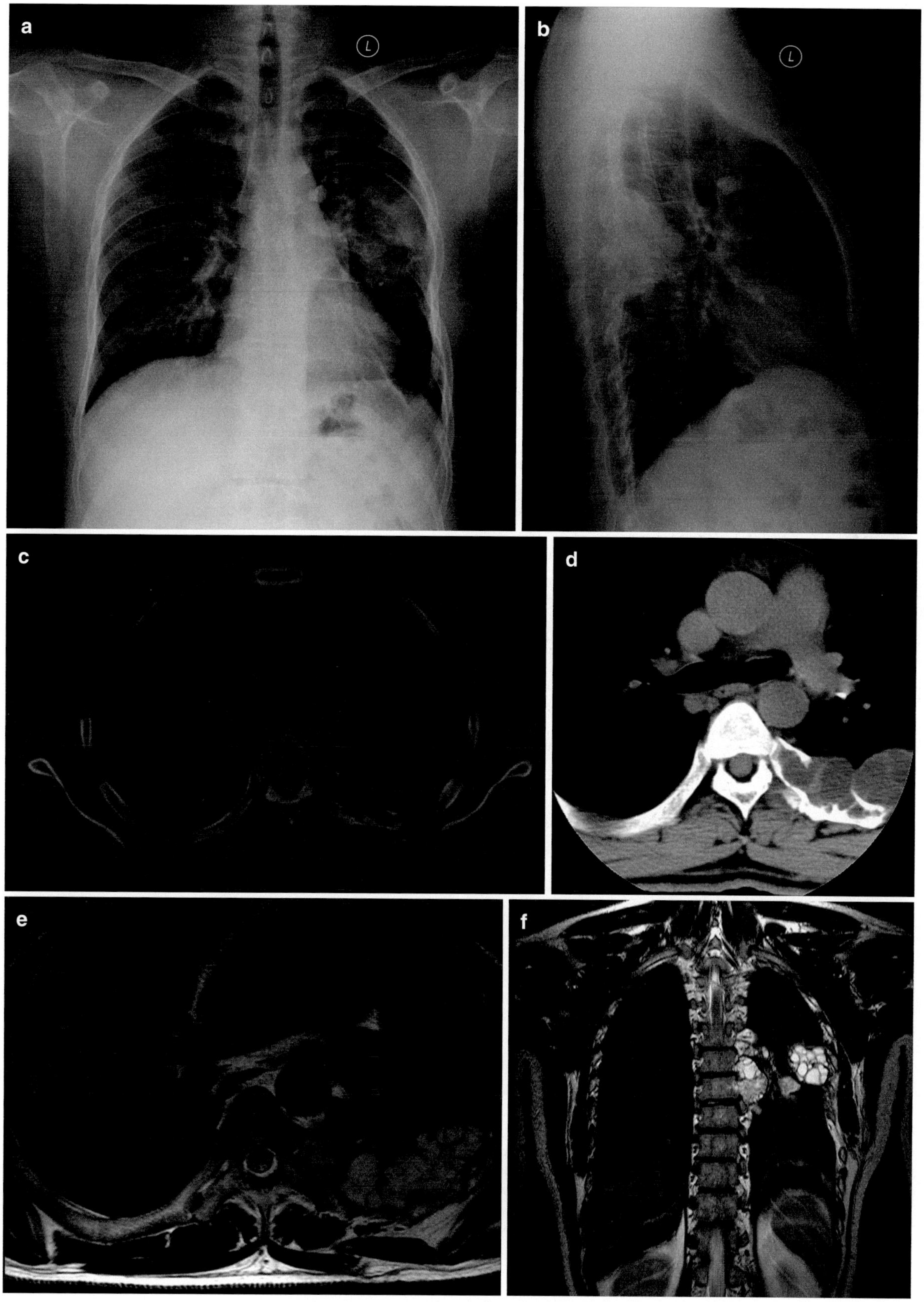

Fig. 8.74 (**a**, **b**) Anterior-posterior and lateral Chest X-rays demonstrated swelling bone destruction in the 6th, 7th and 8th left posterior rib, with uneven internal density and clearly defined boundary. (**c**, **d**) CT plain scan demonstrated multiple cystic lesions in the left posterior chest wall, with multiple small daughter cysts inside. Swelling cystic bone destruction were detected in the 6th, 7th and 8th left rib. The left intervertebral foramen was shown to be enlarged at the 7th thoracic vertebra, with the lesion protruding inwards. Contrast scan demonstrated no obvious enhancement of the lesion and more clearly demonstrated lesion. (**e**, **f**) MR imaging demonstrated lump and flakes of long T2 signal in the left transverse appendiceal area of thoracic 5–7th vertebrae, paravertebral soft tissue, and 6th, 7th and 8th left ribs, with multiple well defined septa opacity with vesicle like change. The lesion was shown in a size of 1.02 cm×4.51 cm×7.06 cm that protruded into the left lung

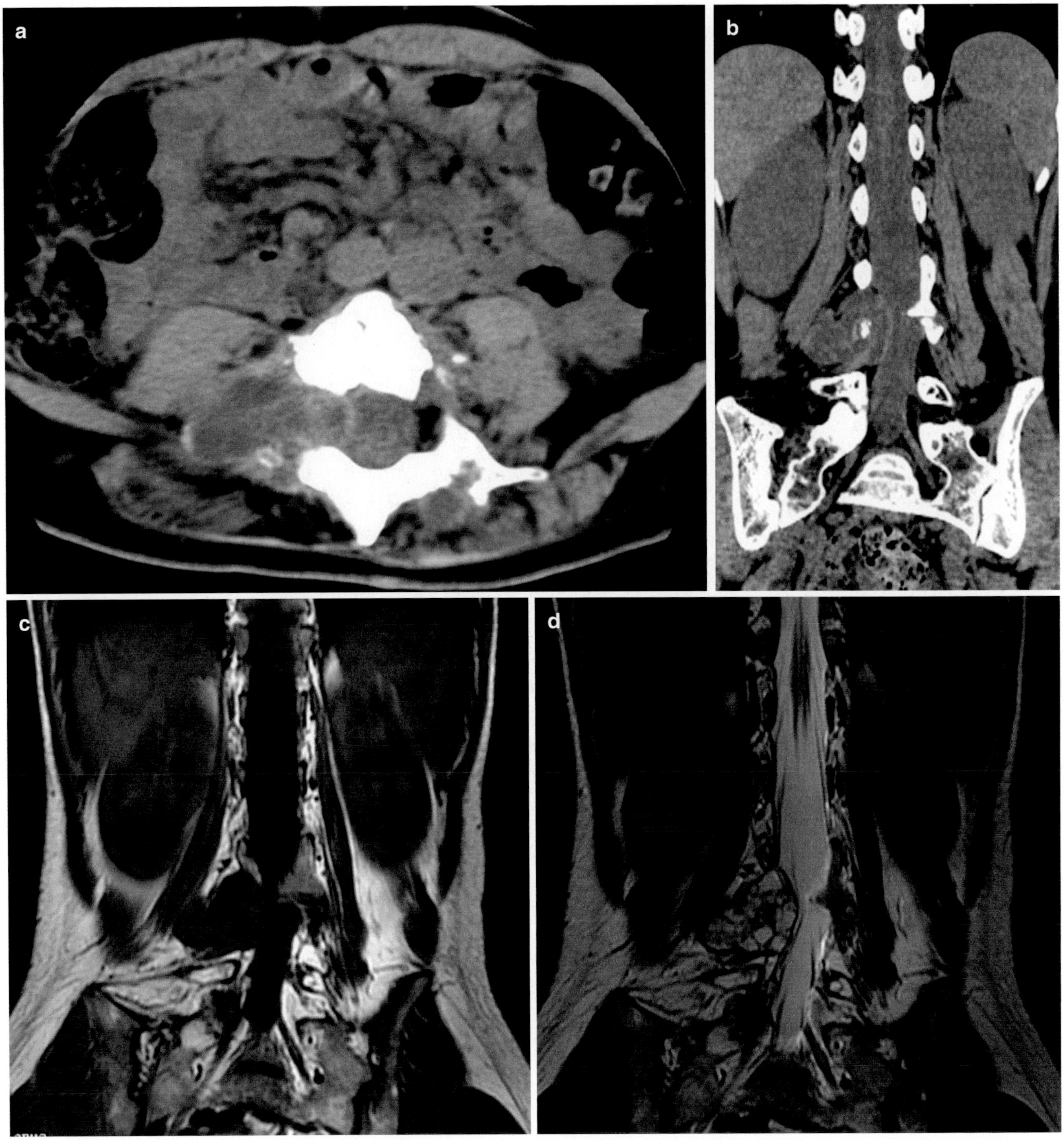

Fig. 8.75 (**a–f**) The appendiceal area of the 4th lumbar vertebra was demonstrated with bone destruction. And the bilateral paravertebral soft tissue was demonstrated with cystic space occupying lesion in irregular shape, which was dominantly water like low density by CT plain scan, with moderate density envelope and multiple moderate density septa opacity. MR plain imaging demonstrated the lesion with predominantly low T1WI signal and multiple equal signal septa opacity inside. MR plain imaging demonstrated the lesion with predominantly high T2WI signal. The lesion was shown to grow along and across the right intervertebral foramen to protrude into the spinal canal. Local dural cyst was compressed to show deformation. Spinal MR hydrography demonstrated the lesion in rose-petals sign

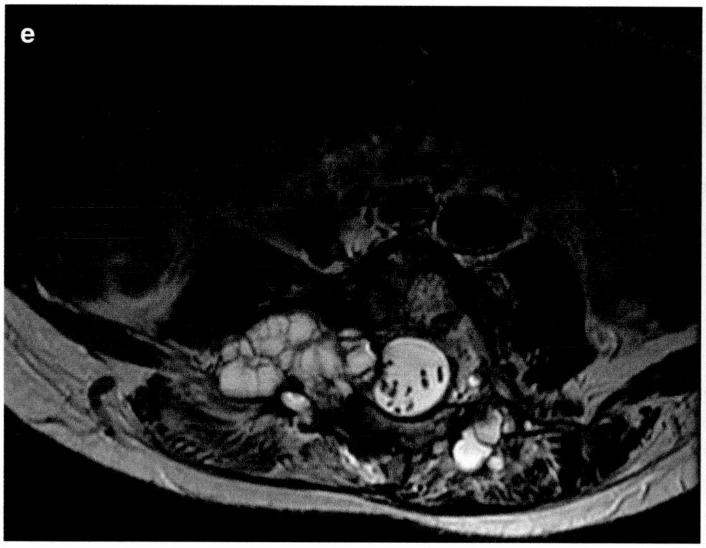

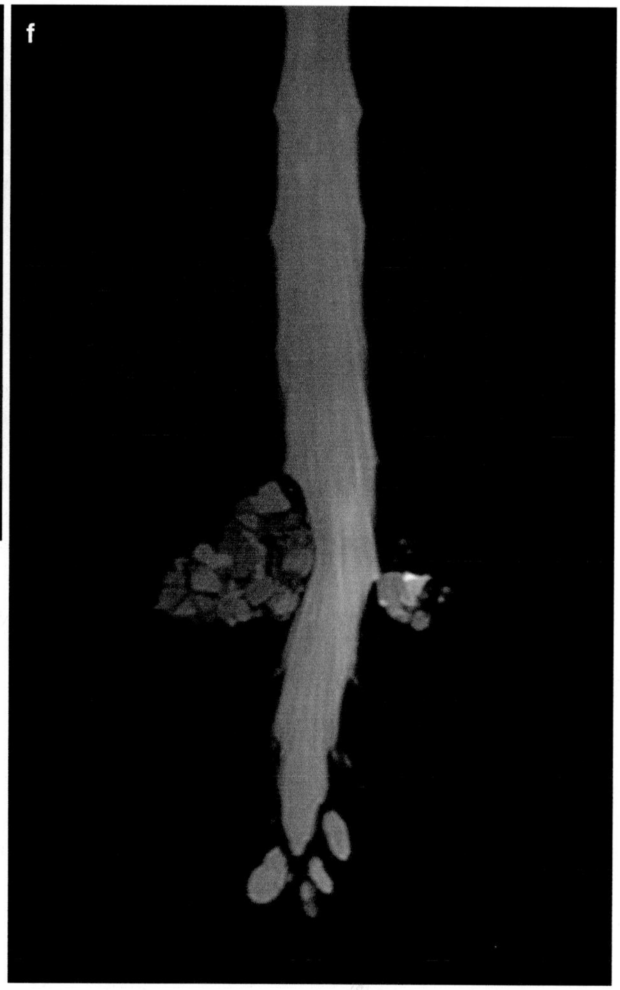

Fig. 8.75 (continued)

Spinal cystic echinococcosis should be differentiated from spinal tuberculosis and metastasis.

Spinal tuberculosis the most commonly occurs in lumbar vertebra, followed by thoracolumbar segment, but rarely occurs in cervical vertebra. Based on the firstly destructed bone, it can be further divided into centrum tuberculosis and adnexal tuberculosis. The incidence rate of centrum tuberculosis is far higher than that of adnexal tuberculosis. With remissive onset, the lesion of spinal tuberculosis progresses slowly, showing slight symptoms such as low grade fever, fatigue and other systemic symptoms. Spinal tuberculosis is commonly manifested as osteolytic bone destruction, narrowed or absent intervertebral space, spinal kyphosis, paravertebral abscess, and soft tissue calcification of at least two vertebral bodies.

Spinal metastasis commonly occurs in middle-aged and elderly populations, with a higher incidence in males. Its clinical manifestation is characterized by pain, and some patients experience pathological fracture and compression symptoms. Spinal metastasis can be further divided into osteolysis type, osteogenesis type and mixed type, showing respective radiological demonstrations. X-ray and CT scan demonstrate the lesion of osteolysis type as irregular osteolytic bone destruction, and MR plain imaging demonstrates it with slightly low T1WI signal and slightly high T2WI signal. However, X-ray and CT scan demonstrate the lesion of osteogenesis type as spots, flakes, nodules or masses of high density opacity in cancellous bone tissue, and MR plain imaging demonstrates it in slightly low T1WI signal and low T2WI signal. The lesion of mixed type shows the radiological demonstrations of both osteolysis and osteogenesis types. Spinal metastasis is characterized by multiple lesions, jumping growth, and early involvement of the vertebral appendix.

Case Study 63

[Brief Medical History]

A 57-year-old man, Han nationality, complained of sacrococcygeal pain and numbness for 40 days. The 4-item examination for echinococcus showed Anti-EgCF antibody negative (–), Anti-EgP antibody negative (–), Anti-EgB antibody negative (–) and Anti-Em2 antibody negative (–).

[Radiological Demonstration] (See Fig. 8.76)

[Diagnosis] Sacral cystic echinococcosis.

[Discussion]

Skeletal echinococcosis occurs with an incidence of 0.5–2 %, with 50 % of the cases to be spinal echinococcosis. Radiological examination is important for the diagnosis of skeletal echinococcosis, especially CT scan and MR imaging, which can more clearly demonstrate characteristic signs of cystic echinococcosis so as to increase its diagnostic accuracy. CT scan can clearly demonstrate the characteristic signs the lesion, including cystic or multicystic swelling bone destruction, intact or destructed bone cortex, hardened margin of the lesion, shell like or strips of calcification, and local soft tissue lump. MR imaging can demonstrate the shape, range and location of the lesion from sagittal, coronal and transverse perspectives, with unique advantages in demonstrating relationships between the vertebral centrum, intervertebral space and spinal cord. Therefore, MR imaging shows the most important value in all radiological modalities for the diagnosis of spinal echinococcosis. In this case, the diagnosis can hardly be made by X-ray. But its combination to transverse CT scan, sagittal CT scan and 3-dimensional reconstruction as well as MR imaging, the lesion can be more clearly demonstrated for the diagnosis of sacral cystic echinococcosis.

Case Study 64

[Brief Medical History]

A 37-year-old herdsman complained of a lump in the left groin for 10 years, with aggravated left leg pain during standing or walking for 1 year.

[Radiological Demonstration] (See Fig. 8.77)

[Diagnosis] Skeletal echinococcosis in the right sacrum and ilium.

[Discussion]

After access to human systemic circulation, the larva of echinococcus may settle and parasitize in bone to cause skeletal echinococcosis. Constrained by hard bone tissue, the larva of echinococcus develops slowly with the cyst growing along the cavity and space with less resistance. Its gradual enlargement cause the sclerotin to be thinner, showing as an incomplete cyst. Several such incomplete cysts cluster together to show grapes like cavity. Since bone tissue cannot form fibrous envelope, the skeletal echinococcus cyst shows no envelope, which is a characteristic sign of skeletal echinococcosis.

Skeletal echinococcosis should be differentiated from the following diseases:

1. Osteoclastoma
   Osteoclastoma commonly occurs at the epiphysis of long bone, manifested as eccentric swelling osteolytic bone destruction, with fibrous bone septa appearing like soap bubbles inside. The echinococcus cyst grows to the areas with the lowest resistance in the bone, often extending from metaphysis to diaphysis with sharp margin of the cystic bone destruction. The lesion is irregular in shape. By MR imaging, the lesion of osteoclastoma is demonstrated with moderate T1WI signal and high or equal T2WI signal.
2. Centrum tuberculosis
   The lesion of vertebral or appendiceal echinococcosis is multilocular, with no collapse of vertebral centrum in patients with mild conditions. Paravertebral soft tissue lump is commonly detected unilaterally, and the intervertebral disc remains intact with no stenosis. The lesion of spinal tuberculosis can invade its neighbouring vertebral centrum, with the intervertebral disc involved. Paravertebral soft tissue lump is detected bilaterally.
3. Bone cyst
   The lesion of bone cyst is demonstrated as well defined oval transparent area, with moderate T1WI signal and high T2WI signal, with no low signaling external echinococcus cyst and daughter cysts.
4. Aneurysmoid bone cyst
   The lesion of aneurysmoid bone cyst is demonstrated as swelling multilocular bone destruction in the vertebral centrum, often with the vertebral arch involved. Although it is difficult to be differentiated from skeletal echinococcosis, their MR imaging demonstrations are totally different. Stairs like fluid-fluid level can be shown in the lesion of aneurysmoid bone cyst, with multiple diverticulum like protrusions in different sizes, which helps for the differential diagnosis.

Case Study 65

[Brief Medical History]

A 19-year-old Kazak young woman was hospitalized due to the chief complaints of repeated left hip pain and upset for 4 years, which aggravated with restricted movement for 1 year. She experienced left hip pain and upset with no known causes 4 years ago, which was intermittent soreness and swell-

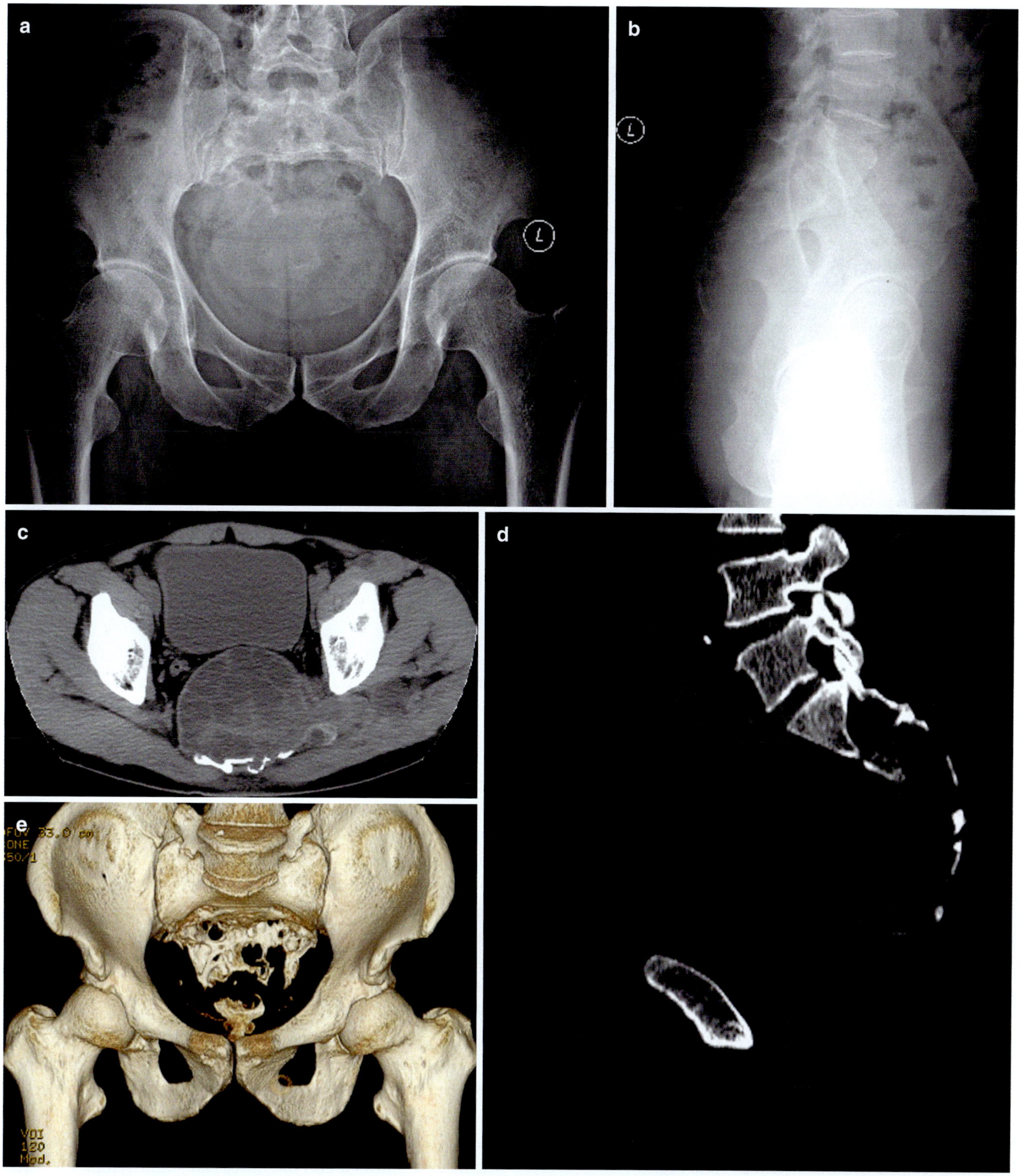

Fig. 8.76 (**a**, **b**) Anterior-posterior and lateral sacral X-rays demonstrated absorption and destruction sacrococcyx, with thinner sclerotin and well defined lesion but no sclerosis. (**c–e**) Pelvic CT scan demonstrated a well defined large round like cystic lesion in the sacral canal and presacral adipose space, with a large section in size of about 8.41 cm × 1.62 cm. The lesion was shown with septa and diffuse nodular opacity inside, with surrounding bone destruction of the sacrum. The rectum was revealed with rightward anterior shift due to compression. The left piriformis, gluteus medius and iliopsoas were demonstrated with multiple round like cystic lesions, sometimes with septa inside and clearly demonstrated surrounding adipose space. (**f**, **g**) Sacral MR imaging demonstrated the sacral vertebrae below S1 and some caudal vertebrae in abnormal shape, with irregular shape long T1 and long T2 mixed signals in the vertebral centrum, its neighbouring soft tissue including left piriformis, space of gluteus medius and pelvic space anterior to the sacral vertebrae. Fat suppression sequence demonstrated high signal, with multiple septa inside, multiple daughter cysts in the mother cyst, and smooth margin. The organs in the pelvic cavity showed shift due to compression of the lesion

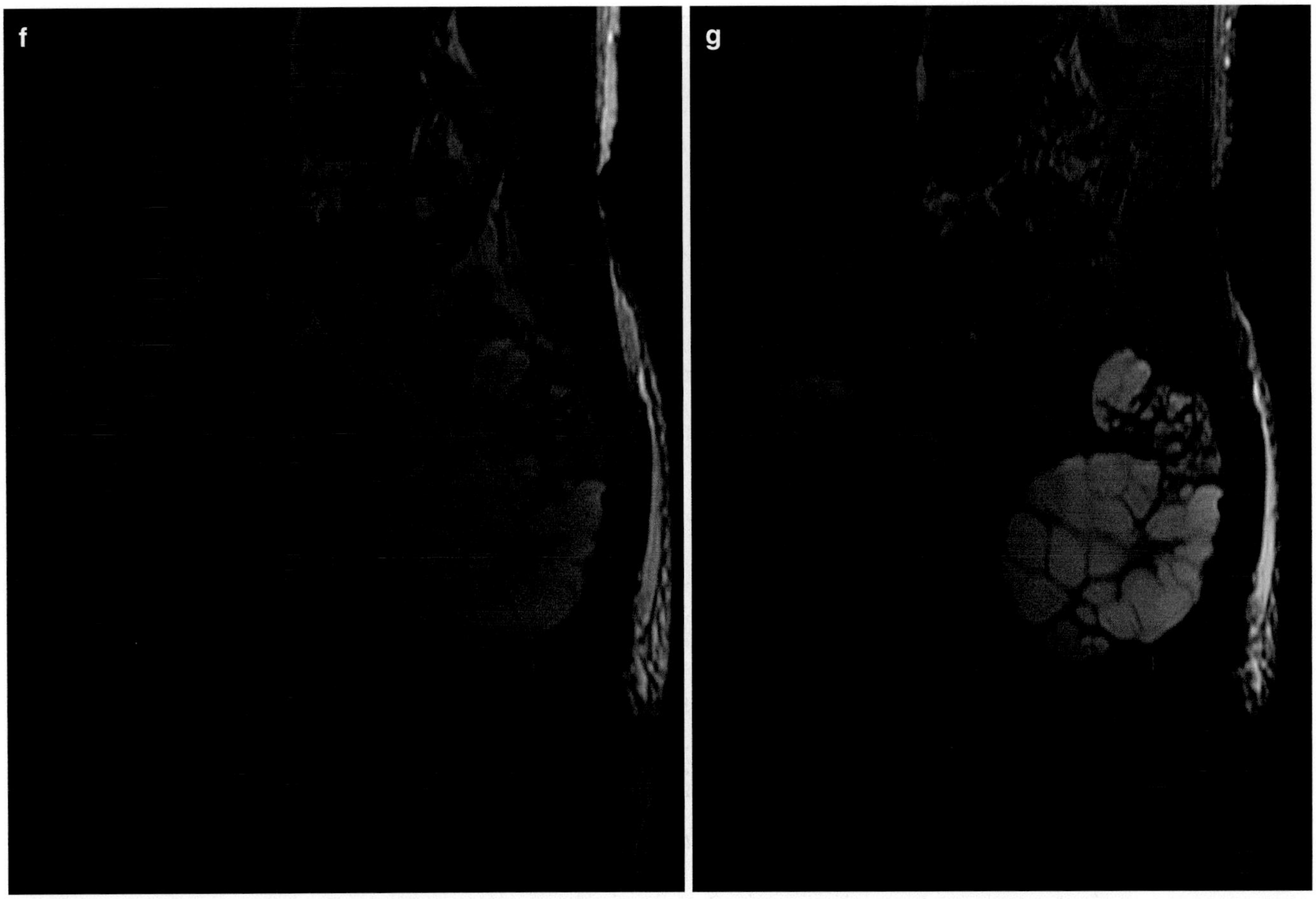

Fig. 8.76 (continued)

ing and more obvious during long-term standing or walking. After that, left hip pain occurred repeatedly and progressively aggravated. By physical examination, no malformation in left hip was observed, with no obvious local redness, swelling, rupture or ulceration. Obvious local tenderness was detected at the anterior and lateral hip joint, with restricted movements of anteflexion, posterior extension, abduction, extorsion, adduction and intorsion. Grade III muscular strength was detected in the biceps and quadriceps femoris of left lower limb, while grade V muscular strength in the triceps surae and anterior tibial muscle of her left lower leg. And her left lower limb was about 2 cm shorter than her right lower limb. The 4-item examination for echinococcus showed Anti-EgCF antibody positive (+), Anti-EgP antibody positive (+), Anti-EgB antibody positive (++) and Anti-Em2 antibody (±).

[**Radiological Demonstration**] (See Fig. 8.78)

[**Diagnosis**] Cystic echinococcosis in left hip joint.

[**Discussion**]

Cystic echinococcosis rarely occurs in joints, with the flat bones in hip or sacroiliac joint more commonly involved. Restrained by hard bone tissue, the lesion of early stage cannot grow into large ball shape cyst, but only grows, infiltrates and destructs along the medullary cavity and loose bone tissue to form multilocular cystic lesion in different sizes. X-ray demonstrates multiple round like cystic bone destructions in different sizes, with irregular boundary. CT scan demonstrates multiple round like cystic low density in different sizes, often with its surrounding soft tissue involved. MR imaging demonstrates multiple round like lesions of different sizes, with smooth and sharp boundary, low T1WI signal and high T2WI signal. If the echinococcus cyst continues to grow, it can penetrate the bone tissue to involve the surrounding soft tissue and form a lump. It can also rupture to form nonhealing fistula, with overflow of daughter cysts in some cases.

Cystic echinococcosis in hip joint should be differentiated from hip tuberculosis and purulent hip arthritis.

Hip tuberculosis commonly occurs in teenagers and children, often with nonbearing joint surface involved. Clinically, the disease shows a chronic onset and slight symptoms, such as joint swelling and pain as well as restricted movement. During its active stage, the patients also experience night sweat, low grade fever, weight loss and other systemic symptoms. X-ray demonstrates narrowed joint space, bone destruction on the nonbearing joint surface, swelling of surrounding soft tissue, increased density, and accompanying osteoporosis. CT scan demonstrations resemble to those by X-ray, but more

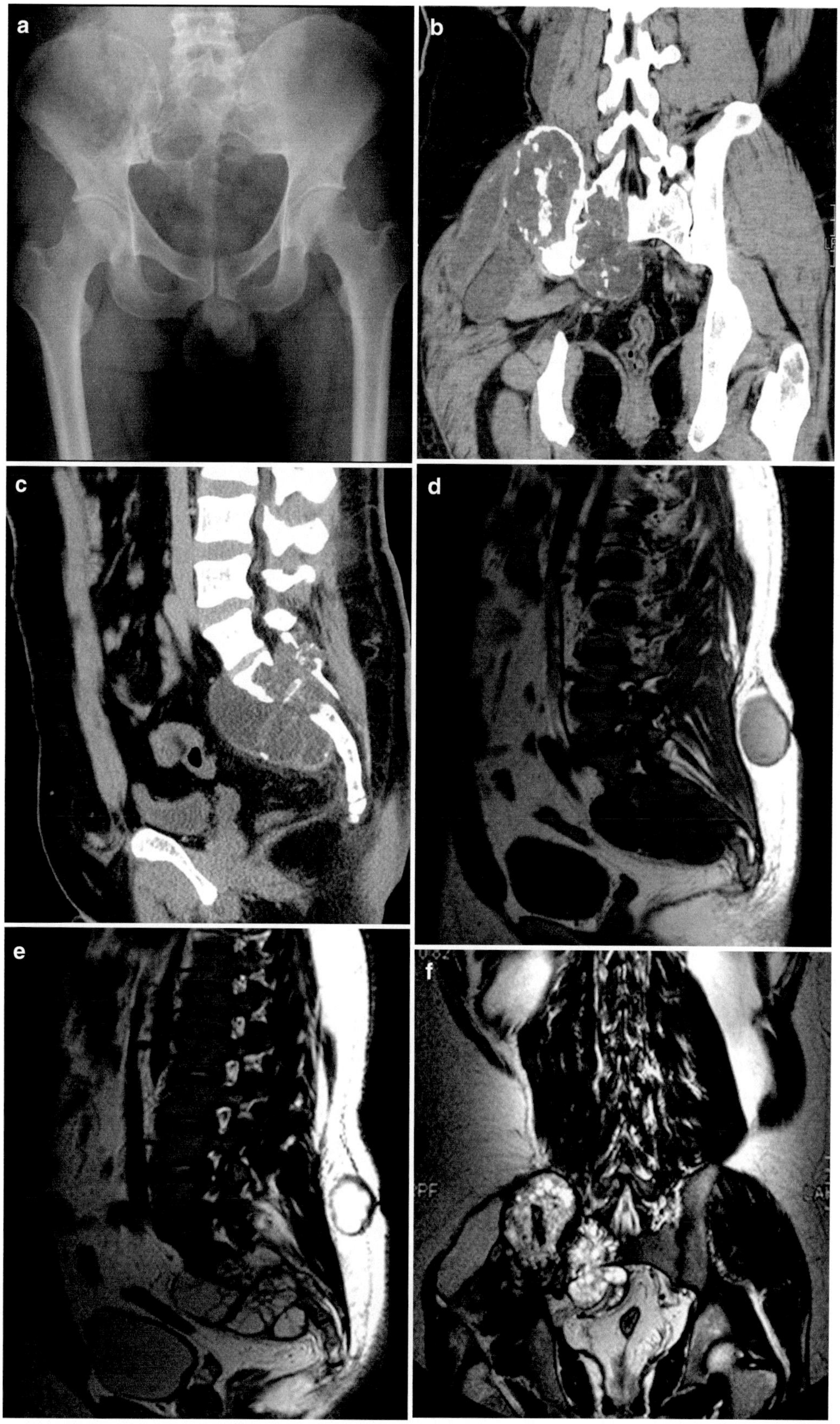

Fig. 8.77 (**a**) X-ray demonstrated well defined bone destruction in the right ilium and sacrum. (**b**, **c**) CT scan demonstrated well defined swelling bone destruction in the ilium and sacrum, with patches of ossification opacity inside. (**d**–**f**) MR imaging demonstrated bone destruction in the sacrum, which is comprised of several vesicles, with low T1WI signal and high T2WI signal. The lesion was shown with septa inside. The dorsal soft tissue at about the level of L4-5 was shown with an oval lesion with long T1 long T2 signals (Reprint with permission from Hongjun Li, Radiology of Infectious diseases, Volume 2, 2015)

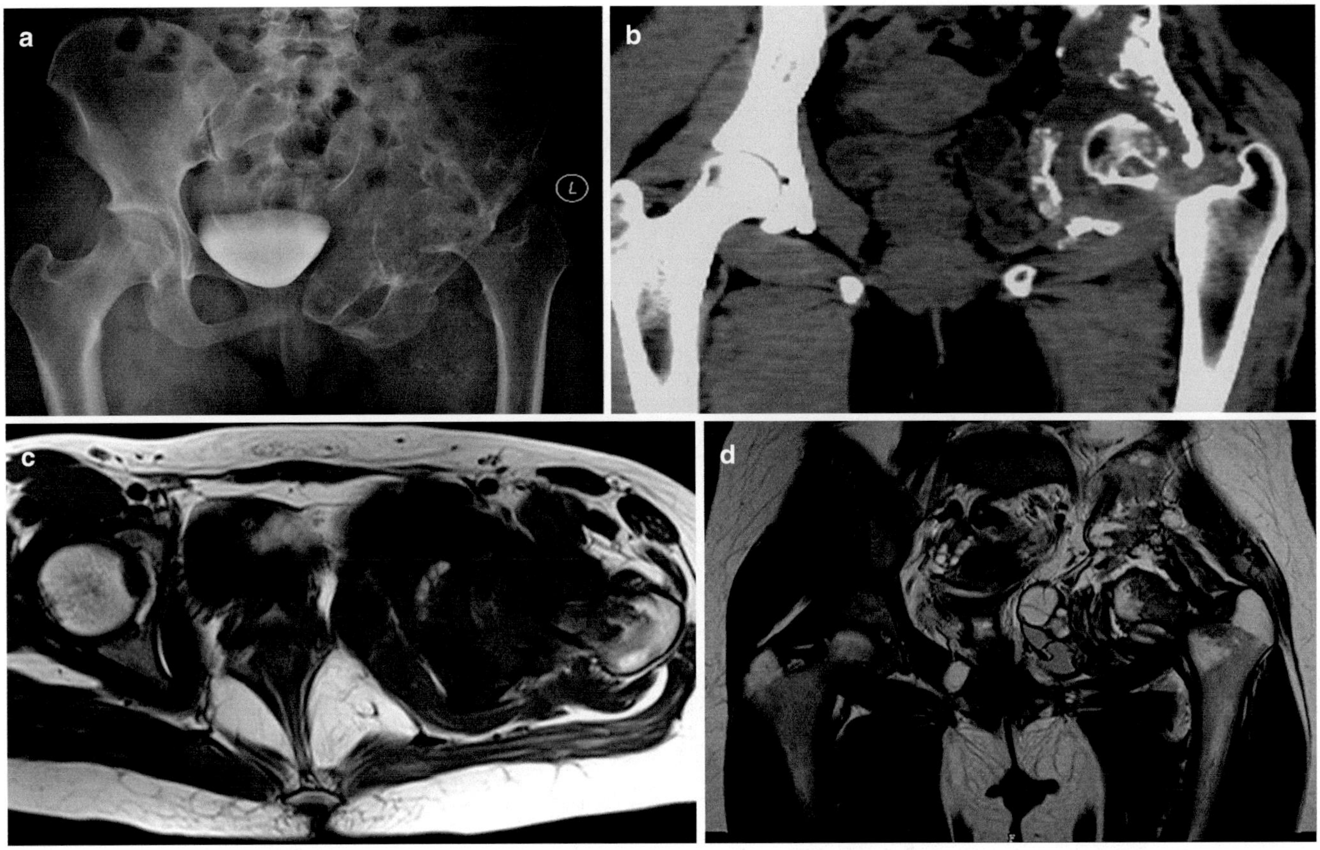

Fig. 8.78 (**a**) X-ray demonstrated asymmetric bilateral hip joints and multiple cystic decreased bone density in the left ilium and left femoral head, collapse of the left humeral head, and shortening of the femoral neck. (**b**) CT plain scan demonstrated left ilium and left femoral head with multiple cystic bone destructions, the soft tissue surrounding the hip joint with cystic low density, with a small quantity of septa opacity in the lesion. (**c**, **d**). MR plain imaging demonstrated low T1WI signal of the lesion and equal signal of the cystic wall and septa; high T2WI signal of the lesion and slightly low T2WI signal of the cystic wall and septa

clearly showing effusion in articular cavity and swollen soft tissue around the joint. In the cases with accompanying abscess, CT scan can define its location and range. Contrast scan demonstrates obvious abnormal enhancements of the joint cyst and the cystic wall. By MR plain imaging, the bone destruction under the joint surface is demonstrated with slightly low T1WI signal and high T2WI signal. If with abscess, the lesion is demonstrated with low T1WI signal of the abscess fluid and equal T1WI signal of the abscess wall; and with high T2WI signal of the abscess fluid and low T2WI signal of the abscess wall. Contrast imaging demonstrates obvious ring shape enhancement of the abscess wall.

Purulent hip arthritis commonly occurs in children and infants, and its pathogenic bacteria is the most commonly staphylococcus aureus. The lesion is commonly singular, and the patients commonly experience symptoms of swollen joints, acute inflammatory signs of redness, swelling, heat and pain, as well as restrained joint movements. X-ray demonstrates, at the early stage, widened joint space, narrowed joint space after local osteoporosis, and bone destruction under the joint surface that is more serious under the strength bearing surface. X-ray demonstrates, in the late stage, bone ankyloses and calcification in surrounding soft tissue. CT scan demonstrations resemble to those by X-ray, but CT scan is more sensitive to bone destruction and abscess than X-ray. MR imaging demonstrates bone destruction on the strength bearing surface, with low T1WI signal and high T2WI signal. Compared to CT scan, MR imaging can more clearly demonstrate the involvement of soft tissues around inflammation.

Case Study 66

[**Brief Medical History**]

A 43-year-old Kazak woman complained of repeated left hip pain for more than 10 years, which aggravated with left hip lump for 2 months. The 4-item examination for echinococcus showed Anti-EgCF antibody positive (+++), Anti-EgP antibody positive (+++), Anti-EgB antibody positive (+++) and Anti-Em2 antibody positive (+).

[**Radiological Demonstration**] (See Fig. 8.79)

[**Diagnosis**] Echinococcosis in the left ilium, sacral vertebrae and lungs.

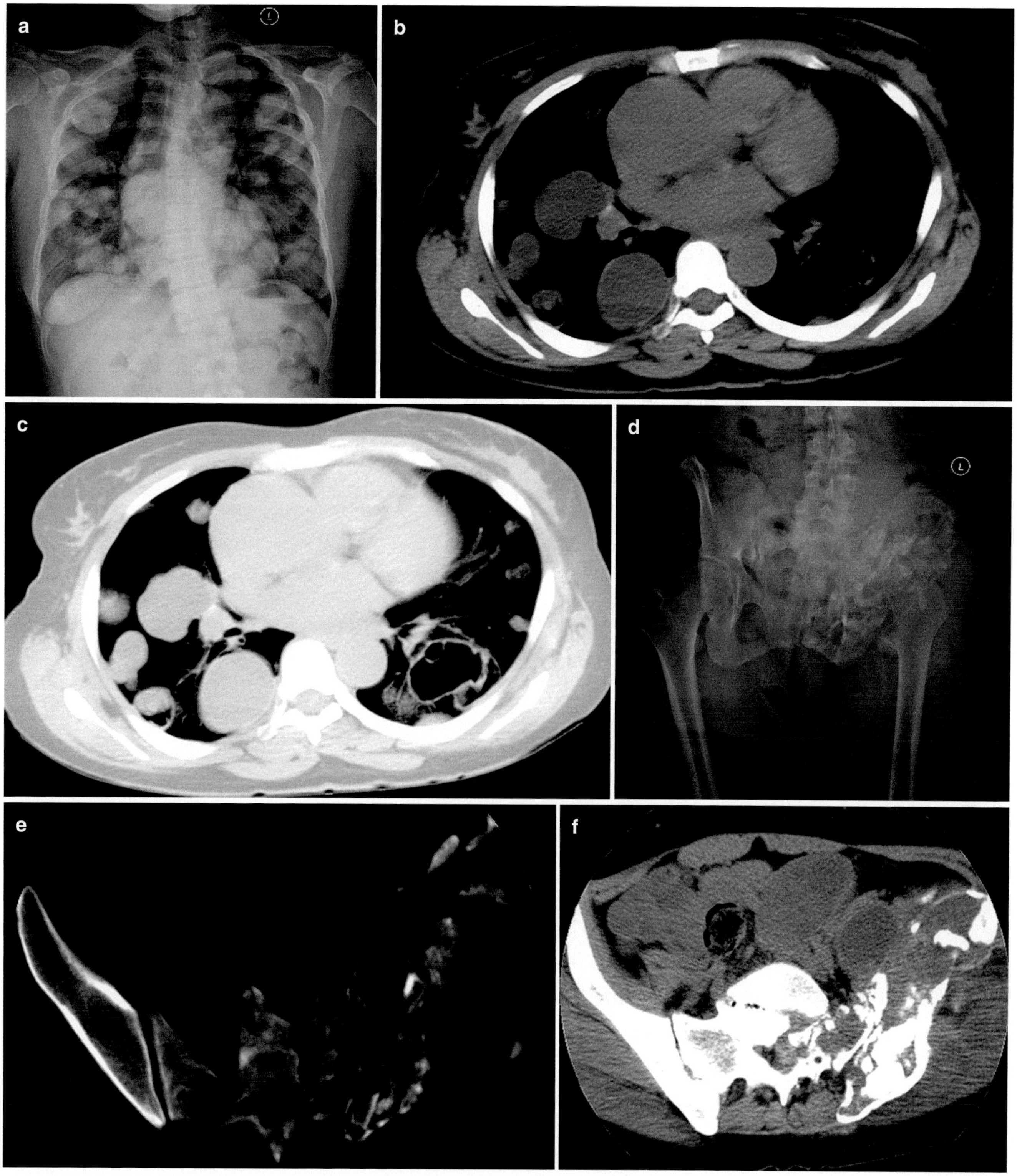

Fig. 8.79 (**a**) Chest X-ray demonstrated both lungs with nodular and ball shape lesions of different sizes, with cavity opacity in some cases. The lesions were shown to be well defined with smooth margin. (**b**, **c**) CT plain scan demonstrated the lesions in both lungs with homogeneous fluid density, with cavity opacity in some cases. The lesions were shown to be well defined with smooth margin. (**d**) Pelvic X-ray demonstrated irregular bone destruction in the left ilium, left phalanx and partial left sacral vertebrae, with uneven density as well as patches of high density opacity and low density area. The left femoral head and hip joint were demonstrated with abnormal shape, and the femoral neck was shortened. (**e**, **f**) CT plain scan demonstrated the 2nd sacral vertebral centrum and left ilium with extensive irregular bone destruction in fence like pattern. The lesions were shown with uneven density, well defined margin and discontinuous iliac bone cortex. (**g**) 3-dimensionally reconstructed CT scan image. (**h**–**j**) Multiple cystic lesions in different sizes with mixed long T1 and long T2 signals were demonstrated in the left paravertebral area of L3 to sacral vertebrae level, lumbar erector spinae, abdominal cavity, left iliac fossa, soft tissue of the left hip as well as the right ilium and the right sacrum. The lesions were shown with multiple septa inside. Some lesions were shown with development into the sacral canal, with mixed high signal by fat suppression sequence

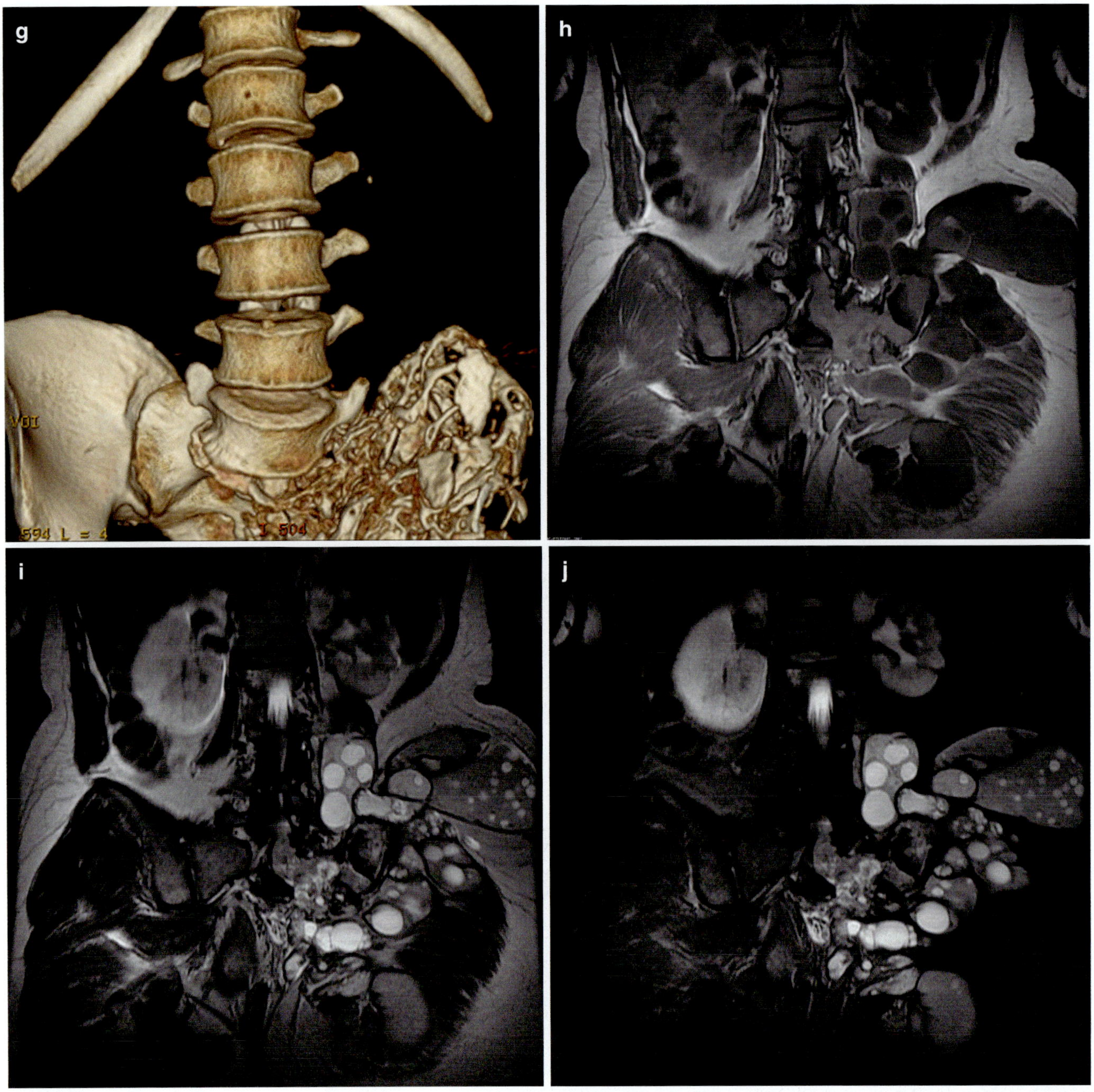

Fig. 8.79 (continued)

[Discussion]

Radiological examination has important diagnostic value for skeletal echinococcosis, with X-ray as the first choice. In terms of CT scan, contrast scan and multiplanar reconstruction (MPR) can more clearly demonstrate the lesions, which show high diagnostic value for more accurate diagnosis of echinococcosis. MR imaging can demonstrate characteristic signs of echinococcosis and shows special value in demonstrating the relationship between the lesions and their neighboring tissue, based on which the diagnosis can be made. In conclusion, CT scan and MR imaging can more clearly demonstrate the echinococcus cyst so as to increase the diagnostic accuracy of skeletal echinococcosis. In this case, the patient reported a medical history of multiple pulmonary echinococcosis. A preliminary diagnosis was made based on the lesion of pelvic bone destruction demonstrated by plain X-ray, but its qualitative diagnosis cannot be made. By multiplanar reconstruction of CT images as well as sagittal, coronal and transverse MR imaging, the skeletal echinococcus was observed in details for its shape, range and location.

Case Study 67

[Brief Medical History]

A 54-year-old woman, Han nationality, complained of repeated swelling of the right lower leg for 13 years that aggravated with pain for 1 year. She reported a painless lump at the right medial lower leg with no known causes 13 years ago, which then gradually enlarged. She experienced swelling and pain of the right lower leg for 1 year that could not relieved by rest or applying hot pack. By physical examination, her right lower leg showed obvious swelling, which was the most obvious at the site 13 cm inferior to the right patella. Her left lower leg showed a perimetre of 31 cm, while the right lower leg 36 cm. Multiple lumps were palpable at the medial and dorsal areas of her right lower leg, with tenderness. The muscular tension and strength of the four limbs were normal.

[Radiological Demonstration] (See Fig. 8.80)

[Diagnosis] Cystic echinococcosis in the muscular soft tissue of right lower leg.

[Discussion]

Cystic echinococcosis in soft tissue rarely occurs, but commonly secondary to echinococcosis at other body parts. Due to the slow development of echinococcus cyst, the illness course is relatively long. As no barrier to restrain the growth of echinococcus cyst in soft tissue, delayed treatment causes a large area of involvement, with typical signs of cystic echinococcosis in the cases of cystic echinococcosis in soft tissue. These typical signs include low T2WI signal of the cystic wall, homogeneous thickness of the cystic wall, rose-petals sign for daughter cysts in mother cyst, and ribbon sign for rupture of internal cyst. In rare cases of cystic echinococcosis in soft tissue, due to its slow development, its hardened envelope, no adhesion with surrounding skin and no communicating duct, the cyst in soft tissue is rarely subject to rupture, infection, degeneration, and calcification, showing atypical radiological signs. Some of the cases show only simplex cyst, singular or multiple, with smooth and intact margin but no cystic rupture, infection, and calcification. Based on the information about its high-prevalence regions, medical history and typical radiological signs, cystic echinococcosis in soft tissue can be differentiated from other cystic lesions in soft tissue.

Cystic echinococcosis in soft tissue should be differentiated from hemangioma, plexiform neurofibromatosis (PNF) and schwannoglioma.

Hemangioma is the most common benign lesion in soft tissue, which commonly occurs in infants and children, with a higher incidence in females. The patients generally experience no obvious symptoms, but possibly intermittent pain and swelling. In some cases, pulse is palpable on the swelling and vascular murmur can be auscultated on the swelling. CT scan demonstrates the lesion as poorly defined soft tissue lump. In some cases, twisted cords like structure can also be detected in its adjacent subcutaneous

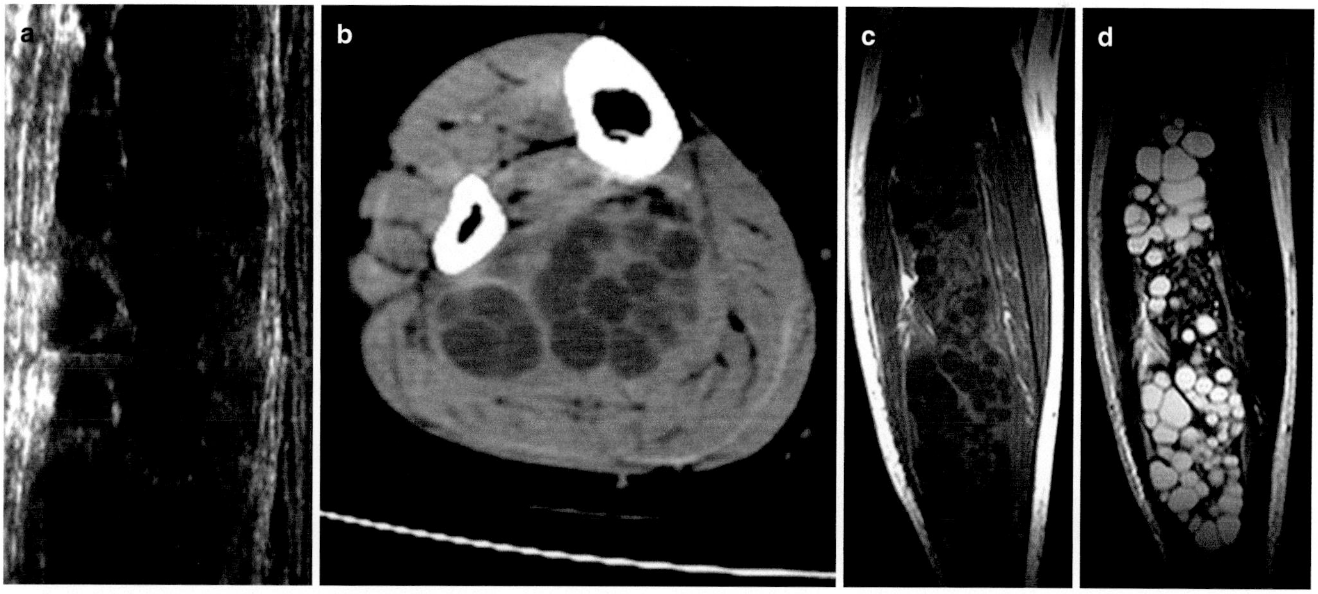

Fig. 8.80 (**a**) Ultrasound demonstrated multiple oval and round like non-echo areas of different sizes in the muscular soft tissue of right lower leg, with dense arrangement, well defined boundary, favorable sound transmission, and corresponding posterior enhancement. (**b–d**) The muscular soft tissue in the right lower leg was demonstrated with multiple round like cystic lesions of different sizes. By CT plain scan, the lesions were shown with water like density, multiple septa opacity inside, and multiple daughter cysts in the mother cyst. MR imaging demonstrated the lesions with low T1WI signal and high T2WI signal, with the cystic wall in homogeneous thickness and partial cystic wall in slight low T2WI signal. The lesions were well defined from their surrounding soft tissue, which were distributed along the muscular space

adipose tissue, which is the feeding artery and drainage vein of the lump. The lump is demonstrated with multiple round or oval calcified phleboliths of different sizes, with obvious abnormal enhancement by contrast scan. MR plain imaging demonstrates low T1WI signal and high T2WI signal, with multiple twisted earthworm like low signal blood flow void opacity on T2WI. If with sub-acute phase of bleeding, high T1WI and T2WI signals are demonstrated, with progressive enhancement along with time lapse by contrast imaging.

Plexiform neurofibromatosis (PNF) is a main type of neurofibromatosis, whose lesion is comprised of Schwann cells, fibroblasts and peripheral neurocytes. The typical lesion is demonstrated with formation of plexiform structure in grapes-cluster pattern that swells into its neighboring soft tissue, growth along the long axis of nerve bundle and its branches, and spread along and around the nerves. MR plain imaging demonstrates that the lesion in grapes-cluster pattern grows in a crawling manner along intermuscular nerve with low T1WI signal and water like high T2WI signal. Within the lesion, septa opacity is demonstrated with well defined boundary. The lesion shows no sign of infiltrative growth but its surrounding soft tissue migrates due to compression.

The lesion of schwannoglioma is a common benign neoplasm in soft tissue, which is more commonly originated from major nerve trunk and may occur at any part of human body, the most commonly lower limbs. Schwannoglioma occurs in any age group, more commonly adults aged 20–40 years. Most of the neoplasms are solitary nodule, showing a slow development and long illness course. The clinical manifestations are related to the location of lesion. The lesion of neoplasm is mostly located in tissue space (e.g. muscular space), demonstrated as well defined oval lump or lobulated nodule. Since the lesion is susceptible to bleeding, necrosis or cystic change, CT plain scan commonly demonstrates the lesion with mixed density, with obvious enhancement of its solid part by contrast scan. MR plain imaging demonstrates the solid part of lesion with moderate signal, necrotic part and cystic change with low T1WI signal and high T2WI signal. If lesion with internal bleeding, the imaging demonstrations vary according to the stage of hematoma, with obvious abnormal enhancement of its solid part by contrast imaging.

Case Study 68

[**Brief Medical History**]

A 54-year-old woman, Han nationality, complained of repeated swelling of her right lower leg for 13 years that aggravated with pain for 1 year. The 4-item examination for echinococcus showed Anti-EgCF antibody (±), Anti-EgP antibody (±), Anti-EgB antibody (±), and Anti-Em2 antibody negative (−).

[**Radiological Demonstration**] (See Fig. 8.81)

[**Diagnosis**] Cystic echinococcosis in the posterior muscles group of right lower leg.

[**Discussion**]

Muscular cystic echinococcosis account for 0.7–3 % of all echinococcosis cases, with B-mode ultrasound, CT scan and MR imaging serving as the main radiological examinations for its diagnosis. B-mode ultrasound is the first choice for its diagnosis, and MR imaging shows characteristic signs of non-calcified lesion of echinococcosis. In this case, plain X-ray showed limited diagnostic value for muscular cystic echinococcosis. Although CT scan demonstrated multiple well-defined multilocular cystic lesions in the posterior muscles group of right lower leg, a qualitative diagnosis cannot be made. MR imaging clearly demonstrated multiple daughter cysts in the large cyst in grapes cluster pattern on T2WI and fat suppression sequence, which are characteristic signs of cystic echinococcosis. Before the surgical operation, examination of the right lower leg by B-mode ultrasound was not ordered.

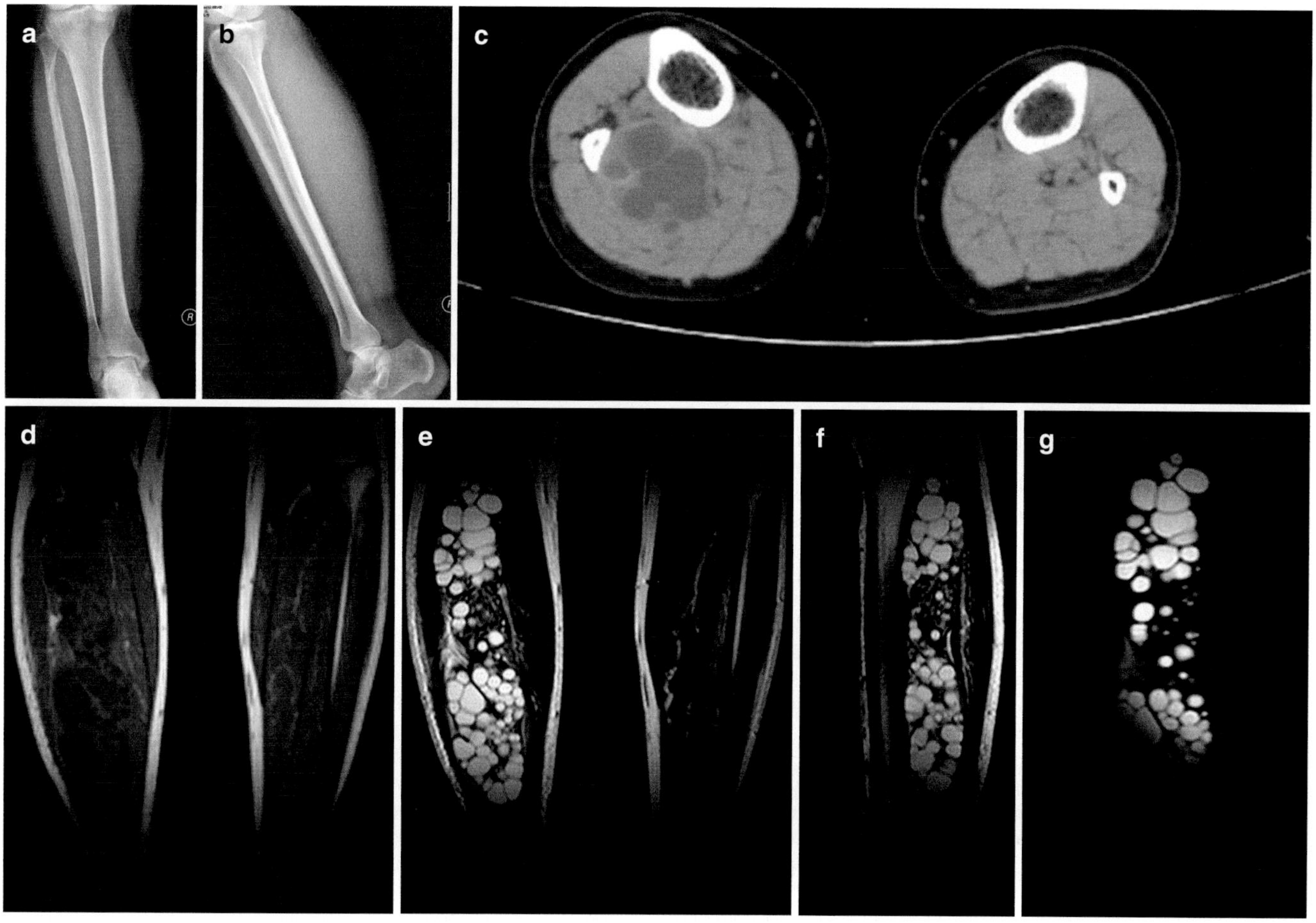

Fig. 8.81 (**a**, **b**) Anterior-posterior and lateral X-rays of tibia and fibula demonstrated narrowing of local bone in the left upper fibula with thinner bone cortex, and increased density of soft tissue in the right lower leg with blurry boundaries. (**c**) CT plain scan demonstrated multiple longitudinal well-defined multilocular lesions in the posterior muscles group of right lower leg, with fibrous septa and poorly-defined boundary from the muscles group. The fibula was shown to be locally compressed, with no obvious bone destruction. (**d**–**g**) MR imaging demonstrated multiple longitudinal strips of alveolar long T1 and long T2 abnormal signals in the posterior muscles group of right lower leg. Fat suppression sequence demonstrated obvious high signal. The small cystic lesions were surrounded by envelope opacity in equal T1 and equal T2 signals, and some in mixed short T2 signal. The lesions were distributed in an area with a size of 23.89 cm × 5.27 cm, which were in grapes-cluster pattern with poorly defined boundaries from the muscles group. Fat suppression imaging demonstrated the surrounding soft tissue with a large quantity of flakes high signal opacity

Further Reading

Abdel Razek AA, El-Shamam O, Abdel Wahab N. Magnetic resonance appearance of cerebral cystic echinococcosis: World Health Organization (WHO) classification. Acta Radiol. 2009;50(5):549–54.

Abdelhakim K, Khalil A, Haroune B, et al. A case of sacral hydatid cyst. Int J Surg Case Rep. 2014;5(7):434–6.

Aderdour L, Harkani A, Nouri H, et al. Hydatid cyst of the thyroid in a child. Rev Stomatol Chir Maxillofac. 2012;113(2):124–6.

Ai T, Hu DY. Biliary ascariasis: report of 1 case. Radiol Pract. 2009;24(10):1167, 1168.

Alonso García ME, Suárez Mansilla P, Mora Cepeda P, et al. Ovarian hydatid disease. Arch Gynecol Obstet. 2014;289(5):1047–51.

Alouini Mekki R, Mhiri Souei M, Allani M, et al. Hydatid cyst of soft tissues: MR imaging findings (Report of three cases). J Radiol. 2005;86(4):421–5.

Arslan F, Zengin K, Mert A, et al. Pelvic and retroperitoneal hydatid cysts superinfected with Brucella sp. and review of infected hydatid cysts. Trop Biomed. 2013;30(1):92–6.

Awuti T, Maihesuti M, Liu WY. CT diagnosis of echinococcus cyst in muscular soft tissue. J Clin Radiol. 2008;27(4):486–8.

Aydinli B, Aydin U, Yazici P, et al. Alveolar echinococcosis of liver presenting with neurological symptoms due to brain metastases with simultaneous lung metastasis: a case report. Turkiye Parazitol Derg. 2008;32(4):371–4.

Azizi A, Blagosklonov O, Lounis A, et al. Alveolar echinococcosis: correlation between hepatic MRI findings and FDG-PET/CT metabolic activity. Abdom Imaging. 2015;40(1):56–63.

Bagheri R, Haghi SZ, Amini M, et al. Pulmonary hydatid cyst: analysis of 1024 cases. Gen Thorac Cardiovasc Surg. 2011;59(2):105–9.

Chafik A, Benjelloun A, El Khadir A, et al. Hydatid cyst of the rib: a new case and review of the literature. Case Rep Med. 2009;2009:817205.

Chang KH, Chi JG, Cho SY, et al. Cerebral sparganosis: analysis of 34 cases with emphasis on CT features. Neuroradiology. 1992;34(1):1–8.
Czermak BV, Akhan O, Hiemetzberger R, et al. Echinococcosis of the liver. Abdom Imaging. 2008;33(2):133–43.
Deng SD, Wei M, Rong J, et al. MRI diagnosis of hepatic cystic echinococcosis: a comparative study to CT diagnosis. J Pract Radiol. 2008;24(11):1504–6.
Ding ZX, Yuan JH, Chong V, et al. 3T MR cholangiopancreatography appearances of biliary ascariasis. Clin Radiol. 2011;66(3):275–7.
Dong Y, Wang L, Jiao TY, et al. Diagnostic value of ultrasound for intestinal ascariasis and biliary ascariasis. Chin J Clin Med. 2013;20(6):835–7.
Dubagunta S, Still CD, Kormar MJ. Acute pancreatitis. J Am Osteopath Assoc. 2001;101(4 Suppl Pt 1):S6–9.
Erol B, Tetik C, Altun E, et al. Hydatid cyst presenting as a soft-tissue calf mass in a child. Eur J Pediatr Surg. 2007;17(1):55–8.
Ertas IE, Gungorduk K, Ozdemir A, et al. Pelvic tuberculosis, echinococcosis, and actinomycosis: great imitators of ovarian cancer. Aust N Z J Obstet Gynaecol. 2014;54(2):166–71.
Gong HJ. CT diagnosis of hepatic alveolar echinococcosis. J Pract Med Technol. 2009;9(16):700.
Gong YP, Pan W, Chen LE. Ultrasonographic findings of hepatic, subcutaneous and muscular cysticercosis: report of 1 case. Chin J Med Ultrasonogr. 2009;25(8):808.
Gougoulias NE, Varitimidis SE, Bargiotas KA, et al. Skeletal muscle hydatid cysts presenting as soft tissue masses. Hippokratia. 2010;14(2):126–30.
Guo H. CT scan for pancreatic echinococcosis: report of 1 case. J Clin Radiol. 2004;23(8):707.
Guo YM, Chen QH. Radiology of respiratory system. Shanghai: Shanghai Science and Technology Press; 2011.
Hamitti M, Ge YJ, Yang F, et al. CT signs frequency in the cases of hepatic echinococcosis from Xinjiang, China. J Xinjiang Med Univ. 2014;37(4):416–8.
Jain S, Chopra P. Cystic echinococcosis of the pelvic bone with recurrences: a case report. Korean J Parasitol. 2011;49(3):277–9.
Jiang DD, Taxipulati, Fang KH. CT diagnosis of pulmonary cystic echinococcosis. J Imaging Diagn Interv Radiol. 1998;7(1):10–2.
Kalpana S, Sridhar K, Murugan B, et al. Hydatid cyst of lung: an uncommon cause of chest pain in young. Lung India. 2014;31(3):262–3.
Kantarci M, Pirimoglu B. Diffusion-weighted MR imaging findings in a growing problem: hepatic alveolar echinococcosis. Eur J Radiol. 2014;83(10):1991–2.
Kireşi DA, Karabacakoğlu A, Odev K, et al. Uncommon locations of hydatid cysts. Acta Radiol. 2003;44(6):622–36.
Kodama Y, Fujita N, Shimizu T, et al. Alveolar echinococcosis: MR findings in the liver. Radiology. 2003;228(1):172–7.
Lantinga MA, Gevers TJ, Drenth JP. Evaluation of hepatic cystic lesions. World J Gastroenterol. 2013;19(23):3543–54.
Lei GQ, Chen Y, Wang XH, et al. CT and MRI diagnosis of hepatic echinococcosis. Chin J Med Imaging Technol. 2010;26(2):291–3.
Lewall DB, McCorkell SJ. Rupture of echinococcal cysts: diagnosis, classification, and clinical implications. Am J Roentgenol. 1986;146(2):391–4.
Li BP, Guo H. Diagnosis and differentiated diagnosis by CT scan for splenic cyst. J Clin Radiol. 2007;26(3):310, 311.
Li CX, Li YF. Hepatic echinococcosis with concurrent splenic echinococcosis. Chin J Digest. 2012;32(6):368.
Li DQ, Yan WY, Jiang HJ, et al. Acetabulum cysticercosis: report of 1 case. Chin J Med Imaging. 2002;10(5):343.
Li FT, Zhou ZL, Luo R, et al. X-ray demonstrations of subcutaneous and muscular cysticercosis: report of 3 cases. J Pract Radiol. 2005;21(1):109–10.
Li H, Qu Y, Jiang J, et al. Diffusion imaging of hepatic cystic echinococcosis and contrastive study of cystic fluid. Chin J Clin Med Radiol. 2012;23(12):845–8.
Li HJ. Practical radiology of infectious diseases. Beijing: People's Medical Publishing House; 2014.
Liu SY, Chen QH, Wu N. Practical chest radiology for diagnosis. Beijing: People's Military Medical Press; 2012.
Liu WY, Han KN, Zhang QN. CT diagnosis of hepatic echinococcosis: report of 138 cases. J Imaging Diagn Interv Radiol. 1995;4(1):22–4.
Liu WY, Lou JR, Xing Y, et al. Hepatic alveolar echinococcosis by multi-slices spinal CT scan. Chin J Radiol. 2005;39(8):860–3.
Liu WY, Shang G, Dang J, et al. Extrahepatic metastasis of alveolar echinococcosis by CT scan: report of 12 cases. Chin J Radiol. 2000;34(4):255–7.
Liu WY, Xie JX, Li L, et al. CT diagnosis of pelvic echinococcosis. Chin J Radiol. 2003;37(1):79–81.
Liu X, Zhao YH, Zhao YF, et al. Splenic alveolar echinococcosis by CT scan: report of 1 case. Chin J Med Imaging Technol. 2014;30(2):316.
Liu YC. Diagnostic value of CT for hepatic echinococcosis and its typing. Med Xinjiang. 2010;40(7):15–7.
Lu GM. Clinical differential diagnosis by CT scan. Nanjing: Jiangsu Science and Technology Press; 2011.
Ma LG, Li WF, Qiao Y, et al. CT diagnosis of hepatic echinococcosis. J Clin Radiol. 1998;17(6):347–9.
Ma SF, Qiao J, Huo Q. Clinical manifestation and surgical treatment of cardiac and pericardial echinococcosis. Chin J Thorac Cardiovasc Surg. 2009;17(3):144, 145.
Ma XX, Zhang XM, Wang Q, et al. Cerebral, pulmonary and hepatic cysticercosis: report of 1 case. Chin J Med Imaging. 2002;12(4):287.
Maihesuti M, Liu WY, Shayiti M. Imaging demonstrations and diagnosis of skeletal echinococcosis. Chin J Radiol. 2007a;41(5):517–9.
Maihesuti M, Liu WY, Shayiti M. Radiological demonstrations and diagnosis of skeletal echinococcosis. Chin J Radiol. 2007b;41(5):517–9.
Meng YF, Pamier, Li KC, et al. CT diagnosis of hepatic cystic echinococcosis. J Chin Clin Imaging. 2000;11(6):408–11.
Oz G, Eroglu M, Gunay E, et al. Aggressive hydatid cysts: characteristics of six cases. Surg Today. 2014(Aug 28).
Ozaydin I, Ozaydin C, Oksuz S, et al. Primary echinococcus cyst of the thyroid: a case report. Acta Med Iran. 2011;49(4):262–4.
Papakonstantinou O, Athanassopoulou A, Passomenos D, et al. Recurrent vertebral hydatid disease: spectrum of MR imaging features. Singapore Med J. 2011;52(6):440–5.
Pei Y, Deng D, Long LL, et al. X-ray demonstrations of pulmonary cysticercosis. Chin J Radiol. 2002;36(5):468–9.
Polat P, Kantarci M, Alper F, et al. Hydatid disease from head to toe. Radiographics. 2003;23(2):475–94.
Shahei A. Radiological diagnosis of hepatic and splenic cystic echinococcosis. Guide Med Medication China. 2013;11(22):489, 490.
Sheng WB, Liu Y, Xu XX. Clinical features and diagnostic examinations for spinal echinococcosis. Chin J Orthop. 2006;26(1):7–12.
Sreeramulu PN, Krishnaprasad, Girish gowda SL. Gluteal region musculoskeletal hydatid cyst: case report and review of literature. Indian J Surg. 2010;72 Suppl 1:302–5.
Subercaseaux VS, Besa CC, Burdiles OA, et al. Retroperitoneal hydatid cyst: a common disease in a rare location. Rev Chilena Infectol. 2010;27(6):556–60.
Sun YM, Liu CC. Diagnosis of biliary ascariasis by CT scan: report of 1 case. J Clin Radiol. 2008;27(8):1028.
Tang GB. Practical radiology of echinococcosis. Beijing: People's Medical Publishing House; 2013.
Wan XJ, Li ZS, Xu GM, et al. Clinical research of recurrent acute pancreatitis. J Surg Theory Pract. 2001;6(5):310–2.
Wang J, Xing Y, Ren B, et al. Alveolar echinococcosis: correlation of imaging type with PNM stage and diameter of lesions. Chin Med J (Engl). 2011;124(18):2824–8.
Wang J, Abudureheman Y, Jiang CH, et al. Characteristic ^{1}H-MRS signs of cerebral alveolar echinococcosis. Chin J Radiol. 2014;48(2):89–92.

Wang J, Jia WX, Chen H, et al. Diagnostic value of MR hydrography in alveolar echinococcosis. Chin J Radiol. 2009;43(4):402–5.
Wani NA, Kousar TL, Gojwari T, et al. Computed tomography findings in cerebral hydatid disease. Turk Neurosurg. 2011;21(3):347–51.
Wen H, Liu WY, Shao YM, et al. Development of radiological diagnosis and surgical treatment for echinococcosis. Int J Parasitosis. 2009;36(5):299–306.
Wen H, Xu MQ. Echinococcosis in clinical practice. Beijing: Science Press; 2007.
Xu S, Yuan XY, Wang YN. Echinococcosis in body parts of children by CT scan. Chin J Med Imaging Technol. 2012;28(1):133–6.
Zhang JT, Chen GA. Diagnosis and treatment of splenic cystic echinococcosis: report of 90 cases. Pract Clin Med. 2010;11(6):125, 126.
Zhang L, Li WF, Wang CW. Imaging diagnosis of spinal echinococcosis. Chin J Clin Med Radiol. 2011;22(1):54–6.
Zhou CF, Zhou F, Ma GM. Diagnostic diagnosis of renal and splenic cystic echinococcosis. Pract Med China. 2009;4(10):72, 73.
Zhou GX, Wu SJ, Shi X. Cysticercosis cellulosae of right ilium and right iliopsoas muscle: report of 1 case. Chin J Bone Neoplasm Bone Dis. 2008;7(6):380.
Zhou ZR. Pathogenic microbiology. 2nd ed. Beijing: Science Press; 2004.

9 Nematodiasis

Wenya Liu, Jian Wang, and Wanjun Xia

Nematodiasis is a common and frequently occurring parasitic disease. Due to its cylindrical body shape, the parasite is nominated as nematode. Nematode has a variety of species distributing extensively in the natural world, and most lead a free life by themselves.

9.1 Morphology

9.1.1 Adult Nematode

Most nematodes have a cylindrical non-segmental body shape, with its anterior part being bluntly round and its posterior part gradually thinner. It is dioecious. In its adulthood, the outer layer is its body wall, with no epithelial cells in the space between the body wall and the alimentary tract. Therefore, the space is known as primary body cavity or pseudo body cavity. The cavity is filled with liquid, which an important medium for exchange of substances with the inner organs immersed in.

9.1.2 Eggs of Nematode

With no operculum but light yellow, yellowish brown or colorless shell, the eggs of nematode is generally oval in shape. Some nematodes lay eggs containing a non-divided ovum such as roundworm, while containing a dividing ovum such as hookworm. Some nematodes lay eggs containing embryo at the tadpoles stage such as pinworm, and others deliver lavae such as placenta-borne filaria and trichina.

W. Liu • J. Wang (✉)
The First Affiliated Hospital, Xinjiang Medical University, Urumqi, Xinjiang Uygur Autonomous Region, China
e-mail: jeanw1265@163.com

W. Xia
The First Affiliated Hospital, Zhengzhou University, Zhengzhou, Henan, China

9.2 Life Circle

According to the need of intermediate host during their life circle, nematodes can be categorized into two types.

9.2.1 Soil-Transmitted Nematode

During the development of nematodes, those directly develop into adulthood with no parasitizing in an intermediate host are known as the directly developed nematodes or soil-transmitted nematodes. The eggs or lavae can gain direct access into human body for their development. Most of the intestinal nematodes can be categorized into this type.

9.2.2 Biologically Transmitted Nematode

Biologically transmitted nematodes need intermediate host for their development, which are also known as indirectly developed nematodes. Most of the nematodes parasitizing in host tissues can be classified into this type. The larvae firstly develop into infective larvae in the intermediate host, followed by their access into human body via skin or mouth for their further development.

9.3 Pathogenicity

The severity of damages to human body by nematodes is related to the type of nematode, the quantity of parasites, the stage that the nematodes develop into, the parasitic site, the mechanical and chemical stimulation of the nematodes to the host as well as the nutritional status and immunity of the host.

9.3.1 Damages Caused by Larva

When larvae gain their access into human body and migrate, they may damage the related tissues or organs. For instance,

H. Li (ed.), *Radiology of Parasitic Diseases*, DOI 10.1007/978-94-024-0911-6_9

the infective larva of hookworm can invade the skin to cause dermatitis. When the larva of roundworm or hookworm migrates into the lung, it may cause lesions in the affected lung or even roundworm/hookworm induced asthma. The larva of trichina can parasitize in the muscle to cause myositis and systematic symptoms.

9.3.2 Damages Caused by Adult Nematode

The adult nematode lives in the parasitic site to cause malnutrition, tissue damage, bleeding, inflammation and other lesions due to its intake of nutrients, mechanical damage, chemical stimulation and immunopathological responses of the host. For instance, trichina can invade the myocardium, a vitally important tissue, to cause myocarditis and pericardial effusion that may progress into cardiac failure and even death. The angiastrongyliasis cantonensis can parasitize in the nervous system of the host to cause server damages to the brain and spinal cord.

9.4 Filariasis

Filariasis is a parasitic disease caused by parasitism of filaria in the lymphatic tissue, subcutaneous tissue and serous cavity of the host. The disease mainly spreads via bites and stings of blood-sucking mosquitoes and insects, which severely threaten the human health and economic development of the affected region. Totally, eight species of filaria can parasitize in human, but only parasitic diseases caused by bancroftian filariasis and malayan filariasis prevail across the world. The two parasitic diseases share similar clinical manifestations. Their early stage is characterized by lymphangitis and lymphnoditis, while their advanced stage is characterized by a series of symptoms and signs caused by lymphatic blockage.

9.4.1 Etiology

Bancroftian filariasis and malayan filariasis are similar in shape, with long and thin body like thread as well as smooth and milk white surface. Both are diecious. Their diagnosis is based on the detection of microfilaria, larva of filaria, which has fine and long body coated by sheath, bluntly round head, fine and sharp tail tip. Both have basically same life circle, including two stages: developing stage of larva in mosquito and insect (intermediate host) as well as developing and reproducing of adult in human body (definitive host). When mosquito bites and stings patient with microfilaria in the peripheral blood, the microfilaria enters into the stomach of mosquito along with blood. After shedding of the sheath, the microfilaria migrates into the pectoralis to develop into sausage and infective stage of larva consecutively. The infective larva then leaves pectoralis to the inferior lip of mosquito. Once the mosquito bites human again, the infective larva gains its chance to infect human body.

9.4.2 Epidemiology

Filariasis prevails in tropical and subtropical regions. Bancroftian filariasis distributes extensively all over Asia, Africa, Latin America, while Malayan filariasis mainly affects Asia. In China, the affected areas include Shangdong province, Henan province, Guizhou province, Sichuan province, Jiangsu province, Zhejiang province, Fujian province, Guangxi Zhuang Autonomous Region, and Guangdong province.

9.4.2.1 Source of Infection

The source of infection includes patients with microfilaria in his/her peripheral blood and asymptomatic carriers of filaria.

9.4.2.2 Route of Transmission

Filariasis spreads via bites and stings of mosquito and insect. The major transmission vector of Bancroftian filariasis includes culex pipiens pallens and culex quinquefasciatus, while that of Malayan filariasis, anopheles sinensis and anopheles anthropophagus.

9.4.2.3 Susceptible Population

Human is the only definitive host of filaria and is generally susceptible to fialariasis.

9.4.3 Pathogenesis and Pathological Changes

The infective larva invades into the lymphatic vessels of human body, and gradually migarates into the major lymphatic vessels or lymph nodes to develop into adult. Malayan filaria commonly parasitizes in the superficial lymphatic system of the four limbs, while Bancroftian filaria parasitizes in not only the superficial lymph nodes but also the deep lymphatic system of such parts as abdominal cavity, spermatic code and lower limbs. After mating, the female adult produces microfilaria, which enters the blood flow along the lymphatic system. During daytime, microfilaria stays at the pulmonary capillaries, while during nights, it migrates into the peripheral blood circulation. The onset and lesions of filariasis are commonly not caused by microfilaria in blood flow but rather by adult filaria and its infective larva. The occurrence and development of filariasis are dependent on the species of filaria, the parasitic site, the quantity of invading larva and the responses of the host.

9.4.3.1 Filarial Fever, Lymphadenitis and Lymphangitis in the Acute Stage

The metabolites of adult and larva filaria as well as the secretions from the uterus of female filaria can cause systemic allergic responses and local responses of lymphatic tissue in the body of the host. The hosts may experience periodical onset of chills and high fever. The lymphatic vessels may sustain intima swelling, endothelial cells proliferation, and the following inflammatory cells infiltration of the surrounding tissue. These responses may further cause thickening of the lymphatic wall, dysfunctional valves and formation of lymphatic thrombus. Due to recurrent lymphatic inflammation, symptoms of chronic lymphatic blockage may occur.

9.4.3.2 Obstructive Diseases in the Chronic Stage

Lymphatic blockage is an important factor inducing chronic signs of filariasis. Due to stimulations from adult filaria, the patients experience dilation of lymphatic vessels, insufficiency of valve, and lymph deposition. These changes are followed by, inflammatory cells infiltration at lymphatic wall, endothelial cells proliferation, and narrowed lymphatic lumen, which further cause lymphatic blockage. With the dead adult filaria and microfilaria as the center, large quantities of inflammatory cells, macrophagocytes, plasma cells, and eosinophils gather to form filarial granuloma, which finally causes lymphatic blockage. The pressure within the lymphatic vessel distal to the blockage, therefore, increases to cause lymphangivarix. Blockage at different sites may induce different clinical manifestations.

9.4.4 Clinic Symptoms and Signs

9.4.4.1 Acute Filariasis

Acute Lymphadenitis and Lymphangitis

Acute lymphadenitis and lymphangitis mostly occur at the lower extremities, with symptoms of inguinal and femoral pain and lymphadenectasis. After that, medial femoral lymphangitis spreads from upside down, which is known as retrograde lymphangitis. When the inflammation further develop to involve intracutaneous capillary, local redness, swelling and tenderness occur, which is known as erysipelas. Lymphadenitis and lymphangitis commonly show periodical onset, mostly when patients are in tiredness or fatigue.

Spermatocystitis, Epididymitis and Testitis

Spermatocystitis, epididymitis and testitis commonly show acute onset, with chills and high fever. The patients also experience unilateral or bilateral groin or scrotum pain, which radiates to medial thigh and even abdomen. By physical examination, the findings may include thickened spermatic cord, swollen epididymis and testis, swollen mass at the spermatic cord, testis and the surface of testis with tenderness. The symptoms are commonly self-limited, persisting only for several days. The swollen mass may be hardened and gradually shrinks into hard nodule in size of a soybean or mung bean.

Filarial Fever

Filarial fever usually show periodical onset with sudden chills and high fever. The symptoms are self-limited, persisting only for 2–3 days. In some cases, the patients may only experience low grade fever but no chills.

Pulmonary Eosinophilic Infiltration Syndrome

The patients experience chills, fever, cough, asthma and lymphadenectasis. The lung can be shown with migrating infiltration, and by X-ray, bronchial and vascular markings are demonstrated to increase with extensive miliary spots of opacity. By sputum examination, eosinophilic granulocytes and Charcot-Leyden crystals can be detected. The peripheral blood can be detected with increased eosinophilic granulocytes and the finding of microfilaria.

9.4.4.2 Chronic Filariasis

Lymphadenectasis and Lymphangivarix

Recurrent lymphadenitis and intralymphatic sinus varicose are the factors contributing to lymphadenectasis. The swollen lymph node and its surrounding centripetal lymphatic varicose form mass, which resembles to spongy lump with a hard core by palpation. By puncture, lymph fluid can be obtained. Inguinal lymphadenectasis is more common, and lymphangivarix occurs more frequently at the groin, spermatic cord, scrotum and medial thigh.

Effusion of Perididymis

Due to lymphatic blockage in both spermatic cord and testis, lymph flows into the cavity of perididymis to cause effusion of perididymis. The patients with a large quantity of effusion show enlarged scrotum with no folds. The patients may experience sensation of dropping but no pain, being positive in transillumination test.

Chyluria

Chyluria is one of the common clinical manifestations of chronic filariasis. Due to lymphatic blockage, lymph flow in the intestinal trunk lymphatic vessel regurgitates into the urinary tract to cause chyluria. The urine turns into milky white in color and sometimes pink when mixed with blood. Before the onset of chyluria, the patients may experience precursory symptoms such as cloudy urine, lumbar pain, inguinal pain and pelvic pain, which are commonly periodical but persistent in severe cases.

Elephantiasis

It is the most common sign at the advanced stage of filariasis. The initial sign of elephantiasis is swollen lymph whose occurrence at the limbs is shown as pitting edema, which can be relieved by elevating the limbs. Subsequently, tissue fibrosis occurs, with occurrence of non-pitting edema which fails to be relieved by elevating the affected limbs. At this time, the skin has no elasticity. The condition finally progresses into elephantiasis, with enlarged limbs, which are formed by large quantities of fibrous tissue and fat as well as dilated lymphatic vessels and retention of lymph. The epithelium of skin is keratinized or shows wart like hypertrophy.

Others

In addition to the above symptoms and signs, women with filiariasis may also experience filarial node at the breast. Occasionally, filaria may cause ophthomic filariasis as well as filarial granuloma at the spleen, chest, back, neck and arm. Filarial pericarditis, chylous pleural effusion, chylous bloody sputum and microfilaria in bone marrow may also occur.

9.4.5 Diagnostic Examinations

9.4.5.1 Laboratory Test

Serological Test

Positive finding of Bancroftian filariasis antigen by rapid immunochromatographic test (ICT) can be used for the serological diagnosis of the disease. Otherwise, positive finding of filarial specific antibody IgG by enzyme-linked immunosorbent assay (ELIZA) can also be used for the diagnosis.

Etiological Examination

Detection of microfilaria in blood or body fluid is the only reliable way to define the diagnosis of filariasis. In addition, the finding of microfilaria in urine, peridídymis effusion, lymph, ascites and chyluria can define the diagnosis. Other findings to define the diagnosis include detected adult filaria in lymphatic vessels and lymph nodes or detected filaria at the pathohistological sections.

9.4.5.2 Diagnostic Imaging

Chest X-Ray

By chest X-ray, the lesions are demonstrated as thickened pulmonary markings, intrapulmonary spots of opacity, pleural effusion and pleural thickening.

CT Scanning

CT scanning is more sensitive to the lesions than chest X-ray, being capable of detecting finer lesions in the lungs, pleural effusion in small quantity as well as pleural and pericardial thickening.

MR Imaging

It is optimal to distinguish soft tissues and can more favorably demonstrate skin thickening caused by filariasis, inflammatory responses of subcutaneous fat, edema of bone marrow, peritendinitis and synovitis.

Case Study 1

[A Brief Introduction to the Case History]

An 83-year-old woman had suffered from arthritis for 20 years. About 1 year ago, she experienced edema of both lower extremities, which aggravated about 2 months ago. In a local hospital, her condition was diagnosed as hypothyroidism and further aggravated after receiving treatment there. After transferred to our hospital, she showed obvious pitting edema at her lower extremities, and extremely distended abdomen. After abdominal puncture for ascites, the routine ascites biopsy indicated chyle positive. The disease was suspectively diagnosed as filariasis.

[Radiological demonstration] (See Fig. 9.1)

[Diagnosis]

Pulmonary filariasis, ascites, and cholecystolithiasis.

[Discussion]

Most of the patients with filariasis are from the prevailing region of filariasis and have lived in the region for years, with detected microfilaria in the blood. Acute filariasis is clinically manifested as lymphangitis, lymphnoditis, and filarial fever. The patients commonly experience periodical chills and high fever. In the cases of chronic filariasis, due to different sites of blockage, clinical symptoms are various, including lymphedema, elephantiasis, or chyluria. In addition, filarial chest granuloma, filarial pericarditis, chylous pleural effusion, or chylous bloody sputum can be occasionally detected. Patients with asymptomatic filariasis commonly experience paroxysmal asthma or cough during nights as well as accompanying fatigue and low-grade fever. The diagnosis of filariasis mainly depends on epidemiological and laboratory examinations. For this case, the patient showed obvious pitting edema at the lower extremities and chylous ascites. CT scanning indicated lung inflammation, which cannot exclude the possibility of filariasis. Enhanced CT scanning is necessary to further examine the lymph nodes and laboratory tests are also necessary for the definitive diagnosis.

Filariasis should be differentiated from: (1) pneumonia, which is radiologically characterized by cloudy opacity with poorly defined boundary and can be treated by routine anti-inflammatory therapy. (2) peripheral lung cancer, which is shown as lobulated mass with irregular boundary by radiological examination and can be differentiated from filariasis by dynamic enhanced scanning. (3) pulmonary tuberculosis, which is characterized by multiple lesions and small caseous

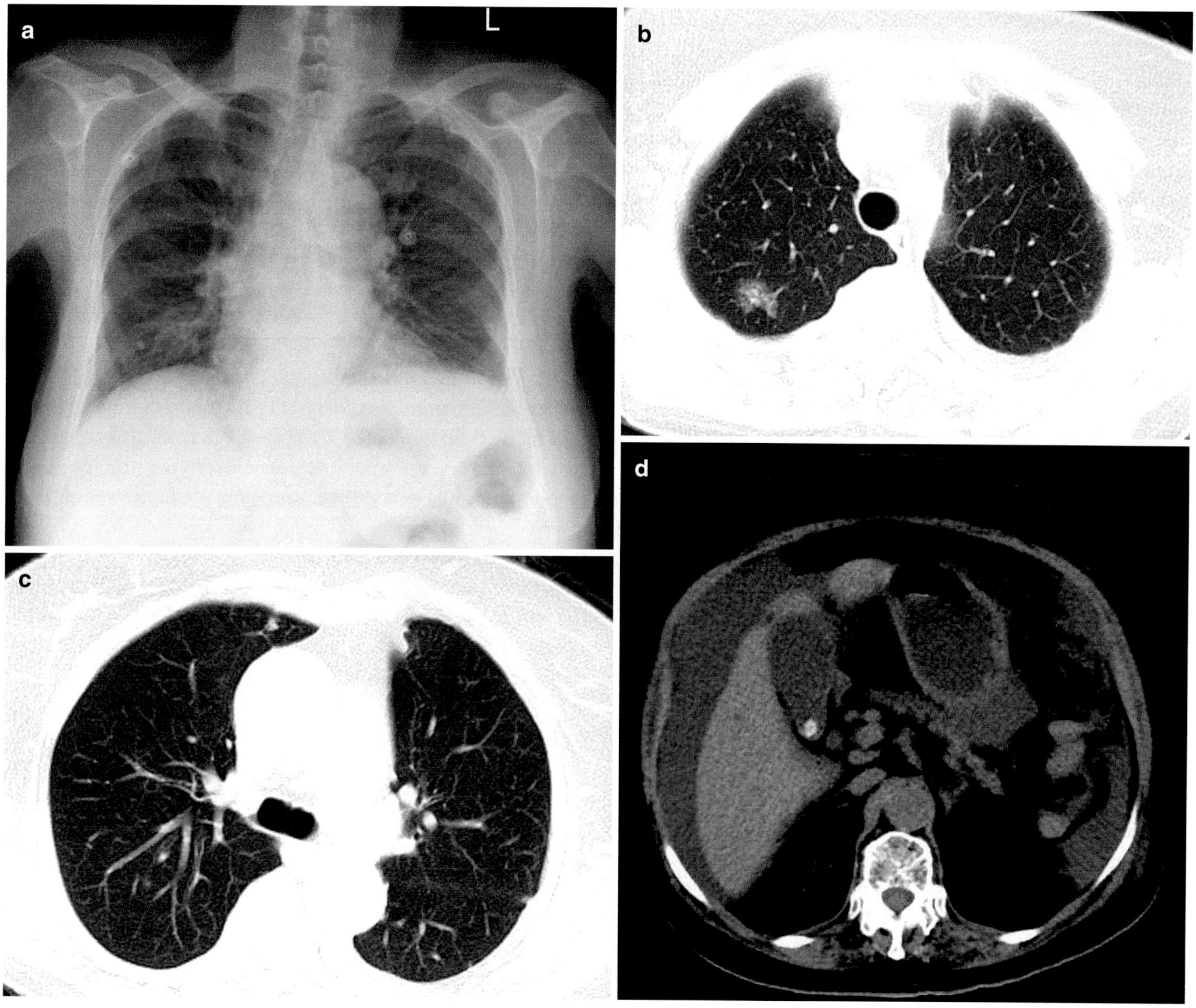

Fig. 9.1 (**a**) Chest X-ray demonstrated thickened and calcified pulmonary markings. (**b**, **c**) CT scanning visualized patches of blurry opacity at the upper right lung lobe with poorly defined boundary, and spots of calcification at the right middle lung lobe. (**d**) Mediastinal CT scanning showed liquid-density opacity around the liver and cholecystolithiasis

necrotic granuloma and is shown with small ring shaped enhancement by enhanced scanning. The patients with pulmonary tuberculosis commonly experience typical tuberculosis poisoning symptoms and can be treated by anti-TB therapy.

9.5 Trichinellosis

Trichinellosis (also known as trichinosis) is caused by parasitism of trichina, which is also nominated as Tricheinella speralis (Owen, 1835) or Railliet (1895), in the body of human or animals, such as wild boars, pigs, bears, and rats. Adult Trichina parasitizes in the small intestine of the host, commonly in the duodenum and jejunum, while its larva commonly parasitizes in striated muscle of the host, such as the jugomaxillary muscle, glossal muscle, diaphragm, intercostal muscle, gastrocnemius and biceps. Trichina is infective to the new host whose adult and larva develop in the body of the host. However, they only can continue the life circle of their next generation after transferring into another host. Therefore, the parasitized host by trichina is either a definitive host or an intermediate host. The host is commonly infected via intake of contaminated meat or meat products by encysted larva of trichina. The clinical symptoms during the acute phase include allergic responses such as fever, skin rash, and eyelid, followed by severe muscular pain, limbs fatigue and pain. In some serious cases, death may occur due to complications.

9.5.1 Pathogenic Process

Larva is mainly pathogenic whose pathogenicity is related to multiple factors, such as the intake quantity of larval cysts, the developing stage of larvae, the invading site of larvae, and immunity of the host. Among these factors, the intake quantity of larval cysts and the developing stage of larvae are more important. For cases of slight infection, such as those taking in 20–30 larval cysts, the patients are commonly asymptomatic. However, intake of thousands of larval cysts causes serious infection.

The pathogenic process of trichina includes three consecutive stages.

9.5.1.1 The Larval Invasion Stage (the Initial 1 Week)

When adults or excysted larvae of trichina gain their access into the intestinal tract, the adults trichina live by intestinal villi to excrete and secrete in large quantities. In addition, they reproduce larvae in a large quantity. The excretions, secretions and larvae stimulate the duodenum and jejunum to cause corresponding inflammatory responses, such as local edema, congestion, focal hemorrhage, and superficial ulceration. The lesions are generally slight.

9.5.1.2 The Larval Migration Stage (the Next 2–3 Weeks)

The newborn larvae penetrate the intestinal mucosa into the blood flow and their toxic metabolites can cause systemic toxic symptoms and allergic responses. When larvae invade muscle, the muscle fibers can be seriously damaged to show swelling and disarrangement as well as absence of transverse striation, mild interstitial edema, and inflammatory cells infiltration of different degrees, which further cause systemic vasculitis and myositis.

9.5.1.3 The Larval Encystment Stage (the Next 4–16 Weeks)

With the growth of larvae, the myocytes that they settle in enlarge into spindle shape. Due to the parasitism of larvae, the myocytes fail to regenerate, and the flat and thin myoblastic nucleus disappears and is transparent to finally form the inner wall of cystic wall, which is known as hyaline layer. The fibrous connective tissue that is around and is directly linked with myolemma is subject to hyperplasia to form an extremely thin outer layer of the cystic wall, which is known as the fibrous layer. Therefore, the outer layer of cystic wall is extremely thin, which is formed by hyperplastic surrounding fibrous connective tissue following responses of the host. The inner layer is relatively thick, which is transformed by myoblasts during the repair of damaged myocytes.

9.5.2 Clinical Manifestations

The incubation period of trichinellosis commonly lasts for 5–15 days, averagely 10 days. The disease shows diverse clinical manifestations and can be divided into three stages based on the staging of pathogenic process.

9.5.2.1 The Intestinal Stage

The trichina invades the intestinal mucosa to cause gastrointestinal upset. During the first week after onset, the patients may experience nausea, abdominal pain, diarrhea or occasionally constipation, and vomiting. Vomiting may occur 2 h after food intake, which may persist up for 4–5 weeks. The symptoms during this stage are commonly mild and tend to be neglected by patients. during this stage, the patients may also experience accompanying fatigue, aversion to cold, low-grade fever and other systemic symptoms.

9.5.2.2 The Acute Stage

The acute stage is characterized by persistent high fever, eyelid and face edema, allergic skin rash, blood eosinophilia, other allergic responses and systemic muscular pain. The patients commonly show persistent high fever 2 weeks after the onset, with a body temperature of 38–40 °C and commonly persisting for 2–4 weeks. In some serious cases, the high fever may persist for up to 6 weeks. The body temperature then gradually decreases. Concurrent to high fever, most patients experience edema around the eyes and face edema. In some serious cases, accompanying lower limbs edema and systemic edema may be found.

9.5.2.3 The Convalescent Stage

Along with the encystment of larvae in muscles, acute inflammation relieves and systemic symptoms also retreat. However, muscular pain may persist up to several months. Patients with serious conditions may experience cachexia, with collapse. Death may even occur due to toxemia or myocarditis.

9.5.3 Diagnosis

9.5.3.1 Etiological Diagnosis

Tissues are firstly harvested from the biceps or gastrocnemius of patient. The sections or imprints are then prepared for microscopic examination. Detected cyst and larvae are the basis for its etiologial diagnosis.

9.5.3.2 Immunological Diagnosis

The trichina shows strong immunogenicity and the commonly used immunological essays include intracutaneous test, circumlarval precipitin test (CPT), bentonite flocculation test (BFT), enzyme linked immunosorbent assay (ELIZA), indirect hemagglutination test, counterimmunoelectrophoresis, and indirect immunoperoxidase staining. Each of these tests has its advantages and the combination of 2 or 3 tests is recommended for more reliable diagnosis.

9.5.3.3 Clinical Manifestations

Typically, the symptoms and signs include fever, eyelid edema, muscular pain and obvious eosinophilia. In combination to epidemiological history, the diagnosis can be clinically made.

Case Study 1

[Brief Medical History]

A 26-year-old young man complained of fever for 3 months, with the highest body temperature being 39.0 °C. He caught a cold 3 months ago. After that, he experienced fever each afternoon, accompanied by slight cough and fatigue. These symptoms relieved after anti-inflammatory treatment. But the symptoms were recurrent. One week ago, persistent swelling and pain occurred at major joints of the limbs. The routine anti-bacterial therapy showed no favorable therapeutic responses. The antibody IgG of trichina was positive.

[Radiological Demonstrations] (See Fig. 9.2)

[Diagnosis]

Trichinellosis related lung lesions.

[Discussion]

Trichinellosis is a zoonosis caused by trichina, with common clinical manifestations of gastrointestinal symptoms, fever, and muscle pain. In the early stage when adults trichina live in intestinal tract of the host, symptoms commonly include nausea, vomiting, abdominal pain and diarrhea, which are generally slight and transient. In acute stage when larvae trichina migrate in blood flow of the host, the onset is acute with symptoms of fever and accompanying aversion to cold. Remittent fever and irregular fever are

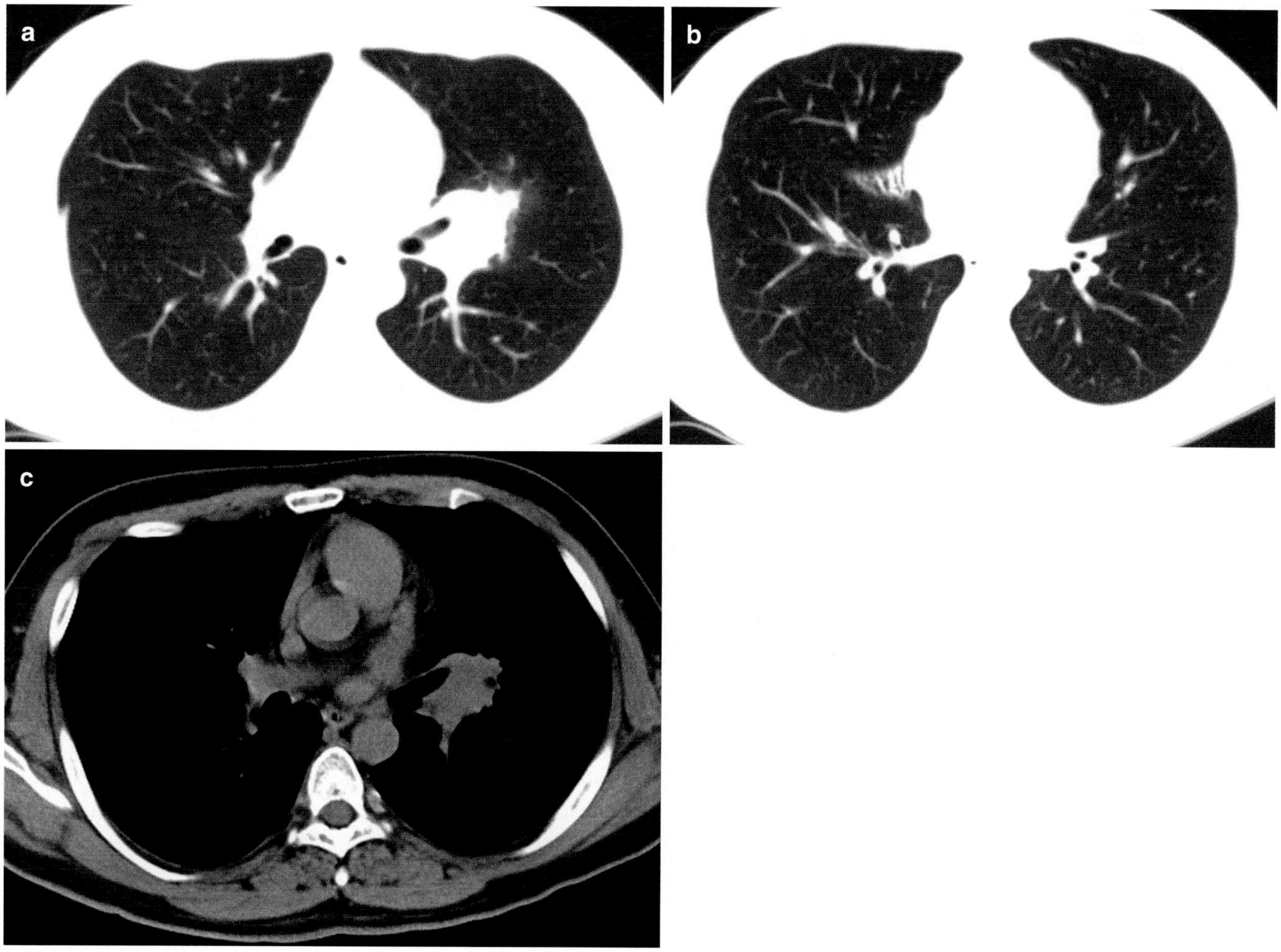

Fig. 9.2 (**a**, **b**) CT scanning demonstrated patchy high density opacity at the right middle lung lobe and nodular opacity at the left pulmonary hilum. (**c**) The left pulmonary hilum was demonstrated to be enlarged, with irregular opacity in soft tissue density and irregular boundary. The bronchus at the left upper lung lobe was shown to be compressed

clinically more common. Meanwhile, about 80 % of the patients experience edema, which is characterized by rapid progress. Skin rash is frequently concurrent with fever, which commonly occurs at the back, chest and limbs. Systemic muscle pain is severe. And the lung lesions may induce cough and lung rales. In the convalescent stage, along with encystment of larvae in muscles, acute symptoms of inflammation subside, with the systemic symptoms alleviating such as fever, edema and muscle pain. In this case, trichinia antibody showed positive, and the patient experienced fever with the highest body temperature of up to 39.0 °C, as well as persistent and recurrent swelling and pain at the major joints of the limbs. The patient also showed slight cough, and fatigue. CT scanning demonstrated enlarged left pulmonary hilum, irregular soft tissue density opacity with irregular boundary, compressed bronchus at the left upper lung lobe, and patchy high density opacity at the right middle lung lobe. Anti-inflammatory therapy showed unfavorable therapeutic efficacy and the symptoms were recurrent. In this case, the radiological findings were consistent with the clinical manifestations, both of which can be interpreted by acute stage of trichinellosis.

Trichinellosis should be differentiatd from: (1) pneumonia with mediastinal lymphadenectasis, which can be treated by routine anti-inflammatory therapy with no relapse and shows trichinia antibody IgG negative. (2) sarcoidosis, which shows bilaterally symmetrical enlargement of lung hilum and mediastinum and Kvein-Siltzbach skin test positive.

Case Study 2

A boy aged 4 years and 6 months complained of fever with a body temperature of around 38.0 °C but the cause was unknown. The conventional antipyretic therapy showed poor therapeutic efficacy and the patient still showed increased WBC count, being up to $15.0 \times 10^9/L$. After 3 days treatment for fever, his body temperature decreased to normal level but the WBC count was still as high as $20.0 \times 10^9/L$. After the patient was transferred to our hospital, bone marrow puncture for biopsy showed eosinophils 0.30, both IgG and IgM antibodies positive (+). His neutrophils count fluctuated and RBC count decreased, with slight anemia.

[Radiological Demonstrations] (See Fig. 9.3)

[Diagnosis]

The diagnosis was highly suspected to be trichinellosis.

[Discussion]

Cui, YX et al. proposed that patients with the following symptoms be highly suspected with trichinellosis and receive immunological assays to define the diagnosis: (1) fever, fatigue, and systemic muscle pain with no known causes; (2) irregular fever that is accompanied by headache, cough, eyelid edema, and eye conjunctival congestion, which fail to relieved by therapies treating upper respiratory tract infection; (3) fever and systemic allergic responses with no known causes; (4) increased eosinophil count with no known causes; (5) increased WBC count that is accompanied by fever, systemic muscle pain, and a dysfunctional organ. In this case, the patient showed 4 of the above 5 conditions. Further detection of trichinia IgG and IgM antibodies positive defined the diagnosis of trichinellosis.

Radiologically, the patient showed blurry flakes of low density opacity at the right liver lobe, which has quite even density but poorly defined boundary. These findings excluded the suspected diagnosis of simple hepatic cyst. Concerning the mesenteric multiple small lymph nodes, due to the absence of a medical history of neoplasm and inflammatory lesions, the possibilities of neoplasm and inflammation-induced lymphadenectasis were also excluded. In combination to the medical history of trichinellosis, these radiological signs are more likely to be induced by invasion of trichina into the liver and mesentery.

Trichinellosis should be differentiated from the following conditions: (1) Liver cyst, which can be demonstrated by CT scans with singular or multiple, round or oval, homogeneous low density opacity with smooth boundary; CT values being close to water; and no enhancement by contrast scans. In the cases with complicating hemorrhage or infection, the lesions may show increased density. (2) Liver metastatic neoplasm, which shows small lesions, obvious edema around the lesions, and obvious space occupying effect. The primary lesions can be detected.. (3) Cholangiocellular carcinoma, which originates from intrahepatic biliary epithelial cells and may occur in each liver lobe. Along with the growth of neoplasm, the lesions are susceptible to ischemic necrosis to show multiple cysts with different sizes and poorly defined boundary, which are commonly accompanied by intrahepatic biliary wall thickening and intrahepatic biliary dilation to invade adjacent vascular and liver tissues. The hepatic portal and abdominal lymph nodes are commonly enlarged. Distant metastasis is likely to occur. (4) mesenteric lymphoid metastasis that the patients commonly report a medical history of neoplasm.

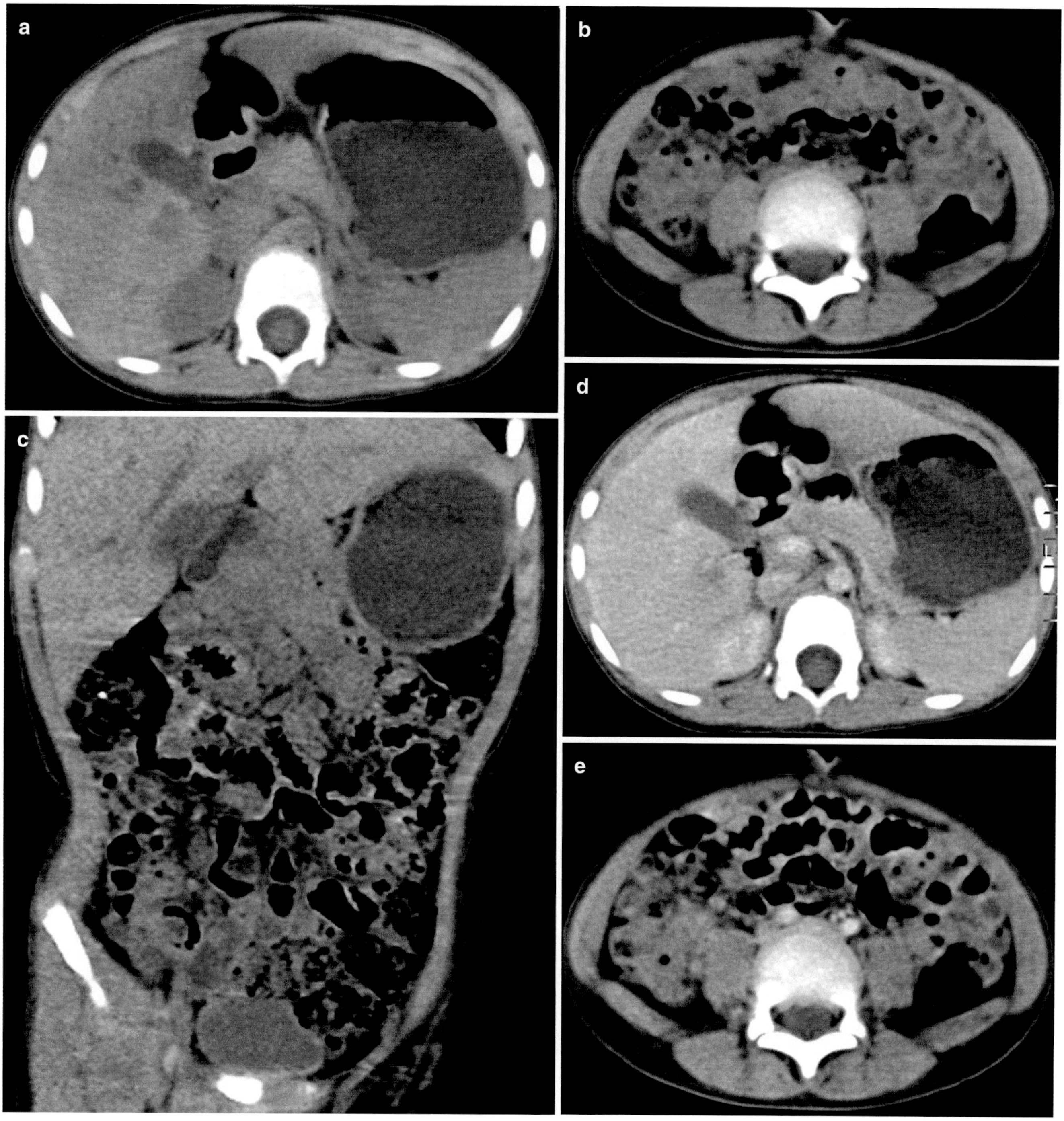

Fig. 9.3 (**a–c**) CT scanning demonstrated blurry flakes of low density opacity in the right liver lobe, multiple nodular lesions at the mesentery root, intestinal dilation and pneumatosis in the abdominal cavity. (**d**, **e**) Enhanced CT scanning showed slight enhancement of the lesions at the right liver lobe and enhancement of multiple small lymph nodes in the mesentery

Further Reading

Blacksin MF, Lin SS, Trofa AF. Filariasis of the ankle: magnetic resonance imaging. Foot Ankle Int. 1999;20(11):738–40.

Cano J, Rebollo MP, Golding N, et al. The global distribution and transmission limits of lymphatic filariasis: past and present. Parasit Vectors. 2014;7(1):466.

Cui YX, Guan JC, Xu XQ. Trichinellosis: a study of 38 cases. Chin J Misdiagnosis. 2002;11(2):1723.

Guo RZ. Atalas of infectious diseases and parasitic diseases in pathology. Guiyang: Guizhou Science and Technology Press; 2012.

Ji HJ, Wang H. Diagnosis and treatment of pleural effusion caused by filariasis. J Clin Pneumotol. 2013;18(1):112,–113.

Mendoza N, Li A, Gill A, et al. Filariasis: diagnosis and treatment. Dermatol Ther. 2009;22(6):475–90.

Wang GQ. Clinical guidelines for infectious diseases department. Beijing: Chinese Medical Science and Technology Press; 2012.

10 Trematodiasis

Bailu Liu, Li Li, Song Shu, Yi Xiao, and Jiangfeng Pan

Trematodiasis is a group of diseases caused by endoparasites and ectoparasites that belong to monogenea, aspidogastraea, and digenea of trematoda, a class within phylum Platyhelminthes. The trematodes of monogenea and aspidogastraea mainly parasitize in fish, amphibians, reptiles, mollusks and crustaceans. And the digenetic trematodes of digenea also parasitize in in mammals and birds. Trematodes the parasitize in human body are categorized into digenea, which is known as digenetic trematodes. Although digenetic trematodes have diversifying species, varying shapes, and complex life cycle, their basic structure and development share similarities.

10.1 Morphology

Most of adult digenetic trematodes have a flat body shape that appears like a leaf or tongue, with varying sizes. They are bilaterally symmetric, which is dorsally and ventrally flat, with oral sucker and ventral sucker. Some species of digenetic trematodes also have a genital sucker.

B. Liu
The Second Affiliated Hospital, Harbin Medical University, Harbin, Heilongjiang, China

L. Li
Beijing You'an Hospital, Affiliated to Capital Medical University, Beijing, China

S. Shu
Tianmen First People's Hospital, Tianmen, Hubei, China

Y. Xiao (✉)
Shanghai Changzheng Hospital, Shanghai, China
e-mail: xiaoyi@188.com

J. Pan
Jinhua Municipal Central Hospital, Jinhua, Zhejiang, China

10.2 Life Cycle

The digenetic trematodes have a complex life cycle, with generations and host transfer. In their life cycle, the digenetic trematodes necessarily go through alternation of sexual generation and asexual generation. The asexual generation generally parasitizes in mollusk, an intermediate host which is commonly gastropods (e.g. snails) or pelecypoda (e.g. mussel). The sexual generation mostly parasitizes in vertebrate, a definitive host. Despite of their complex life cycle with variance in different species, the life cycle is basically composed of egg, myracidium, sporocyst, redia, cercaria, metacercaria, excysted metacercaria (polypide in the cyst with no tail) and adult.

10.3 Pathogenicity

In China, trematodiasis that is caused by parasitism of trematodes in huaman body includes schistosomiasis japonica, clonorchiasis sinensis, paragonimiasis westermani, and fasciolopsiasis.

10.4 Schistosomiasis

Schistosomiasis is a disease caused by parasitism of schistosomes in human body. A total of six species of schistosomes can parasitize in human body, with schistosomiasis japonica prevailing in China. After access of cercaria of schistosoma japonicum to human body via skin or mucosa contacts to contaminated water, schistosomiasis japonica is caused by parasitism of schistosoma japonicum and its eggs in the portal system. The main pathological change is fomration of granuloma induced by deposition of eggs of schistosoma japonicum in large intestines, liver and the other tissues. In this section, we mainly introduce schistosomiasis japonica.

H. Li (ed.), *Radiology of Parasitic Diseases*, DOI 10.1007/978-94-024-0911-6_10

10.4.1 Epidemiology

10.4.1.1 Source of Infection

Its source of infection includes patients and animals with schistosomiasis japonica, which varies in different prevailing regions. Epidemiologically, patients and cattles with schistosomiasis japonica are the most important sources of its infection.

10.4.1.2 Route of Transmission

Schistosomiasis japonica is transmitted based on three necessary conditions: eggs of schistosoma japonicum in water; existence and multiplication of oncomelania, human or animal contact to contaminated water.

10.4.1.3 Susceptible Population

People are generally susceptible to the disease. Certain immunity can be acquired after its infection. Sometimes, the disease occurs collectively, showing an outbreak.

10.4.2 Pathogenesis and Pathological Change

All cercaria, larva, adult and egg within the life cycle of schistosoma japonicum can impair organisms to cause complex immunopathological responses.

Invasion of its cercaria into skin can cause dermatitis, with local occurrence of papules and pruritus, dilation and congestion of vascular vessels with hemorrhage and edema as well as peripheral infiltrations of neutrophils and monocytes. When larva migrate in human body, organs that they pass through develop vasculitis as well as capiliary embolism and rupture to produce local cellular infiltration and spots of hemorrhage. Commonly, its adult is not pathogenic, but may cause slight mechanical damages. Its egg is the major pathogenic factor of schistosomiasis, whose depisition in tissues of human body causes granuloma and fibrosis that are the main lesions of schistosomiasis.

Schistosoma japonicum mainly parasitizes in the inferior mesenteric vein and the vena superior of rectal hemorrhoids, with its eggs depositing in the submucosa of intestinal wall. The eggs then flow along with blood in the portal vein to reach the intrahepatic venous branches. Therefore, the lesions are more commonly located in the liver and colon. In addition, the deposition of eggs out of the portal system can also induce formation of lesions, namely ectopic lesion, which is more commonly located in the lung and brain.

10.4.3 Clinical Manifestation

According to the severity of infection, course of the disease, clinical symptoms, immunity of the host, different regions of egg deposition, schistosomiasis japonica can be categorized into acute schistosomiasis, chronic schistosomiasis, advanced schistosomiasis and ectopic schistosomiaosis.

10.4.3.1 Acute Schistosomiasis

Acute schistosomiasis is common in patients with primary infection and the patients usually have a definite history of contact to contaminated water. The patients may experience intermittent fever 1–2 months after contact to contaminated water, as well as poor appetite, nausea, abdominal pain, diarrhea, bloody purulent stool and other gastrointestinal symptoms. More than 90% of the patients with acute schistosomiasis show hepatalgia and hepatomegaly.

10.4.3.2 Chronic Schistosomiasi

Chronic schistosomiasi account for majority of the cases of schistosomiasis japonica in the affected regions. It commonly occurs in untreated patients after the sympotoms of acute schistosomiasis retreat or in patients with acquired partial immunity after repeated slight infections in the affected regions. The course usually lasts for above 6 months. The asymptomatic patients often show no obvious symptom and the disease is commonly detected occasionally. The symptomatic patients usually experience abdominal pain and diarrhea with accompanying hepatosplenomegaly.

10.4.3.3 Advanced Schistosomiasis

The course of advanced schistosomiasis often lasts for 5–15 years. According to clinical symptoms, advanced schistosomiasis is categorized into megalosplenia type, ascites type, colon granuloma type and dwarf type. A patient may concurrently show symptoms of two to three types.

10.4.3.4 Ectopic Schistosomiasis

Ectopic schistosomiasis mainly include pulmonary schistosomiasis, cerebral schistosomiasis and spinal schistosomiasis.

10.4.4 Detection

10.4.4.1 Laboratory Test

Etiological Examination

The findings of eggs and hatched myracidium in stool are the direct evidence supporting the diagnosis of schistomiasis.

Biopsy of rectal mucosa can also provide etiological evidence for the diagnosis.

Immunoassay

Intracutaneous test is usually used for field screening to define the suspected cases. The cases with intracutaneous test positive should receive further examination.

Indirect hemagglutination test is one of the immunoassays for the screen of this disease.

Circunoval precipitin test and ELISA can also be used for comprehensive diagnosis of the disease.

10.4.4.2 Radiological Examination

Ultrasound

Ultrasound is an important examination in detecting hepatic schistosomiasis. B-mode ultrasound can demonstrate morphological changes of lesion in the liver, help in assessing the severity of hepatic fibrosis, and locate for liver tissue puncture and biopsy. Color doppler ultrasonography can be chosen to observe the conditions of blood flow and portal hypertension.

X-Ray

X-ray is mainly applied for the diagnosis of pulmonary schistosomiasis.

CT Scan

CT scan can help in assessing the severity of hepatic fibrosis and it can favorably demonstrate calcification of the liver and intestinal wall. CT scan also shows important value in the diagnosis of pulmonary and cerebral schistosomiasis.

MR Imaging

MR imaging is mainly applied to detect the lesions of schistosomiasis in the central nervous system and abdomen.

Case Study 1

[Brief Medical History]

A 45-year-old man complained of abdominal distension for more than 2 months. Circunoval precipitin test for schistosoma showed positive.

[Radiological Demonstration] (See Fig. 10.1)

[Diagnosis]

Hepatic schistosomiasis.

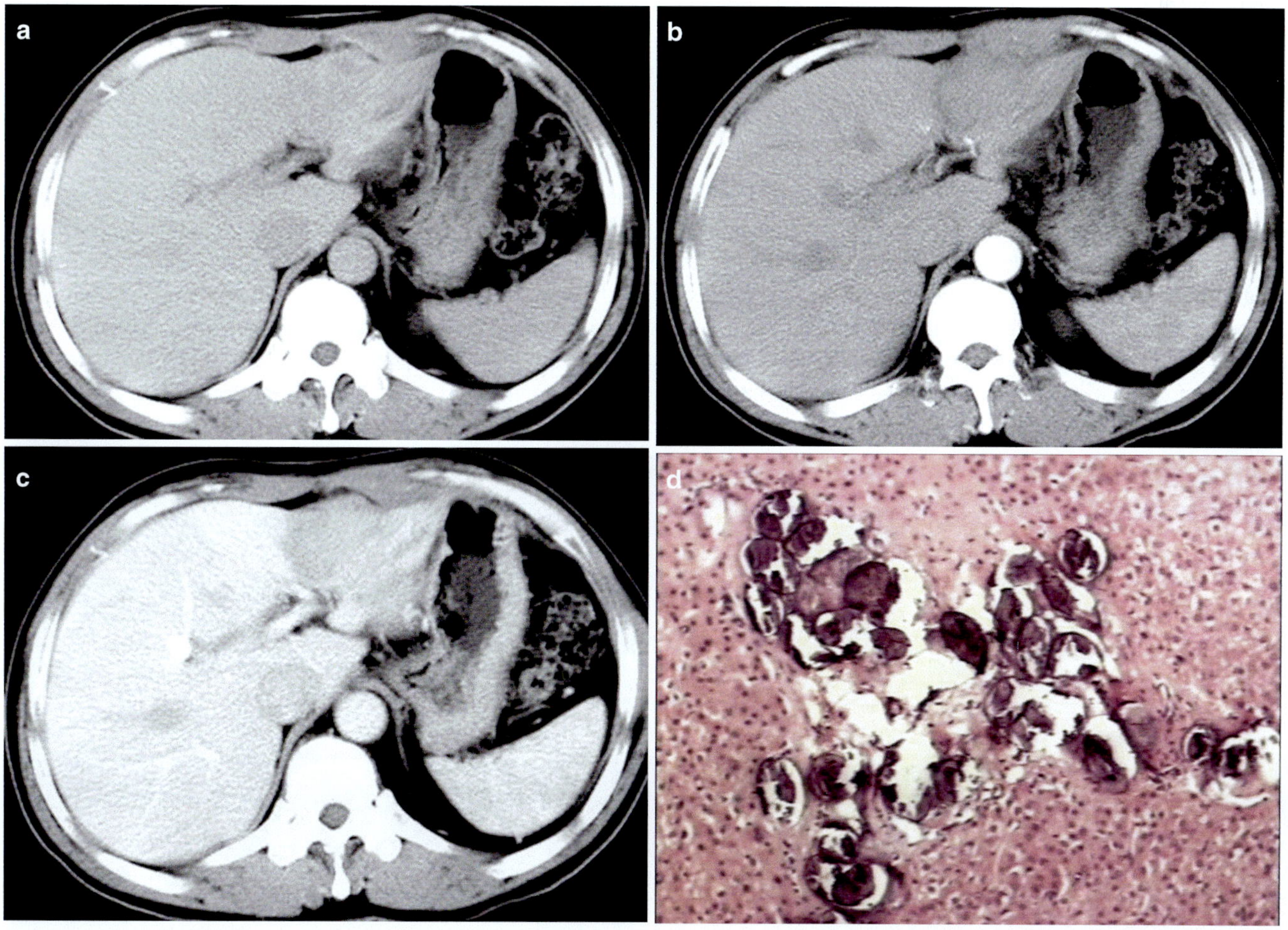

Fig. 10.1 (**a**) Plain CT scan demonstrated poorly defined heterogeneous low density opacity in the left liver. (**b**, **c**) Contrast scan demonstrated the lesion with mild heterogeneous enhancement at the arterial and portal vein phases. (**d**) Postopertive pathology demonstrated granuloma with deposition of schistosoma eggs and calcification (Reprint with permission from Hongjun Li, Radiology of Infectious diseases, Volume 2, 2015)

[Discussion]

In the advanced stage, also known as hepatocirrhosis stage, schistosomiasis is manifested with intrahepatic nodular low density opacity due to the formation of inflammatory granuloma. And by contrast scan, ring enhancement is demonstrated. Hepatic schistosomiasis with low density lesion in the liver should be differentiated from amebic liver abscess. Clinically, the patients with amebic liver abscess experience fever, hepatalgia and other symptoms. Ultrasound or CT scan demonstrates space occupying lesion with fluid density in the liver. And by liver puncture, typical chocolate like pus is shown. Amoeba trophozoite can be detected in the stool.

Case Study 2

[Brief Medical History]

A 55-year-old woman complained of upper abdominal upset for 1 week, hematochezia for 1 day and hematemesis for 5 h. Laboratory tests revealed WBC 7.34×10^9/L, and Hb 46 g/L. Eggs of schistosoma were detected in the stool. The patient had a history of living in a schistosomisis affected region.

[Radiological Demonstrations] (See Fig. 10.2)

[Diagnosis]

Chronic hepatic schistosomiasi.

[Discussion]

Chronic hepatic schistosomiasi is mainly caused by deposition of schistosoma eggs in the liver with manifestations of hepatocirrhosis and portal hypertension due to terminal blockage of the intrahepatic portal vein, hyperplasia of fibrous tissue, and a large quantity of calcified egg nodules. CT scan is the most commonly applied for its diagnosis, with intrahepatic calcification as the characteristic sign of advanced schistosomiasis. Generally, the calcification of intralobular fibrous septa is demonstrated to be linear, the calcification of liver capsule to be curved, and the calcification around portal vein in the portal area to be mass like and crab like. These calcifications may be demonstrated as map like change. Due to recurrent formation of egg nodules in a large quantity and hyperplasia of fibrous tissue, a series of signs are shown to be liver cirrhosis, such as disproportional liver lobes commonly with enlarged left lobe and shrun right lobe. Contrast scan demonstrates the liver with septa like enhancement, which may be related to formation of egg granuloma. In some cases, capsular enhancement and amorphous enhancement can be demonstrated Calcification in portal vein system may be caused by deposition of some eggs on the vascular wall of the splenic vein and the superior mesenteric vein due to portal hypertension. The spleen is shown to be obviously enlarged, which further develops into megalosplenia in advanced stage. MR imaging is usually applied to show liver cirrhosis of hepatic schistosomiasis. However, MR imaging can unfavorably demonstrate calcification in the liver. Therefore, MR imaging is not recommended for the cases of suspected hepatic schistosomiasis.

Chronic schistosomiasis with liver cirrhosis should mainly be differentiated from liver cirrhosis after hepatitis. Liver cirrhosis after hepatitis is demonstrated with shrinkage of liver, disproportional liver lobes and wave like changes of the hepatic margin but no linear calcification in the subcapsular and parenchymal areas. CT plain scan demonstrates schistosomiasis with liver cirrhosis with characteristic signs of linear, crab-like and map-like calcifications. In addition, the liver may be subject to enlargement or shrinkage with enlarged left lobe and shrunk right lobe. Otherwise the liver keeps its normal size. In some rare cases, imaging

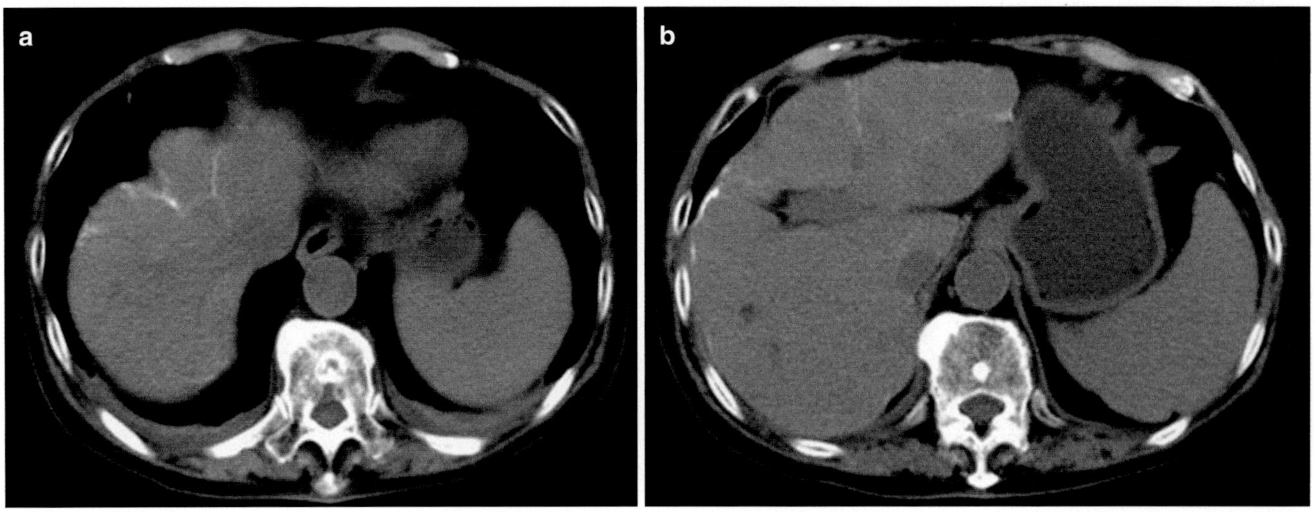

Fig. 10.2 (**a**) CT plain scan demonstrated cirrhosis of liver, wave like changes of the hepatic margin, linear calcification in liver parenchyma, and bilateral pleural effusion. (**b**) The liver was shown with calcification of liver capsule and multiple calcifications of the fibrous septa

demonstrations are atypical, showing only fine linear and spots of capsular calcification.

Case Study 3

[Brief Medical History]

A 64-year-old man had a history of schistosomiasis for 30 years.

[Radiological Demonstrations] (See Fig. 10.3)

[Diagnosis]

Hepatic schistosomiasis.

[Discussion]

In the advanced stage of schistosomiasis, the liver is shown with liver cirrhosis of different severity, namely schistosomiasis with liver cirrhosis. Calcification of the liver is a basic pathological change of schistosomiasis with liver cirrhosis and the major evidence for its diagnosis by CT scan. Liver cirrhosis commonly occurs in the right liver lobe, with capsular calcification as its the earliest and the most obvious sign. Pathologically, calcification in the liver occurs due to deposition of eggs in a large quantity in the subcapsular area and intervenous space, formation of fibrous scar tissue and degeneration of the eggs. The calcifications are morphologically diversified, including linear, reticular, crab-like, map-like, mass-like or subcapsular, which are more serious in the peripheral liver than in the central liver. These features pathologi-

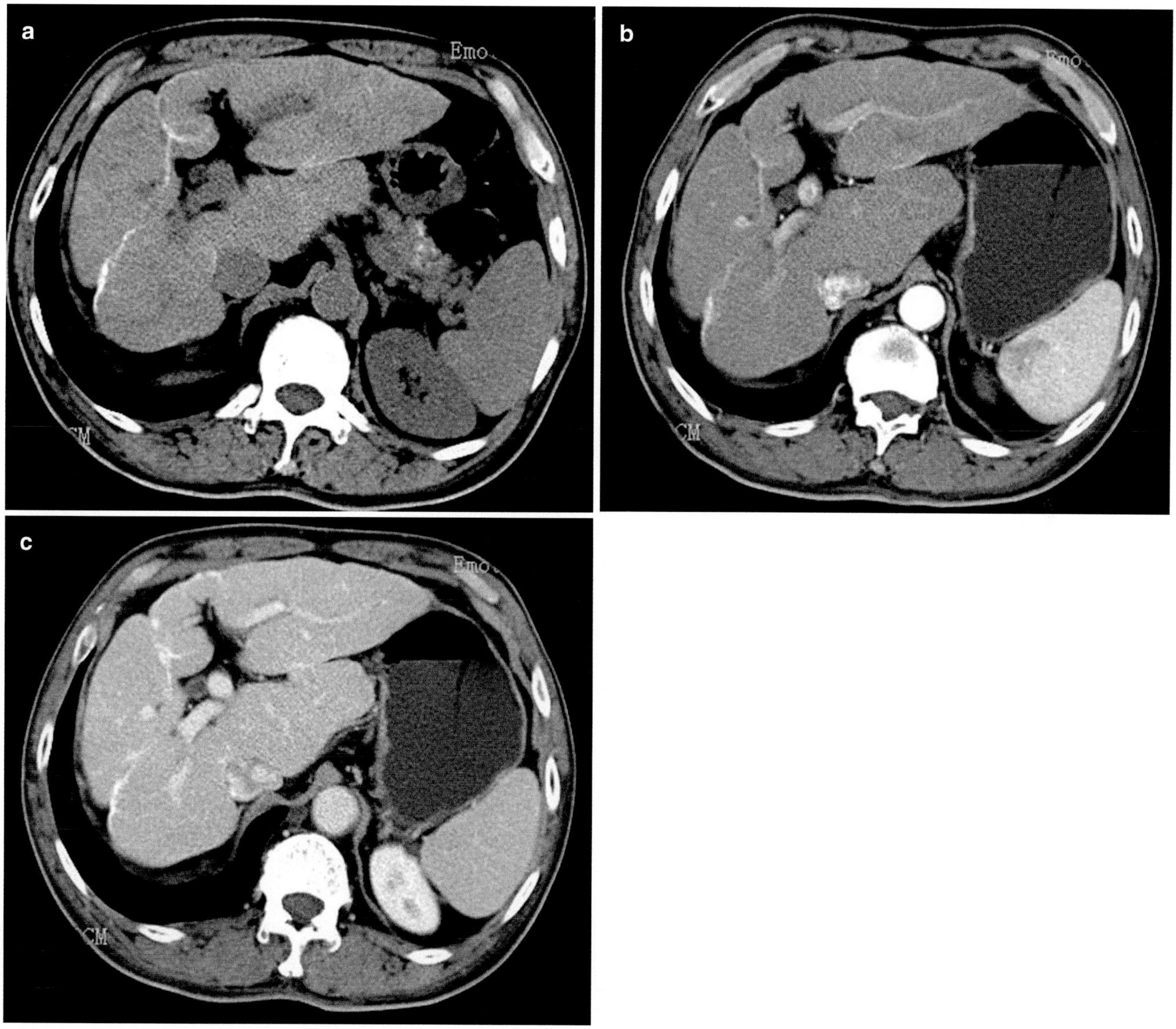

Fig. 10.3 (**a**) Plain CT scan demonstrated malformation of the liver, enlarged left and caudal lobes, increased density of liver parenchyma and strips of calcification in liver and its capsule. (**b**, **c**) Contrast scan demonstrated relatively homogeneous enhancement of the liver parenchyma, and linear enhancement of the intrahepatic septa (Reprint with permission from Hongjun Li, Radiology of Infectious diseases, Volume 2, 2015)

cally related to trunk liver cirrhosis, that it is, branches like distribution of fiber hyperplasia along the portal vein. Calcification at different sites has their own features, with subcapsular calcification often to be linear, and calcification around the portal vein in portal area to be mass-like. If calcification extends along the surface of lobule, crab-like calcification is shown. If interlobular calcifications connect, curved calcification is revealed. Extensive linear calcifications may intertwine into a map like or reticular appearance, which is the most serious calcification. Multiple calcifications with diversified appearance may be concurrent.

Case Study 4
[Brief Medical History]

A 30-year-old man complained of cough and chest pain for 2 days. Laboratory tests revealed obviously increased eosinophils in the peripheral blood, and eggs of schistosoma in stool.

[Radiological Demonstrations] (See Fig. 10.4)

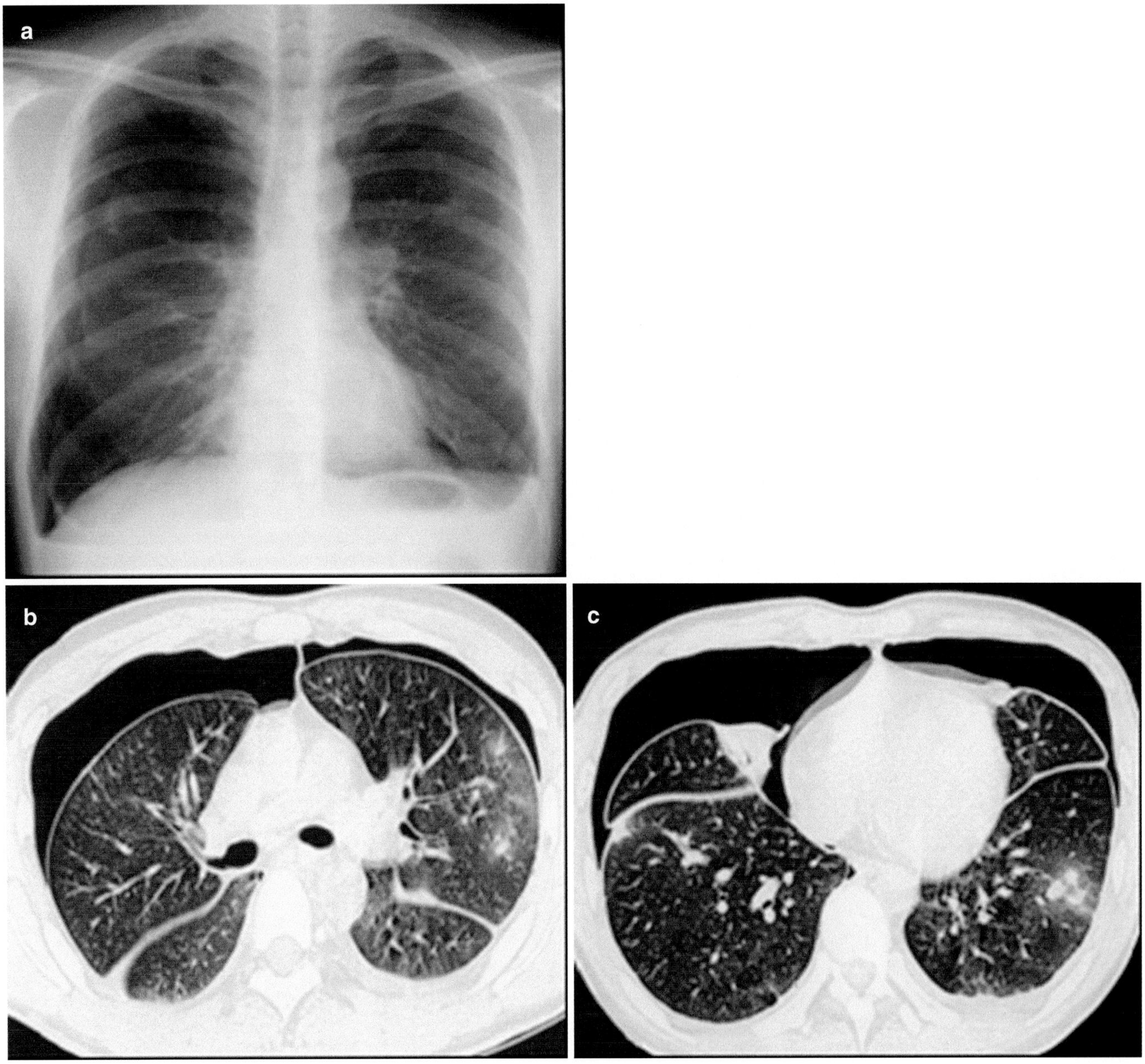

Fig. 10.4 (**a**) X-ray demonstrated scattered poorly defined patches of opacity in both lungs, and bilateral pneumothorax. (**b**, **c**) Plain CT scan demonstrated scattered nodular opacity in both lungs, with higher central density and surrounding ground-glass opacity in a halo sign. Cords like opacity were shown to connect with pleura. And bilateral pneumothorax was revealed (Reprint with permission from Hongjun Li, Radiology of Infectious diseases, Volume 2, 2015)

[Diagnosis]

Chronic pulmonary schistosomiasi, and bilateral pneumothorax.

[Discussion]

Pulmonary schistosomiasi commonly occurs in patients with acute schistosomiasi, which is caused by pulmonary interstitial lesion due to deposition of eggs. The respiratory symptoms are often mild and are usually overlapped by systemic symptoms, such as mild cough and dull chest pain. The pulmonary signs are often not obvious. The pathogenesis of pulmonary schistosomiasi is mainly mechanical damage caused by penetration of larva through pulmonary tissue and delayed cell-mediated allergic responses caused by egg granuloma. The pulmonary lesions include interstitial egg granuloma and inflammatory infiltration of peripheral alveoli.

The pulmonary lesions caused by acute schistosomiasis can be divided into early stage lesions and late stage lesions. In the early stage, the lesions are caused by mechanical damage due to access of cercariae and adults schistosoma to the pulmonary tissue and allergic responses to their metabolites. Most of the patients only experience increased and thickened lung markings with early onset and rapid retreat that usually persist for 2–3 weeks. About 2–3 months after infection, the disease develops into its late stage with deposition of eggs in the pulmonary interstitium to form predominately pseudo-nodules. During the late stage, the lungs are shown with scattered heterogeneous density poorly defined miliary opacity with a diameter of 2–5 mm. CT scan demonstrates chronic pulmonary schistosomiasis with fissure like exudative opacity in lung fields, multiple fibrous cords opacity in lungs, and typical nodular or micro-nodular opacity in lungs. The nodules commonly distribute in middle lower lung field, under the pleura or bifurcating point of trachea, with central high density, poorly defined margin and surrounding ground-glass opacity in a halo sign. CT scan can also demonstrate pulmonary interstitial fibrosis and pulmonary hypertension along with prolongation of the illness course.

Case Study 5

[Brief Medical History]

A 45-year-old man complained of intermittent right low abdominal pain and irregular bowel movements for 4 months. The stool was loose paste like with mucus. And he experienced no hematochezia and tenesmus. The abdominal pain was relieved after bowel movements. The patient had a medical history of schistosomiasis and received etiological treatment.

[Radiological Demonstrations] (See Fig. 10.5)

[Diagnosis]

Intestinal schistosomiasis.

[Discussion]

Colon is the passage for discharge of schistosoma eggs, therefore, it is commonly involved by schistosomiasis. Pathologically, the lesion shows submucosal hyperplasia of connective tissue in a large quantity and diffuse fibrosis with calcified eggs, hyperplasia and thickening of mucosa. CT scan is the most common used radiological examination for its diagnosis, which can demonstrate thickening of the intestinal wall and calcification of the colonic wall. Hyperplasia of colonic mucosa in the cases of chronic schistosomiasis may be cancerated. In the cases with relatively large granuloma of intestinal schistosomiasis, the intestinal lumen may be narrowed to cause intestinal obstruction, which is the main reason for prevailing large intestinal obstruction in schistosomiasis affected regions. Deposition of eggs in the mesenteric root induces fibrosis responses to cause thickening and shrinkage of mesentery, and fan-shaped shrinkage of intestinal tube toward the mesenteric root. Generally speaking, calcification of the intestinal wall is a typical sign of intestinal schistosomiasis. In combination to positive findings of eggs and cercaria in stool, the diagnosis can be defined.

Intestinal schistosomiasis is mainly manifested as granuloma formed by deposited eggs in intestinal submucosa and consequent calcification, thickened intestinal wall and narrowed intestinal lumen. Therefore, it should be differentiated from the following diseases.

1. **Intestinal tuberculosis**
   The lesion of intestinal tuberculosis is commonly located in the ileocecal junction and terminal ileum, and the disease is secondary to pulmonary tuberculosis. Ulcerative signs are relatively rare but the intestinal wall slightly thickens with a long range of involvement. Inflammatory granuloma is restricted and smooth, with narrowed, malformed and shortened intestinal tube. The mesenteric lymph nodes are subject to enlargement and calcification.
2. **Crohn's disease**
   The lesion of Crohn's disease is mainly located in the right colon, with the rectum generally not involved. Its characteristic signs include segmental and continual mucosa with thickening like paving stones and formation of fistula in its advanced stage.

Case Study 6

[Brief Medical History]

A 67-year-old man experience right lower abdominal pain with no known causes 1 month ago, which was persistent dull pain. But he experienced no fever, jaundice, pantothenic acid, eructation, diarrhea, hematemesis, bloody stool and anal exsufflation but discontinued bowel movement. So he paid no special attention to the condition. However, the

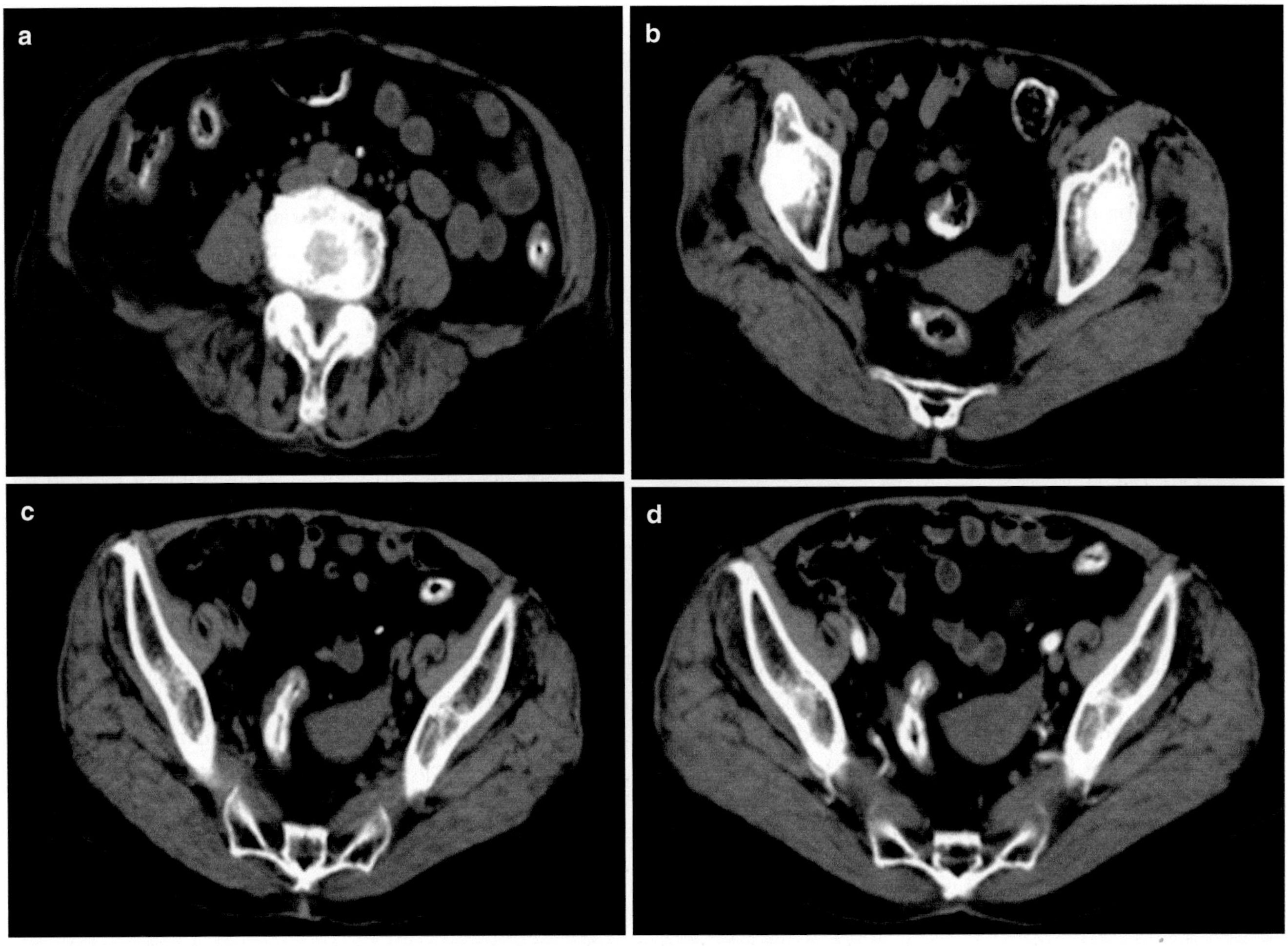

Fig. 10.5 (**a**) Plain CT scan demonstrated wall calcification of the ascending colon, transverse colon and descending colon. (**b**) Irregular wall calcification in the rectum and sigmoid colon. (**c**) Thickening of the sigmoid colon wall with calcification and luminal stenosis. (**d**) Contrast scan demonstrated wall thickening and calcification of the sigmoid colon, and enhancement of the thickened wall

abdominal pain aggravated 1 week ago and he paid clinic visit in a local hospital for treatment. Anti-inflammatory fluid infusion was administrated and his conditions were relieved. Abdominal CT scan demonstrated lump in peritoneum. For further diagnosis and treatment, the patient was hospitalized into our hospital. The patient had a medical history of schistosomiasis 40 years ago and the condition was improved after treatment. Laboratory tests on January 17th, 2007 revealed WBC 5.2×10^9/L, neutrophil 0.572, RBC 3.06×10^{12}/L. The patient received more tests and examinations during his hospitalization. An exploratory laparotomy was performed on January 23th, 2007, which revealed lump inferior to the terminal ileal mesentery in size of 4×3 cm. The lump was hard with poorly defined boundary and intact root, which was suspected to be appendiceal mass. Intraoperative frozen pathology showed appendiceal schistosomiasis. Paraffin pathology on January 26th, 2007 revealed acute appendiceal cellulitis with accompanying deposition of schistosome eggs in tissue.

[Radiological Demonstrations] (See Fig. 10.6)

[Diagnosis]

Appendiceal schistosomiasis with acute appendiceal cellulitis and intestinal schistosomiasis.

[Discussion]

Human is the definitive host of schistosoma japonicum and can be infected via contact to its cercaria on the surface of water. After the cercaria gains its access into human body, it flows along with blood to settle in the superior and inferior mesenteric veins where the male and female copulate, followed by their oviposition in venules of intestinal wall. Along with increased intravenous pressure, the venules may rupture to release the eggs to appendiceal submucosa. In addition to liver and colon, appendix is also a common region for the eggs deposition in patient with chronic or advanced schistosomiasis, manifested as appendiceal schistosomiasis. In this case, the disease was complicated by appendicitis, with an underlying mechanism that enzymes produced by eggs induce necrosis of appendical mucosa

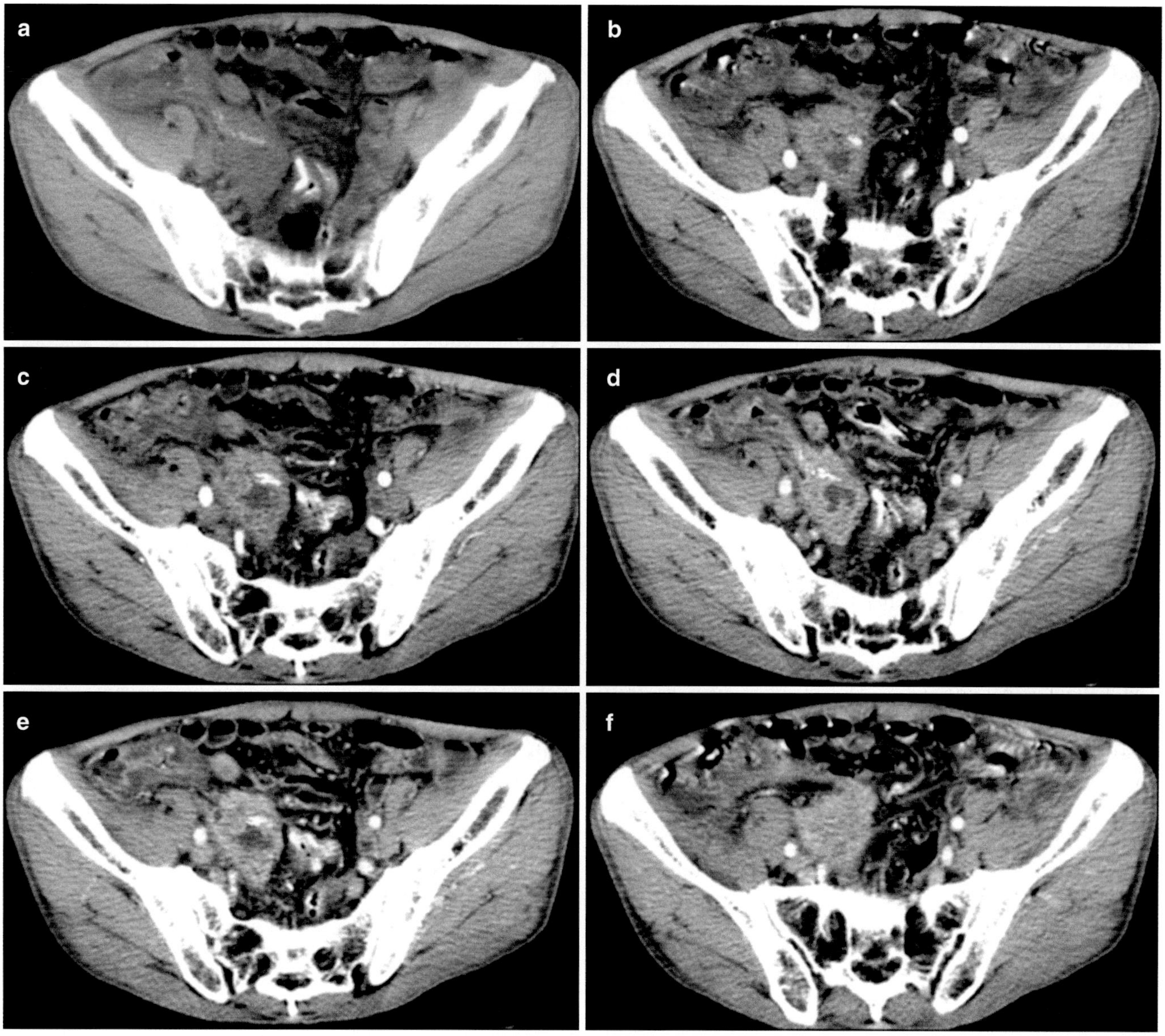

Fig. 10.6 (**a**-**f**) Plain CT scan demonstrated a lump in the terminal ileal mesentery of right lower quadrant, thickened intestinal wall of ileocecal junction, linear calcification of the colonic wall. Contrast scan demonstrated heterogeneous enhancement of the lesion with central low density and marginal moderate enhancement. The lump was shown at the bifurcation of right common iliac artery, with its posterior wall being closely related to the vascular vessels lateral and medial to the right ilium and its anterior part being poorly defined from some intestinal tubes (Reprint with permission from Hongjun Li, Radiology of Infectious diseases, Volume 2, 2015)

whose falling into enteric lumen finally causes obstruction of appendiceal cavity. The matured eggs contain myracidium whose cephalic gland can produce toxins to cause necrosis of appendiceal wall and subsequent eosinophilic abscess. Its further progression induces formation of scar fibrosis or calcification that finally narrowed and obstructed appendiceal cavity. Bacterial infection is then induced with occurrence of appendicitis. Therefore, when appendiceal schistosomiasis is complicated by acute appendicitis, the vascular lumen of appendix is subject to deposition of schistosoma eggs or embolism to cause angiosclerosis and mucosal atrophy. In addition to the occurrence of fibrous change in appendiceal wall, the appendiceal cavity is partially obstructed to cause appendicular perforation and subsequent peritonitis. The patient of this case suffered from schistosomiasis 40 years ago and the condition was improved after treatment. And presently, the disease lasted for about 1 month with aggravated abdominal pain for 1 week. The abdominal pain was relieved after anti-inflammatory fluid infusion. Preoperative CT scan demonstrated lump with abscess under the terminal ileal mesentery in the right lower quadrant. Intraoperative frozen pathology indicated a diagnosis of appendiceal schistosomiasis and paraffin pathology indicated a diagnosis of acute appendiceal cellulitis with deposition of eggs in tissue. Based on these findings, the disease was finally diagnosed as appendiceal schistosomiasis with acute cellulitis. The disease is mainly differentiated from different mesenteric neoplasms occurring in this area. Generally speaking, such neoplasms are larger in size than the lesion of appendiceal schistosomiasis, with more obvious enhancement by contrast scan, blurrier boundary and aggressiveness. Moreover, literature reports have demonstrated that the lesion of appendiceal schistosomiasis may be cancerated. The clinically suspected cases based on radiological signs should received surgical operation as early as possible.

Case Study 7

[Brief Medical History]

A 70-year-old male peasant who lived in the suburb of Shanghai, China, was detected with occult blood in stool positive in a community health survey on 10th last month. The patient experienced no stool changes such as fresh bloody stool, tenesmus and poor appetite. He then paid his clinic visit in Jinshan Hospital affiliated to Fudan University, Shanghai, China, and received endoscopy which revealed multiple polyps in colon. He then was hospitalized in our hospital due to a suspected diagnosis of hemorrhage in the lower gastrointestinal tract. The laboratory tests on admission showed PA 108 g/L, TP 60 g/L, ALb 29 g/L, A/G 0.94, Hb128g/L, and HCT 0.376. An exploratory surgery revealed a shrunk hard lump with a diameter of 2.5 cm in the middle part of sigmoid colon and the serosa involved. The sigmoid colon was showen with obvious adhesion to the lateral abdominal wall and the adhesion was surgically treated to restore their original anatomic relationship. During the surgery, the mesentery of sigmoid colon and some anterior sacral fascia were dissected; and the corresponding vascular vessels in the mesenteric area were ligated and transfixed. The mesentery was crispy with edema. After the respective vascular vessels in the mesenteric area were ligated, the intestine tube was transected about 5 cm proximal and distal to the lump and the intact specimen was sent for pathological examination. And the pathological diagnosis was grade II adenocarcinoma (ulcerative type) with the malignancy infiltrating extraserosal fibrous adipose tissue, obvious adhesion of sigmoid colon to lateral abdominal wall, and sporadic deposition of schistosoma eggs in the intestinal wall.

[Radiological Demonstrations] (See Fig. 10.7)

[Diagnosis]

Malignancy in the sigmoid colon with intestinal schistosomiasis and schistosomiasis live cirrhosis.

[Discussion]

Pathological studies have demonstrated that granuloma and fibrosis are the predominate pathological changes of schistosomiasis. Chronic schistosomiasis can induce damages to multiple abdominal organs, especially liver and colon. Human is the definitive host of schistosoma japonicum and can be infected by contacts to the surface of water with cercaria of schistosome. After cercaria gains its access into human body via skin, it migrates to the superior and inferior mesenteric veins along with blood flow, where the male and female copulate. After that, they lay eggs in venule of intestinal wall. Due to the anatomic shunt, the eggs deposit in the hepatic portal area and the branches of portal vein after passing through the portal system, which induces liver cirrhosis. Liver cirrhosis after schistosomiasis is common manifested with compensatory enlargement of the left liver lobe, which is different from liver cirrhosis after hepatitis with shrunk right liver lobe and simplex enlargement of the caudate lobe. In combination to the case history, such a difference serves as a characteristic sign for their differential diagnosis. In addition, calcification with diversified shapes in the liver is another important evidence of schistosomal hepatopathy. Schistosome eggs are discharged via colon whose deposition on the colonic wall causes fibrosis of the intestinal wall. CT scan demonstrates thickening of the intestinal wall with linear and track like calcifications, which is a characteristic sign of chronic intestinal schistosomiasis. In this case of chronic intestinal schistosomiasis complicated by colonic adenocarcinoma, CT scan demonstrated obviously irregular thickening of the intestinal and soft tissue lump with calcification in addition to signs of chronic intestinal schistosomiasis, which indicates correlationship between their occurrences. Literature reports have demonstrated that, compared to simplex colorectal cancer, schistosomal enteritis related colorectal cancer occurs at an earlier age, with higher pathological staging, obviously higher percentage of stages III and IV, and higher percentage of mucous adenocarcinoma. These findings further proved that the incidence of colorectal cancer is high in the schistosomiasis affected region and endemic schistosomiasis is a high risk factor of colorectal cancer. Therefore, in clinical practice, follow-ups of the clinical, radiological and endoscopic data should be paid focused attention for patient with schistosomal enteropathy. The lesion of schistosomal enteritis related colorectal cancer often grows in a polyp like manner or protruding manner, which is consistent with the enteroscopic demonstrations of the case we reported. In this case, the patient was an elderly peasant who lived in the suburb of Shanghai, China, where schistosomiasis occurred with a high incidence. In a population health investigation for colorectal cancer, he was detected with stool occult blood positive and CT scan demonstrated

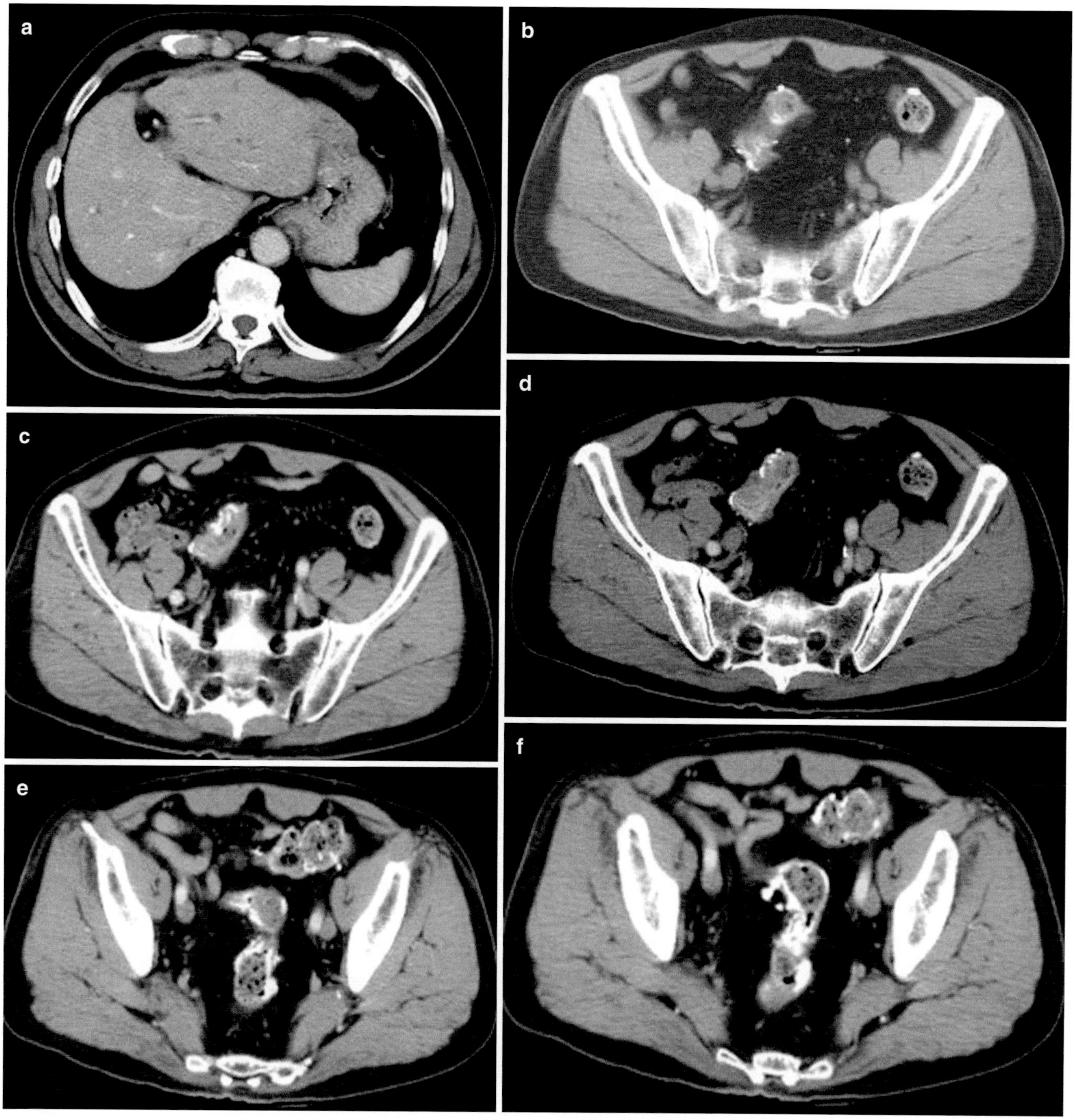

Fig. 10.7 (**a**) Abdominal Contrast CT scan demonstrated irregular shape of the liver, linear calcification at the margin, and disproportional liver lobes with enlarged caudate and left liver lobes, which indicated schistosomiasis with liver cirrhosis. (**b–f**) Pelvic CT scan demonstrated multiple linear and track like high density calcification opacity in the colon and rectum, with accompanying multiple cystic diverticulum. The wall of sigmoid colon was show to be obviously thickened with luminal stenosis, stiff morephology, adhesion to the right abdominal wall, soft tissue lump at the intestinal wall with accompanying enhancement. The lesion was revealed with poorly defined serosal margin, which was considered to be penetration of the lesion into serosa

typical signs of schistosomal liver cirrhosis and enteropathy, with the lesion being located in the sigmoid colon with soft tissue lump in the intestinal wall and its enhancement. Therefore, the preoperative diagnosis was defined to be schistosomal enteropathy with space occupying lesion.

Case Study 8
[Brief Medical History]

A 51-year-old man complained of headache and dizziness for more than 20 days, and hand convulsion for 4 days. The circumoval precipitin test (CODT) for schistosomiasis showed titer 1:40 positive (+). The patient lived in a schistosomisis affected region and had a history of contact to contaminated water. Surgical tissue harvest for biopsy indicated cerebral granuloma of schistosomiasis.

[Radiological Demonstrations] (See Fig. 10.8)

[Diagnosis]

Cerebral schistosomiasis with schistosomal granuloma in the left parietal lobe.

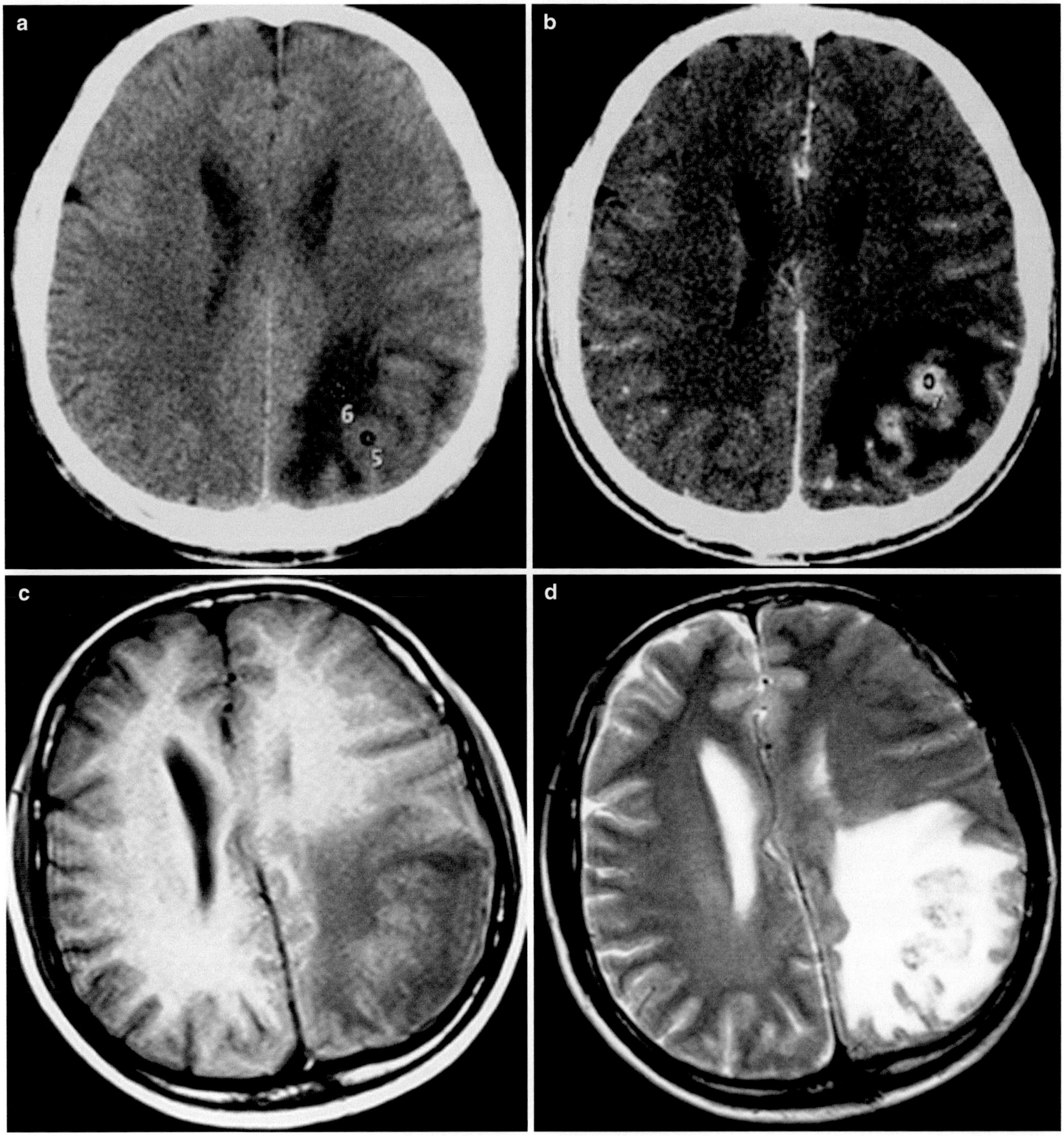

Fig. 10.8 (**a**) Plain CT scan demonstrated the lesion in the left parietal lobe, with equal density to the cerebral cortex and surrounding obvious fingerstall like edema. (**b**) Contrast scan demonstrated several spots of enhanced nodules that clustered together. (**c**, **d**) Plain MR imaging demonstrated the lesion with equal signal and surrounding obvious fingerstall like edema. (**e**, **f**) Contract MR imaging demonstrated several spots of enhanced nodules that clustered together in the left parietal lobe with central linear enhancement, showing a tree-branches sign

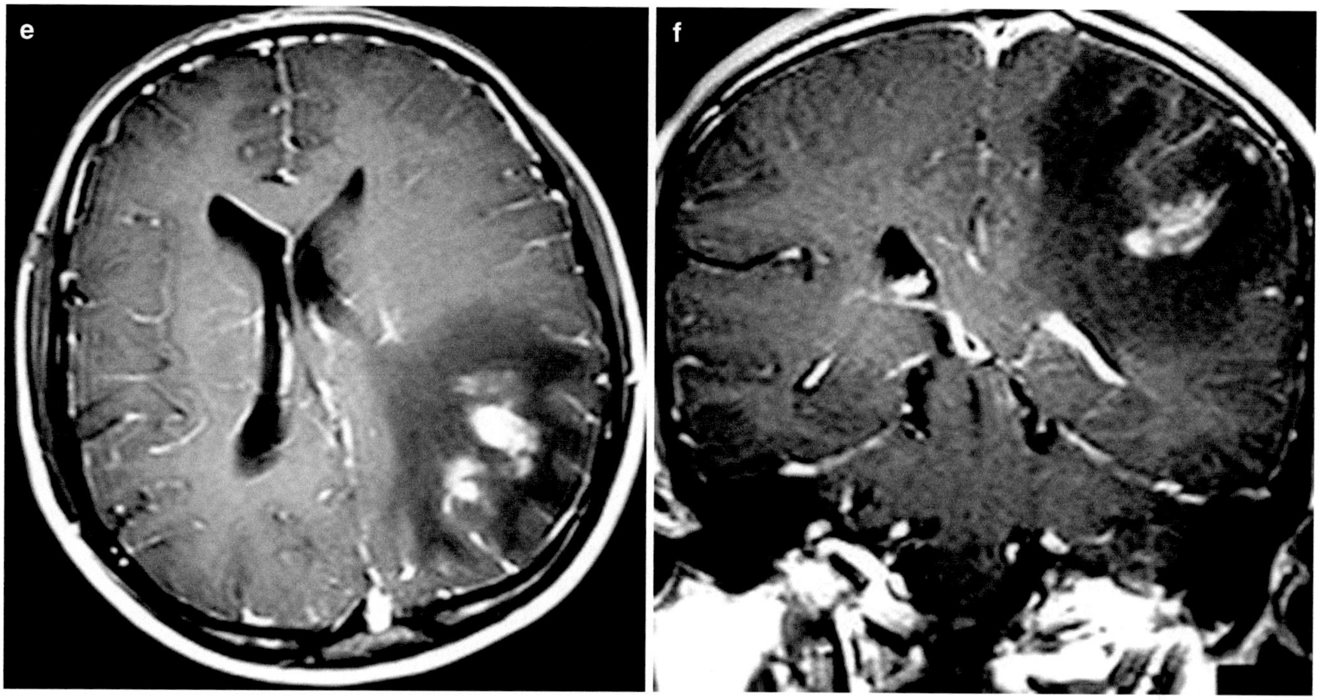

Fig. 10.8 (continued)

[Discussion]

The patients with cerebral schistosomiasis have a history of contact to contaminated water and the disease is common in young adults with more male patients than female patients. Contract MR imaging has important diagnostic value, showing a characteristic sign that spots of enhanced nodules cluster together. If with central linear enhancement showing a tree-branches sign, in combination to the history of contact to contaminated water by schistosoma, the diagnosis can be basically defined.

Cerebral schistosomiasis should be differentiated form the following diseases.

1. Other common parasitic encephalopathy
   Other common parasitic encephalopathy includes cerebral cysticercosis and cerebral cenuriasis, which commonly occur in brain cortex and subcortex. The lesion of cerebral cysticercosis is shown as small vesicles with spot of eccentric scolex inside in a typical small-target sign. The lesion of cerebral cenuriasis is demonstrated with characteristic cyst-in-cyst sign.
2. Cerebral tuberculoma
   The lesion of tubercular meningitis is often located near the basal cistern, showing small nodular enhancement by contrast scan, rarely with aggregation of nodular lesions but commonly with hydrocephaly. Tuberculoma may occur in any part of the intracranial region, with the lesion demonstrated with ring enhancement by contract scan. The radiological finding of rare target sign strongly indicates a diagnosis of tuberculoma.
3. Intracerebral malignant glioma and metastatic neoplasm
   The lesion of intracerebral malignant glioma is commonly located in the deep white matter and is commonly subject to necrosis and liquefaction. Contract MR imaging generally demonstrates wreth like enhancement of the lesion, which is inconsistent with characteristic MR sign of schistosomal granuloma that spots of enhanced nodules cluster together. Intracerebral metastasis generally shows multiple lesions in the brain, commonly subcortical area, with obvious edema around lesion. The lesion may show nodular enhancements of different sizes, but no aggregated spots of enhanced nodules. A concurrent medical history of primary carcinoma also supports the diagnosis of intracerebral metastasis.
4. Cerebral infarction
   Hypertensive cerebral infarction generally occurs at a late age and the patients usually have a medical history of hypertension. However, schistosomal cerebral infarction generally occurs at a young age and the patients usually have a medical history of schistosomiasis.

Case Study 9

[Brief Medical History]

A 15-year-old teenager boy complained of intermittent dull headache with no known causes for more than 6 months. The pain persisted for a short period of time and could be relieved by itself. During the recent 1 month, the headache aggravated. About 6 h ago, he experienced convulsion of the limbs, vomiting whitish foams, and no response to calling, which persisted for about 1 h. Routine blood test showed normal WBC count and increased eosinophil count.

Abdominal B-mode ultrasound demonstrated hepatic schistosomiasis. The patient reported a definite history of contact to contaminated water by schistosome cercaria 1 year ago.

[Radiological Demonstrations] (See Fig. 10.9)

[Diagnosis]

Granuloma type of cerebral schistosomiasis.

[Discussion]

The lesion of cerebral schistosomiasis is caused by atopic deposition of schistosomal eggs in brain tissue, which induces a series of neurological symptoms. The granuloma type, as the most common type of cerebral schistosomiasis, shows the lesion of granuloma caused by egg deposition and immune response to foreign substance. Currently, CT scan and MR imaging serve as the main diagnostic examination for its diagnosis. CT plain scan demonstrates the nodules in the cortex and subcortex at the margin of edema area, with equal or slightly high density. The nodules show diversified sizes and shapes, with homogeneous enhancement by contrast scan. Edema around the lesion is obvious, commonly demonstrated as flakes or fingerstall like in appearance. Plain MR imaging demonstrates singular or multiple nodular opacity in the cortex and subcortex of the brain with equal or slightly long T_1 signal and slightly long T_2 signal and obvious surrounding edema. These MR signs are shown because inflammatory responses and other pathological changes occur in the diseased area. Contract imaging demonstrates obvious enhancement of the nodules, with some of them fuse into a large nodule that is homogeneously enhanced. Before formation of granuloma by eggs deposition, spots, patches and sand like enhancement opacity are demonstrated.

Commonly, the lesion of granuloma type of cerebral schistosomiasis resembles to intracranial space occupying lesion, which should be differentiated from the following diseases.

1. Glioma
   The lesion of glioma is often located in the deep white matter of the brain and shows obvious space occupying effect. By contrast scan, nodular, ring and lace like enhancements are demonstrated at the arterial phase due to abundant blood supply to the lesion. At the delayed phase of contrast scan, enhancement of the lesion gradually weakens. In addition, serum immunoassay negative faciliates their differential diagnosis.
2. Metastasis
   The lesion is commonly located in the subcortex with singular or multiple nodules and obvious space occupying effect. Contract scan demonstrates ring or irregular enhancement with central necrosis or cystic change but no tendency of infusion. Clinically, the patients often have a medical history of primary neoplasm. Granuloma of cerebral schistosomiasis is the result of aggregation of mutiple nodules with different sizes that may infuse into a mass.

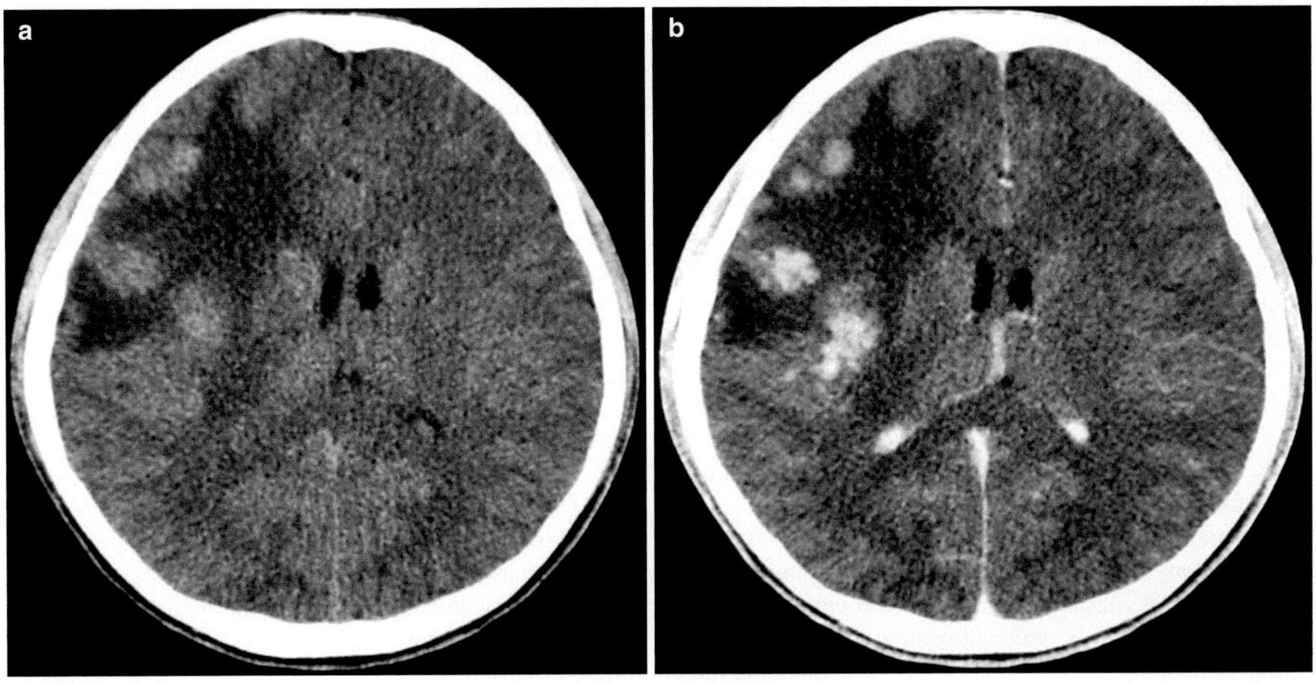

Fig. 10.9 (**a**) Plain CT scan demonstrated larger flakes of hypo-intense opacity in the right temporal lobe, and multiple slightly hyper-intense nodular opacity with different sizes in the cortex and subcortex. (**b**) Contrast scan demonstrated nodular enhancement in the cortex and subcortex at the margin of edema area in the right temporal lobe, and widened adjacent gyrus

3. Tuberculoma
   The lesion may be located at any part of brain parenchyma that is caused by deposition of tubercle bacillus in the brain, and is subject to calcification. By contrast scan, the lesion shows thick-wall ring or nodular enhancement with relatively mild edema around the lesion.
4. Cerebral cysticercosis
   Cerebral cysticercosis is the most common parasitic disease occurring in the brain, with its lesion commonly located in the white mater, ventricles and subarachnoid space of the brain. Commonly, the lesion is multiple sporadic small vesicles with small nodular dense opacity inside, namely scolex of cysticercus. Accompanying edema commonly occurs with mild space occupying effect. Serum complement test for cysticercosis is positive which faciliates the differential diagnosis.

Briefly, CT scan and MR imaging demonstrate the lesion of cerebral schistosomiasis with characteristic multiple nodular opacity in the cortex and subcortex of brain, which is in a cluster with obvious enhancement. Otherwise, mulitple small nodules may infuse into a large nodule, with a large area of finger like edema around the large nodule. MR imaging has a higher resolution and shows higher sensitivity to patches and sands like enhancements of the lesion at the acute stage of cerebral schistosomiasis as well as the adjacent meningeal enhancement. When the lesion involves the brain tissue at the cranial base, MR imaging is superior to CT scan in demonstrating the shape and range of lesions at the top of the brain, with no negative impacts by the skull artifacts.

Case Study 10
[Brief Medical History]

A 14-year-old teenager boy was hospitalized due to chief complaints of persistent fever for 3 days. The patient experienced fever with no known causes 3 days ago, with the highest body temperature of 38.8 °C. He also had symptoms of headache, vomiting, fatigue, cough, jaundice, hepatalgia, abdominal distension, diarrhea, hepatomegaly with tenderness, and yellowish urine but no convulsion of limbs. Physical examination revealed T 38.5 °C, P 86 times/min, R 23 times/min, and BP 126/85 mmHg. The patient showed acute complexion, yellowish sclera, hepatomegaly with palpable liver inferior to the right ribs and palpably moderate in texture. The abdomen was examinted to be soft, with splenomegaly as well as multiple systemic superficial lymphadenectasis. The laboratory tests revealed RBC 4.9×10^{12}/L, WBC 13.8×10^9/L, Hb 146 g/L, EOS 0.018, PLT 198×10^9/L, ALT 1152U/L, AST 712U/L, and TBil 9 μmol/L. Serum examination for schistosomiasis showed positive (+).

[Radiological Demonstrations] (See Fig. 10.10)

[Diagnosis]

Cerebral schistosomiasis

[Discussion]

Schistosomiasis is a parasitic disease caused by invasion of schistosoma into the human blood system. Ectopic deposition of schistosoma eggs in the brain causes diseases of the central nervous system disease, which is known as cerebral schistosomiasis. Cerebral schistosomiasis is the most serious ectopic schistosomiasis and commonly occurs in young and middle-aged adults in schistosomiasis affected regions. Schistosomiasis japonica prevails in China and schistosoma mansoni prevails in Africa. Schistosoma japonicum is a common pathogenic parasite for the cerebral type of central nervous schistosomiasis and schistosoma mansoni is a common pathogenic parasite for the spinal type of central nervous schistosomiasis. MR imaging demonstrates the lesion of cerebral schistosomiasis as multiple small nodules or nodules of different sizes in the brain cortex and subcortex that infuse in most cases. In the cases with multiple nodules, a main lesion is located in a certain area of brain. Contrast scan demonstrates the lesion with obviously homogeneous enhancement. Cerebral schistosomiasis rarely occurs with diversified clinical manifestations and radiological demonstrations. Clinically, it tends to be misdiagnosed. Its diagnosis can be defined based on the following key points.

The disease commonly occurs in young and middle-aged adults from schistosomiasis affected regions. Serologic examination is commonly applied for diagnosis.

MR imaging shows characteristic signs of multiple lesions, infusion of nodules, relatively concentrated distribution of the main lesions, and obvious homogeneous enhancement.

Medication of praziquantel shows favorable therapeutic responses. The lesions shrink or disappear after 15–30 days of its administration.

Case Study 11
[Brief Medical History]

A 40-year-old man was hospitalized due to complaints of dizziness, blurry vision and fever for 3 weeks. The patient had a medical history of schistosomiasis and the disease was cured, with a generally good physical condition. He experienced no symptoms of fever, nausea, vomiting, abdominal pain and diarrhea. The cerebrospinal fluid examination revealed cerebrospinal pressure 240 mmH_2O and the fluid was colorless and clear. Routine laboratory tests and biochemical examination of the cerebrospinal fluid were negative. No bacterial growth was revealed by culture of the cerebrospinal fluid. And the cerebrospinal fluid examination for parasite showed negative. Routine blood test revealed WBC 12.4×10^9/L, NEUT 0.796, EOS 0.08, Hb 110 g/L, and PLT 203×10^9/L. Serological test for schistosomiasis showed (±).

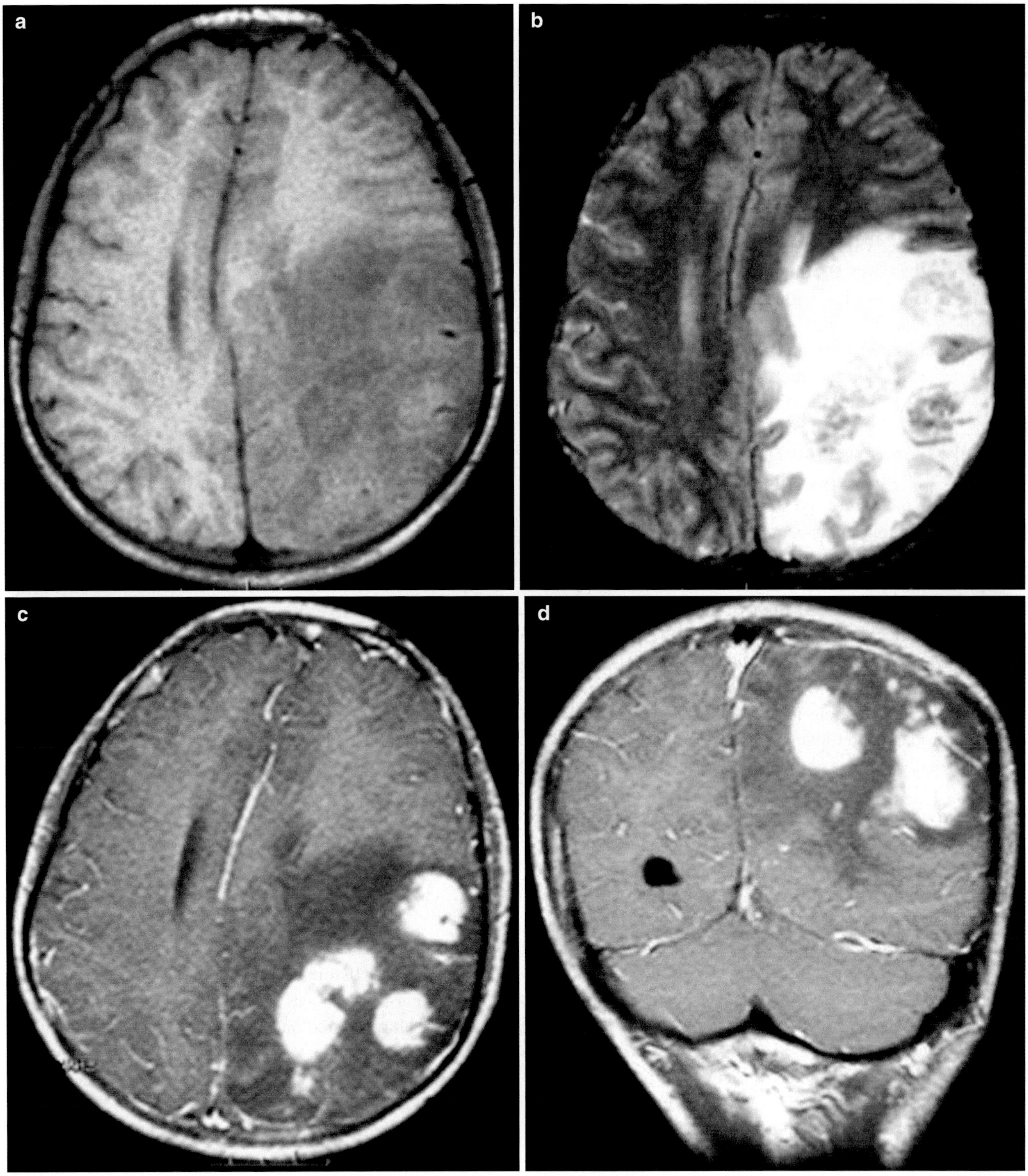

Fig. 10.10 (**a**, **b**) Plain MR imaging demonstrated a large area of long T_1 and long T_2 signal opacity in the left parietal-occipital region, rightward migration of the midline structure and space occupying effect. (**c**, **d**) Contract scan demonstrated the lesion in the left parietal-occipital region with cluster like or mass like enhancement, a large area of surrounding edema, and compressed lateral ventricles

[Radiological Demonstration] (See Fig. 10.11)

[Diagnosis]

Cerebral schistosomiasis

[Discussion]

Schistosomiasis is a parasitic disease caused by invasion of schistosoma into human blood system. Ectopic deposition of its eggs in the brain can cause diseased condition of the central nervous system, which is known as cerebral schistosomiasis, the most serious ectopic type of schistosomiasis. Cerebral schistosomiasis commonly occurs in young and middle-aged adults in schistosomiasis affected regions. Schistosomiasis japonica prevails in China and schistosomiasis mansoni prevails in Africa. Schistosoma japonicum commonly causes central nervous schistosomiasis and schistosoma mansoni commonly causes spinal schistosomiasis.

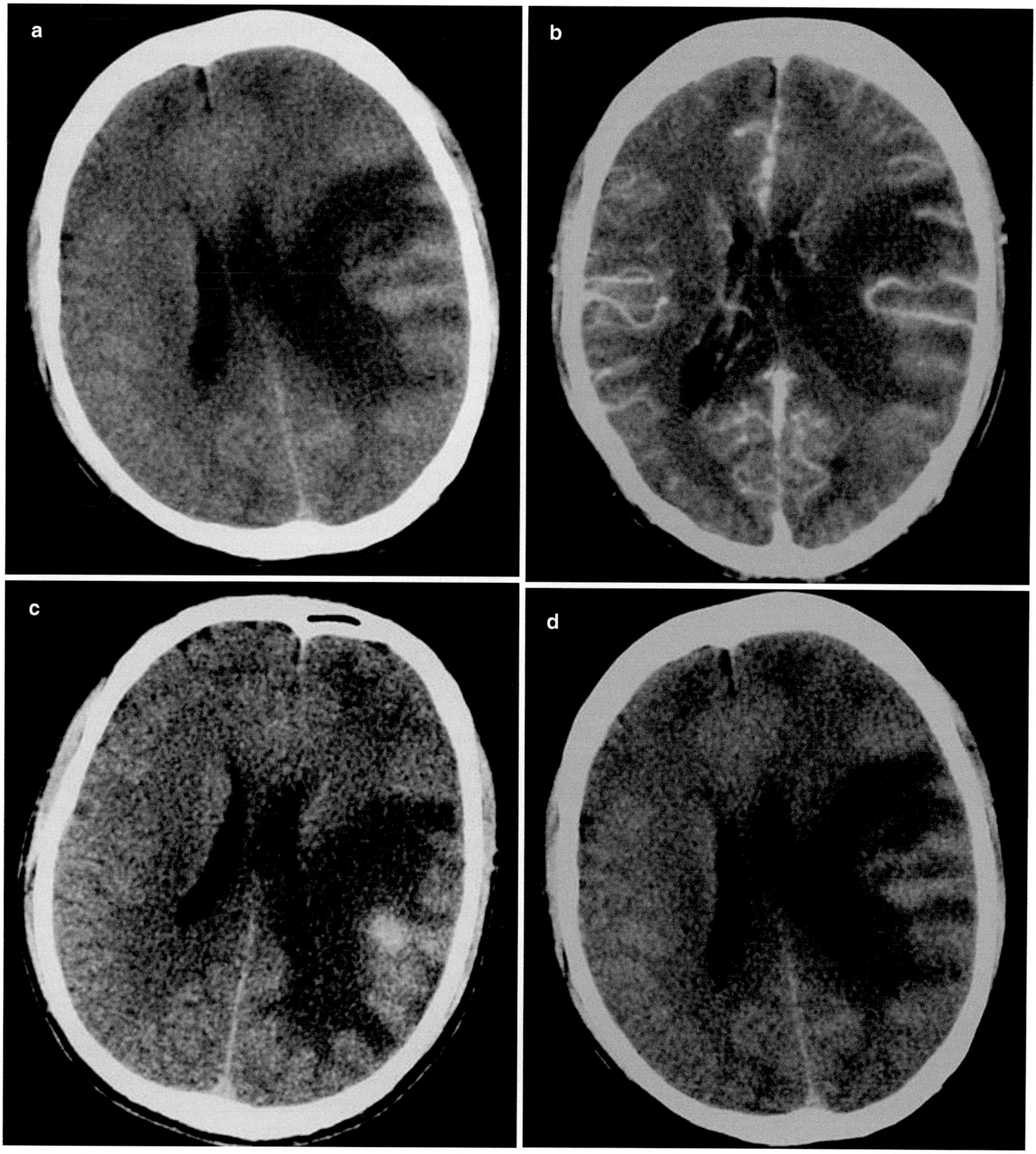

Fig. 10.11 (**a**) Plain CT scan demonstrated a large area of hypointense edema in the left parietal lobe. (**b**) Contract scan demonstrations 1 minute after scanning. (**c**) Delayed scan for 20 minutes demonstrated mass like enhancement in the left parietal lobe and a large area of surrounding edema. (**d**) Delayed scan for 25 minutes demonstrated absence of enhancement (Reprint with permission from Hongjun Li, Radiology of Infectious diseases, Volume 2, 2015)

Cerebral schistosomiasis rarely occurs and can be categorized into two stages based on its CT demonstrations.

1. Acute cerebral ischemia and edema stage
   In the cortex and subcortex, flakes of hypo-intense opacity is demonstrated with mild space occupying effect, which often involve 1–3 lobes, especially the temporal and parietal lobes. The opacity is actually toxins secreted by eggs and their metabolites, allergy caused by vascular embolism induced by eggs deposition, and toxic encephalitis.
2. Granulomatous nodules formation stage
   The lesion is demonstrated with calcification and surrounding large area of fingerstall like or irregular edema with obvious accompanying space occupying effect. The lesion is revealed with masses of different sizes or nodular opacity with mixed density. Contract scan demonstrates the solid part of lesion with flakes, spots or nodular enhancement in different degrees, with no enhancement of the surrounding low density area. Delayed scan for 10–20 min demonstrates the most obvious enhancement of the lesion, but delayed scan for 25 min shows no enhancement. The surrounding area of nodules is revealed to be coarse and unsmooth, which is related to formation of collagen fiber in a large quantity. During this stage, cerebral schistosomiasis is characterized by patches, sands like, nodular, and cluster like (fusion of small nodules) homogeneous enhancement at the interface between gray and white matters in the brain cortex and subcortex.

10.5 Clonorchiasis

Clonorchiasis, also known as hepatic distomiasis, is a parasitic zoonosis caused by parasitism of Clonorchis sinensis in the intrahepatic bile duct of human beings or animals. Human infection is commonly induced by oral intake of uncooked or unthoroughly cooked freshwater fish or shrimp containing its metacercaria. Clinically, it is manifested as chronic gastrointestinal dysfunction and hepatobiliary diseases.

10.5.1 Epidemiology

10.5.1.1 Source of Infection

Humans and mammals (e.g. cat, dog, and pig) infected by clonorchis sinensis are the main sources of infection.

10.5.1.2 Route of Transmission

Its infection is mainly via oral intake of uncooked or unthoroughly cooked freshwater fish or shrimp containing alive metacercaria of Clonorchis sinensis.

10.5.1.3 Susceptible Population

People are generally susceptible to clonorchiasis with no significant differences in terms of age, gender and race.

10.5.2 Pathogenesis and Pathological Changes

The occurrence of clonorchiasis is usually related to mechanical obstruction of the biliary duct by polypide, and blood-sucking polypide with cholangiocytes as its food, which cause damages to local biliary duct and drop of the mucosa. Its occurrence is also related to metabolites of polypide as well as local biliary inflammation and secondary bacterial infection caused by direct stimulation of polypide to the local biliary duct. In addition, the occurrence of clonorchiasis is related to the age, nutrition and immunity of the host as well as occurrence of other diseases.

The lesion is mainly located in the intrahepatic biliary ductule. During its early stage or when infection is mild, no obvious pathological change is shown. But when the infection is serious, the biliary duct may be subject to cystic or cylindrical dilation, with thickening of the biliary duct and surrounding hyperplasia of fibrous tissue. In the cases with severe infection, the biliary lumen is filled with clonorchis sinensis and static bile. The lesion is rather obvious in the left liver lobe, which may be related to the comparatively straight biliary duct in the left liver lobe that is more vulnerable to invasion of laval clonorchis sinensis.

10.5.3 Clinical Manifestation

The patients with mild infection experience no symptoms or only epigastric fullness after intake of food, anorexia or mild abdominal pain, and common feeling of fatigue. The patients with more serious infection experience anorexia, epigastric fullness, mild diarrhea, and dull pain at the hepatic region. Hepatomegaly occurs with more obviously enlarged left liver lobe with tenderness or percussion pain. The patients may also experience neurasthenic symptoms such as dizziness, insomnia, low spirits and decline of memory.

When chronic and repeated serious infection progresses into liver cirrhosis and portal hypertension, the patients experience emaciation, anemia, abdominal venous varicose, hepatosplenomegaly, ascites, and jaundice. Children with serious infection may develop malnutrition and growth disturbance or even dwarfism.

The most common complications include biliary tract infection, cholangitis and cholelithiasis.

10.5.4 Diagnostic Examination

10.5.4.1 Laboratory Test

Etiological Examination

By stool examination or duodenal drainage of bile for examination, the finding of eggs of clonorchis sinensis is the direct evidence for the definitive diagnosis of clonorchiasis.

Immunoassay

Immunoassay is applied in patients with mild infection for diagnosis. Otherwise, it is applied for epidemiological study.

10.5.4.2 Radiological Examination

Ultrasound, CT scan and MR imaging can demonstrate polypide and other changes in the dilated biliary duct. But the finding cannot provide direct evidence for its definite diagnosis.

Case Study 1

[Brief Medical History]

A 40-year-old man complained of poor appetite, epigastric dull pain and distension as well as dull pain at the hepatic region for more than 1 month. He reported a past history of frequent intake of uncooked fish. ELISA showed positive (+).

[Radiological Demonstrations] (See Fig. 10.12)

[Diagnosis] Clonorchiasis

[Discussion]

Clonorchiasis is a parasitic disease caused by parasitism of clonorchis sinensis in human biliary tract, with lesions in the liver and biliary duct. Its infection is mainly via oral intake of uncooked freshwater fish containing its larva. The

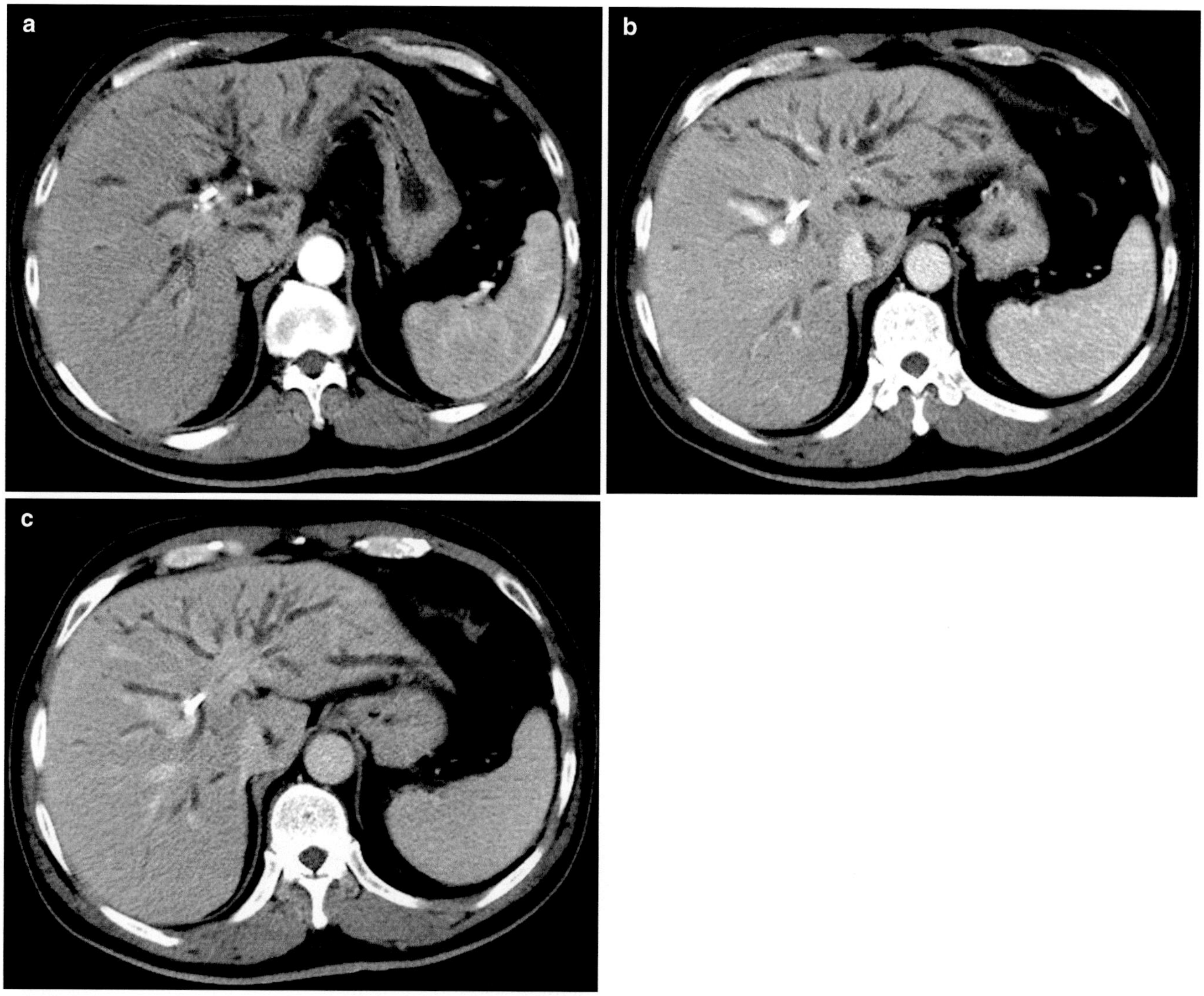

Fig. 10.12 (a–c) Contrast CT scan demonstrated small cystic and fine branches like dilation of the intrahepatic biliary duct, with clustering like distribution

manifestation of clinical symptoms depends on the severity and duration of infection, with mild cases possibly showing no obvious symptoms. But the paitents with serious infection experience systemic fatigue, abdominal upset or pain, diarrhea, hepatomegaly, jaundice, cholecystitis, and hepatitis. In the advanced stage of serious cases, hepatocirrhosis, splenomegaly, ascites and edema may occur, which may further induce calculi, cholecystitis, cholangiocarcinoma, cholecystocarcinoma, and pancreatitis. Eggs of clonorchis sinenis can be detected in the stool or duodenal fluid from the patients.

Based on CT scan demonstrations, clonorchiasis can be classified into three types:

1. Peripheral liver type
   The dilation of intrahepatic bile duct is predominantly dilation of intrahepatic ductule in the peripheral liver but rarely dilation of intrahepatic biliary duct in the hilar area. The lesions are characterized by small cystic and fine branches like dilation of intrahepatic terminal biliary ductule, which are mainly distributed as a cluster in the right posterior liver lobe. The small cyst is commonly in a diameter of below 2 cm, which communicates with the dilated biliary ductile, showing a comma sign. In some rare cases, contrast scan demonsntrates small granulomatous nodule opacity of trematode with well defined boundary and slight enhancement.
2. Hepatic hilum type
   The hepatic hilum type is characterized by branches like dilation of the intrahepatic biliary duct beside the liver hilum. Most of the intrahepatic biliary ducts are subject to slight and proportionate dilation. However, the moderately and seriously dilated biliary ducts show homogeneous dilation from the lateral hepatic hilumm outwards to the hepatic capsule, with the diameter of distal biliary duct being almost the same as that of proximal biliary duct.
3. Mixed type
   Most of the intrahepatic biliary ducts are subject to diffuse dilation, showing signs of both peripheral liver type and hepatic hilum type. The dilation of subcapsular biliary ductules is commonly more serious than that of large biliary ducts in the hilar region. The biliary ducts dilate disproportionately, rarely with extrahepatic biliarly duct dilation. The cases with gall bladder involved show thickening of the gall bladder wall, and heterogeneous density within the gall bladder.

Clonorchiasis should be differentiated from the following diseases.

1. Small liver abscess
   In the cases of clonorchiasis, small cystic dilation of the intrahepatic biliary ducts commonly occurs subcapsularly, with a cluster like distribution and possibly comma sign. However, the lesion of small liver abscess is commonly round in shape with well defined boundary but no communication to the biliary ducts.
2. Obstructive jaundice
   The patients with obstructive jaundice show obvious symptoms of jaundice that progressively aggravate. CT scan shows moderately and seriously dilated intrahepatic biliary ducts extending from the hilum as the center and being gradually thinner. Lithiasis and lump are observable on the plane of obstruction. And in most cases, obstructive jaundice is complicated by extrahepatic biliary dilation. However, the patients with clonorchiasis show no symptoms of jaundice or just show transient mild jaundice, which does not progressively aggravate. The intrahepatic biliary ducts mainly show small cystic or fine branches like dilation of the terminal biliary ducts. The diameter of terminal biliary ducts is larger than or almost the same as the diameter of hilar biliary ducts. And the extrahepatic biliary ducts commonly show no dilation.

In clonorchiasis affected area, the cases with subcapsular terminal biliary ducts dilation in the liver should be firstly suspected with clonorchiasis. In the cases with complicating small calculi or calcification in the intrahepatic terminal biliary ducts, a diagnosis of clonorchiasis is further supported.

10.6 Paragonimiasis

Paragonimiasis, also known as lung fluke disease, is an acute or chronic endemic parasitic disease caused by parasitism of paragonimus (lung fluke). In China, two types of paragonimiasis are commonly detected, namely paragonimiasis westermani and paragonimiasis skrjabini. Paragonimus westermani mainly parasitizes in human lungs to cause clinical symptoms of cough and expectoration of brownish red phlegm. It can also parasitizes in multiple human organs and tissues, such as brain, spinal cord, gastrointestinal tract, abdominal cavity and subcutaneous tissue, to cause corresponding symptoms. Paragonimiasis skrjabin is mainly manifested as migrating subcutaneous lump and exudative pleuritis.

10.6.1 Life Cycle of Paragonimus

Paragonimus finishes its development after experiencing three alternations of its host. The first intermediate host is simisalcospira calculus and the second intermediate host is freshwater crabs (e.g. sinopotamon) and crawfish. And its definite host may be human, livestock or wild cats and dogs. And the adult paragonimus commonly parasitizes in human or animal lung cysts.

10.6.2 Epidemiology

10.6.2.1 Source of Infection

Human or carnivorous mammal that can discharge eggs of paragonimus are the source of its infection.

10.6.2.2 Route of Transmission

The disease is mainly transmitted via intake of uncooked or unthoroughly cooked second intermediate host carrrying paragonimus metacercaria, such as freshwater crabs and crawfish. Pig, wild boar, rabbit, chicken, Paa boulengeri, rat and bird may act as the transport host of paragonimus. Humans or animals may also be infected after the intake of meat of such transport host. After death of its intermediate host, metacercaria fall into water to infect humans or animals after their drinking the contaminated water.

10.6.2.3 Susceptible Population

People are generally susceptible to the disease, with higher incidences in children and adolescents.

10.6.3 Clinical Manifestation

10.6.3.1 Acute Paragonimiasi

Clinically, acute paragonimiasi caused by paragonimiasis westermani is characterized by a short incubation period, acute onset and obvious systemic symptoms. The patients commonly experience the symptoms of poor appetite, abdominal pain, diarrhea, fever, fatigue, night sweat, skin rash (e.g. repeated occurrence of urticaria). And the secondary symptoms include chest pain, chest distress, shortness of breath and cough.

10.6.3.2 Chronic Paragonimiasi

Acute paragonimiasis rarely occurs, but chronic paragonimiasis is more common. The symptoms of paragonimiasis westermani mainly include cough, hemoptysis, and chest pain. Extra-pulmonary symptoms may occur if paragonimus westermani invades the brain, spinal cord, liver and subcutaneous tissue. However, paragonimiasis skrjabin is clinically manifested as symptoms related to migration of its larva, such as migrating subcutaneous nodule. Extra-pulmonary symptoms may also occur if paragonimus skrjabin invades the liver, pericardium, eyes, brain and spinal cord.

Chest-Lung Type

The patients with paragonimiasis westermani commonly experience the symptoms of cough, expectoration, chest pain and hemoptysis. The pleura is commonly involved to show pleural adhesion or thickening and rarely pleural effusion. The patients of chest-lung type often experience initial symptoms of dry cough, followed by expectoration of whitish thick sputum with fish smell. Subsequently, the sputum turns to be typically rusty bloody or jam-like bloody, and sometimes rotten peach like. The patients commonly show more severe cough in varying quantity in mornings. In such characteristic bloody sputum, eggs of paragonimiasis westermani, Charcot-Leyden crystals and eosinophils are detectable.

Cerebrospinal Type

The cerebrospinal type is more common in patient with paragonimiasis westermani, especially pediatric patients, with more commonly involved brain than the spinal cord. The main symptoms and physical signs include increased intracranial pressure, stimulated cerebral cortex, destructed brain tissue, and meningitis.

Skin Type

The skin type is the most common clinical type of paragonimiasis skrjabin, with an incidence rate of up to 50–80 %. The lesions mainly include subcutaneous nodule and lump that is characterized by its migratability. Subcutaneous lump is commonly detected in the abdomen, whose biopsy demonstrates tunnel like changes caused by migration of larva or adult paragonimus skrjabin.

Abdomen-Liver Type

The abdomen-liver type is mainly manifested with abdominal pain, which is commonly dull pain with no fixed site. The patients may experience bloody stools and diarrhea, extensive inflammation and adhesion of abdominal organs, and even peritoneal inflammation. Eosinophilic abscess occurs in the liver when the liver is invaded by paragonimus. The migrating paragonimus can also damage the vascular vessels to cause hemorrhagic lesion in the liver tissue. In addition to abdominal symptoms, the patients often experience accompanying symptoms of fatigue, poor appetite, and fever.

Subclinical Type

The patients with subclinical type of paragonimiasis commonly experience no obvious clinical symptoms and physical signs. The skin test and serological test for paragonimiasis show positive with increased eosnophils but no obvious impairments to organs. The infection may be mild or in its early stage. Otherwise, the patients with subclinical type of paragonimiasis are infected but the parasitic paragonimus has already been absent from their bodies.

10.6.4 Diagnostic Examination

10.6.4.1 Laboratory Test

The findings of eggs of paragonimus in sputum, stool and body fluid or the finding of polypide of paragonimus in subcutaneous nodule are the evidence for definite diagnosis. The findings by serological test and immunoassay faciliate the diagnosis.

10.6.4.2 Radiological Examination

Chest X-ray and CT scan have important diagnostic value for the chest-lung type of paragonimiasis. Brain CT or MR imaging demonstrates the lesions of cerebrospinal type in terms of their location, shape, quality or obstructed region.

Case Study 1

[Brief Medical History]

A 32-year-old woman complained of cough, chest pain and rusty colored sputum. The patient had a history of eating uncooked river crab 1 month ago. The laboratory test revealed WBC 20×10^9/L, EOS 2.0×10^9/L, and ESR 70 mm/h. ELISA for paragonimiasis showed positive (+). And the sputum examination detected eggs of paragonimus.

[Radiological Domonstration] (See Fig. 10.13)

[Diagnosis] Paragonimiasis of the left upper lung lobe.

[Discussion]

Paragonimus westermani is categorized into the family of troglotrema salmincola. In addition to its parasitism in human lungs, it can also parasitize in subcutaneous tissue, liver, brain, spinal cord, muscle and orbit to cause systemic paragonimiasis. Its first intermediate host is simisalcospira calculus and its second intermediate host is freshwater crabs and crawfish. Humans can be infected by intake of uncooked, drunken and unthoroughly cooked crabs and then suffer from paragonimiasis. The main symptoms include mechanical damages to organs caused by migration or settle of polypides in the tissue as well as allergic reaction induced by metabolites produced by polypides.

Radiologically, the lesion of paragonimiasis westermani is characterized by the following demonstrations.

1. **X-ray**
   X-ray demonstrates the lesion of paragonimiasis with different signs in different stages, with poorly defined opacity with cotton like boundary in the tissue destruction stage and multilocular cystic opacity in the cyst stage. The lesions in the scar stage are demonstrated to be hard nodular, with spots of calcification. And the pleura is demonstrated with cystoid opacity in the pleural adhesion and thickening stage.
2. **CT scan**
   CT scan shows signs of the lesion corresponding to the findings by X-ray. In the tissue destruction stage, in addition to the poorly defined patches of opacity with cotton like boundary, the lesion may be demonstrated to be movable along with migration of polypides and possible track sign in the patches of opacity. In the cyst stage, nodular lump like opacity is mainly found with singular or multilocular cystic change. In this case, the patient showed typical signs of the cyst stage, with multiple poorly-defined lesions that were multilocular cystic lump like. In the scar stage, the allergic responses gradually weaken and the cyst is absorbed to develop into fibrous scar. In the pleural adhesion and thickening stage, pleural effusion and thickening may occur as a result of polypides moving back into the pleural cavity during their development. Sometimes, polypides may penetrate through the pleura to cause pneumothorax. The characteristic sign is alternative occurrence or repeated occurrence of pleural effusion in the left and right thoracic cavity.

Clinically, the patients who experienced corresponding symptoms following eating uncooked crabs or drinking unboiled water with increased eosinophils are highly suspected with paragonimiasis. Immunoassay is an important

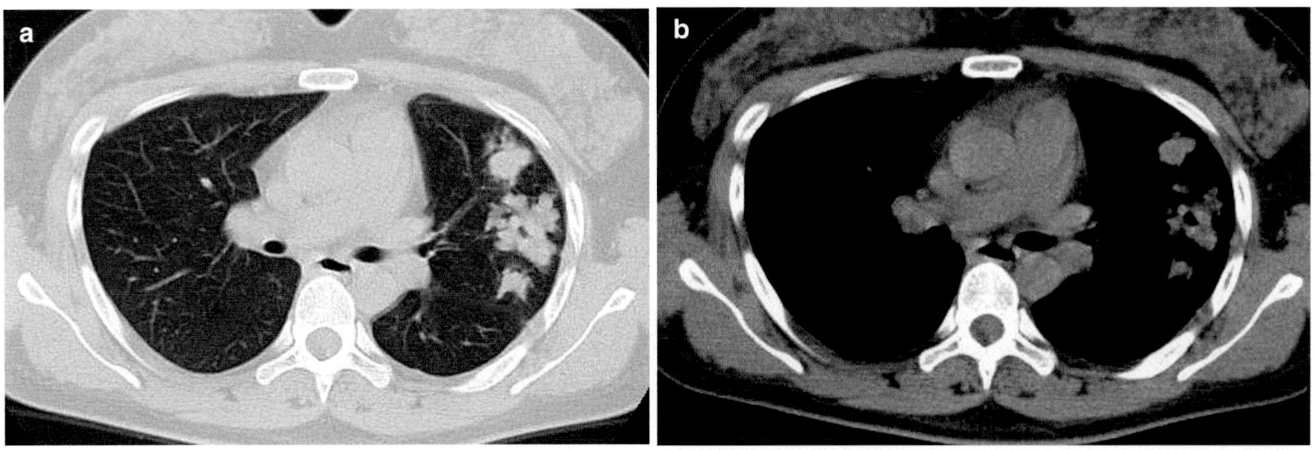

Fig. 10.13 (**a**) CT scan demonstrated multiple poorly defined patches of opacity in the left upper lung lobe. (**b**) The multiple patches of opacity in the left upper lung lobe was shown with mixed high and low density, with multilocular cystic change

way for the diagnosis of paragonimiasis, such as ELISA, Dot-ELISA, and IEST.

The disease shows complex demonstrations due to their varying parasitic site and different stages, whose radiological signs are overlapping with those of many other pulmonary diseases. The diagnosis can be defined in combination to clinical demonstration and immunoassay.

Paragonimiasis should be differentiated from the following diseases.

1. **Pulmonary tuberculosis**
   At its early stage, paragonimiasis shows similar symptoms to pulmonary tuberculosis with many overlapping radiological signs that is most likely to be misdiagnosed as pulmonary tuberculosis. However, the lesion of pulmonary tuberculosis is characterized by surrounding multiple satellite lesions and calcification. The patients with paragonimiasis that is predominantly manifested as pleural effusion are also likely to be misdiagnosed with tuberculous pleuritis. The differential diagnosis can be made based on the related labortary tests such as PPD.
2. **Lobar pneumonia**
   Lobar pneumonia may develop into chronic pneumonia with radiological signs of irregualr consolidation opacity, relatively poorly-defined margin of the lesion, and obvious thickening of adjacent pleura. Lobar pneumonia commonly shows intra-pulmonary lesions.
3. **Fungal infection**
   Halo sign is commonly demonstrated in the cases of fungal infection, with thrombotic necrosis of the vascular coagulation type at its center. Such signs are relatively rare in the cases of paragonimiasis.
4. **Lung cancer**
   The lump is well defined and commonly isolated, commmonly with neoplasm signs of lobulation and spike. Based on these signs, the differential diagnosis can be made. In addition, Three-dimensionally reconstructed CT scan images with post-processing techniques can more clearly demonstrate corresponding lesions. In combination to the case history and laboratory tests, the differential diagnosis can be made.

Case Study 2

[Brief Medical History]

A 47-year-old man reported a parabronchial lump like lesion in the right lung that had been detected by chest CT scan in a physical examination in 2001. But he experienced no cough, expectoration, chest distress, dyspnea, fever and night sweat. Then he was hospitalized in Peking Union Hospital, Beijing, China and received bronchoscopy, lung puncture, and examination for parasite antibody, which showed no abnormality. After being discharged, the patient received following-up chest CT scan each year, which demonstrated no enlargement of the lesion. Reexamination of CT scan in 2010 showed progression of the lesion in the right lung, with newly emerged round like lesion and patches of opacity adjacent to the pleura. On August 17th, 2010, the patient received PET/CT in Huashan Hospital, Shanghai, China, which indicated paratracheal soft tissue opacity in the right upper lung, with increased metabolism of FDG. By delayed contrast scan, the maximal value of SUV was shown with slight increase and the possibility of malignancy should be excluded And the multiple nodules in both lungs are suspected to be inflammatory lesions. On May 6th, 2011, the patient received examination for schistosome antibody in Shanghai Institute of Parasitosis, which showed positive (+). On May 16th, 2011, the total IgE level was examined in Huashan Hospital, which elevated to 2592 ng/ml; on May 10th, 2011, T-SPOT strong positive. The diagnosis was suspected to be pulmonary tuberculosis and anti-tuberculosis therapy was prescribed with isoniazide, rifampicin, ethambutol and levofloxacin. Reexamination by chest CT scan in August, 2011 demonstrated that the lesions in both lungs were almost the same as those demonstrated by previous examinations, only with relatively enlarged lymph nodes in the mediastinum. By inquiry of his medical history, the patient reported a suspected history of contact to contaminated water 20 or 30 years ago and a history of eating uncooked drunk shrimps, crabs and fish. The patient was then hospitalized and received more examinations on admission. Praziquantel was administered to treat pulmonary schistosomiasis

[Radiological Demonstration] (See Fig. 10.14)

[Diagnosis] Paragonimiasis.

[Discussion]

Paragonimiasis prevails in Southeast Asia, East Asia, Africa and Latin America. And the infection is commonly caused by intake of uncooked or unthoroughly cooked crabs carrying metacercaria of paragonimus, or by contact to contaminated water. Paragonimiasis is a disease with its lesions induced by mechanical damages to local tissue due to migration and parasitism of larva and adult paragonimus in human body as well as immune responses to metabolites (antigentic substances) produced by polypides of paragonimus. Metacercarias of paragonimus is infected via oral intake, which then ruptures in human stomach and duodenum to release the larva. The released larva then penetrates the intestinal wall into the abdominal cavity, or penetrates the diaphragm into the thoracic cavity and lungs. In lungs, the larva develops into adult, and some larva and adult paragonimus further migrate into other body parts along soft tissue space and vascular system. Therefore, although paragonimiasis primarily occurs in the lung, ectopic infection is likely to occur in other body parts such as brain, subcutaneous tissue, muscle, liver, omentum, mesentery and retroperitoneum.

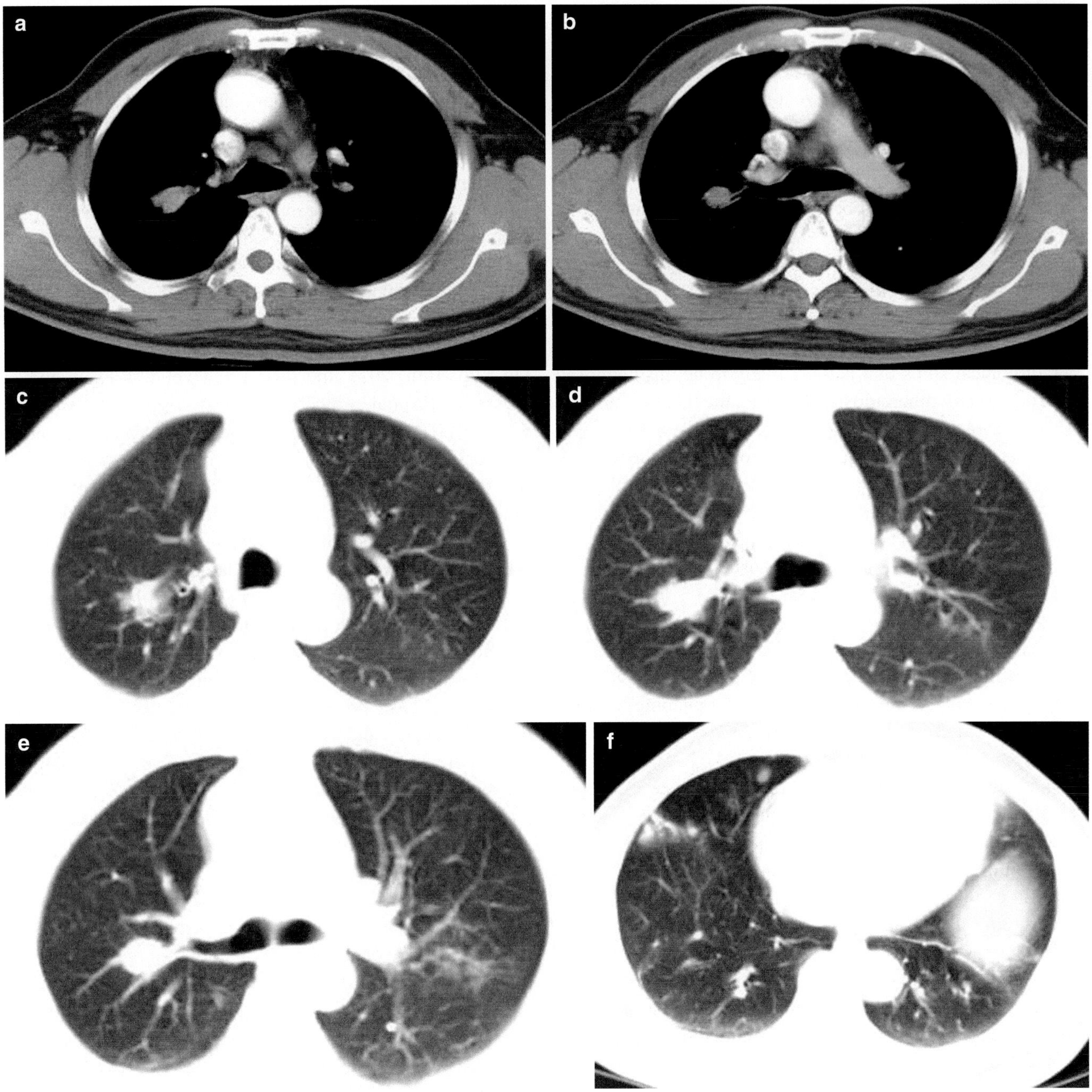

Fig. 10.14 (**a**–**f**) CT scan demonstrated parabronchial round like soft tissue density opacity in the right upper lung, with quite smooth margin and visible incisura. The lesion was shown with a diameter of about 1.6 cm. The lymph nodes in the mediastinum were revealed with enlargement and multiple patches, nodular and cords like opacity were shown in both lungs

Primary thoracic paragonimiasis shows diversified lesions in the lung according to its onset time. Based on literature reports, the major changes are as following:

(1) Infiltrative change, with characteristic radiological demonstration of tunnel sign. (2) Peribronchial inflammatory change. (3) Changes in the cyst stage, with thick-wall cavity in mediastinal window. (4) Hollow cavity with nodules attached to the wall. (5) Pleural lesion, such as pleural effusion, spontaneous pneumothorax and hydropneumothorax, which facilitate its diagnosis. (6) Formation of fibrous scar, with nodular opacity with sharp margin or patches and cords like dense opacity.

The first two types occur in the early stage. The change in the cyst stage is shown as liquefaction and necrosis during the development of infiltrate lesion, with no communication to bronchus. The hollow cavity with nodules-attached wall is caused by communication of the cyst to the bronchus with accompanying hyperplasia of granulation tissue in the cavity.

The formation of fibrous scar occurs in the late stage. In this case, the patients experience atypical symptoms. In 2001, chest CT scan demonstrated parabronchial lump like lesion in the right lung, enlarged mediastinal lymph nodes as well as multiple patches, nodules and cords like opacity in both lungs. However, bronchoscopy, lung puncture and examination for parasite antibody showed no abnormality. Subsequent follow-ups of chest CT scan each year showed no enlargement of the lesion. In 2010, reexamination by chest CT scan demonstrated progression of the lesion in the right lung, newly emerged round like lesion and patches of opacity adjacent to the pleura. In August, 2010, the patient received PET/CT in Huashan Hospital, Shanghai, China, which indicated paratracheal soft tissue opacity in the right upper lung, increased FDG metabolism, with slight increase of the maximum value of SUV by delayed scan. Before a diagnosis can be made, malignancy should be excluded. And the multiple nodules in both lungs were suspected to be lesions of inflammation. By further inquiry about the medical history, the patient reported a suspected history of contact to contaminated water 20 or 30 years ago and a history of eating uncooked drunk shrimps, crabs and fish. On May 6th, 2011, an examination for schistosome antibody in Shanghai Institute of Parasitic Diseases showed positive, and pulmonary schistosomiasis was suspectively diagnosed. Praziquantel was then administered. The radiological demonstrations were also atypical, with no typical tunnel sign, thick-wall cavity and hollow cavity with nodules-attached wall. Based on the finding of T-SPOT strong positive, tuberculosis was suspectively diagnosed, but the anti-tuberculosis thearpy was ineffective. The key points for differential diagnosis of paragonimiasis from tuberculosis are as the following:

The lesions of tuberculosis are commonly located in the apex of both lungs and the dorsal segment of lower lung lobes. The lesions are diversifed, including infiltrative proliferation lesion, thin-wall cavity and fibrosis. CT scan demonstrates different lesions from those of paragonimiasis.

Case Study 3

[Brief Medical History]

A 3-year-old boy experienced cough, hemoptysis, yellowish sputum with blood and low grade fever, with the highest body temperature of 38.0 °C about 1 year ago. Anti-infection therapy was administed and the symptoms were relieved. Subsequently, the symptoms relapsed and showed no improvement after anti-tuberculosis therapy was adminstered. A full set of parasite examination revealed cysticercus antibody positive (+). Bone marrow examination revealed obviously increased ratio of eosinophils, active bone marrow hyperplasia, quite active erythroid hyperplasia and active granulocyte hyperplasia.

[Radiological Demonstrations] (See Fig. 10.15)

[Diagnosis] Paragonimiasis.

[Discussion]

Infiltrative lesions of paragonimiasis commonly emerge in the early stage, which is mixed with hemorrhagic lesion caused by migration and penetration of paragonimus in the lung tissue and exudative lesion caused by local allergic reaction. Radiologically, the lesion was shown as poorly defined large area of cotton like or fusion of small patches opacity that is amorphous. The inflammatory lesion shows poorly defined small vesicles. And based on such signs, the differential from other lung infection can hardly be made.

Parabronchial inflammatory changes tend to be missed, which are related to the development of larva paragonimus around the bronchus into adult after its migration between the liver and abdominal wall and then penetration through diaphragm into the thoracic cavity. On plain film, the lesions are shown as increased and blurry lung markings with the hilum as the center or in bilateral middle and lower lungs. CT scan demonstated small patches and spots of opacity along lung markings, with poorly defined margin, which resemble to those of bronchopneumonia.

The hollow cavity with nodules attached wall is a demonstraton of evacuation after trematode cyst communicates to the bronchus, which is shown as isolated thin-wall cavity containing air and almost complete absence of its surrounding inflammatory lesions. However, well-defined cords like opacity are observable. The observable cavity with nodules attached wall is a characteristic sign, resembling to watermelon seeds that attach to the inner wall of the cavity and resembling to mould mass in the lung cavity. However, the nodules attached to the cavity inner wall are relatively small, while the mould mass in lung cavity often occupies a large area of the cavity, with only arc-like space left in the cavity. The nodules attached to the inner cavity wall in the cases of paragonimiasis are caused by paragonimus, its eggs or hyperplasia of granulation tissue. Such a demonstration is rarely found in lung cavity of other diseases. Therefore, it plays an indicative role in the diagnosis of paragonimiasis.

Paragonimiasis should be differentiated from the following diseases.

1. Pulmonary tuberculosis

   The lesions of pulmonary tuberculosis are multiple small caseous necrotic granuloma, showing ring enhancement by contract scan. The patients commonly experience typical tubercular poisoning symptoms, which show therapeutic responses to anti-tuberculosis therapy.
2. Pneumonia

   The lesion of pneumonia is demonstrated as poorly defined cloudy opacity, and shows improvement after administration of routine anti-inflammatory therapy.
3. Peripheral lung cancer

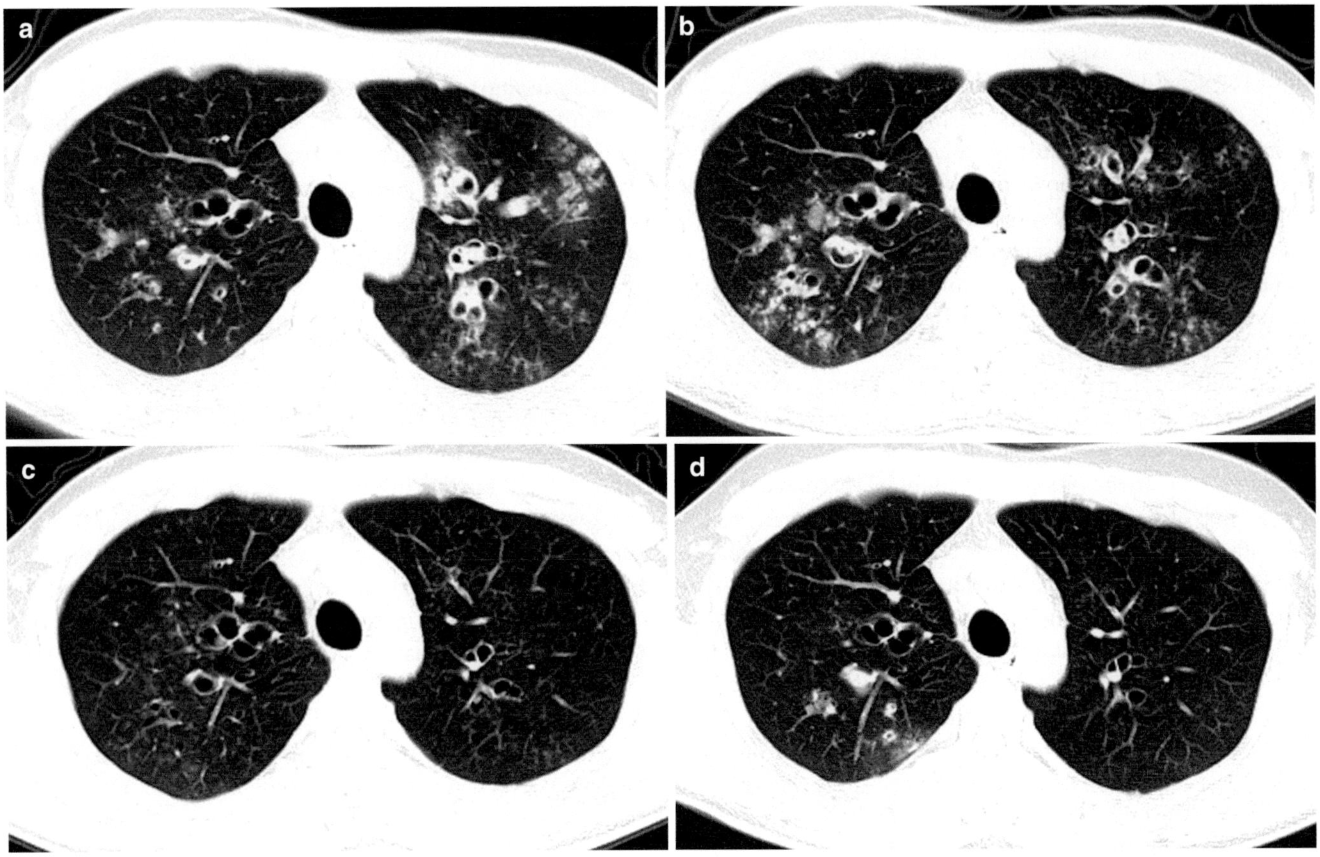

Fig. 10.15 (**a**) CT scan demonstrated multiple patches and cords like hyper-intense opacity in both lungs, columnar dilation of local bronchi, obviously thickened bronchial wall with poorly defined margin. The mediastinum was shown with multiple small lymph nodes opacity. (**b**) CT scan showed no obvious decrease in quantity and absorption of the lesions in both lungs after treatment for 1 week with anti-inflammatory therapy. (**c**) The lesions were shown to be decreased in quantity and shrinkage after treatment for 1 month with anti-inflammatory therapy and anti-parasite therapy. (**d**) The old lesions were shown with decrease in quantity and shrinkage, but with emergence of new lesions after treatment for 4 months with anti-inflammatory therapy and anti-parasite therapy. (**e**) The lesions were shown with decrease in quanity and shrinkage after continued treatment for 1 month with anti-inflammatory therapy and anti-parasite therapy. (**f**) Reexamination after 1 year demonstrated relapse of the lesions. (**g**) Pathologically, lung smear demonstrated neutrophils in a moderate quantity as well as phagocytes, lymphocytes and eosnophils in small quantities

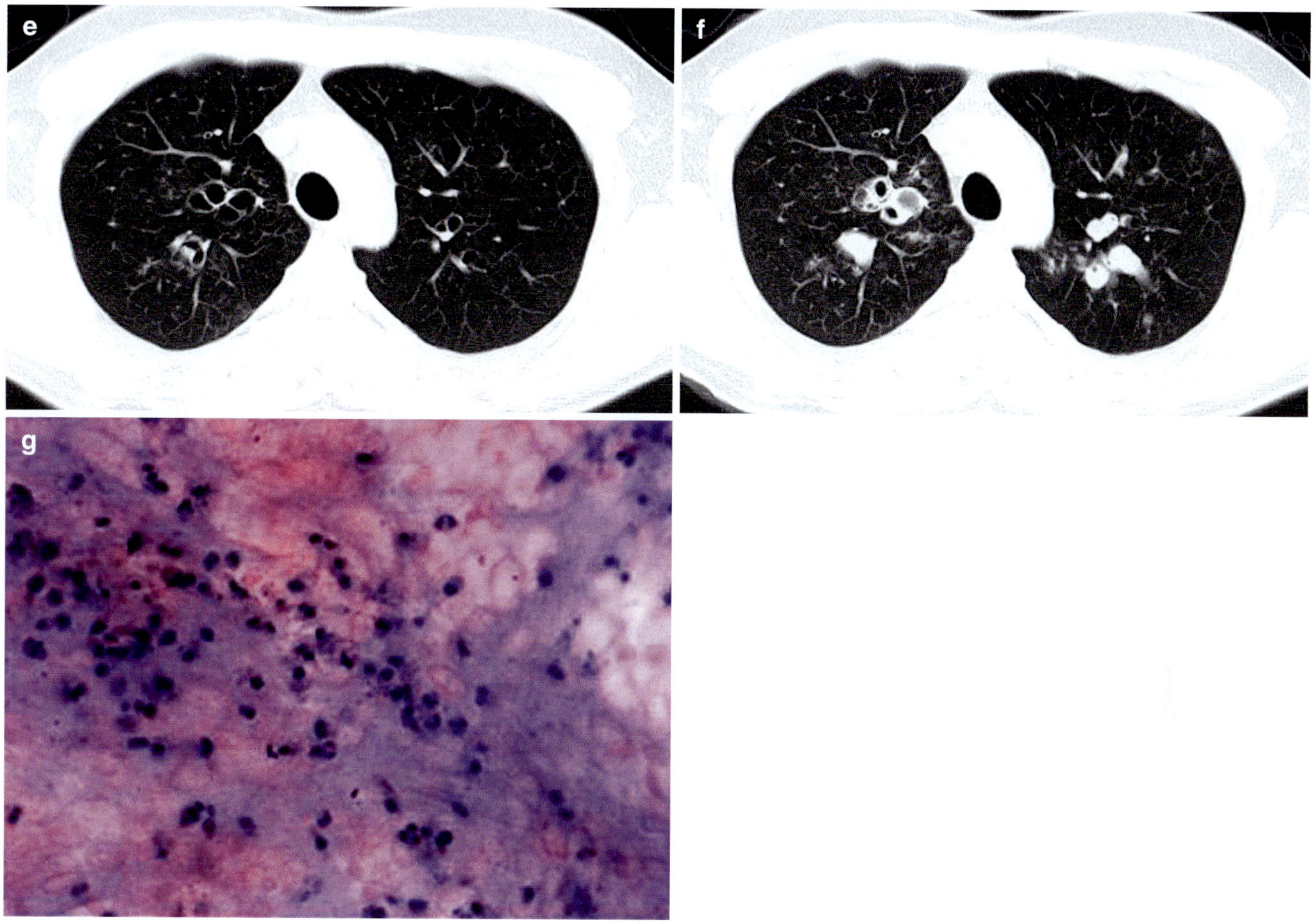

Fig. 10.15 (continued)

The lump is demonstrated with irregular margin, and lobulation. Dynamic contract scan facilitates their differential diagnosis.

Case Study 4
[Brief Medical History]

A 33-year-old man showed HBsAg positive for 17 years. He complained of fatigue, abdominal distension, and poor appetite for 10 days. Laboratory tests revealed increased eosinophil count and monocyte count as well as inverted ratio of albumin to globulin. Pathological report demonstrated non-caseous necrotic granulomatous inflammation in the liver, infiltration of eosinophil in a large quantity, focal fat degeneration of surrounding liver tissue.

[Radiological Demonstrations] (See Fig. 10.16)

[Diagnosis] Paragonimiasis.

[Discussion]

Based on the major organs the parasite invades, paragonimiasis can be clinically categorized into 4 types.

1. **Chest-lung type**
   The chest-lung type is the most common, and the patients experience symptoms of cough, bloody sputum and chest pain. The migration of paragonimus into the thoracic cavity causes chest pain, exudative pleuritis or pleural thickening.
2. **Abdomen type**
   The patients of this type experience abdominal pain, diarrhea, and hepatomegaly, with formation of eosinophilic granuloma in the liver.
3. **Nodule type**
   The lesions are the most commonly subcutaneous or muscular nodules.
4. **Brain type**
   The brain type is common in patients of children and adolescents, with symptoms of headache, vomiting, meningeal irritation sign and intracranial hypertension.

By chest CT scan the lesions of paragonimiasis can be categorized into five types, infiltration, parabronchial inflammation, cystoid opacity, pleural effusion and cavity. Based on the chronological development of the lesion, infiltration and parabronchial inflammation are early lesions of paragonimiasis. Cystoid opacity emerges in the sub-early or abscess stage, and cavity with nodules attached wall is a sign of the paracmasis stage. Pleural effusion may exist alone or along with lung lesions.

In this case, chest CT scan demonstrated infiltration, shown as mixed hemorrhagic lesion and exudative lesion caused by migration and penetration of paragonimus in lung tissue and local allergic responses. The infiltration was singular or multiple large flakes or small-patches-integration cotton like opacity, with poorly defined boundary, uncertain shape, and different sizes. The infiltration is commonly detected in the marginal part of bilateral middle and lower lung fields, with poorly defined small vesicles in inflammatory lesion. Based on such demonstrations, its differential diagosis from other lung infections can be hardly made. In addition, polypides of paragonimus can penetrate the diaphragm into the pleural cavity to cause pleural lesions, which are especially common on the diaphragmatic and mediastinal surfaces with accompanying pleural effusion. Pleural adhesion with pericardial membrane and shell like pleural calcification also occur.

Paragonimiasis should be differentiated from the following diseases.

1. Pulmonary tuberculosis
   The lesions of pulmonary tuberculosis are multiple small caseous necrotic granuloma, showing ring enhancement by contract scan. The patients commonly experience typical tubercular poisoning symptoms, which show therapeutic responses to anti-tuberculosis therapy.
2. Pneumonia
   The lesion of pneumonia is demonstrated as poorly defined cloudy opacity, and shows improvement after administration of routine anti-inflammatory therapy.
3. Peripheral lung cancer

The lump is demonstrated with irregular margin, and lobulation. Dynamic contract scan facilitates their differential diagnosis.

Case Study 5
[Brief Medical History]

A 21-year-old man complained of abdominal upset with no known causes for 10 days, with accompanying hiccup, anorexia, and right lumbar upset. The patient experienced chest distress, cough and night sweat with no known causes 4 days ago. During the recent 8 years, he had intermittent skin eczema in anterior chest and the condition relapsed following medications. He had medical histories of pulmonary empyema 15 years ago that was healed, and hepatic abscess more than 10 years ago that was healed. Pathologically, biopsy following laparoscopy revealed nodules in the omentum and abdominal wall with eggs like substance, which are consistent to granulomatous inflammation. Laboratory tests revealed increased eosnophils count, monocyte count and WBC count as well as increased C reactive protein; increased mucus in urine sediment; and qualitative DBDX protein positive (+).

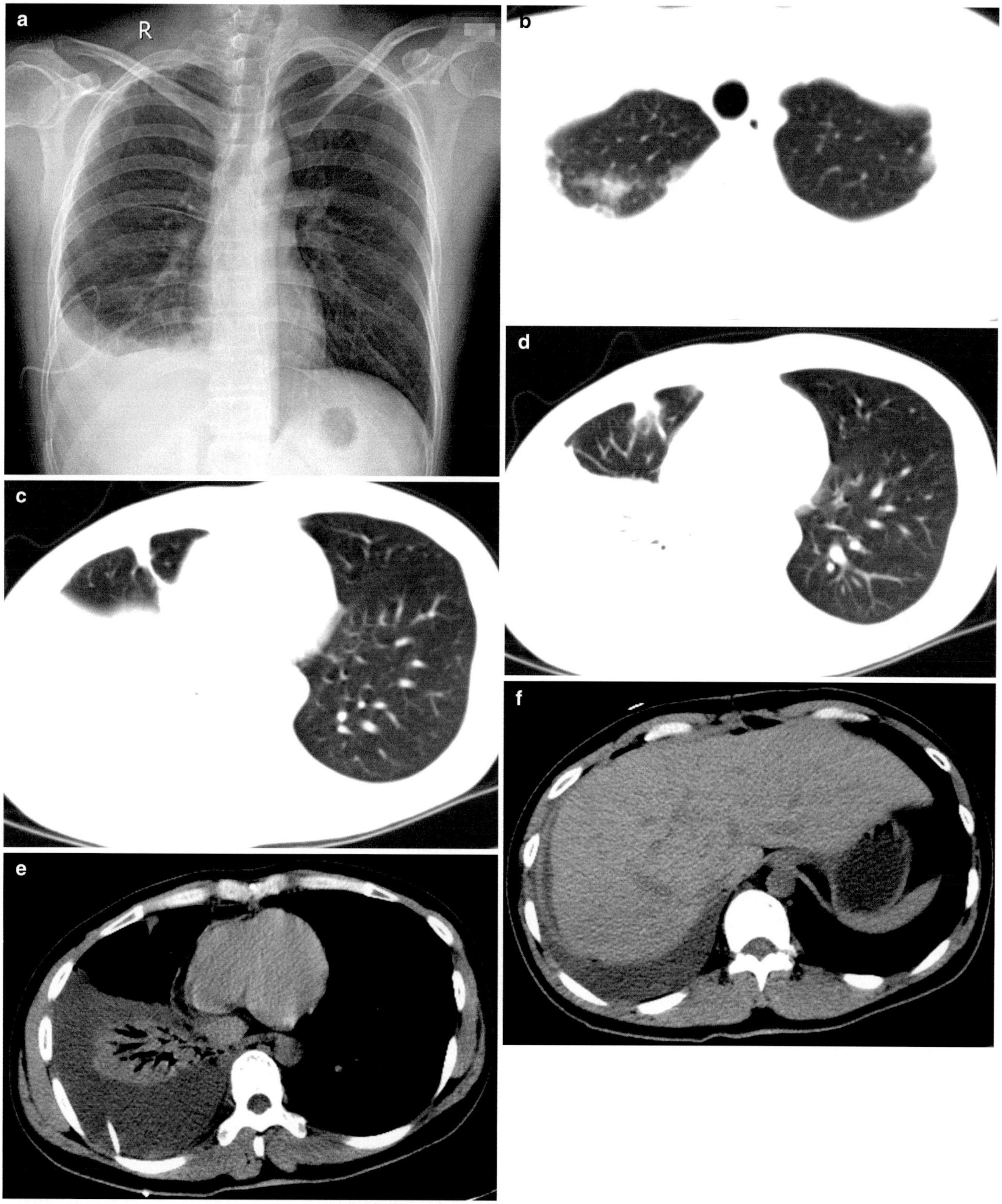

Fig. 10.16 (**a**) Chest X-ray demonstrated absence of right costophrenic angle, and dense arc-like liquid opacity. (**b**) CT plain scan demonstrated poorly defined flake of opacity in the right upper lung lobe. (**c**, **d**) The right middle lung lobe was shown with patches of hyper-intense opacity and cords like opacity, and the right thoracic cavity was shown with band like liquid density opacity. (**e**, **f**) CT scan demonstrated the right lower lung to be compressed with consolidation. The right thoracic cavity and tissue around the liver were revealed with arch like liquid density opacity. (**g**) The resected specimen from the liver was observed by naked eyes, with nodules in a diameter of 1.5 cm, grayish white or grayish yellow section and moderate texture. (**h**) Microscopic examination demonstrated non-caseous necrotic granulomatous inflammation in the liver, infiltration of eosinophil in a large quantity, and focal fat degeneration of the surrounding liver tissue

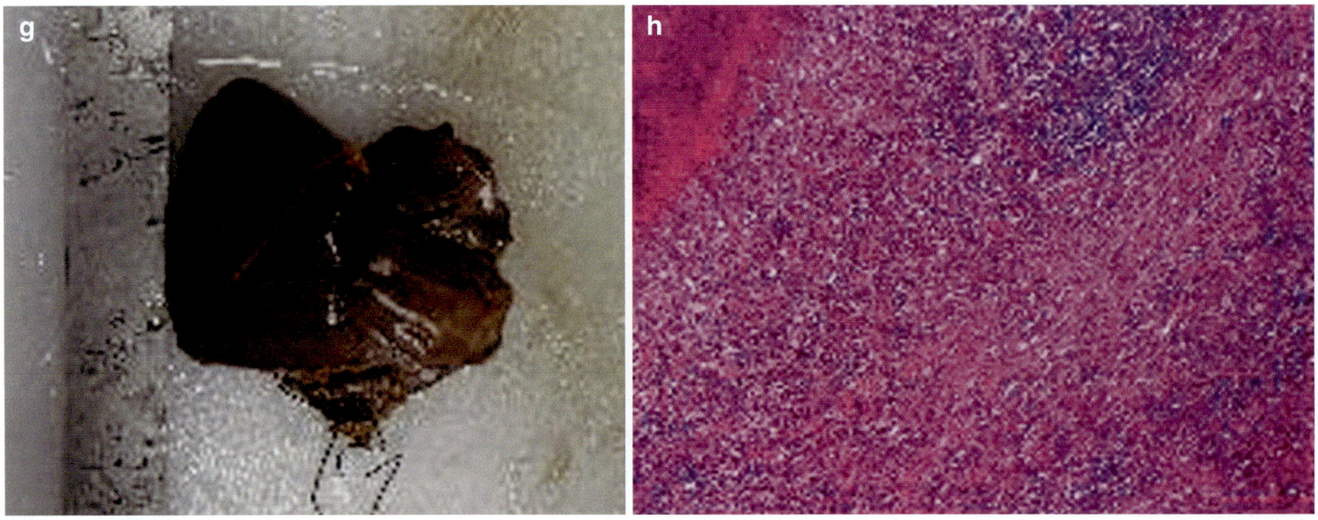

Fig. 10.16 (continued)

[Radiological Demonstrations] (See Fig. 10.17)

[Diagnosis] Paragonimiasis (the liver type).

[Discussion]

Kim et al. reported radiological demonstrations of hepatic paragonimiasis, including cluster like, tubular (multiple cysts) or lobulated low-density lesion in the left and/or right liver lobe by contrast CT scan in its stage, typically as the tunnel sign; no enhancement in the cyst cavity but singular or reticular enhancement of the cyst wall; and nodular or wedge like enhancement of the adjacent liver tissue. These radiological signs are pathologically based on invasion of metacercariae of paragonimus into the liver via the liver capsule and its constant migration in the liver parenchyma. The mixed old and new lesions are various in size due to varying quantity of invading polypides, varying duration of their migration and individual differences of the host. However, the damage of liver tissue caused by polypides is relatively mild with formation of small thin-wall abscess that can be slightly enhanced by contrast scan. The abscesses are communicated via sinus to develop into multilocular abscess. In the cases with retention of polypides and necrotic tissue in intrahepatic bile duct, chronic inflammation of the bile duct may occur to cause adjacent intrahepatic cholangiectasis. The abscess cavity may communicate to the involved dilated bile duct, and their differentiation is sometimes difficult. Portal hypertension and splenomegaly may occur in its middle and advanced stages. In the cases with the omentum invaded by paragonimus, the lesions are demonstrated as multiple pancake like nodules. In addition, infiltration of lung tissue in patient with paragonimiasis may be demonstrated as singular or multiple flakes or cotton-like opacity in different size, with poorly defined boundary, which is commonly detected in the marginal lung fields of bilateral middle and lower lung and resembles to those of common pneumonic infections. Radiologically, pleural effusion is demonstrated as crescent shape liquid density opacity along the lateral posterior wall of thoracic cavity.

Radiological demonstrations of paragonimiasis resemble to those of atypical bacterial liver abscess and primary cholangiocellular hepatocarcinoma with poor blood supply. The lesions of the three diseases are all demonstrated as singular or multiple cystic lesions with sligtly enhanced nodules. The key points for differential diagnosis are as the following:

1. Bacterial liver abscess
   The lesion is commonly located in the right liver lobe. In the early stage of the disease, it is often demonstrated as singular or multiple small abscesses. If untreated, the small abscesses integrate into large one along with constant disintegration and necrosis of liver tissue. The abscess wall is uneven in thickness with obvious enhancement by contrast scan, with edema of the surrounding liver tissue, which can be typically demonstrated as the target sign. In additon, adjacent liver parenchyma may show flake enhancement due to inflammatory response.
2. Cholangiocellular hepatocarcinoma
   The lesion of cholangiocellular hepatocarcinoma is orignated from intrahepatic biliary epithelium, which may occur in any liver lobe. Along with its growth, the lesion is subject to ischemia and necrosis to develop into multilocular, which is poorly defined in different sizes and is accompanied by thickening of intrahepatic biliary wall and dilation of intrahepatic biliary lumen to invade adjacent vascular vessels and liver tissue. The hepatic hilar and abdominal lymph nodes are suject to enlargement. And the lesion may show metastasis into other organs.

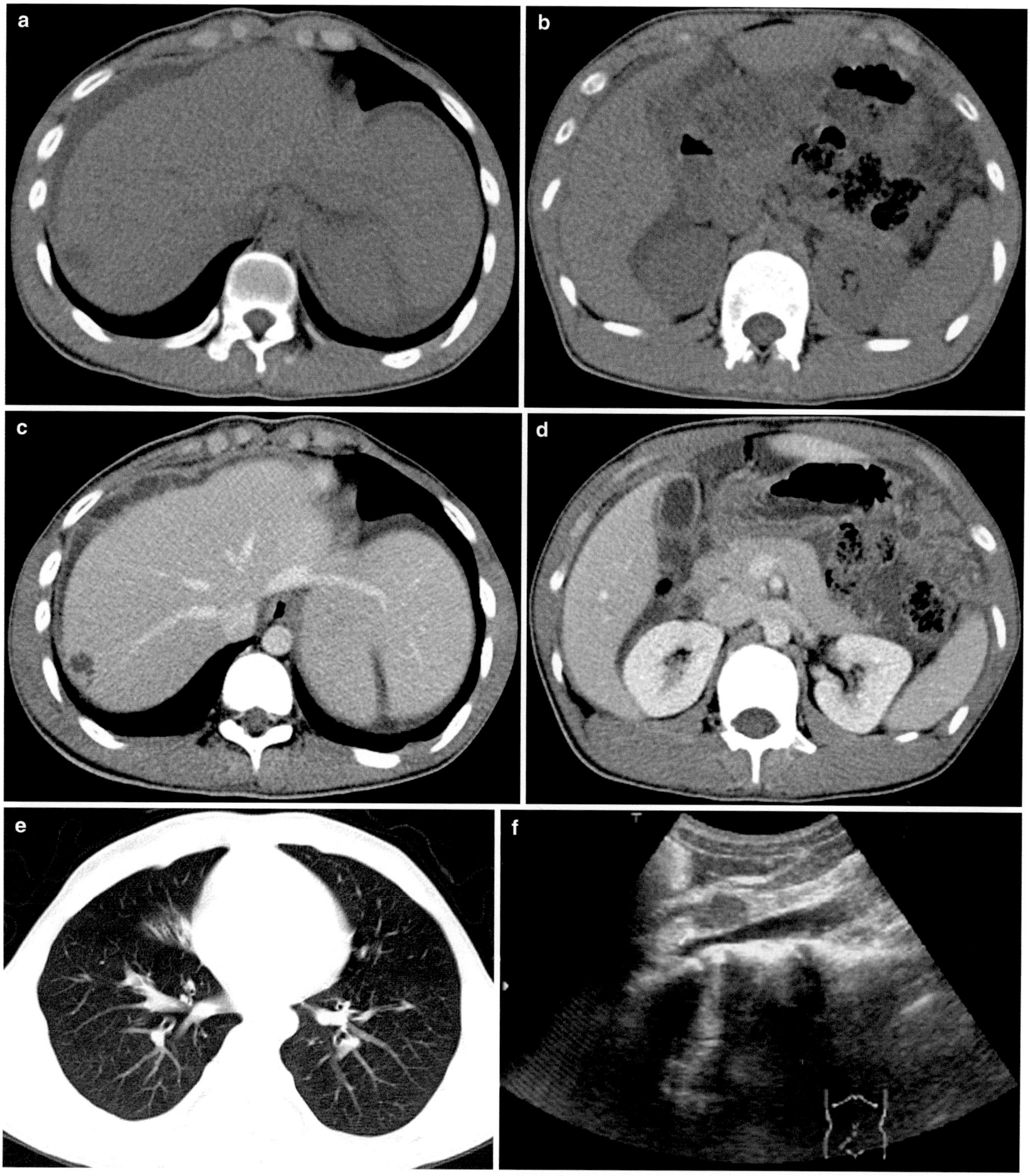

Fig. 10.17 (**a**, **b**) Plain CT scan demonstrated hypo-intense opacity in the right liver lobe, multiple nodules in the omentum, and ascites. (**c**, **d**) Contract CT scan demonstrated enhancement of the lesion in the superior segment of right liver lobe, and pancake like enhancement of the multiple nodules in the omentum. (**e**) the right middle lung lobe was shown with patches of hyper-intense opacity with poorly defined boundary. (**f**) Ultrasound demonstrated echoes from multiple abdominal lymph nodes, with the large one in size of 2.0×1.5 cm; and ascites. (**g**) By naked eyes observation, the nodules in the omentum and abdominal wall with grayish white and grayish yellow fragments. (**h**) Microscopically, centrifuged sediment smear of ascites demonstrated neutrophils and eosinophil in large quantities as well as mesothelial cells in a small quantity. (**i**) Microscopically, the fibrous adipose tissue was shown with focal necrosis, infiltration of neutrophil, lymphocyte, eosinophil and plasmocyte, reactive hyperplasia of multinuclear giant cell, and eggs like substance. These demonstrations are consistent to granulomatous inflammation. Purulent inflammation and abscess were also detected

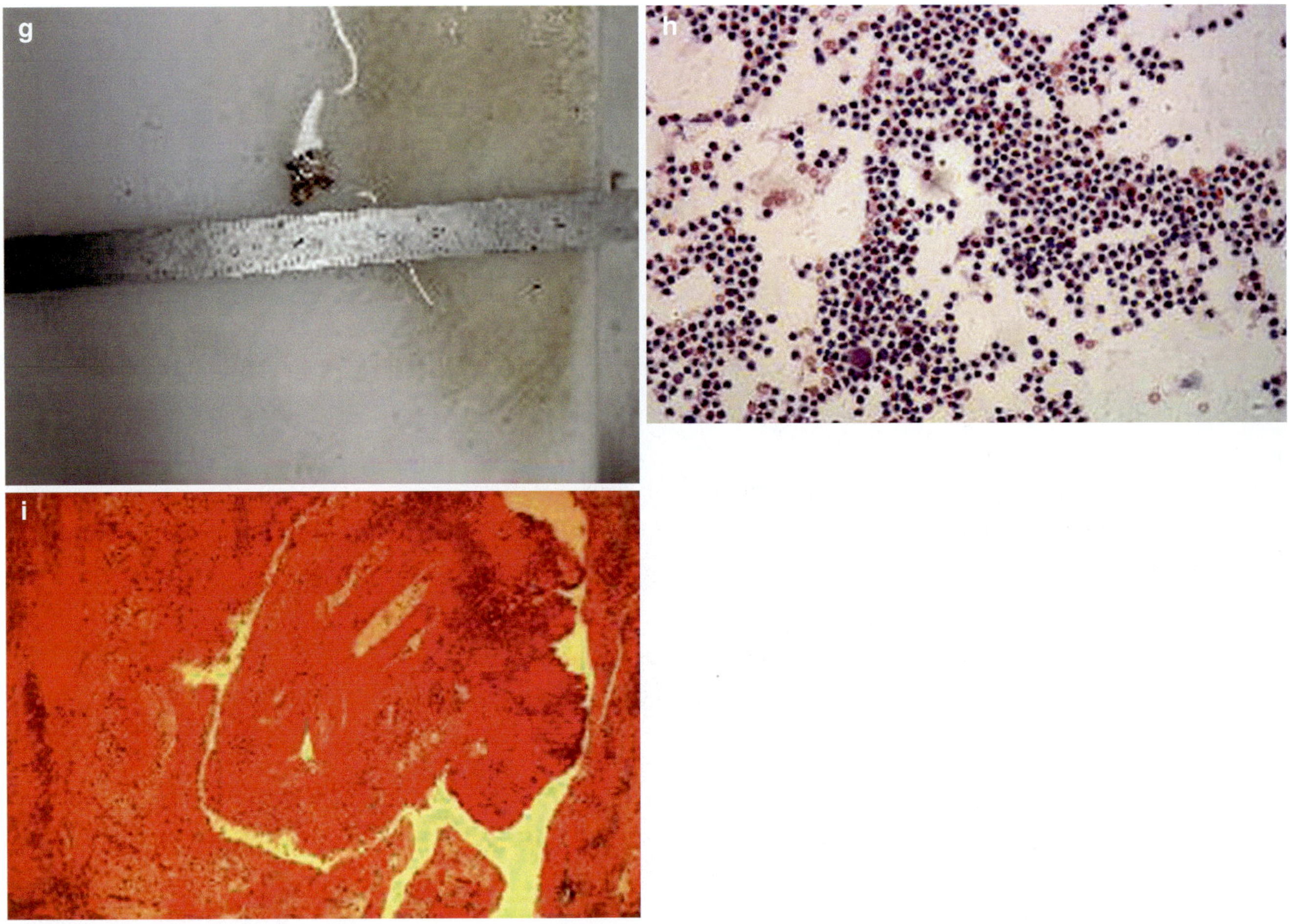

Fig. 10.17 (continued)

Case Study 6

[Brief Medical History]

A 8-year-old girl experienced icteric hepatitis 1 year ago and she complained of systemic yellowish skin mucosa for 1 month with right abdominal pain. The diagnosis was primarily gastric disease for treatment. After that, the abdominal pain was relieved but the yellowish skin mucosa aggravated with yellowish sclera, yellowish urine and kaolinic like stool. In recent several days, her abdominal pain aggravated, with a palpable lump in the right upper quadrant. Laboratory tests revealed increased WBC count and eosnophils count. Pathological examination following surgery indicated inflammatory lesion and suspected parasitic infection. Blood test revealed IgG of paragonimus antibody positive (+), indicating a past history of paragonimiasis.

[Radiological Demonstrations] (See Fig. 10.18)

[Diagnosis] Paragonimiasis granuloma.

[Discussion]

The patient had a past medical history of paragonimiasis. CT scan demonstrated the retroperitoneum, head of pancreas and hepatic hilar region with round like soft tissue lumps. Pathological examination following surgery indicated granuloma. In combination to clinical manifestations and laboratory tests, the diagnosis was considered to be recurrence of paragonimiasis. The abnormal space occupying lesions were suspected to be ascites or extensive inflammatory adhesion and granuloma due to migration of polypides in the abdominal cavity.

Paragonimiasis should be differentiated from the following diseases.

1. Pancreatic head carcinoma

   The lesion of pancreatic head carcinoma is demonstrated with double-line sign, with gradual enhancement by delayed scan if the lesion has poor blood supply.
2. Retroperitoneal neoplasm

   Retroperitoneal neoplasm is commonly benign teratoma, and the lesion is demonstrated as equal density lesions in the adipose tissue and bone by CT scan. Among retroperitoneal malignancies, liposarcoma is relatively common, with adipose density lesion that shows obvious uneven enhancement by contrast scan.

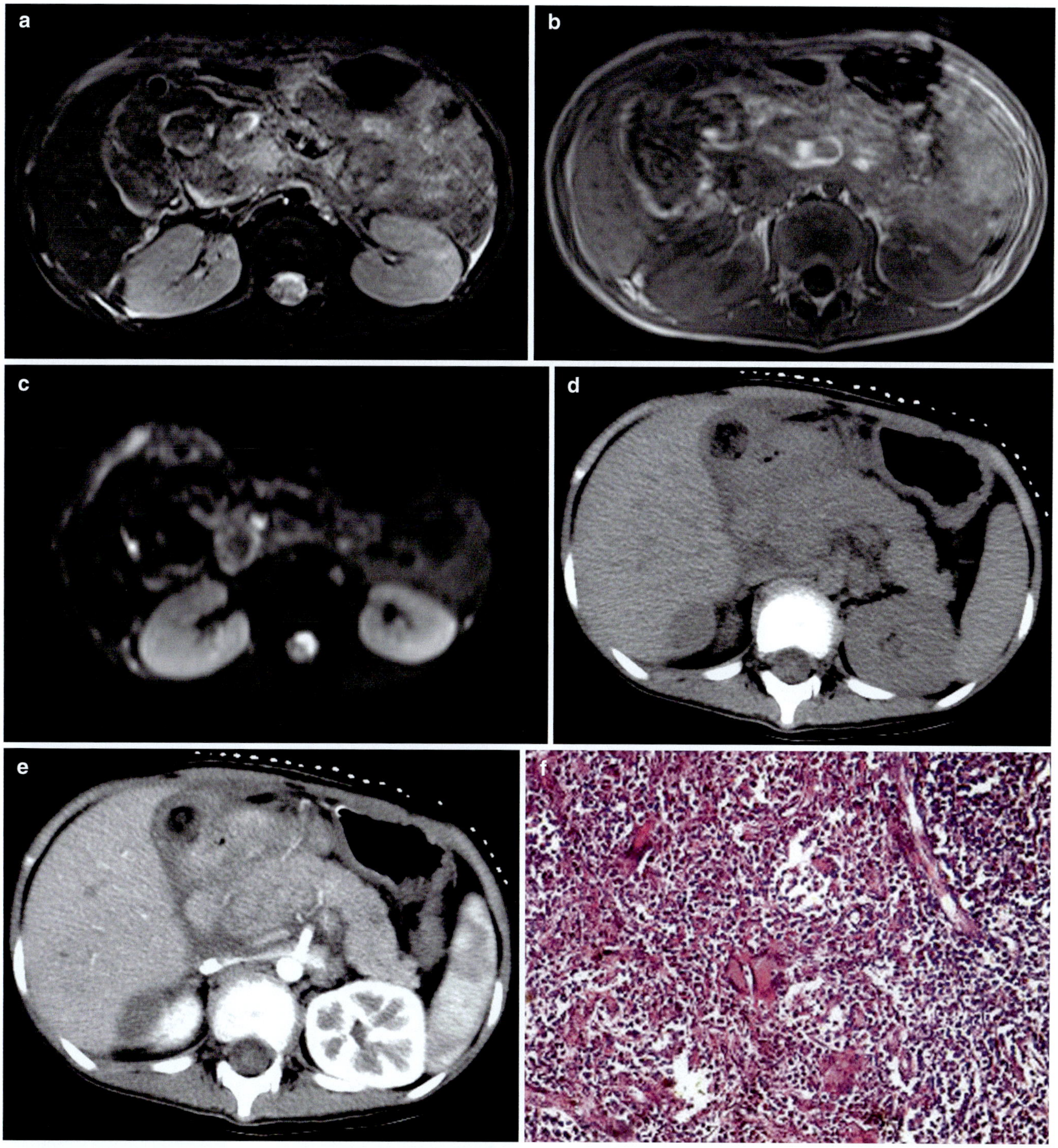

Fig. 10.18 (**a**–**c**) MR imaging demonstrated round like soft tissue lumps in the retroperitoneum, head of pancreas and hepatic hilar region, with long T_1 and mixed long T_2 signals, relatively high DWI signal. The adjacent structure was shown to be compressed, with slight backward migrations of inferior vena cava and abdominal aorta, forward and leftward migration of the superior mesenteric artery and vein. (**d**, **e**) CT scan demonstrated deranged structures between liver and stomach as well as posterior to the head of pancreas, with poorly defined irregular lump like opacity, with moderate uneven enhancement by contrast scan. The lump like opacity was shown to be surrounded by flakes of exudation and effusion. The right hepatorenal recess was revealed with effusion. (**f**) Pathology indicated granuloma (histocytes, epithelial cells and multinuclear giant cells)

Case Study 7

[Brief Medical History]

A 35-year-old man experienced sudden blurry vision of both eyes with no known causes 3 months ago, with subsequent severe pulsing headache at the left tempus that persisted for several minutes as well as slow response, consciousness disturbance and dizziness when headache occurred. He also experienced intermittent fever 1 month ago, commonly after noons, with the body temperatures fluctuating between 38 and 39 °C that persisted for several hours. The patient experienced relapse of headache 3 days ago with accompanying pain at the right tempus, dizziness and slow response dizziness that persisted for about 4 h, but no consciousness disturbance. Laboratory tests revealed increased eosnophil count and monocyte count as well as increased C reactive protein. The patient had a history of being scratched by a cat and a history of eating uncooked crabs.

[Radiological Demonstrations] (See Fig. 10.19)

[Diagnosis] Paragonimiasis.

[Discussion]

Infiltration of paragonimiasis is demonstrated as singular or multiple flakes or cloud like opacity and pleural effusion. Radiologically, hepatic paragonimiasis is characterized by typical tunnel signs, no enhancement in the cystic cavity but singular or reticular enhancement of the cystic wall as well as nodular or wedged like enhancement of the adjacent liver tissue.

Cerebral paragonimiasis is caused by upward migration of adult paragonimus in abdominal or thoracic cavity from mediastinum into the brain, and then migrating, laying eggs, disintegration and metabolism in the brain tissue. The major pathological changes include infiltrative aseptic inflammation, cerebral hemorrhage or cerebral infarction. Long-term stay of the polypides in the brain tissue or aggregation of eggs in a large quantity causes abscess, cyst, granuloma, fibroatrophy or calcification, with clinical symptoms of headache, vomiting and epileptic seizure.

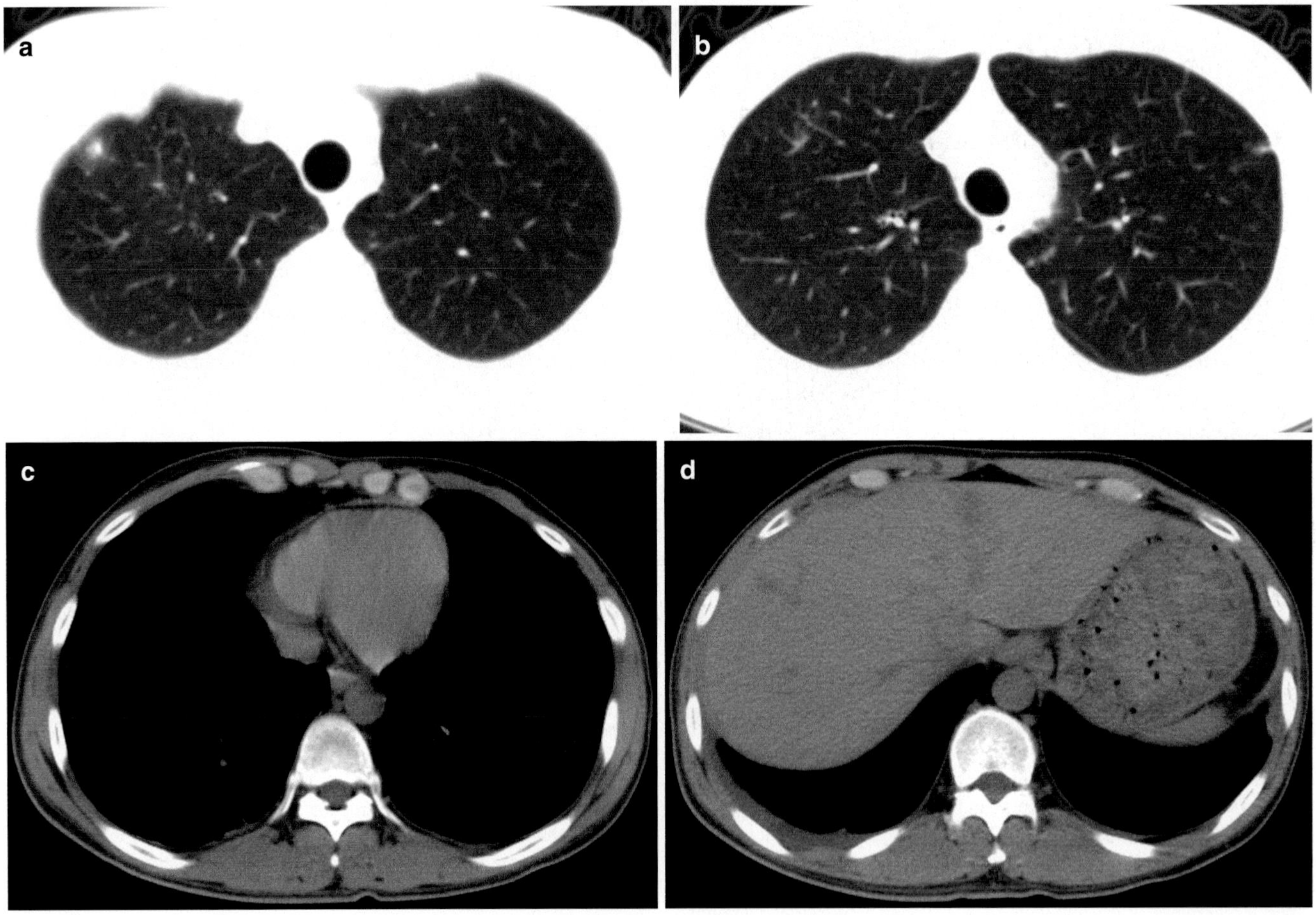

Fig. 10.19 (**a**, **b**) CT scan demonstrated multiple small flakes of hyper-intense opacity in both lungs with poorly defined boundary. (**c**) CT scan demonstrated effusion opacity in bilateral thoracic cavity in a small quantity. (**d**) CT scan demonstrated hypo-intense opacity in the liver. (**e**, **f**) MR imaging demonstrated flakes of mixed long T_1 long T_2 signals in the left frontal lobe. (**g**) Contract MR imaging demonstrated patches of mild delayed enhancement of the lesion. (**h**) DWI demonstrated slightly high signal of the lesion. (**i**) MR imaging demonstrated slightly low NAA, and high Cho, which did not support the diagnosis of typical neoplasm

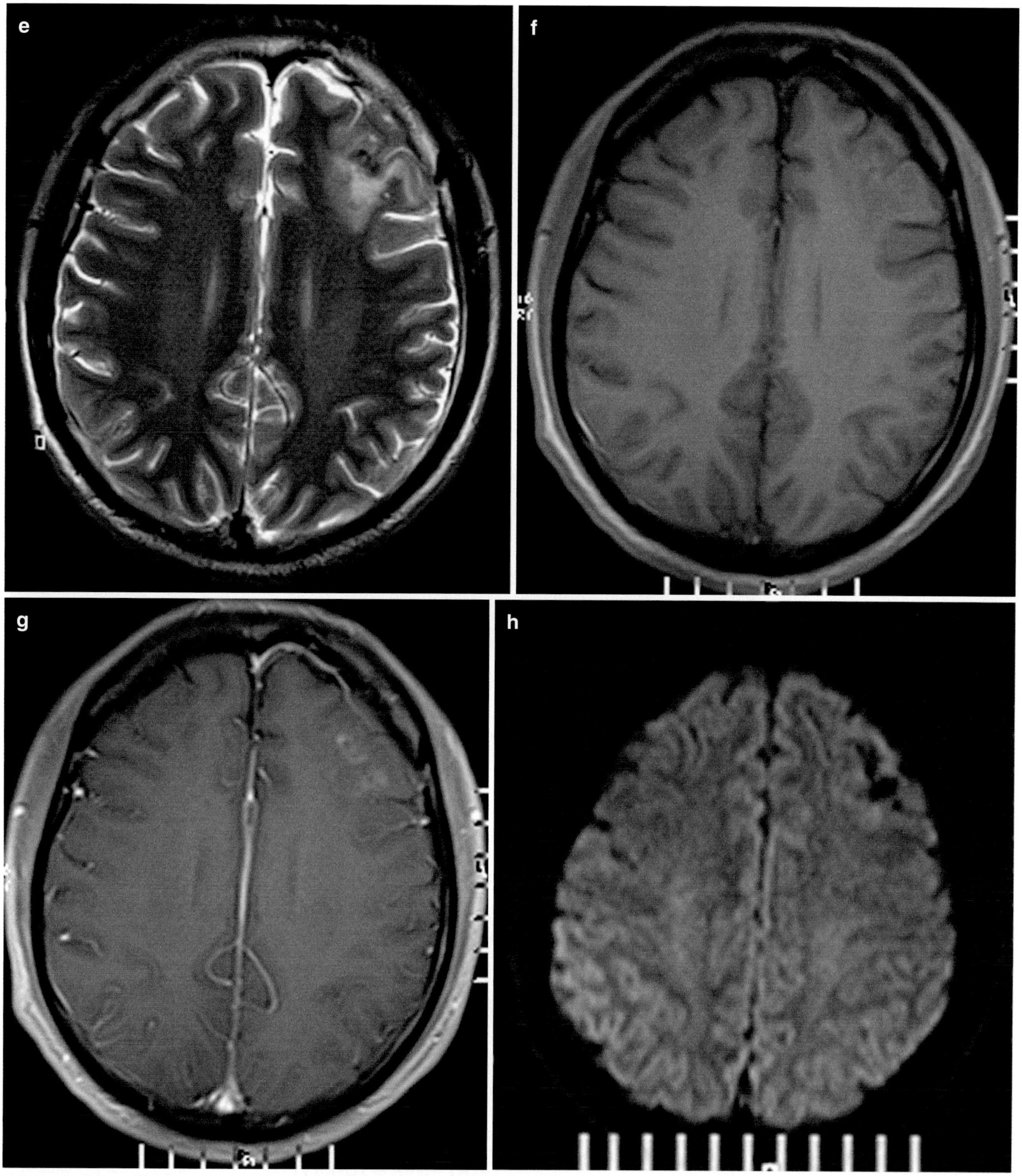

Fig. 10.19 (continued)

Fig. 10.19 (continued)

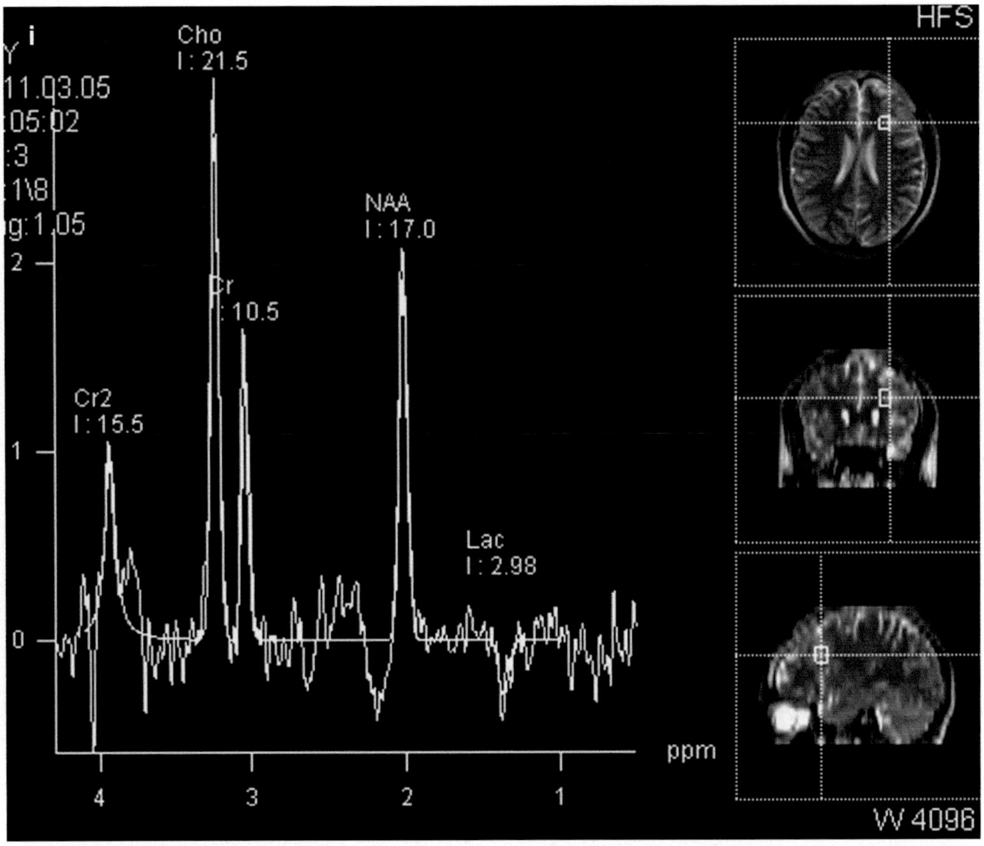

Radiologically, cerebral paragonimiasis can be further typed into encephalitis type, cyst type and atrophy type. Plain MR imaging demonstrates hemorrhagic lesion and large edema area around the hemorrhagic lesion, namely small hemorrhagic lesion with large edema area. The lesions distribute randomly in aggregating or migrating pattern, with detectable tunnel sign. Contract imaging demonstrates patches, nodular, small ring enhancement, with more obvious tunnel sign. The advantages of MR imaging in its diagnosis are as the following: (1) multiple irregular hemorrhages in different degrees; (2) tunnel sign with long T1 signal after absorption of hemorrhage; (3) relatively aggregating and migrating appearance of the lesions that is consistent with inflammation; (4) irregular edema signal (including inflammatory edema and edema around hemorrhagic lesion).

Paragonimiasis should be differentiated from the following diseases.

1. Cerebral abscess
   The abscess wall is commonly regular with high tension. And the patients with cerebral abscess commonly show obvious symptoms and signs of acute infection.
2. Pulmonary tuberculosis
   The patients with pulmonary tuberculosis commonly experience typical tubercular poisoning symptoms and anti-tuberculosis therapy is effective. The lesions are demonstrated as multiple small caseous necrotic granulomas. Contract scan often demonstrates small ring enhancement of the lesion.
3. Metastatic liver neoplasm
   The lesion of metastatic liver neoplasm is small with obvious surrounding edema. The space occupying effect is obvious and the primary lesion can be found.
4. Hepatic cyst
   The singular or multiple lesions of hepatic cyst are round or oval in shape, with homogenous low density and smooth margin. Its CT value is close to water, and the lesion cannot be enhanced by contract scan. The density may increase when the disease is complicated by hemorrhage or infection.

Case Study 8
[Brief Medical History]

A 22-year-old man complained of persistent headache for 20 days with lethargy for 3 days. Routine blood test revealed increased eosinophils, skin test for paragonimus antigen positive (+). The patient came from a paragonimiasis affected region and had a history of eating uncooked crabs.

[Radiological Demonstrations] (See Fig. 10.20)

[Diagnosis] Paragonimiasis (right basal ganglia area).

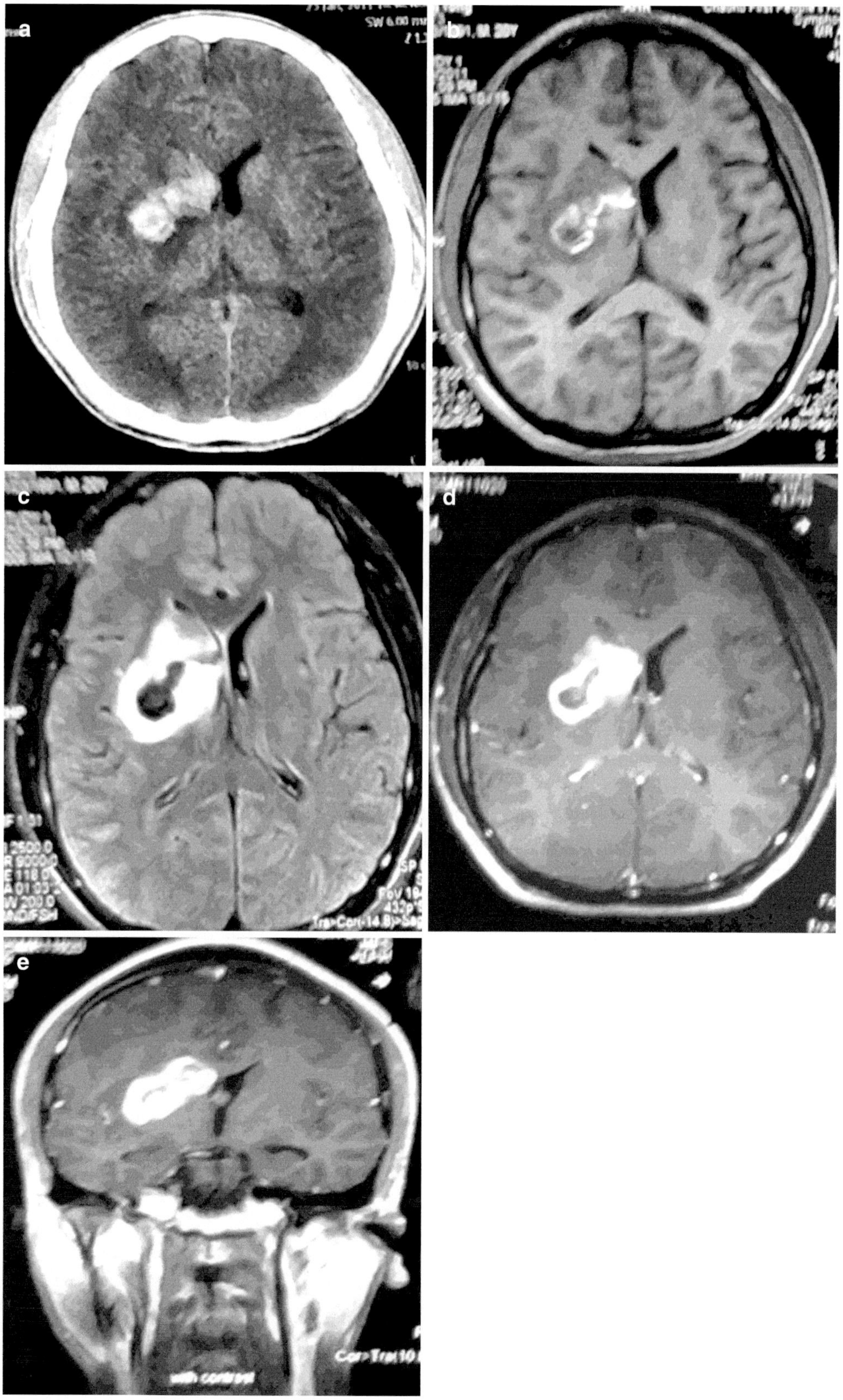

Fig. 10.20 (**a**) Plain CT scan demonstrated hyper-intense lesion in the right basal ganglia area with mild surrounding edema. (**b**) Plain MR imaging demonstrated the lesion with central tunnel sign in long T1 signal following absorption of hemorrhage. (**c**) FLAIR demonstrated low central signal of the lesion. (**d**, **e**) Contract MR imaging demonstrated no central enhancement of the tunnel sign but irregular enhancement around the lesion

[Discussion]

Head CT scan or MR imaging demonstrates multiple irregular lesions and hemorrhage in the brain. Contract scanning or imaging demonstrates multiple aggregating ring enhancement of the lesion and surrounding edema if the lesion shows tunnel sign. The patient should be highly suspected with paragonimiasis base on a history of living in an affected region, a history of eating uncooked crabs and drinking unboilded contaminated water. And immunoassays should be ordered to define the diagnosis.

Cerebral paragonimiasis should be differentiated from the following diseases.

1. Cerebral tuberculosis
   The patients with tuberculosis often experience typical tubercular poisoning symptoms, and the anti-tuberculosis therapy is effective. The lesion are demonstrated to be multiple small caseous necrotic granulomas, with small ring or chain like enhancement by contract scan. If complicated by meningeal granuloma, contract scan demonstrations of cast like enhancement and hydrocephaly are its characteristic signs.
2. Brain abscess
   The abscess wall is commonly regular with high tension. The patients experience obvious symptoms and signs of acute infection.
3. Cerebral echinococcosis
   The patients with cerebral echnicoccosis should have definite epidemiological history. And formation of daughter cysts or granddaughter cysts is detected in the lesion.
4. Cerebral cysticercosis
   The brain parenchyma type is demonstrated with multiple lesions, which are demonstrated as multiple mixed signal lesions by plain imaging and nodular or ring enhancement by contrasts imaging. The ring enhancement and wall-nodules enhancement (scolex of cysticercus) greatly facilitate the differential diagnosis. Immunoassay of cerebrospinal fluid for antibody facilitates its qualitative diagnosis.
5. Nodular sclerosis
   Clinical manifestations of nodular sclerosis include epilepsy, intelligence disturbance, steatadenoma, hamartoma characterized by its formation in different organs, and commonly occurring subependymal nodules. The nodules may also distribute sporadically in the cortex, subcortex, basal ganglia area and occasionally in brainstem and cerebellum.
6. Cerebrovascular malformation
   When the lesion is subject to hemorrhage, paragonimiasis should be differentiated from cerebrovascular malformation. MR imaging characteristically demonstrates arteriovenous malformation with wool-ball like or honeycomb like vascular void opacity. MRA can directly demonstrate the supplying artery and the draining vein of arteriovenous malformation. Cerebral angiography is recommended to define the diagnosis.
7. Metastatic neoplasm
   The lesion of metastatic neoplasm is small with obvious surrounding edema. In addition, its space occupying effect is obvious and the primary lesion can be found.

Further Reading

Cha SH, Chang KH, Cho SY, et al. Cerebral paragonimiasis in early active stage: CT and MR features. AJR Am J Roentgenol. 1994;162(1):141–5.

Chen WJ, Wu EF, Hu ZT, et al. Chest CT scan of paragonimiasis. Chin J Radiol. 2004;38(4):407–9.

Ghosh K, Ghosh K. Pulmonary hydatidosis, strongyloidiasis and paragonimiasis in India. J Assoc Physicians India. 2013;61(12):954–5.

Gu XH, Huang YX. Clinical diagnosis of schistosomiasis by B-mode ultrasound and X-ray barium enteroscopy. Chin J Schistosomiasis Control. 2007;19(6):456,457.

Guo YM, Chen QH, Wang W. Radiology of respiratory system. Shanghai: Shanghai Science and Technology Press; 2011.

Hu XP, Cui GC, Yin B. Diagnostic value of 16-slice spiral CT and its post processing technology for large intestinal schistosomiasis. Radiol Pract. 2010;25(9):1030–3.

Huang XB, Shen YD. Diagnostic value of CT scan for clonorchiasis. Chin Int Med Res. 2014;12(28):59–60.

Kim HJ. Pneumothorax induced by pulmonary paragonimiasis: two cases report. Korean J Thorac Cardiovasc Surg. 2014;47(3):310–2.

Kim EA, Juhng SK, Kim HW, et al. Imaging findings of hepatic paragonimiasis: a case report. J Korean Med Sci. 2004a;19(5):759–62.

Kim EA, Juhng SK, Kim HW, et al. Imaging findings of hepatic paragonimiasis: a case report. J Korean Med Sci. 2004b;19(5):759–69.

Li L, Wang RK, Jiang CD, et al. Clinical data analysis of paragonimiasis: a report of 62 cases. J Parasitosis Infect Dis. 2010a;8(3):166,167.

Li Y, Sun L, Chen K. Misdiagnosis of paragonimiasis: a report of 199 cases. J Parasitosis Infect Dis. 2010b;8(1):46–7.

Liang DR. CT scan demonstrations of clonorchiasis. J Guangxi Med. 2011;33(5):638–40.

Liu HQ, Chen ZJ. MR imaging diagnosis of cerebral schistosomiasis. Chin J Radiol. 2002;36(9):821–3.

Liu ZZ, Zhang LS. CT and MRI for paragonimiasis in its acute stage. J Chuanbei Med Coll. 2009;24(6):580–2.

Liu T, Song MF, Dong JS, et al. Abdominal CT scan for chronic schistosomiasis: a comparative study to pathological examination. Chin J Radiol. 2005;39(11):1188–91.

Liu H, Lim CC, Feng X, et al. MRI in cerebral schistosomiasis: characteristic nodular enhancement in 33 patients. AJR Am J Roentgenol. 2008a;191(2):582–8.

Liu H, Lim CC, Feng X, et al. MRI in cerebral schistosomiasis: characteristic nodular enhancement in 33 patients. AJR Am J Roentgenol. 2008b;191(2):582–8.

Liu BL, Liang WQ, Chen BL, et al. CT diagnosis of clonorchiasis. J Chin Medication Guide. 2009;7(22):132–3.

Liu ZB, Wang L, Yang Y, et al. Clinical pathology and prognosis of schistosomiasis related colorectal cancer in the elderly from the petrochemical area of Shanghai, China. Chin J Geriatr. 2011;30(10):836–8.

Lu GM, Xu J, Chen JK. Guide in reading CT scan images. Nanjing: Jiangsu Science and Technology Press; 2006.

Mei CL, Li ZS, Zhu L. Manual of internal medicine. 7th ed. Beijing: People's Medical Publishing House; 2010.

Nascimento-Carvalho CM, Moreno-Carvalho OA. Clinical and cerebrospinal fluid findings in patients less than 20 years old with a presumptive diagnosis of neuroschistosomiasis. J Trop Pediatr. 2004a;50(2):98–100.

Nascimento-Carvalho CM, Moreno-Carvalho OA. Clinical and cerebrospinal fluid findings in patients less than 20 years old with a presumptive diagnosis of neuroschistosomiasis. J Trop Pediatr. 2004b;50(2):98–100.

Roberts M, Cross J, Pohl U, et al. Cerebral schistosomiasis. Lancet Infect Dis. 2006;6(12):820.

Ross AG, McManus DP, Farrar J, et al. Neuroschistosomiasis. J Neurol. 2012;259(1):22–32.

Salim OEH, Hamid HK, Mekki SO, et al. Colorectal carcinoma associated with schistosomiasis: a possible causal relationship. World J Surg Oncol. 2010;13(8):68–70.

Sanelli PC, Lev MH, Gonzalea RG, et al. Unique linear and nodular MR enhancement pattern in schistosomiasis of the central nervous system. AJR Am J Roentgenol. 2001a;177(6):1471–4.

Sanelli PC, Lev MH, Gonzalez RG, et al. Unique linear and nodular MR enhancement pattern in schistosomiasis of the central nervous system. AJR Am J Roentgenol. 2001b;177(6):1471–4.

Shen X, Zhang W, Wang PJ. Abdominal CT scan and pathological analysis of chronic schistosomiasis. Chin J Schistosomiasis Control. 2012;24(2):200–2.

Si CW, Jia FZ, Li JT. Study of infectious diseases. Beijing: People's Medical Publishing House; 2004.

Sun X, Li CJ, Zhang JS. Practical medical parasitology. Beijing: People's Medical Publishing House; 2005.

Waldman AD, Day JH, Shaw P, et al. Subacute pulmonary granulomatous schistosomiasi: high resolution CT appearances-another cause of the halo sign. Br J Radiol. 2001;74(887):1052–5.

Wang WL, Chen WJ. Chest CT scan of paragonimiasis westermani: a report of 32 cases. J Radiol Pract. 2003;18(7):489–90.

Wu L, Wu M, Tian D, et al. Clinical and imaging characteristics of cerebral schistosomiasis. Cell Biochem Biophys. 2012;62(2):289–95.

Ye S, Wang WL, Zhao K. F-18 FDG hypermetabolism in mass-forming focal pancreatitis and old hepatic schistosomiasis with granulomatous inflammation misdiagnosed by PET/CT imaging. Int J Clin Exp Pathol. 2014;7(9):6339–44.

Yu HL, Jiang TJ, Meng QR, et al. CT scan demonstrations of clonorchiasis: report of 48 cases. J Chin Pract Med. 2011;6(8):58–9.

Zeng JW, Zhang JH, Ma JP, et al. Analysis of paragonimiasis: a report of 2 cases. Chin J Endemiol. 2013;32(6):700.

Zhang W, Wang PJ, Shen X, et al. CT presentations of colorectal cancer with chronic schistosomiasis: a comparative study with pathological findings. Eur J Radiol. 2012a;81(8):e835–43.

Zhang H, Zeng HH, Xie ZW, et al. Diagnostic value of multi-slice spiral CT scan for clonorchiasis. Mod Hosp. 2012b;12(4):67–9.

Zhao DM, Chen D, Han FG, et al. CT and MRI diagnosis of cerebral paragonimiasis. J Pract Radiol. 2007;23(11):1445–8.

Zhou LP. Modern neurosurgery. Shanghai: Fudan University Press; 2001.

Zhu WZ, Gong CY, Zhou YC, et al. MR imaging demonstrations and pathological findings in the cases of cerebral schistosomiasis. Chin J Radiol. 2000;34(10):701–3.

Zhu GY, Luo LK, Zu DL. Clinical and pathological analysis of appendiceal schistosomiasis compliated by infection. Chin J Zoonoses. 2005;21(6):540, 541.

MIX
Papier aus verantwortungsvollen Quellen
Paper from responsible sources
FSC® C105338

If you have any concerns about our products,
you can contact us on
ProductSafety@springernature.com

In case Publisher is established outside the EU,
the EU authorized representative is:
Springer Nature Customer Service Center GmbH
Europaplatz 3, 69115 Heidelberg, Germany

Printed by Libri Plureos GmbH
in Hamburg, Germany